Saas-Fee Advanced Course 21
Lecture Notes 1991

W. B. Burton   B. G. Elmegreen
R. Genzel

The Galactic Interstellar Medium

Saas-Fee Advanced Course 21
Lecture Notes 1991
Swiss Society for Astrophysics and Astronomy

Edited by D. Pfenniger and P. Bartholdi

With 118 Figures

Springer-Verlag
Berlin Heidelberg New York
London Paris Tokyo
Hong Kong Barcelona
Budapest

Professor W. B. Burton
Sterrewacht Leiden, PO Box 9513, NL-2300 RA Leiden, The Netherlands

Professor B. G. Elmegreen
IBM, Thomas J. Watson Research Center, PO Box 218, Yorktown Heights, NY 10598, USA

Professor R. Genzel
Max-Planck-Institut für Physik und Astrophysik, Karl-Schwarzschild-Strasse 1,
W-8046 Garching, Fed. Rep. of Germany

Volume Editors:

Dr. D. Pfenniger
Dr. P. Bartholdi
Observatoire de Genève, ch. des Maillettes 51, CH-1290 Sauverny, Switzerland

This series is edited by Professor T. J.-L. Courvoisier on behalf of the
Swiss Society for Astrophysics and Astronomy:

Société Suisse d'Astrophysique et d'Astronomie
Observatoire de Genève, ch. des Maillettes 51, CH-1290 Sauverny, Switzerland

ISBN 3-540-55805-5 Springer-Verlag Berlin Heidelberg New York
ISBN 0-387-55805-5 Springer-Verlag New York Berlin Heidelberg

Library of Congress Cataloging-in-Publication Data. Burton, W. B. (William Butler), 1940- . The galactic interstellar medium / W. B. Burton, B. G. Elmegreen, R. Genzel; edited by P. Bartholdi and D. Pfenniger. p. cm. – (Saas-Fee advanced course 21 lecture notes ; 1991). Includes bibliographical references and index. ISBN 3-540-55805-5 (Berlin). – ISBN 0-387-55805-5 (New York). 1. Milky Way. 2. Interstellar matter. 3. Molecular clouds. 4. Astrophysics. I. Elmegreen, Bruce G. II. Genzel, R. III. Bartholdi, P. IV. Pfenniger, D. V. Title. VI. Series: Saas-Fee advanced course ... lecture notes ; 1991. QB857.7.B87 1992 523.1'125–dc20 92-30271

Typesetting: Camera ready copy from the author/editor
Production Editor: P. Treiber
55/3140 - 5 4 3 2 1 0 - Printed on acid-free paper

Foreword

The previous Saas-Fee Advanced Course dedicated to the interstellar medium took place in 1972. The tremendous scientific advances that have occurred in this field since then, in particular owing to the availability of receivers working at completely unexplored wavelength bands, fully justified a new set of lectures. As a consequence, the members of the Swiss Society for Astrophysics and Astronomy voted that "The Galactic Interstellar Medium" should be the subject of the 1991 course. The 21st Saas Fee Advanced Course took place in Les Diablerets from 18 to 23 March 1991, gathering together about 80 participants from all over the world, but mostly from Europe.

According to a rule that has proved to lead to success, but also to challenge the lecturers' energy, the format of a Saas-Fee Advanced Course consists traditionally of 28 lectures of 45 minutes which take place in the morning and late afternoon, leaving ample time for discussions, self-study, hiking or skiing. Despite the inordinate work load imposed, this year's lecturers felt that the subject was sufficiently dense to increase the lecture time by 1/3! This proved judicious and left more time for questions and discussions during the lectures.

This volume contains the written form of the delivered lectures, which fascinated many in the audience not only because of the width and depth of the subject, but because of the high quality of the transmitted information. The course topics were divided up as follows: Butler Burton spoke about "Distribution and Observational Properties of the ISM", Bruce Elmegreen about "Large Scale Dynamics of the ISM", and Reinhard Genzel about "Physics and Chemistry of Molecular Clouds". Although a complete coverage of the Galactic ISM is nowadays quite impossible in a one-week course, these written lectures give probably one of the most up-to-date presentations of the essential facts about the interstellar medium. Over the coming years this volume should remain invaluable to students and active or new researchers in the field.

The Saas-Fee Courses are financed in large part by the Swiss Academy of Natural Sciences, to which the Swiss Society for Astrophysics and Astronomy belongs. Thanks to continuous support this series of successful lectures has continued uninterrupted for more than 20 years.

The practical organisation of this year's course was greatly facilitated by our secretary, Mrs Irène Scheffre, and our assistants, Daniel Friedli and Stéphane Udry. We thank them for their generous efforts.

Geneva, June 1992 D. Pfenniger and P. Bartholdi

Contents

Distribution and Observational Properties of the ISM

By W.B. Burton (With 72 figures)
Chapters 5 and 6 are co-authored, respectively, with E.R. Deul & H.S. Liszt

Large Scale Dynamics of the Interstellar Medium

By Bruce G. Elmegreen (With 12 figures)

Physics and Chemistry of Molecular Clouds

By Reinhard Genzel (With 34 figures)

List of Previous Saas-Fee Advanced Courses:

---

* Out of print

Books up to 1989 may be ordered from: SAAS-FEE COURSES
                                        GENEVA OBSERVATORY
                                        ch. des Maillettes 51
                                        CH-1290 Sauverny
                                        Switzerland

Books from 1990 on may be ordered from Springer-Verlag.

Distribution and Observational Properties of the ISM

W.B. Burton[1]

Chapters 5 and 6 are co-authored, respectively, with *E.R. Deul*[1] *& H.S. Liszt*[2]

[1] Sterrewacht Leiden, Leiden, The Netherlands
[2] National Radio Astronomy Observatory, Charlottesville, USA

1 Introduction

1.1 Some Qualifying Remarks

The six chapters which follow deal with various aspects of the large-scale distribution and observational properties of the cooler constituents of the interstellar medium in our Galaxy. The discussions are based on the lectures given at the 21st "Saas-Fee" course organized by the Swiss Society for Astrophysics and Astronomy and held in Les Diablerets, Switzerland, during March, 1991. Some qualifying remarks are in order. These chapters should be viewed as collected notes on selected topics, not as a comprehensive review or formal monograph covering the subject of the galactic interstellar medium. The Saas-Fee course affords a pleasant opportunity to select topics which fall within the lecturer's own field of view. The risk of topics selected in this way is that the collected lecture notes may not — and in this case certainly *do* not — provide a comprehensive summary of the subject. The chapters presented here focus on the general subject of the morphology, or shape, of the layer containing the cooler, denser constituents of the interstellar medium as observed at lower galactic latitudes.

Such a focus implies important limitations. Discussions of galactic morphology require attention to material at low latitudes, where the lines of sight sample the Galaxy at large. These individual lines of sight typically intercept several tens of kpc of interstellar space. Low-latitude observations of essentially all tracers of the medium, but certainly of the prime tracers HI and CO, are unavoidably heavily blended because the tracers are distributed so generally throughout the gas layer. Blending due to the ubiquity of a single constituent is only one aspect of the problem, however; each low-latitude line of sight also samples a great variety of different interstellar environments, representing a large range of temperatures, densities, pressures, etc. These physical properties reveal themselves in the low-latitude observations as some sort of average, often difficult to interpret in a quantitative way.

Because an emphasis on the morphology of the gas layer in the Milky Way requires low-latitude data, this emphasis has as a consequence that but little information is provided on *individual* structures. The concept of an individual interstellar cloud is only of quite limited use at low latitudes; individual structures can generally only be isolated as such at latitudes greater than some 15°. At these higher latitudes, lines of sight typically sample lengths of only a few hundred pc. It is perhaps wise to remind ourselves that much of our knowledge of the detailed physical state of individual structures making up the interstellar gas layer refers more to the local neighborhood than to the Galaxy at large. (Similarly, much of our knowledge of the detailed kinematics of the interstellar medium is also derived from higher latitudes, and thus refers to motions largely perpendicular to the galactic plane.)

Some of the lecture-note chapters which follow review material already published; some others contain fresh results, currently being worked up for publication. The fresh material represents joint efforts with colleagues, so that the chapters reporting it are coauthored. Chapter 5, dealing with the morphology of the galactic dust layer derived from a radial unfolding of IRAS infrared emissivities, is based on some of the work done in the Leiden Ph.D. thesis of Erik Deul, and not previously published. Chapter 6 summarizes work being done together with Harvey Liszt, of the National Radio Astronomy Observatory, on the gas distribution in the part of the gas layer encompassed by the galactic bulge and which evidently enjoys substantial independence from the gas layer in the Galaxy at large.

The lectures given at this Saas-Fee course by Bruce Elmegreen and Reinhard Genzel complement the ones given by me in such a way as to restore some of the imbalance inherent in my emphasis on the heavily-blended, low-latitude data. My emphasis on the *morphology* of the gas layer neglects, in particular, matters of dynamics and of the physical and chemical details of the interstellar medium, as well as detailed properties of individual features. Dr. Genzel's lectures deal with the *physical state* of the interstellar gas, including that of gas species revealing much more extreme conditions than the morphological tracers HI and CO, and with the conditions within individual clouds. Dr. Elmegreen's lectures deal with the *dynamics* governing interstellar clouds, and with the dynamics governing the gas layer as a whole.

1.2 Some Suggested General References on Various Aspects of Galactic Structure

The selected-topic nature of these lecture notes makes it all the more appropriate to mention some general background reading, and to point out recent reviews which have stressed matters not covered, or covered only superficially, in these notes.

The development of galactic astronomy during the decades between 1900 and 1950 is intriguingly summarized in the book by Berendzen, Hart, & Seeley (1976) entitled "Man Discovers the Galaxies". Summarized in that book are the efforts leading to the establishment of the fact of galactic rotation by B.

Lindblad and J.H. Oort, and the observational verification of the far-reaching consequences of this rotation as predicted by Oort; the arguments of R.J. Trumpler and others leading to the recognition of the general nature of the extinction of starlight during its passage through a pervasive interstellar medium; the observational evidence collected by E.E. Barnard and others of specific examples of the obscuring matter; and the path which led to the establishment of the scale of the Milky Way and its proper placement in the firmament of galaxies. Many of the advancements in galactic studies which occurred in the decades after 1950 were based on studies in the radio regime, in particular of the 21-cm line of neutral hydrogen. The initial development of the radio epoch is the subject of Sullivan's (1984) book on "The Early Years of Radio Astronomy".

The textbook by Binney & Tremaine (1987) on "Galactic Dynamics" is the standard current reference on this general subject. The 1989 Saas-Fee course resulted in a treatment of "The Milky Way as a Galaxy", by Gilmore, King, & van der Kruit which reviews many aspects of the dynamics of our system, as well as many observational aspects. Elmegreen's contribution to this volume represents a general treatment of the dynamics of the interstellar component of the Galaxy. The dynamics of the stellar component of the Galaxy is emphasized in the review by Gilmore, Wyse, & Kuijken (1989). The textbook by Mihalas & Binney (1981) on "Galactic Astronomy" focuses on the observational foundation of matters of galactic structure.

Many of the observational aspects of galactic structure are effectively summarized in the proceedings of the NATO Advanced Study Institute on the Galaxy, edited by Gilmore & Carswell (1987), and in the proceedings of IAU Symposium No. 106 on "The Milky Way Galaxy", edited by van Woerden, Allen, & Burton (1985). The proceedings of IAU Symposium No. 144 on "The Interstellar Disk-Halo Connection in Galaxies", edited by Bloemen (1991), is a rich source of information on aspects of the gas layer which remain largely hidden at lower latitudes. The proceedings of the symposium held on the occasion of F.J. Kerr's retirement, edited by Blitz & Lockman, contains contributions dealing with the special structural and physical circumstances pertaining at large galactocentric distances.

Several older reviews still remain current on a large number of general aspects of galactic structure. The volume of the Chicago compendium on "Galactic Structure" edited by Blaauw & Schmidt (1965) contains generally useful information on the kinematic and other distinctions separating different stellar populations. The review by Kerr (1968) in volume VII of this compendium on "Nebulae and Interstellar Matter" summarizes the physical background of the 21-cm emission line. The more recent review by Kulkarni & Heiles (1988) gives an important treatment of the differing temperature regimes which are represented in 21-cm spectra.

Mention should also be made of two comprehensive reviews which have recently appeared in *Annual Review of Astronomy and Astrophysics*. The article by Dickey & Lockman (1990) on "HI in the Galaxy" discusses observational aspects of HI, as displayed over a range of temperatures and densities, especially regarding its vertical distribution; the detailed information which they give for the higher-latitude gas is particularly relevant here in view of the fact that just this sort of information is not yet generally accessible at lower latitudes. The article by Combes (1991) on the "Distribution of CO in the Milky Way" reviews the galactic-structure information provided by observations of the CO molecule, in a context which includes not only our own Galaxy but also nearby external systems. Burton's (1988) review of the HI morphology in the Milky Way stresses that many, but certainly not all, of the problems initially posed in our own system will ultimately be solved in other systems because they are viewed under more advantageous circumstances than pertain for our embedded perspective on our own Galaxy.

References

1.1 Berendzen, R., Hart, R., Seeley, D. 1976, *Man Discovers the Galaxies*, Science History Pub.
1.2 Binney, J., Tremaine, S. 1987, *Galactic Dynamics*, Princeton University Press
1.3 Blaauw, A., Schmidt, M., (eds.) 1965, *Stars and Stellar Systems* V, *"Galactic Structure"*, University of Chicago Press
1.4 Blitz, L., Lockman, F.J., (eds.) 1988, *The Outer Galaxy*, Springer-Verlag, New York
1.5 Bloemen, J.G.B.M., (ed.) 1991, *Proceedings IAU Symp.* 144, *"The Interstellar Disk-Halo Connection in Galaxies"*, Kluwer Acad. Pub.
1.6 Burton, W.B. 1988, in *Galactic and Extragalactic Radio Astronomy*, G.L. Verschuur, K.I. Kellermann, (eds.), Springer-Verlag, New York, p. 295
1.7 Combes, F. 1991, *Ann. Rev. Astron. Astrophys.* **29**, 195
1.8 Dickey, J.M., Lockman, F.J. 1990, *Ann. Rev. Astron. Astrophys.* **28**, 215
1.9 Gilmore, G., Carswell, B., (eds.) 1987, *The Galaxy*, Reidel Pub. Co.
1.10 Gilmore, G., King, I., van der Kruit, P. 1989, *The Milky Way as a Galaxy*, R. Buser, I. King, (eds.), Geneva Observatory
1.11 Gilmore, G., Wyse, R.F.G., Kuijken, K. 1989, *Ann. Rev. Astron. Astrophys.* **27**, 553
1.12 Kerr, F.J. 1968, in *Stars and Stellar Systems* VII, *"Nebulae and Interstellar Matter"*, B.M. Middlehurst, L.H. Aller, (eds.), University of Chicago Press, 575
1.13 Kulkarni, S.R., Heiles, C. 1988, in *Galactic and Extragalactic Radio Astronomy*, G.L. Verschuur, K.I. Kellermann, (eds.), Springer-Verlag, New York, p. 95
1.14 Mihalas, D., Binney, J. 1981, *Galactic Astronomy*, W.H. Freeman & Co.
1.15 Sullivan, W.T. 1984, *The Early Years of Radio Astronomy*, Cambridge University Press
1.16 van Woerden, H., Allen, R.J., Burton, W.B., (eds.) 1985, *Proceedings IAU Symp.* **106**, *"Milky Way Galaxy"*, Kluwer Acad. Pub.

2 Observations of HI and Other Tracers of the Low-Latitude Morphology of the Interstellar Medium in Our Galaxy

Abstract. Observations of atomic neutral hydrogen, supplemented by those of other tracers of the galactic gas layer at low latitudes, in particular of CO, provide information on a broad range of physical and morphological characteristics of the interstellar medium. Measurement noise need not be an important general limitation to the usefulness of HI survey data; not does it seem necessary to increase the kinematic sharpness of the observations beyond about $1\,\mathrm{km\,s^{-1}}$ or to expect substantially more information in observations made, at low latitudes, with an angular resolution greater than the $0°.5$ currently available. The large areal covering factor and the consequently severe blending in low-latitude HI profiles lead to situations in some broad swaths of longitude where the effects of non-negligible optical depth can be identified. The partial saturation effects follow those predicted by the behavior of the macroscopic velocity-crowding parameter $|\Delta v/\Delta r|$. These effects can be identified in the variations across the sky of the maximum temperature and of the integrated properties of HI 21-cm data. The effects of partial saturation can be demonstrated without resort to a population of cold HI clouds. Nevertheless, HI self-absorption due to relatively cold, discrete, HI features, heavily blended in the data but evidently distributed in much the same global way as the entities comprising the galactic ensemble of molecular clouds, contributes some of the partial saturation.

2.1 Observations of Pervasive, Heavily Blended HI Emission from the Low-Latitude Galactic Gas Layer

Study of the radio emission from the neutral hydrogen component of the interstellar medium is important for several reasons. Neutral atomic hydrogen contributes more mass than any other observed interstellar constituent; because it is so widely spread, the temperature, density, and motion of HI give important information on the whole gamut of physical circumstances which can be found in the complicated galactic environment. It is, of course, fortunate that the interstellar medium is generally transparent enough to emission at the 21-cm radio wavelength that HI investigations can be carried out over much, but not all, of the Galaxy, well beyond the optical horizon; the exceptional circumstances that lead to non-negligible optical depths in the 21-cm line also yield information on temperatures and volume- and column-densities.

HI data provide information on a broad range of physical and morphological characteristics of the interstellar medium. For a few phenomena, HI remains essentially the *only* source of information. The 21-cm line is, for example, the only tracer yet identified with the high-velocity-cloud objects. Another example in which HI information largely stands alone concerns the gaseous counterparts found in association with many interstellar cirrus dust clouds but with highly anomalous velocities; excepting three cases in which these counterparts have been seen in CO emission, the anomalous cirrus velocities are uniquely traced by HI emission. HI is also particularly important for studies of the outer Galaxy; until a few years ago, it was the only directly observable tracer, and it remains the most general one, of the form of the gravitational potential and thus of the overall mass in the outer Galaxy.

Our knowledge of most of the physical and morphological characteristics of the galactic interstellar medium is, however, augmented by information from many other observations. Studies of the Milky Way must specify its morphology in a rather intricate manner. Thus interstellar molecules are mostly concentrated in an annulus lying interior to the Sun's location and hosting most newly-formed stars, but interstellar atomic hydrogen shows no preferential concentration to this annulus; the vertical thickness of the Milky Way gas layer is much thinner in some tracers than in others; the gas layer is especially thin in the inner few kpc of the system, just in the region where the stellar population shows that the gravitational potential describes a thick bulge. Different tracers likewise reveal different kinematic situations. Studies of morphology depend strongly on kinematic information, not only because distances are often derived from velocities, but, more importantly, because velocities reflect important aspects of both the local and global physical as well as dynamical environment.

Hydrogen emission in the radio regime is caused by a hyperfine transition in the ground state of the atom. All energy levels of the hydrogen atom allow two relative orientations of the spins of the electron and the proton. The energy difference in the ground state between the situation when the spins are parallel and when they are antiparallel results, upon spin flip, in a quantum of radiation with a natural frequency — accurately measured in the laboratory using hydrogen masers — of 1420.4058 MHz (see Kerr 1968). This frequency corresponds to a wavelength of 21.106 cm; at this wavelength, a Doppler shift of $1\,\mathrm{km\,s^{-1}}$ corresponds to a frequency shift of $-4.74\,\mathrm{kHz}$. Physical structures with widths as narrow as $1\,\mathrm{km\,s^{-1}}$ are rarely (although occasionally) encountered in HI observations. The natural width of the spin-flip transition is $10^{-16}\,\mathrm{km\,s^{-1}}$, which is negligibly small in any astrophysical context. The probability that the transition from the upper to the lower hyperfine ground state sublevel will occur spontaneously is small: the Einstein emission coefficient for the transition is $2.85 \times 10^{-15}\,\mathrm{s^{-1}}$, corresponding to a spontaneous transition only after some 11 million years.

The prediction published by H.C. van de Hulst in 1945 that the 21-cm line would, in fact, be observable from interstellar space was based on his recognition that the total number of hydrogen atoms in the ground state expected along a typical line of sight traversing the entire Galaxy would be very large and that, in any case, energy-level transitions caused by encounter collisions would occur much more frequently than spontaneous transitions, and would typically result in reorientations of the hydrogen-atom spins after only a few hundred years. The 1945 van de Hulst paper is written in Dutch; an English translation (but unfortunately only of part of the paper which foresees several important

6

aspects of interstellar radiation, later confirmed, in addition to the 21-cm line) is given in the resource compendium compiled by Lang & Gingerich (1979).

The 21-cm line was first detected by H.I. Ewen & E.M. Purcell in 1951 at Harvard University. Their detection was confirmed within a few weeks by C.A. Muller & J.H. Oort in the Netherlands and by J.L. Pawsey in Australia. The detection and the two confirmations were published in the same, September 1, 1951, issue of Nature. A historical account of the circumstances associated with the detection of this first spectral line in radio astronomy is given by Sullivan (1984). A replica of Ewen & Purcell's small horn antenna, which had been perched on a shelf outside of their laboratory window when the line was detected, is now on display at the National Radio Astronomy Observatory in Green Bank, West Virginia (together with replicas of Jansky's antenna and of Reber's). The early Dutch work was carried out using one of the so-called Wurzburg radar dishes which had been abandoned in the dunes at the end of the second World War. The Netherlands Foundation for Radio Astronomy has just this year donated its Wurzburg antenna to the *Deutches Museum von Meisterwerken der Naturwissenschaften und Technik* in Munich, where it will form an important part of the new exhibition on radio astronomy currently being installed.

Figure 1 shows the distribution of HI emission observed in the galactic equator over a full 360° swath of longitude. Interpretation of data such as these in terms of the global properties of the galactic gaseous disk is the subject of these Saas-Fee lecture notes.

Table 1 lists the observational parameters of surveys of the HI 21-cm line which cover enough of the sky and which are sensitive enough by current standards to be suitable for general investigations of galactic morphology. Earlier and more specialized surveys are summarized by Kerr (1968), Burton (1974), Burton & Liszt (1983), and Bajaja (1983). A limitation of all HI survey data concerns the rather coarse angular resolution provided by the telescopes of modest size which are the only ones practicable for time-consuming survey work. Although single-dish telescopes (rather than interferometric arrays) must be used for most morphological studies of the HI in our Galaxy (because interferometers do not respond usefully to emission spread so smoothly as that from galactic hydrogen), the higher resolution provided by the arrays is crucial in some specific Milky Way studies and in all work on the HI distribution in external galaxies. It is, in fact, difficult to defend the need for resolution higher than that afforded by a 25-m telescope for most low-latitude investigations of the Milky Way. At higher latitudes, where line blending is not so dominant as it is near the galactic equator, higher angular resolution *does* reveal additional information, but is not practical except in regions of quite limited extent. The Leiden/Dwingeloo survey, currently being completed, has involved more than two years of full-time observing on the 25-m telescope. The principal merits of this survey lie in its velocity coverage and resolution, and in its total sky coverage.

Data reduction of material intended for galactic investigations commonly involves converting the measured frequencies to radial velocities, expressed in

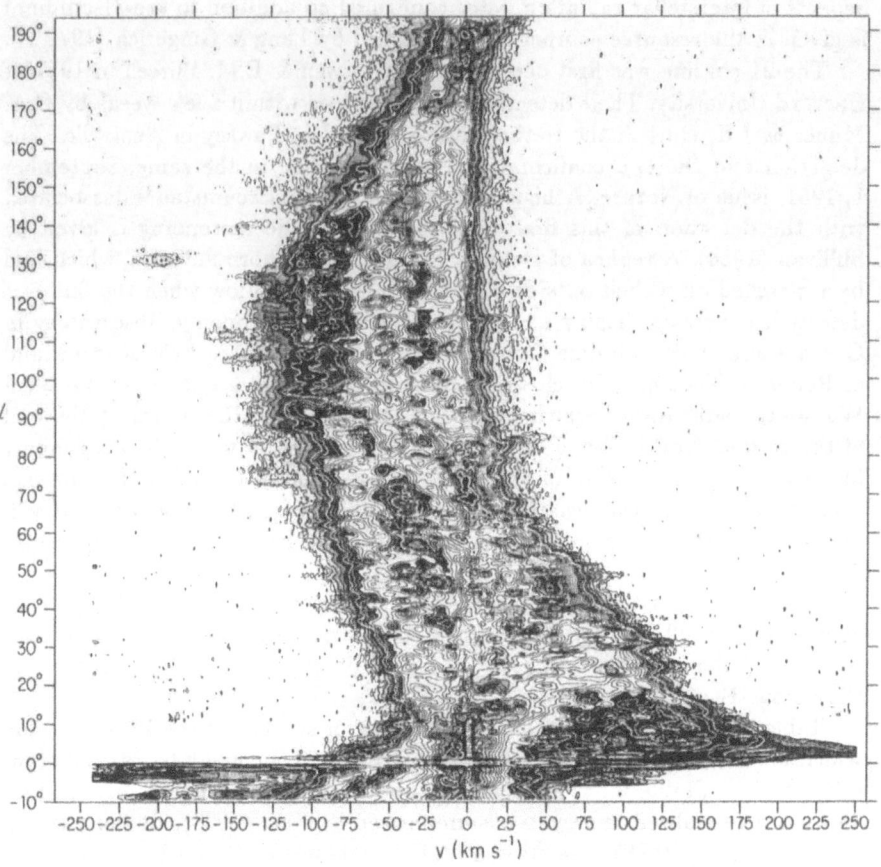

Fig. 1a. Longitude, velocity arrangement of neutral hydrogen emission intensities observed along the galactic equator, $b = 0°$, in the first and second longitude quadrants. The observations are from Burton (1985). The contours represent brightness temperatures at levels $T_b = 0.4, 0.8, 1.5, 2.5, 4, 7, 10, 15, \ldots, 35, 45, \ldots$ K; the grey-scaling darkens until $T_b = 20$ K, and resumes at the 60 K level

the Local Standard of Rest reference frame (see review by Delhaye 1965) which accounts for the motion of the Sun with respect to its neighbors. It is not unlikely that uncertainties in the determination of this reference frame involve systematic errors of the order of $5\,\mathrm{km\,s^{-1}}$ (see, for differing interpretations of the possible error: Shuter 1981; Clube 1985; Blitz & Spergel 1991). Such an error is large compared to the typical internal instrumental accuracy of the data, which can be made very small. It is also not negligible compared to some of the broadening mechanisms intrinsic to the galactic interstellar medium. For this reason, and because of differences in the definitions of the reference frame used in the literature, it is important that the matter be explicitly stated. The most commonly used definition of the LSR, and the one used here, involves the standard solar motion of $20\,\mathrm{km\,s^{-1}}$ toward $(\alpha, \delta) = (18^{\mathrm{h}}, 30°)$, epoch 1900.

8

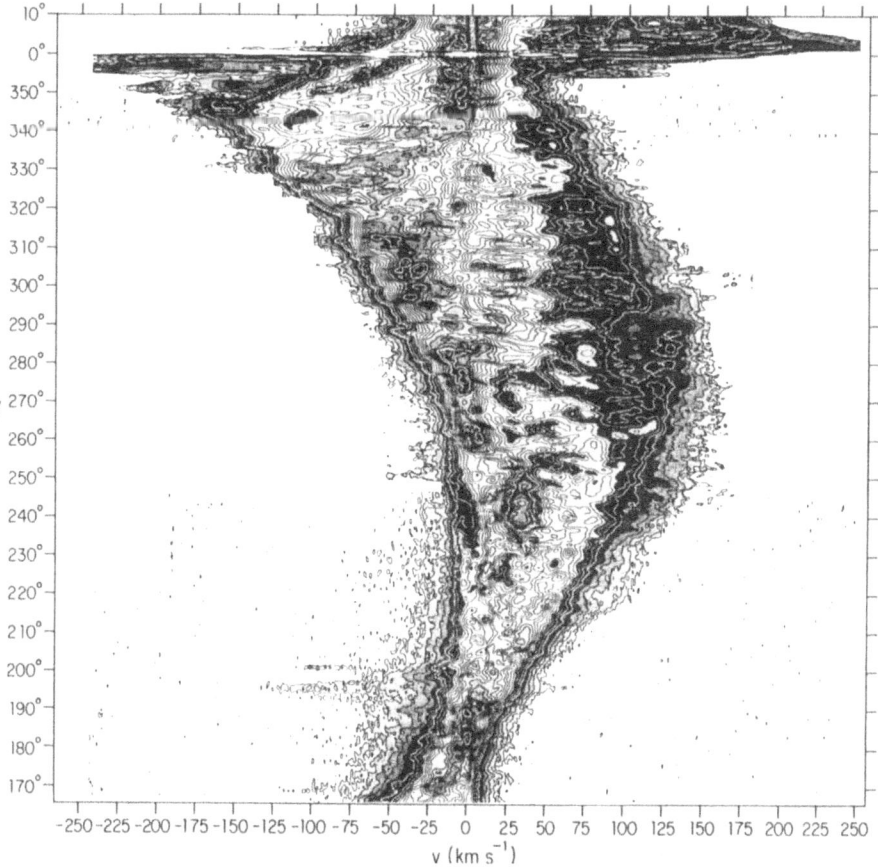

Fig. 1b. Longitude, velocity arrangement of neutral hydrogen emission intensities observed along the galactic equator, $b = 0°$, in the third and fourth longitude quadrants. The observations in the range $240° < l < 350°$ are from Kerr et al. (1986); outside this range they are from Burton (1985). The representation of the brightness temperatures is as in Fig. 1a

Measurement noise need not be an important limitation to the usefulness of HI survey data. Integration times of a few minutes per spectrum are sufficient to produce an r.m.s. noise level of less than 0.1 K; there is no motivation known for substantially greater sensitivity for most low-latitude studies, except those dealing with the high-velocity-cloud phenomenon. To reach this sensitivity, however, it is important that the observations be corrected for contamination by stray radiation. Such contamination will arise for all radio telescopes of the conventional parabolic design, involving a receiver or subreflector supported in the optical path. The problem is complicated by reflection of unwanted radiation from the ground and off of the receiver support structure, and entering the sidelobes of the antenna. The HI survey of Stark et al. (1992) was the first to be able to achieve uniform sensitivity at the 0.1 K level: the Bell Laboratories

Table 1. General surveys of the galactic distribution of HI and of CO

Authors	Reference	Beam [arcmin]	l-coverage [degree]	b-coverage [degree]	v-coverage [km s^{-1}]	Sensitivity [K]
Surveys of HI emission:						
Weaver & Williams	1973	36	10 to 250 $\Delta l = 0.5$	-10 to 10 $\Delta b = 0.25$	$v_0 \pm 100$ $\Delta v = 2.1$ (6.3)	1.7 (1.0)
Heiles & Habing	1974	36	all l at $\|b\| > 10$, $\delta > -30$; $\Delta l = 0.3/\cos b$	all $\|b\| > 10$ at $\delta > -30$; $\Delta b = 0.6$	± 50 at $\Delta v = 2.1$; to $-92, +75$ at $\Delta v = 6.3$	1.2
Burton	1985	21	all l at $\delta > -46$, $\|b\| < 20$ (33); $\Delta l = 1$	$\|b\| < 20$ (33) at $\delta > -46$; $\Delta b = 1$	$v_0 \pm 250$; $\Delta v = 1$ (2.1)	0.2 (0.14)
Kerr et al.	1986	48	$240 < l < 350$; $\Delta l = 0.5$	$\|b\| < 10$; $\Delta b = 0.25$	$v_0 \pm 150$; $\Delta v = 2$	0.8
Stark et al.	1992	150	all l at $\delta > -40$; $\Delta l \geq 2$	all b at $\delta > -40$; $\Delta b = 2$	$v_0 \pm 300$; $\Delta v = 5.3$	0.1
Leiden/ Dwingeloo	1992	36	all l at $\delta > -30$; $\Delta l = 0.5/\cos b$	all b at $\delta > -30$; $\Delta b = 0.5$	$v_0 \pm 460$; $\Delta v = 1$	0.07
Surveys of CO emission:						
Stark et al.	1988	0.6	$-5 < l < 122$ $\Delta l = 3'$	$\|b\| < 1$ $\Delta b = 3'$	$v_0 \pm 80$; $\Delta v = 0.7$	^{13}CO: 0.15 ^{12}CO: 0.3
Dame et al.	1987	8.7	all l; $\Delta l = 0.5$	$\|b\| < 3.3$; $\Delta b = 0.5$	$v_0 \pm 80$; $\Delta v = 1.3$	0.2
Clemens et al.	1986	0.8	$8 < l < 90$ $\Delta l = 0.125$	$\|b\| < 1.05$ $\Delta b = 0.05$	$-50 < v < 150$ $\Delta v = 1$	1.2

horn antenna used for this survey has an unblocked aperture and thus does not suffer badly from stray radiation. (This was an important consideration when the same horn antenna was used by Penzias & Wilson for the discovery in 1965 of the 3 K cosmic background radiation.)

The Leiden/Dwingeloo survey is being corrected for effects of stray radiation by P. Kalberla, at the University of Bonn, using the algorithm described by Kalberla et al. (1982). Figure 2 illustrates the necessity of the correction for stray-radiation contaminating emission entering the sidelobes of the Dwingeloo antenna, by a comparison of two observations made with the telescope pointed toward a single direction, but at differing times. Lockman et al. (1986), in particular, have stressed the importance of stray-radiation contamination, which can be important at intensities up to about $T_b = 0.5$ K, and, of course, over a velocity swath as broad as that corresponding to the part of the Milky Way which might happen to be above the horizon and contributing to the sidelobes at the time of observation.

It is important to ask if additional information would be gained by observing with better velocity resolution, or with better angular resolution, than is the case for the surveys listed in Table 1. Insofar as this question is asked with a view to the derivation of galactic morphology from low-latitude emission

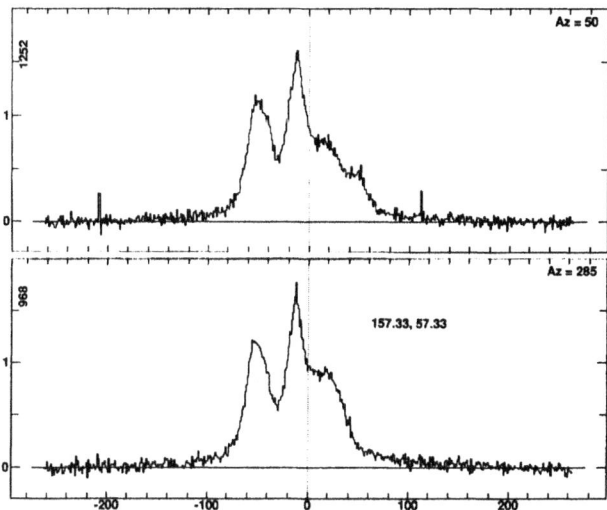

Fig. 2. Contamination from stray radiation demonstrated by comparing two spectra from the Leiden/Dwingeloo survey made towards the same direction but at differing hour angles, when emission from differing regions of the Galaxy enter the antenna's sidelobes. Single-dish 21-cm spectra must be routinely corrected for stray-radiation contamination. This contamination seldom reaches intensities as high as $T_b = 0.5\,\mathrm{K}$, but it is typically broad in velocity so that in some directions the stray radiation contributes some 50% of the integrated emission. Lockman et al. (1986) estimate uncertainties in N_{HI} of up to about $5 \times 10^{19}\,\mathrm{cm}^{-2}$

data, one may answer, albeit cautiously, that very little additional information is provided by velocity resolution better than about $1\,\mathrm{km\,s}^{-1}$ or angular resolution better than about 10 arcminutes. The effective limit to the necessary velocity resolution follows from the *intrinsic* physical properties of individual HI structures; the effective limit to the useful angular resolution, on the other hand, is a consequence of the severe blending which characterizes low-latitude HI spectra.

Regarding the effective limit to the necessary angular resolution, we stress that this is *not* determined by the intrinsic structural length scales, but by the pervasiveness of HI emission at low latitudes. The situation is illustrated by Fig. 3. Emission spectra measured at *low* latitudes show but little additional structure if measured with the 4-arcminute beam of the Arecibo 305-m telescope rather than with the 36-arcminute beam of the Dwingeloo or Hat Creek 25-m telescope. At *high* latitudes, say at $|b| > 10°$, the situation is different because there the covering factor, or areal filling factor, is not typically unity over the 36-arcminute beam as it is at the lower latitudes. Although, certainly, high-resolution observations at high latitudes will give important information on length-scale and covering-factor matters, it is sobering to note that high-latitude observations only sample the local neighborhood, and encompass a total volume amounting to only some few % of the total volume of the galactic gas layer. Studies with the goal of determining morphological aspects of the Galaxy as a whole must make use of the low-latitude material.

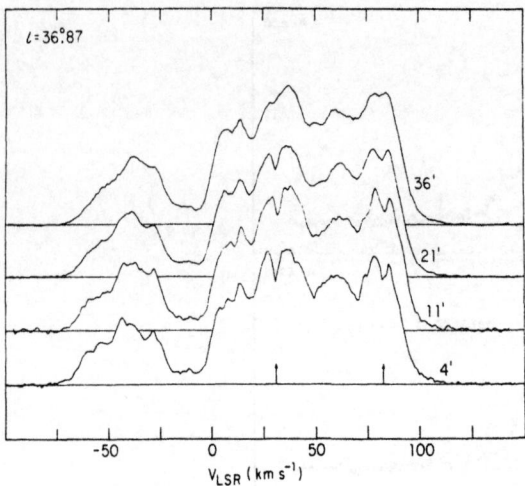

Fig. 3. Change in appearance of low-latitude HI spectra with increasing angular resolution (from Bania & Lockman 1984). The telescopes were each pointed toward $l = 36°.87, b = 0°.0$, but provided the differing angular resolutions marked beside each profile. Most of the increase in detail seen under higher resolution may be attributed to localized self-absorption features; the velocities of two such features are marked by arrows near 30 and $83\,\mathrm{km\,s^{-1}}$. At low $|b|$, increased angular (or, for that matter, kinematic) resolution results in little change in the measured column depth or maximum temperature. The perceived areal covering factor remains high

Regarding the velocity resolution, we note that kinematic variations which are unresolved at a channel separation of $1\,\mathrm{km\,s^{-1}}$ occur only very rarely. Although emission profiles contaminated by self-absorption show sharper structures than shown by uncontaminated ones, even these sharper structures are generally resolved at the $1\,\mathrm{km\,s^{-1}}$ level. Thus there seems to be no general motivation to seek data with kinematic resolution sharper than about $1\,\mathrm{km\,s^{-1}}$. There are several arguments that this is a physically imposed limit. Although isolated physical structures can only rarely be identified as such in low-latitude HI data (and then only in the self-absorption case, not in the pure-emission case), such individual features *can* be identified at high latitudes. Even isolated high-latitude features also show, however, little structure kinematically sharper than about $1\,\mathrm{km\,s^{-1}}$.

A more robust reason for concluding that the kinematic limit of about $1\,\mathrm{km\,s^{-1}}$ is physically relevant is given by absorption measurements, taken against a suitable background source of continuum radiation. The effective angular resolution of such absorption data is determined by the angular size of the background source, and if this source is, for example, an external galaxy, this effective resolution can be vanishingly small, minimizing the role played by blending of unrelated structures within the voluminous sampling element of emission observations. Although absorption observations reveal somewhat narrower line widths than seen in emission, these widths typically involve velocity dispersions of at least $2\,\mathrm{km\,s^{-1}}$ (Dickey et al. 1978; Dickey & Garwood 1989).

This conclusion holds also for the HI gas layer in other galaxies, including those presenting minimum line blending by virtue of their more or less face-on appearance.

The broadening of the 21-cm spectral line reveals information on a variety of physical mechanisms, although most global mechanisms contribute velocity differences greater than the $1\,\mathrm{km\,s^{-1}}$ resolution limit under discussion. Spectra observed at latitudes closer than about $10°$ to the galactic equator typically cover a radial velocity span of more than $100\,\mathrm{km\,s^{-1}}$. Most of this broadening stems from the differential rotation of the Galaxy. Broadening amounting to 5 or $10\,\mathrm{km\,s^{-1}}$ (but sometimes more) reflects *mass motions* of gas structures, which can have a variety of causes. Mass motions within an individual structure typically amount to at least several $\mathrm{km\,s^{-1}}$.

Smaller linewidth contributions are expected from thermal broadening. Such broadening, if due to the random velocities of hydrogen atoms within a single concentration of gas characterized by a single temperature, would produce a spectral component with a Gaussian shape specified by a dispersion $\sigma_v = 0.09\sqrt{T_k}\,\mathrm{km\,s^{-1}}$. For a realistic kinetic temperature of $100\,\mathrm{K}$, σ_v is $0.9\,\mathrm{km\,s^{-1}}$. This dispersion corresponds to the commonly-used measure of full-width to half-maximum intensity, where $\mathrm{FWHM} = 2.36\,\sigma_v$, of $2.1\,\mathrm{km\,s^{-1}}$. HI lines as narrow as this are rarely measured, either in emission or in absorption, either at high latitudes or at low ones, either in our Galaxy or in external ones. This statement may be generalized: spectral features as narrow as what would be expected only from thermal broadening are rarely, if ever, observed in radio studies of *any* tracer of the interstellar medium. In 21-cm work, only a few cases have been observed of spectral features narrow enough that interesting upper limits may be placed on kinetic temperatures.

There are several reasons that lines are in fact broader than the thermal case; some of these are discussed in Chap. 3. But whatever the details, *blending* is always a factor in low-latitude HI data. It is important to distinguish between the *areal* filling factor, or *covering* factor, and the *volume* filling factor. Both are important parameters, but their values have not yet been quantitatively specified for the HI gas in the Galaxy. At low latitudes, the HI covering factor is near unity: no empty line of sight has ever been found, and not even an empty velocity interval *within* a single HI spectrum. Blending in the HI profiles is so severe at low latitudes that it has not yet proven possible to determine the volume filling factor with any certainty. It would be very difficult to argue directly against a volume filling factor of unity on the basis of low-latitude HI data alone; in fact, important insights into the morphology of the HI gas layer can be gained by considering the gas distribution as uniform and space-filling, in a plane-parallel layer. But there are, of course, a gamut of reasons for concluding that HI emission does *not* uniformly fill space. Many of these reasons are based on quite plausible theoretical arguments, although, as Bregman & Ashe (1991) have pointed out, the theoretical predictions relevant to filling-factor questions do not yet adequately account for the HI observations. A large covering factor is not, of course, incompatible with a volume filling factor substantially less than

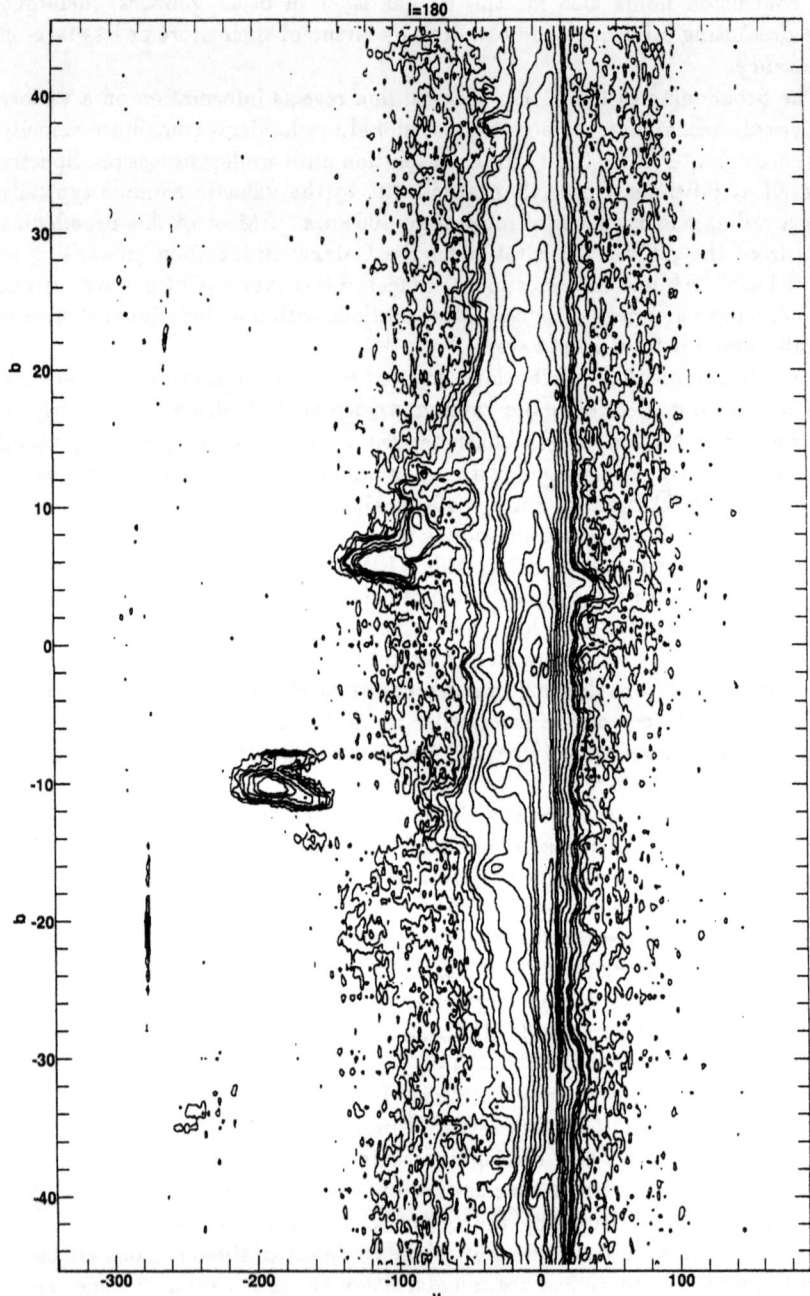

Fig. 4. Neutral hydrogen emission intensities displayed along a representative cut perpendicular to the galactic equator along the meridian plane $l = 180°$. (The data are from the Leiden/Dwingeloo survey of Hartmann & Burton)

unity; such indeed seems to be the situation relevant to low-latitude galactic HI.

It is natural to investigate both the areal and the volume filling factors at high latitudes, where line-of-sight blending is less severe than at lower latitudes. It appears that at $|b| > 20°$ lines of sight through the gas layer sample path lengths short enough that the filling factors may become accessible; the volume filling factor is nowhere unity, but the areal filling factor is revealed in some directions at high $|b|$ to be less than unity (see Burton et al. 1992).

Figure 4 shows HI emission from a representative cut made perpendicular to the galactic equator along the meridian plane $l = 180°$. Empty directions are also never observed at high latitudes, but it appears possible to unravel structure in some cases along individual lines of sight; such unravelling has not yet been possible at low latitudes.

If it has proven difficult to argue directly from the observations against the uniform, plane-parallel case at *low* latitudes (even though the indirect evidence to the contrary is utterly convincing), the situation is more directly revealed at high latitudes, say at $|b| > 20°$. If the gas were uniformly distributed, i.e. with volume filling factor unity, and at a single temperature, and without any mass motions other than galactic rotation, then high latitude profiles would only slightly deviate from Gaussian form, in accordance with the geometric perception of the rotation motions. Under that circumstance, an HI profile observed at high latitude would show a single spectral feature, centered within one $\mathrm{km\,s}^{-1}$ or two from $v = 0\,\mathrm{km\,s}^{-1}$, and only slightly broader than the width expected for the thermal case. Observations such as those shown in Fig. 4 reflect, of course, a much more complex situation. If, on the other hand, the gas layer were still ubiquitously filled and kinematically well behaved (which is the zero-order approximation we wish to test at low latitude), but contained gas components at different temperatures, for example a component at a temperature of about a hundred degrees confined to a layer thinner than that containing a second component at a higher temperature extending into the lower halo, then high-latitude profiles would show a broader wing superimposed on a narrower core, but would otherwise be much better behaved than the observed situation. Thus even allowing for a mixture of temperature regimes, as the observations indeed require (see Kulkarni & Heiles 1988; Dickey & Lockman 1990), the plane-parallel assumption is too simple to be of much use at all at high latitudes.

2.2 Observations of CO and Other Tracers of the Gas Layer Morphology

Although crucial to determination of many morphological aspects of the galactic interstellar gas, 21-cm observations do not, by any means, reveal the situation completely. There are other widely distributed constituents of the interstellar medium which have, like HI, generally observable spectral lines, but which reveal different aspects of galactic structure. Although in the temperature range from a few tens of K to a few thousands, the interstellar gas is dominated by atomic hydrogen, hotter, ionized, gas can be followed using, for example, hydrogen recombination lines, and colder, molecular, gas can be followed using a variety of molecular species.

In interstellar regions with temperatures colder than a few tens of K, hydrogen in the molecular form predominates by mass over all other gaseous material. H_2 molecules at these temperatures do not, however, emit in the radio or optical windows because H_2 lacks a permanent electronic dipole moment. Information from the ultraviolet H_2 absorption bands, first observed by Carruthers (1970), is limited to that from short lengths of path by severe extinction in the UV due to interstellar dust. Observations of the CO molecule have been used widely as a *surrogate* for observations of molecular hydrogen: CO is thought to trace H_2 because the CO rotational transitions are predominantly excited by collisions with H_2. These rotational transitions lead to radiation at mm- and sub-mm wavelengths, where the interstellar medium is largely transparent. Like H_2, CO is stable at low temperatures. It is several orders of magnitude more abundant than any molecule other than H_2.

The ^{12}CO $(1 \rightarrow 0)$ line at a wavelength of 2.6 mm was first observed by Wilson et al. in 1970. Since then, large efforts have been directed toward observations of this line as a tracer of galactic structure, first in the Milky Way, and later, when improved receiver sensitivities made it feasible, in external systems, and also as a tracer of a variety of chemical and physical properties of the interstellar medium. There have been three surveys of CO emission done on a large-enough scale to remain important in galactic-structure work, but a large amount of other, detailed, CO material is available for specific purposes. Some of the observational parameters are given in Table 1. The surveys have generally been published piecemeal: the specific references shown in Table 1 give the appropriate detailed references. Single-beam mapping is always time consuming: the added aspect of small telescope beams in the mm-regime has, so far, precluded complete sky coverage.

Figure 5 shows the distribution of ^{12}CO emission from the inner part of the galactic equator. This distribution can be viewed as the molecular analog to that shown in the Fig. 1 plots for the HI emission. The molecular gas is largely marshalled in interstellar clouds of volume densities typically several orders of magnitude greater than the densities characterizing the pervasive HI gas. Unlike the heavily-blended HI situation at low latitudes, individual molecular features may be isolated, although rarely fully resolved. The interstellar medium outside of the molecular clouds is largely transparent at mm wavelengths. Self-absorption of the ^{12}CO $(1 \rightarrow 0)$ line *within* a cloud is, however, substantial. Emission from the less abundant, more transparent, ^{13}CO isotope is a better probe of the inner regions of clouds.

Viewed as a surrogate of molecular hydrogen, it is important to establish the conversion factor which would allow the column density of H_2 to be derived from the integrated CO emissivity. This conversion has remained controversial (see Combes's review 1991). The issue involves assumptions concerning the [C/H] abundance ratio, the fraction of C confined to CO, the $[^{12}CO/^{13}CO]$ abundance ratio, and the degree of constancy of these measures over the Galaxy, as well as simplifications pertaining to the evidently quite high degree of saturation in the observed CO lines. Although the conversion is crucial for many galactic-morphology purposes, including the obvious goal of determining the total amount of interstellar material in the Milky Way and the relative amount of that mass which resides in dense molecular clouds (and perhaps available for star formation) compared to that which resides in the more diffuse, more pervasive, atomic medium, there remain important global matters to which the details of the $\int T_b(CO)\,dv \to N(H_2)$ conversion are not particularly important. Data such as that in Fig. 5 reveal the kinematics of the molecular clouds, and thus have allowed determination of the shape of the ensemble of clouds. Even if the opacity of the CO lines hides the interiors of the clouds, the clouds may at least be *counted*.

There are additional gaseous constituents of the interstellar medium which reveal information on galactic morphology, being widely spread throughout the Galaxy and emitting or absorbing with sufficient intensity at radio wavelengths. Emission at radio wavelengths from the higher-quantum-level recombination lines of ionized hydrogen are particularly useful tracers of the gas at the lower density, higher temperature, ends of the range of interstellar gas. Most of the recombination line emission is contributed by hot, ionized, material confined to discrete, large, HII regions; the data are relevant to the physical details of these regions with their embedded ionizing heat sources associated with ongoing star formation. The morphological interest of the data lies in the information on the overall galactic distribution and kinematics of the HII regions (see e.g. Lockman 1984). Although emission from *discrete* HII regions dominates the intensities of most hydrogen recombination lines, there does seem to be a contribution from a diffuse, distributed medium. At low latitudes, this contribution is difficult to separate from what might be expected from a superposition of many low-density HII regions. The evidence at higher latitudes for a layer of diffuse ionized hydrogen is more convincing. Reynolds (1991, and references there) has emphasized that diffuse ionized hydrogen at temperatures of about 10,000 K is the dominant state of the interstellar matter in the lower galactic halo, at z-heights of about 1 kpc.

Recent observations of the far-infrared fine-structure line at $158\,\mu$m from singly-ionized interstellar carbon have shown that about 10% of the total mass of the neutral interstellar medium is located near the interfaces between neutral, relatively dense, interstellar clouds and lower density, fully-ionized, regions (see e.g. Stacey et al. 1991). The [CII] line is associated with relatively warm molecular gas, and therefore promises to be a good diagnostic of the energetics of star formation from molecular clouds.

Several widespread constituents of the galactic interstellar medium are principally observed in absorption, rather than in emission. These tracers include formaldehyde, ammonia, and the hydroxyl radical. At low latitudes, the usefulness of lines usually measured in absorption is limited by the distribution of the sources of absorbed background radiation; the vagaries of the distribution of background radiation prejudice the absorption tracers for many morphological purposes. At higher latitudes, however, absorption tracers are particularly

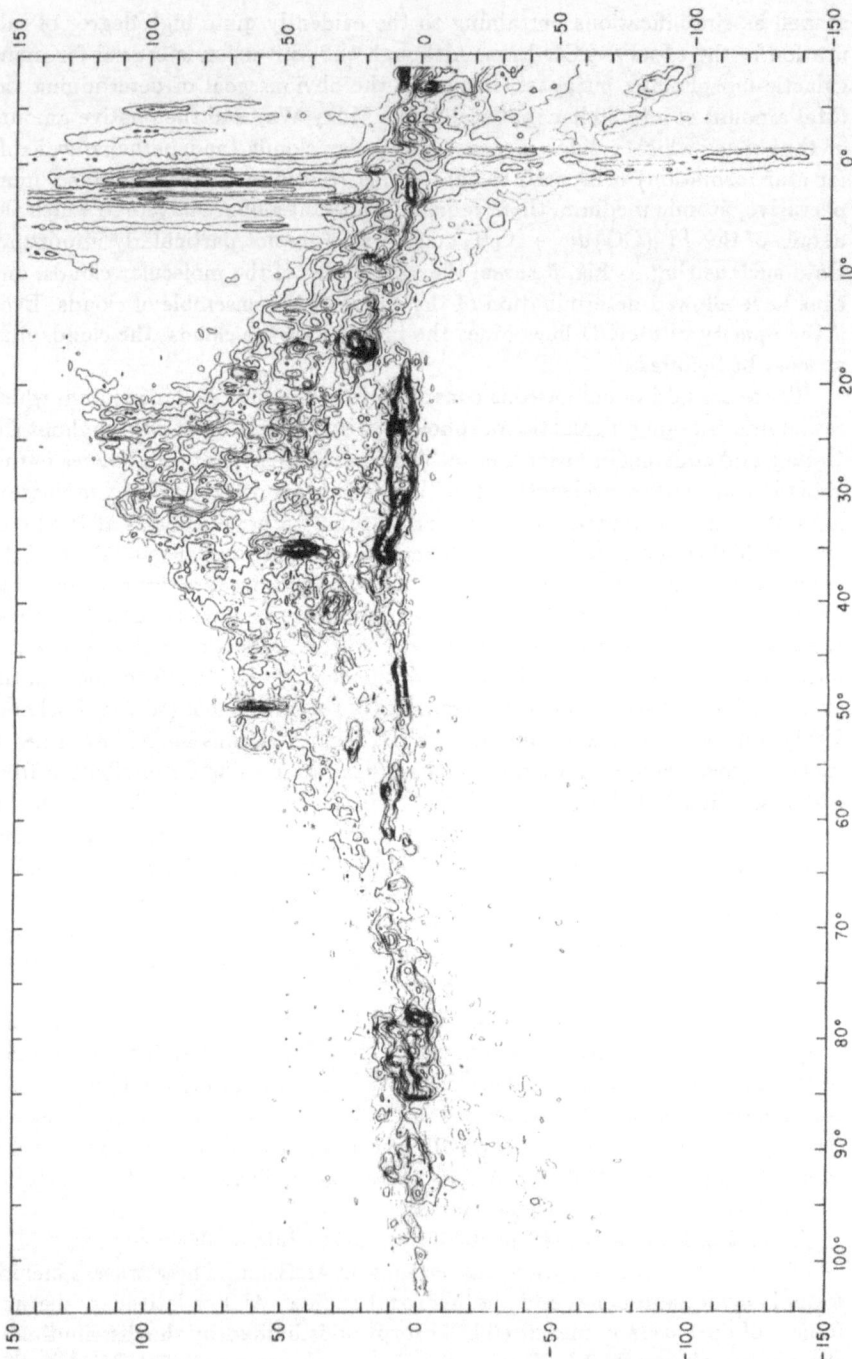

Fig. 5a. Longitude, velocity arrangement of ^{12}CO emission intensities observed in the first longitude quadrant between latitudes $-3°.25$ and $+3°.25$ and displayed projected onto the galactic equator (from Dame et al. 1987)

18

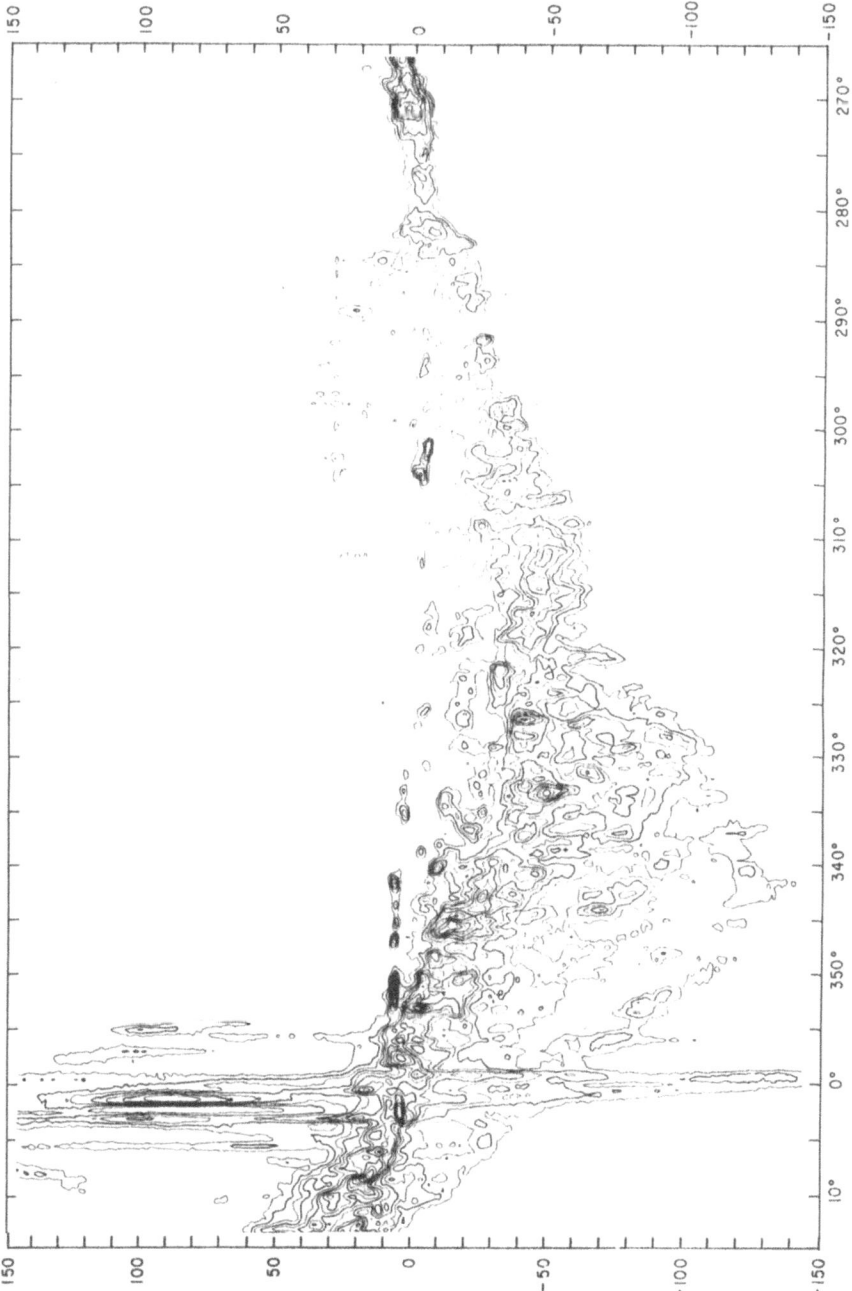

Fig. 5b. Longitude, velocity arrangement of ^{12}CO emission intensities observed in the fourth longitude quadrant between latitudes $-3°.25$ and $+3°.25$ and displayed projected onto the galactic equator (from Dame et al. 1987)

valuable because they can provide limits on distances, if the distance of the background source of radiation — often a star — is known: kinematic distances are, of course, generally unreliable from higher latitudes. Danly (1989, 1991) has emphasized that absorption data, furthermore, provides information from species which would be too weak to study in emission.

Mention should also be made of the deuterium isotope of hydrogen. Most of the deuterium formed during the Big Bang reacted to form helium; the amount left is thought to measure the amount of matter in the familiar baryonic form in the early Universe (see Wagoner 1973). Deuterium produces, as hydrogen does, a spectral line by a hyperfine spin-flip transition. This analogue to the HI 21-cm line occurs at a wavelength of 92 cm, or a frequency of 327 MHz. Substantial efforts to detect the DI line, both in emission as well as absorption, are currently being made by several groups. Recently Heiles (1992) announced a tentative detection of this important spectral line.

2.3 Global Optical-Depth Effects and the Interpretation of Measured HI Intensities

An observation of the 21-cm line gives a spectrum showing intensity, as a function of frequency or velocity, Doppler-shifted from the rest frequency of the hyperfine transition by the radial motion of the gas with respect to the observer. Some of the circumstances which influence the measured intensities, in particular the effects of optical depth, are discussed below. The circumstances which influence the measured kinematics are the subject of Chap. 3.

The intensity of an HI line profile is usually expressed as a brightness temperature, which, in the Rayleigh-Jeans approximation valid at radio wavelengths, is linearly proportional to the specific intensity (see reviews by Field 1958; Kerr 1968; and Kulkarni & Heiles 1988). Brightness temperature is related to kinetic temperature, measuring the energy of the gas, if atomic encounters determine the populations in the energy levels. Interpretation of the measured intensities is complicated because the telescope beam samples a large volume of space for all spectra; each spectrum typically embraces a wide range of interstellar temperatures, densities, and optical depths, as well as a range of kinematics. Under these conditions, it is not easy to decipher the circumstances of radiation transfer.

In the simplest possible case of an isolated cloud, homogeneous and at a single temperature, the equation of radiative transfer yields

$$T_b(v) = T_{bg}(v) \exp(-\tau(v)) + T_k [1 - \exp(-\tau(v))]. \tag{1}$$

What is in fact measured is the difference,

$$T_b(v) - T_{bg}(v) = (T_k - T_{bg}(v)) [1 - \exp(-\tau(v))], \tag{2}$$

between the contribution from the cloud and the contribution from background radiation incident on the cloud. Note that if $T_{bg}(v) < T_k$ then the line is observed in *emission*; if $T_{bg}(v) > T_k$ then the line is observed in *absorption*. If the background temperature and the gas kinetic temperature are equal, there is no net emission; this situation probably is not of much relevance in Milky Way studies, where the gas temperatures are greater than the 3 K cosmic background, but may well be important for tenuous intergalactic HI. In galactic studies it is common to ignore the general continuum background radiation, assuming that $T_{bg}(v) \ll T_k$. Cases of absorption against discrete background sources, or cases of self-absorption by a foreground HI feature superposed on a warmer HI background, do not allow $T_{bg}(v)$ to be ignored at all velocities.

The HI column density specifies the number of atoms at a particular velocity in a cylinder of cross-sectional area $1 \, \text{cm}^2$. At a particular velocity the optical depth and column density are related by $N_{HI}(v) = 1.823 \times 10^{18} \, T_k \, \tau(v)$. The column density contributed by the portion of a line of sight contributing emission to a specified velocity range, Δv, is thus given by

$$N_{HI}(\Delta v) = 1.823 \times 10^{18} \int T_k \, \tau(v) \, dv \quad [\text{cm}^{-2}]. \tag{3}$$

This is in general not a measured quantity. The limiting cases of optically thinness and of optical thickness are both interesting, and both occur in interstellar HI work.

If all the HI contributing at a particular velocity is *transparent*, then $\tau(v) \ll 1$. (The velocity dependence expressed here for the various parameters is not trivial in galactic studies, because the long lines of sight at low latitudes can typically involve a plethora of structures contributing differently at different velocities.) In the optically thin case, $T_b(v) = T_k \left[1 - \exp(-\tau(v))\right] \simeq T_k \, \tau(v)$, so that the column density becomes measurable through the integrated intensity observed, because all photons emitted can escape the medium:

$$N_{HI} = 1.823 \times 10^{18} \int T_b(v) \, dv \quad [\text{cm}^{-2}]. \tag{4}$$

A measure of the *volume* density smoothed over a length of path, Δr, corresponding to a specified velocity range, Δv is given by

$$n_{HI}(\Delta v) \propto \int T_b(v) \, \frac{dv}{\Delta r} \quad [\text{cm}^{-3}]. \tag{5}$$

The condition of optical thinness is frequently invoked, but (as discussed below) it is probably valid only for certain velocity segments of low-latitude profiles; at higher latitudes, the condition is much more generally valid.

If all the HI contributing at a particular velocity is *opaque*, then $\tau(v) \gg 1$. In this case the observed brightness temperature does not reveal the number of atoms emitting in the relevant element, but it does indicate the gas temperature, because $T_b(v) = T_k \left[1 - \exp(-\tau(v))\right] \simeq T_k$. At lower latitudes, it is not uncommon for $\tau(v)$ to approach or even exceed unity (over some, but not all, velocities on a particular line of sight). Velocity blending and beam smearing remain problems, however: so many different environments will generally contribute to each velocity range, that the *measured intensities represent an averaging difficult to quantify.* The profile integral does lead to an interesting measure, however; in the case of substantial optical depths,

$$\int T_b(v) \frac{dv}{\Delta r} \simeq T_k \left| \frac{\Delta v}{\Delta r} \right|, \tag{6}$$

where Δv is the velocity extent of the portion of the profile considered. The "velocity-crowding" parameter $|dv/dr|$, which is discussed below and further in Chap. 3 of these lecture notes, dominates the behavior of the observed profile integrals in cases where saturation effects cannot be ignored. In the optically thin case, the left-hand side of the above equation gives a measure of volume density; in the thick case, it follows the behavior of the velocity-crowding parameter. Examining the *actual* behavior of the measurable left-hand quantity provides an estimate of the eventual importance of saturation effects.

The wide variety of conditions found in the interstellar environment preclude the validity of the single-temperature assumption. Consequently, interpretation of measured HI intensities in terms of kinetic temperatures is usually not straightforward. We note in this regard that the volume of interstellar space sampled by a typical survey-telescope beam of $0°.5$ angular width characteristically involves an approximately cylindrical volume whose linear dimension on the plane of the sky is about 100 pc. The length parallel to the line of sight of this characteristic volume depends on the galactic kinematics sampled along the direction in question, but for a typical direction the representative sampling element of, say, $2 \, \mathrm{km \, s^{-1}}$, might be several hundred pc long.

Although 21-cm observations which measure the *emission* properties of the gas are most commonly used for studies of galactic morphology work, the line can also be measured in *absorption*, under some circumstances; such measurements play a crucial role in investigations of certain physical parameters of the interstellar medium (see e.g. Dickey et al. 1983; Dickey & Garwood 1989; Heiles & Kulkarni 1988). Particularly important is the information which absorption data provide regarding the optical depth of the absorbing gas. Absorption measurements require a suitable source of background emission at 21 cm; the background emission typically comes from a discrete source of continuum radiation, which is commonly extragalactic and therefore subtends a very small angle. The advantage of this situation lies in the very narrow cylindrical element of space which is sampled; the disadvantage lies in the vagaries of the distribution of suitable background sources.

2.4 Interpretation of the HI Integrated Emission and the Maximum Brightness Observed: Is the Low-$|b|$ HI Gas Layer Globally Transparent?

The total number of hydrogen atoms in the line of sight at all velocities is a measure of obvious interest. Interpretation of the integrated emission requires establishing the extent to which the HI gas layer is transparent. The question must be posed globally, with reference to the material as it is perceived, severely blended, in single-dish emission spectra. The answer to the question seems not a simple one: there are substantial sectors of longitude and of velocity where the optical depth seems to hover around unity.

2.4.1 An Indication that the HI Gas Layer is Globally Thin, and a Contrary Indication that it is Thick.

An indication that the HI gas layer viewed near $|b| = 0°$ is optically *thin* is given by the following argument. Because of the well-known distance ambiguity, a line of sight through the *inner* Galaxy samples a given velocity *twice*, once on the near side of the tangent-point circle (where the line of sight is tangent to a galactocentric circle), and once on the far side; the two regions contribute the same velocity because, for any plausibly well-behaved kinematics, the regions are equally distant from the galactic center. Where the same line of sight traverses the *outer* Galaxy, however, a given velocity is contributed by a *single* spatial region because the velocity-to-distance transformation is single valued. For a line of sight traversing both the inner and the outer Galaxy, this situation means that, typically, twice as many HI atoms are emitting per velocity element at velocities corresponding to those expected from the inner Galaxy compared to the number emitting from outer-Galaxy velocities.

Transparency is required for photons emitted from the far distance to continue, unabsorbed, their passage through material at the near distance. The details of the number of atoms contributed per unit velocity depend on the manner in which the kinematic parameter $|dv/dr|$ is perceived along a given line of sight, but to a first approximation the expectation holds that, if the gas is transparent, column depths measured over a velocity interval at positive velocities (in the first longitude quadrant) will be typically about *twice* what they are when measured over negative velocities. The distinction between the inner- and outer-Galaxy occurs at zero velocity, so the expectation is that, if the gas is transparent, there will then be a drop of emission at $v = 0 \, \mathrm{km \, s^{-1}}$, with higher intensities expected at $v > 0 \, \mathrm{km \, s^{-1}}$ in the first quadrant, and at $v < 0 \, \mathrm{km \, s^{-1}}$ in the fourth. Inspection of the HI emission maps in Fig. 1 shows that this expectation is observed, roughly over the longitude range $20° < |l| < 70°$. This is an indication that the HI gas involved is globally transparent.

But a contrary indication that the HI gas layer viewed near $|b| = 0°$ is optically *thick* in some broad sectors can also be given. The total length of path sampled at low latitudes along a line of sight in the general direction of $l = 180°$ is much less than that sampled in the general direction of $l = 0°$. If the total galactic HI layer has a diameter of, say, 50 kpc, and if $R_0 = 10 \, \mathrm{kpc}$, then the anticenter path is some 15 kpc long whereas the path towards the center is some 35 kpc long. If the HI gas were characteristically transparent, and more or less smoothly distributed, the integrated intensities and peak temperatures found generally toward the center would be more than twice the values found toward the anticenter. This situation is not observed; in fact, the profile integrals are approximately equal in the two opposed general directions; these two directions also show similar peak temperatures. Furthermore, similar values of these measures are found in the complementary general directions near $l = 75°$

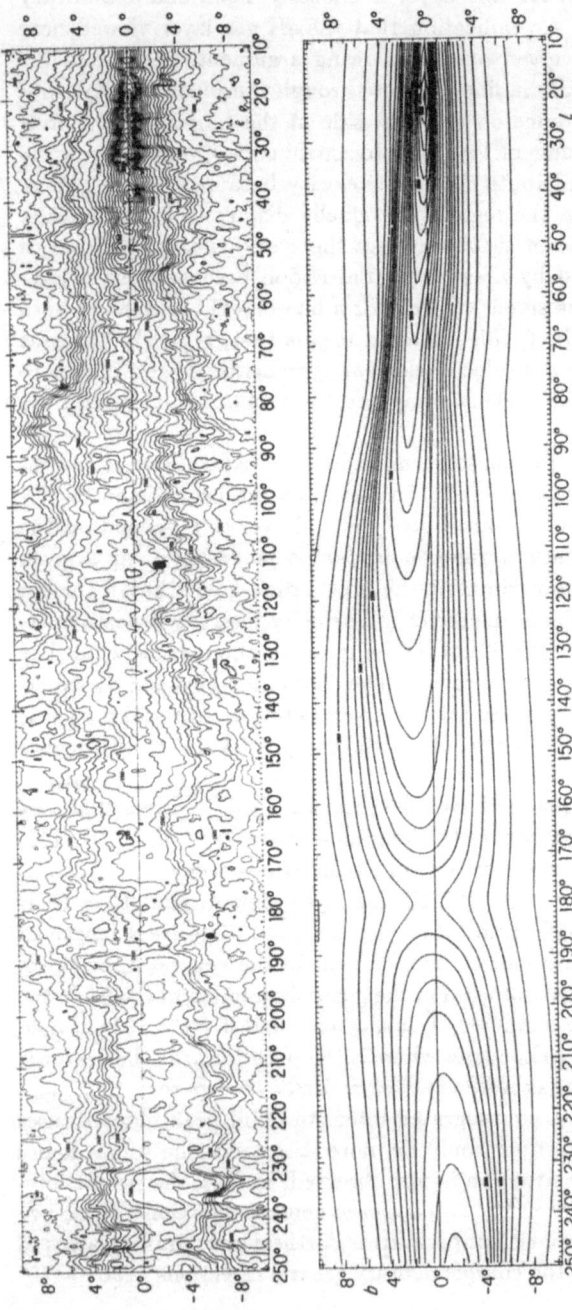

Fig. 8. Arrangement on the plane of the sky of the total-velocity integrated HI emission as observed and as simulated in a simple model. The upper panel shows the integrated antenna temperatures calculated from the observations of Weaver & Williams (1973); this integral gives the HI column density only if the gas is optically thin at all velocities. The lower panel (from Burton 1976) shows the total-velocity integrals for a model of the gas layer which incorporates a description of the observed warp of the outer-Galaxy HI layer but which is otherwise featureless: the relative weakness of both the observed and modelled integrated intensities near $l \simeq 0°$ and $l \simeq 180°$ may be attributed to non-negligible saturation effects

and $l = 285°$, where velocity crowding is responsible for lengths of path of order 5 kpc contributing to a narrow velocity range centered near $v \simeq +5 \, \mathrm{km \, s^{-1}}$ at $l \simeq 75°$ and near $v \simeq -5 \, \mathrm{km \, s^{-1}}$ at the complementary fourth-quadrant longitude. The similarity of the peak-temperature and integrated-emission measures is consistent with a gas layer emitting under conditions of substantial optical depth.

The above two paradoxical arguments imply that over some broad swaths of longitude at low latitudes the HI gas layer is largely transparent, but over others it is not. It appears that $\tau(v)$ hovers near unity over substantial portions of the spectra from the low-latitude HI gas layer.

2.4.2 Considerations Regarding the Integrated HI Emission. The upper panel of Fig. 6 shows the distribution on the plane of the sky of HI total-velocity integrated intensities, calculated from the data of Weaver & Williams (1973). At latitudes more than about 5° from the galactic equator, there is no evidence that the volumes sampled by HI emission spectra are not generally transparent, although evidence is commonly found in absorption data for small, localized features with substantial opacities. Along lines of sight lying closer than a few degrees to the galactic equator, optical thinness may generally pertain, or may pertain over only limited portions of the profile. When deriving the layer thickness, for example by scanning vertically through the gas layer, it is important to be aware of changes of the effective opacity of the line of sight with latitude.

The velocity range of the integrals entering the upper panel of Fig. 6 embraces most of the HI associated with the Galaxy, excepting high-velocity clouds. For most purposes, recognition of individual features requires careful discrimination of the velocity range appropriate to the feature under consideration. Burton & te Lintel Hekkert (1985) have published a series of maps showing the HI column densities in $2.5 \, \mathrm{km \, s^{-1}}$ bins, in which some individual features can be more easily identified; generally, however, individual features are best displayed by custom-tailored integration ranges.

The lower panel of Fig. 6 shows the distribution of HI total-velocity integrated intensities calculated from simulated profiles representing the radiative transfer of HI emission through a smooth distribution of gas, at a single temperature, at a constant midplane density, and partaking in galactic rotation in a conventional manner. A number of aspects of the observed situation are accounted for by the simulation. In particular, the decrease in the integrated emission levels near the galactic anticenter and center directions can be easily identified in the controlled conditions inherent in modelling with saturation effects as the $|dv/dr|$ parameter crowds the 21-cm photons together over a short velocity interval centered near $0 \, \mathrm{km \, s^{-1}}$. The lower-panel distribution shows that near the anticenter non-negligible optical depth effects persist to latitudes of about $\pm 10°$; at longitudes away from the anticenter, including those near the center where saturation is important near $b = 0°$, the effects are not very important beyond $|b| \simeq 5°$.

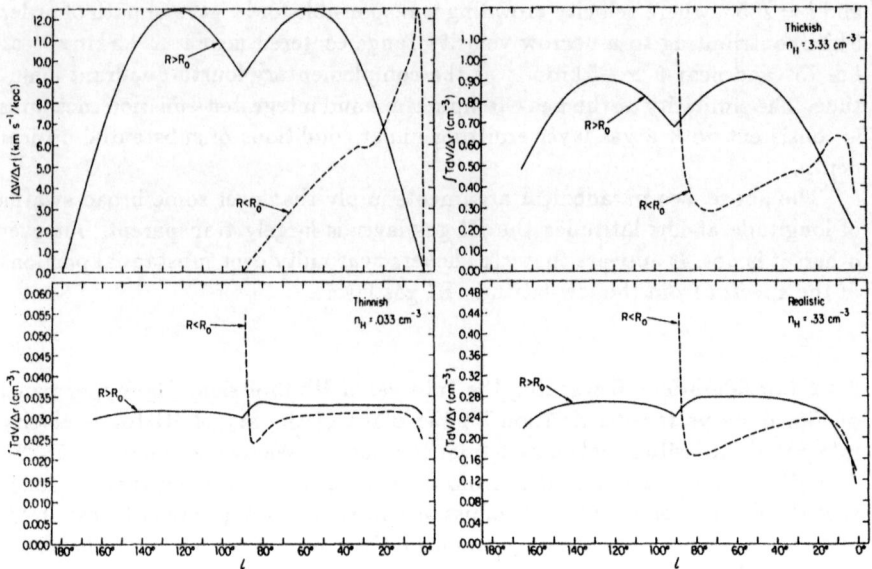

Fig. 7a. Demonstration under the controlled circumstances inherent in modelling that the integrated HI emission depends on the global optical depth through the circumstances of velocity crowding. The upper-left panel shows the line-of-sight behavior of the macroscopic velocity-crowding parameter $|\Delta v/\Delta r|$. The remaining panels show, as explained in the text, how this measure influences the integrated intensities calculated for a model Galaxy with three different values of the number density of HI atoms

Having identified the signature of global optical-depth effects, it is reasonable to attempt to correct the integrated-intensity measure for its underestimate of the total HI column depth. An attempt can be made (see also Burton 1976) to correct for the effects of partial saturation by determining the consequences of saturation which pertain for synthetic spectra, calculated under the controlled conditions possible when modelling but utilizing physical parameters suggested by observations of different sorts.

Figure 7a illustrates the influence of the velocity-crowding parameter on the global properties of the integrated emission. The macroscopic variation of the parameter $|\Delta v/\Delta r|$ is plotted in the upper-left panel of the figure. This parameter represents the total velocity range sampled in each direction, divided by the total length of path from which these velocities are contributed. Its calculation requires, of course, adopting a simple rotation curve for the Galaxy, such as the one given in the following chapter. The parameter was calculated separately for the portion of each line of sight traversing, at $b = 0°$, the inner Galaxy and the outer. Where this macroscopic kinematic parameter is relatively large, the emission from a length of path will be spread over a relatively large range of velocity, so that saturation effects due to velocity crowding will be relatively minor. The plot shows that the outer Galaxy will, in general, suffer less global saturation than the inner: because of the double-valuedness of the

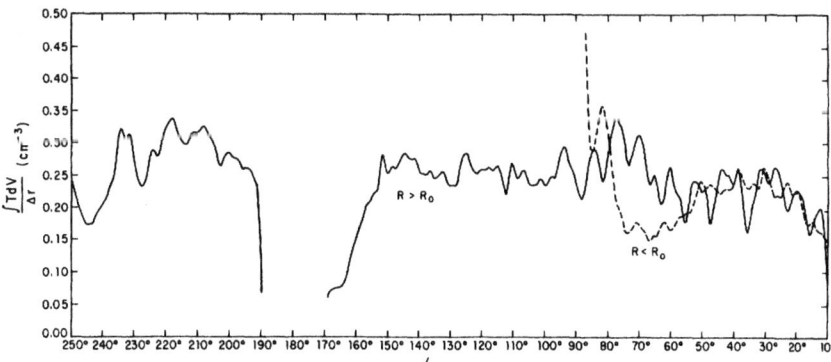

Fig. 7b. Demonstration that the observed integrated HI emission depends on the global optical depth through the circumstances of velocity crowding. The integrated emission was determined from the observations of Weaver & Williams (1973), discriminating the inner- and outer-Galaxy from each other on the basis of velocity; the integral was determined for each longitude at the latitude of the emission centroid (which is usually at $|b| \leq 2°$). Each integral was divided by the total length of path through the inner- or outer-Galaxy, as appropriate, found from the relevant velocity extent of the profile at $v > 0$ or at $v < 0\,\mathrm{km\,s^{-1}}$. Within about 15° of both $l = 180°$ and $l = 0°$, the variation plotted with the full-drawn line shows saturation effects as predicted, otherwise the situation at $R > R_0$ is quite transparent. The effects of the accumulated optical depth hovering near unity are more important at $R < R_0$, where the dashed line shows the predicted influence of partial saturation not only near $l = 0°$ but also between 60° and 80°

velocity-to-distance transformation on lines of sight through the inner Galaxy, emission from there is crowded.

The remaining three panels of Fig. 7a illustrate the consequences of the crowding. Simulated HI spectra were calculated for the simplest-possible, one-component, situation, with the gas density, temperature, and dispersion held at constant values throughout the modelled Galaxy. Three cases were considered, in which only the volume density was varied. The three panels show the calculated measure $\int T_b(v)\,dv/\Delta r$, which, as mentioned above, should return the input volume density *if* the condition of optical thinness holds; deviations from optical thinness are revealed by the manner and extent in which this returned measure *deviates* from the input. For the case in which the input density $n_{\mathrm{HI}} = 3.33\,\mathrm{cm^{-3}}$ is too high to be realistic, the measure $\int T_b\,dv/\Delta r$ does not return the input density. Instead, it shows variations which mimic those of the saturation-dominating $|\Delta v/\Delta r|$ parameter plotted in the upper-left panel. For the case in which the input density is too low, $0.033\,\mathrm{cm^{-3}}$, the gas layer is essentially transparent and the integral measure does return a meaningful value of the input density and at the same time gives an indication of the small correction which would have to be applied to recover the input density and thereby to account for macroscopic saturation.

For the case represented in the lower right-hand panel of Fig. 7a, the input volume density was 0.33 HI atoms $\mathrm{cm^{-3}}$. The behavior of the $\int T_b(v)\,dv/\Delta r$ measure modelled resembles that observed. Figure 7b shows the observed integral measure of the volume density, derived from the HI data of Weaver

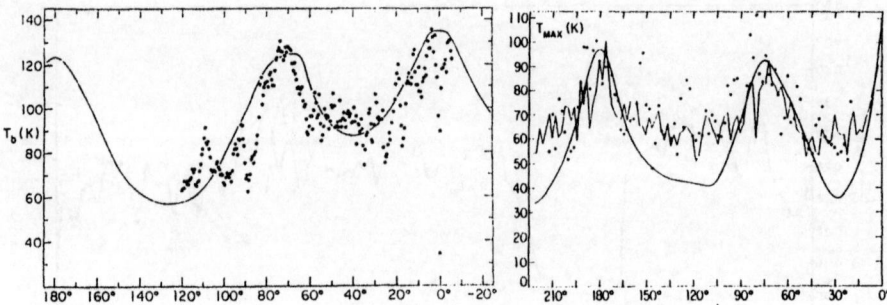

Fig. 8. Demonstration of the dependence of maximum temperature on the global optical depth of low-latitude HI spectra. The left-hand panel (from Burton 1972) compares the maximum T_b read from HI observations at $b = 0°$ with that pertaining to simulated spectra calculated for the simplest possible description of the HI layer, namely for circular rotation and constant density, temperature, and dispersion. The general run of the observed maximum T_b with l varies as expected from the longitude-dependent velocity-crowding influences on opacity. The smaller-scale details of the variation are real (measurement noise in HI observations being not important for most Milky Way purposes) and contain information on the variation of the saturation effects. The right-hand panel (from Burton 1976) shows the same comparison for observations and simulations at $b = +2°$. In addition to the T_b-variation for the simplest single-component case, shown by the smooth line, this variation was determined for a two-component model involving cooler, denser clouds immersed in a warmer, less-dense background. The characteristic distance at which the accumulated optical depth becomes non-negligible depends on the areal filling factor, and can be successfully modelled

& Williams (1973). The variation in this plot follows that predicted by the $|\Delta v/\Delta r|$ parameter. Macroscopic optical depth effects are greatest within about $\pm 15°$ of the center and anticenter directions; they are also substantial along the inner-Galaxy portions of the lines of sight within about $\pm 15°$ of $l = 70°$. In general, the observed variations follow those modelled for the case $n_{HI} = 0.33\,\mathrm{cm}^{-3}$, represented in the lower right-hand panel of Fig. 7a. This argument leads to an estimate of the global correction for macroscopic saturation effects, amounting to about 15% in total. The correction varies as a function of longitude. A realistic value of the line-of-sight-averaged volume density of the HI dominating the emission profiles is about 0.4 atoms cm^{-3}.

2.4.3 Considerations Regarding the Maximum HI Brightness Temperatures Observed. The maximum brightness temperature observed in low-latitude HI observations is about $T_b = 135\,\mathrm{K}$; this maximum value is reached on spectra observed towards directions where velocity-crowding effects are dominant, namely towards the anticenter, the center, and at the appropriate velocities in the complementary directions $l = 75°$ and $l = 285°$. The left-hand panel of Fig. 8 shows a plot of the maximum brightnesses reached in $b = 0°$ observations over a range of longitude.

The smooth line drawn through the data points in this plot represents the variation of maximum temperature displayed by simulated HI spectra calculated, as above, for a single-component gas layer at constant temperature (135 K) and density (0.33 atoms cm^{-3}). The variation in the model reflects the optical-depth characteristics, and follows the variation observed. If the optical

28

depth were in fact everywhere quite large, the variation would be flat; on the other hand, if the optical depth were everywhere quite low, the variation would show more structure. This plot is consistent with a global optical depth which hovers around unity.

The scatter of the observed points around the mean line plotted in the left-hand panel of Fig. 8 is quite large. We have said above that galactic HI spectra can rather easily be gotten with negligible measurement noise. The scatter in the plot thus has a physical significance. In the following section we seek this significance *also* in the general characteristics of absorption.

2.5 Small-Scale Structure in Heavily-Blended Low-|b| HI Emission Spectra

The discussion in the preceding section regarding effects of partial saturation in low-latitude HI observations involved demonstrating aspects of the observed situation which could be understood in terms of a simulation of a smooth HI medium at a single temperature. In fact, of course, it is well established that interstellar HI embodies a range of different temperatures and densities (see, e.g., the review by Kulkarni & Heiles 1988). That fact does not invalidate the demonstrations given above for macroscopic optical depth effects, but it does suggest several additional questions. For example, does the optical depth simply increase monotonically at a given velocity as the length of path at that velocity through a rather uniform medium increases? (This is the situation which pertains for the models represented in Figs. 6, 7, and 8.) Or, on the other hand, does the optical depth increase in increments due to the self-absorption influences of discrete, cold, elements being successively traversed on the line of sight?

It has long been recognized that separate structures of relatively cold HI occur commonly at low latitudes. Heeschen (1955) and Davies (1956) found discrete emission regions with indicated temperatures about half that of their surroundings. Shuter & Verschuur (1964) and Clark (1965) showed that the temperatures observed in HI absorption spectra resembled those of the discrete, cold features appearing in emission. Because the cold gas is largely opaque, the higher temperatures observed generally require that a warmer, but optically thinner, gas be present also. Clark (1965) suggested a "raisin-pudding" model, in which cool, largely opaque, clouds are immersed in a warm, largely transparent, background.

Since Clark's "raisin-pudding" model was introduced there have been a large number of more intricate multi-component models of the interstellar HI, motivated by a wealth of new information. Much of that information has necessarily been generalized from data representing only the immediate neighborhood of the Sun. Confrontation with the collective, blended, behavior of low-latitude observations sampling long lengths of path has received little attention (but see Bregman & Ashe 1991). The term "raisin-pudding" is, of course, a subjective one, but nevertheless suggestive. Other descriptions of the HI layer have

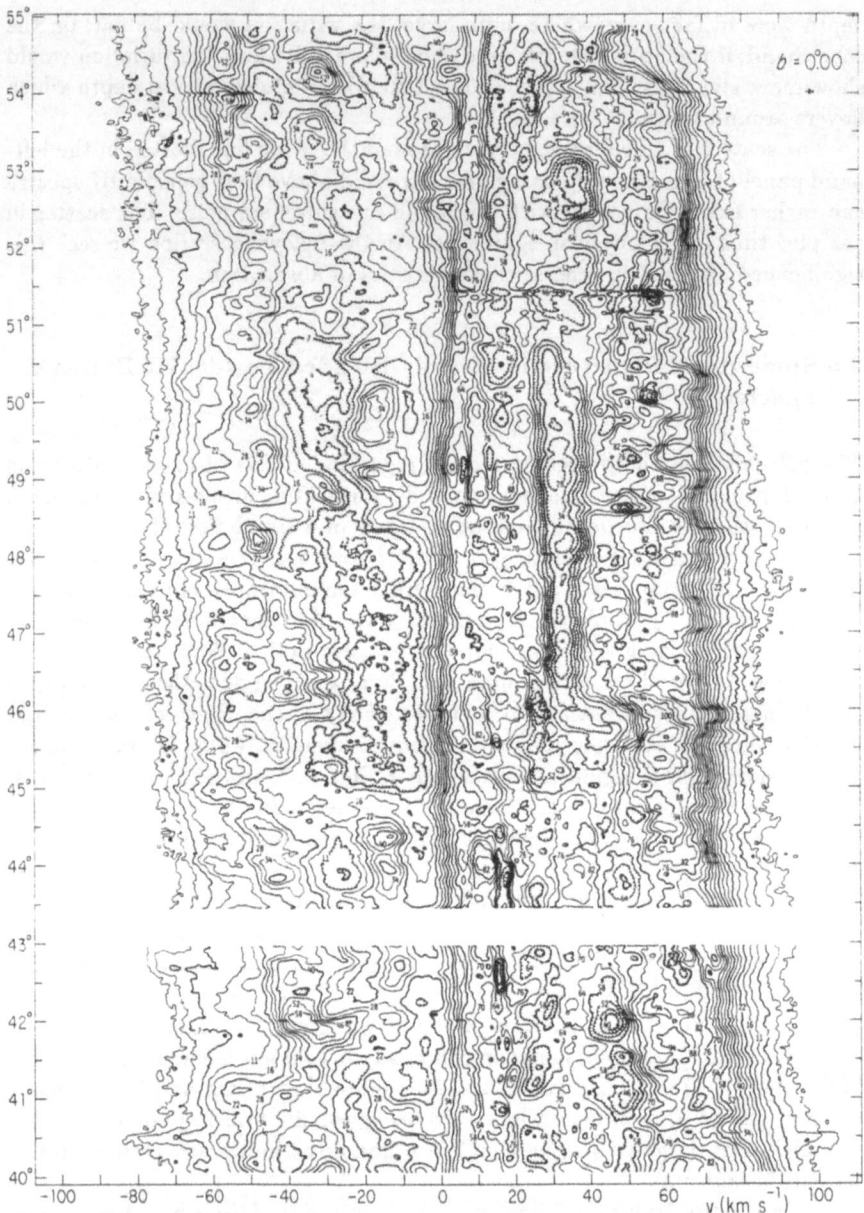

Fig. 9. Longitude, velocity map of HI emission observed at $b = 0°$ at the high angular resolution, $\sim 4'$, afforded by the Arecibo 305-m telescope (from Baker & Burton 1979). Low-latitude HI maps such as this are peppered with small-angular-scale intensity minima due to self-absorption in cold clouds; each of the HI cold clouds have molecular counterparts identifiable in CO observations. The collective influence of these features amounts to only about 5% of the total integrated emission and to about 15% or the derived mean densities

used terms such as "spongy", and "frothy". A "spongy" or "frothy" interstellar medium may be pictured as pervaded by regions of low density and high temperatures, encased in connected regions of higher density and lower temperature. The confrontation between the "raisin" and the "froth" pictures is not an entirely meaningless one. The predictions which the two pictures make regarding areal and volume filling factors are very distinct, for example. In the "raisin" model (where the "raisin" would, in view of more modern material, have a very filamentary, twisted shape), the areal filling factor of the colder material would not have to be unity, whereas it *would* typically be unity in the "froth" view.

Figure 9 shows a map of HI emission observed at $b = 0°$ at high angular resolution using the Arecibo radio telescope. (Note how definite the zero-velocity ridge appears in this map, indicative — as mentioned in Sect. 2.4.1 — of thinness where it appears.) The map is peppered with intensity minima of small angular size, which have been identified with self-absorption in cold clouds. The existence of residual HI in optically identified, dark dust clouds was demonstrated by Knapp (1972) through such self-absorption features contributed from rather near the Solar neighborhood. The features such as those shown in the Arecibo map are generally too small to be resolved in maps made with the half-degree resolution which necessarily characterizes the larger-scale HI surveys.

Insight into the overall optical-depth characteristics of the galactic gas layer is strengthened by the identification of these low-latitude self-absorption HI features with the common occurrence of residual, relatively cold, atomic hydrogen in galactic molecular clouds. Figure 10 allows comparison of a short slice along the galactic equator observed in HI at high, Arecibo, resolution with a map of ^{12}CO emission observed along the same slice. The direction of the map samples the outer portion of the ensemble of molecular clouds. Each of the HI self-absorption features has a counterpart in the molecular map; on the other hand, the requirement that the absorbing HI be located on the *near* side of the tangential point, so that HI emission from the *far* side may be absorbed, indicates that, statistically, only about half of the CO clouds will have HI self-absorption counterparts. This is evidently the situation, although the observational material remains rather meager. It is not yet clear — although deciphering the physical situation is obviously important — if the cold HI associated with the molecular clouds resides on the outer envelopes or if it pervades the cloud well mixed with the CO; interferometric HI studies at very high angular resolution are necessary in this regard, but the requirement of background emission complicates such studies, and suggests that low-latitude regions will have to be involved.

Several aspects of the global saturation effects can be understood if the interstellar medium is pervaded by relatively cold HI features, distributed throughout the galactic plane in a manner similar to that observed for the members of the ensemble of molecular clouds. The features will be individually largely unresolved in the $0°.5$ telescope beams. Collectively, however, they will cause global modulations of the HI emission characteristics resulting from the

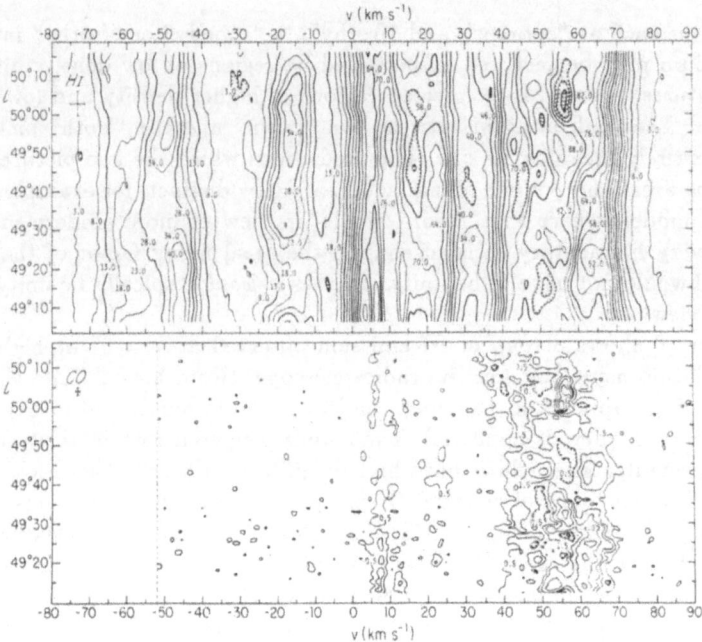

Fig. 10. Comparison of HI and CO observations, made with approximately comparable resolutions, of a field centered on $l = 50°.1$, $b = 0°.0$ (from Liszt et al. 1981). Comparisons of this sort show that, generally, *all* HI self-absorption features have a molecular counterpart; the requirement of suitable background emission restricts most such self-absorption features to the near side of the terminal-velocity locus, so that statistically only about *half* of the molecular features will have an HI self-absorption counterpart

accumulated effects of opaque clouds which are individually non-overlapping in space and which are individually severely diluted in the telescope beam.

Figure 11 demonstrates the influence of individually-unresolved cold HI features on a typical HI emission spectrum. The modelled cold features are distributed as those comprising the ensemble of galactic molecular clouds. The effects are strongest near the terminal-velocity cutoff, because that is where velocity crowding is greatest. The effects are much less at lower velocities. The distribution of molecular clouds with galactocentric radius is such that relatively few clouds occur at velocities corresponding to $R \simeq R_0$, so that fewer associated cold HI clouds will be expected at these velocities; furthermore, at velocities less than zero (in the first quadrant), the HI profiles may be expected to be very smooth *even if* cold features should exist at $R \geq R_0$, because the requirement of background emission will not be met, generally, for a system which is kinematically as well behaved as ours seems to be. Figure 9, for example, shows that these expectations regarding the HI profiles are in accordance with the observed situation.

Figure 11 shows that the effects of individual clouds, largely unresolved, will not be large and that the collective behavior of such clouds will in many ways mimic (in $0°.5$ telescope beams) the low-$|b|$ behavior of a single-component gas;

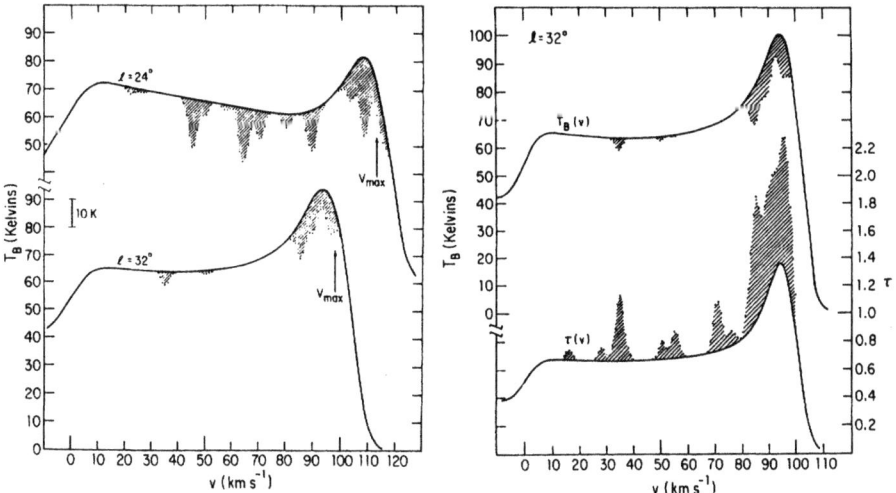

Fig. 11. Demonstration of the consequences to 21-cm emission data resulting from self-absorption provided by cold HI associated with molecular clouds (from Liszt et al. 1984). The left-hand panel shows HI brightness-temperature and optical-depth profiles simulated at $l = 24°$ and $32°$ in the galactic equator. The hatched area in the T_b-profile represents the emission lost by the localized high opacity; the hatched region in the τ-profile indicates how much opacity is added by the cold HI. The right-hand panel shows the radial distribution of HI column densities at $b = 0°$ derived after correcting for a general opacity; the hatched area shows the contribution lost when cold HI clouds are included in the analysis

the manner in which the effects of these clouds accumulate will, in particular, be dominated by the $|\Delta v/\Delta r|$ parameter. This mimicking provides the principal justification for paying any attention at all to the one-component case.

Although optical-depth effects along long lines of sight will accumulate in the absence of absorbing entities, just as they will in the presence of such entities, they will accumulate in a more ragged fashion when the absorbing clouds are present than would be the case in their absence. In this regard it is interesting to return to Fig. 8, and notice in the right-hand panel the difference in the peak temperature which is predicted for the case of a smooth, one-component, gas, and that which is predicted for the ragged accumulation of optical depth. The two-component modelled situation accounts for the scatter in the $T_{\max}$ observed.

The latitude represented in the right-hand panel of Fig. 8, $b = 2°$, was chosen away from the galactic equator so that the effects of decreased areal filling factor would be relevant; areal filling factor of unity at this latitude will be reached at a larger distance from the Sun than is the case at $b = 0°$, where complete beam coverage is expected at short range. Dickey & Brinks (1992a,b) have recently offered a suggestion regarding the cloud population of M31 which is interesting in the present context. They note that the individual molecular and atomic clouds in M31 are quite similar to those observed, individually, in the Milky Way. Yet they notice that the *peak* HI temperature is some 50% higher in M31 than it is in the Milky Way (see also Braun 1991). Dickey &

33

Brinks argue that if the Milky Way were to be seen under the same inclination as M31, the accumulated effects of the colder gas might be sufficiently smeared out that a higher T_{max} would result. Their interpretation utilizes the perceived velocity gradients in M31 in a manner analogous to that discussed above for the global optical-depth and maximum-temperature characteristics of our Galaxy.

References

2.1 Bajaja, E. 1983, in *Surveys of the Southern Galaxy*, W.B. Burton, F.P. Israel, Reidel Pub. Co., p. 49

2.2 Baker, P.L., Burton, W.B. 1979, *Astron. Astrophys. Suppl. Ser.* **35**, 129

2.3 Bania, T.M., Lockman, F.J. 1984, *Astrophys. J. Suppl. Ser.* **54**, 513

2.4 Blitz, L., Spergel, D.N. 1991, *Astrophys. J.* **370**, 205

2.5 Braun, R. 1991, *Astrophys. J.* **372**, 54

2.6 Bregman, J.N., Ashe, G.A. 1991, in *Proceedings IAU Symp. 144*, *"The Interstellar Disk-Halo Connection in Galaxies"*, J.B.G.M. Bloemen, (ed.), Kluwer Acad. Pub., p. 387

2.7 Burton, W.B. 1972, *Astron. Astrophys.* **16**, 158

2.8 Burton, W.B. 1974, in *Galactic and Extragalactic Radio Astronomy*, 1st edition, G.L. Verschuur, K.I. Kellermann, (eds.), Springer-Verlag, New York, p. 82

2.9 Burton, W.B. 1976, *Ann. Rev. Astron. Astrophys.* **14**, 275

2.10 Burton, W.B. 1985, *Astron. Astrophys. Suppl. Ser.* **62**, 365

2.11 Burton, W.B., Bania, T.M., Hartmann, D., Tang Yuan 1992, in *Proc. CTS Workshop No. 1. "Evolution of Interstellar Matter and Dynamics of Galaxies"*, J. Palouš, W.B. Burton, P.O. Lindblad, (eds.), Cambridge University Press, in press

2.12 Burton, W.B., Liszt, H.S. 1983, *Astron. Astrophys. Suppl. Ser.* **52**, 63

2.13 Burton, W.B., te Lintel Hekkert, P. 1985, *Astron. Astrophys. Suppl. Ser.* **62**, 645

2.14 Carruthers, G.P. 1970, *Astrophys. J.* **161**, L81

2.15 Clark, B.G. 1965, *Astrophys. J.* **142**, 1398

2.16 Clemens, D.P., Sanders, D.B., Scoville, N.Z., Solomon, P.M. 1986, *Astrophys. J. Suppl. Ser.* **60**, 297

2.17 Clube, S.V.M. 1985, in *Proceedings IAU Symp. 106*, *"The Milky Way Galaxy"*, H. van Woerden, R.J. Allen, W.B. Burton, (eds.), Reidel Pub. Co., p. 145

2.18 Combes, F. 1991, *Ann. Rev. Astron. Astrophys.* **29**, 195

2.19 Dame, T., Ungerechts, H., Cohen, R.S., de Geus, E., Grenier, I., May, J., Murphy, D.C., Nyman, L.-Å., Thaddeus, P. 1987, *Astrophys. J.* **322**, 706

2.20 Danly, L. 1989, *Astrophys. J.* **342**, 785

2.21 Danly, L. 1991, in *Proceedings IAU Symp. 144*, *"The Interstellar Disk-Halo Connection in Galaxies"*, J.B.G.M. Bloemen, (ed.), Kluwer Acad. Pub., p. 53

2.22 Davies, R.D. 1956, *Monthly Notices Roy. Astron. Soc.* **116**, 443

2.23 Delhaye, J. 1965, in *Stars and Stellar Systems V*, *"Galactic Structure"*, A. Blaauw, M. Schmidt, (eds.), University of Chicago Press, p. 61

2.24 Dickey, J.M., Brinks, E. 1992a, in *Proceedings CTS Workshop 1*, *"Evolution of Interstellar Matter and Dynamics of Galaxies"*, J. Palouš, W.B. Burton, P.O. Lindblad, (eds.), Cambridge University Press, in press

2.25 Dickey, J.M., Brinks, E. 1992b, preprint

2.26 Dickey, J.M., Garwood, R.W. 1989, *Astrophys. J.* **341**, 201

2.27 Dickey, J.M., Kulkarni, S.R., van Gorkom, J.H., Heiles, C.E. 1983, *Astrophys. J. Suppl. Ser.* **53**, 591

2.28 Dickey, J.M., Lockman, F.J. 1990, *Ann. Rev. Astron. Astrophys.* **28**, 215

2.29 Dickey, J.M., Salpeter, E.E., Terzian, Y. 1978, *Astron. Astrophys. Suppl. Ser.* **36**, 77

2.30 Field, G.B. 1958, *Proceedings IRE* **46**, 240

2.31 Heeschen, D.S. 1955, *Astrophys. J.* **121**, 569

2.32   Heiles, C. 1992, in *Proceedings CTS Workshop 1*, *"Evolution of Interstellar Matter and Dynamics of Galaxies"*, J. Palouš, W.B. Burton, P.O. Lindblad, (eds.), Cambridge University Press, in press

2.33   Heiles, C, Habing, H.J. 1974, *Astron. Astrophys. Suppl. Ser.* **14**, 1

2.34   Kalberla, P.M.W., Mebold, U., Reif, K. 1982, *Astron. Astrophys.* **106**, 100

2.35   Kerr, F.J. 1968, in *Stars and Stellar Systems* **VII**, *"Nebulae and Interstellar Matter"*, B.M. Middlehurst, L.H. Aller, (eds.), University of Chicago Press, 575

2.36   Kerr, F.J., Bowers, P.F. Jackson, P.D., Kerr, M. 1986, *Astron. Astrophys. Suppl. Ser.* **66**, 373

2.37   Kulkarni, S.R., Heiles, C. 1988, in *Galactic and Extragalactic Radio Astronomy*, G.L. Verschuur, K.I. Kellermann, (eds.), Springer-Verlag, New York, p. 95

2.38   Knapp, G.R. 1972, Ph.D. Thesis, University of Maryland

2.39   Lang, K.R., Gingerich, O. 1979, *A Source Book in Astronomy and Astrophysics 1900-1975*, Harvard University Press: see p. 627 for a translation into English of van de Hulst, H.C. (1945)

2.40   Liszt, H.S., Burton, W.B., Bania, T.M. 1981, *Astrophys. J.* **246**, 74

2.41   Liszt, H.S., Burton, W.B., Xiang, D.-L. 1984, *Astron. Astrophys.* **140**, 303

2.42   Lockman, F.J. 1984, *Astrophys. J.* **283**, 90

2.43   Lockman, F.J., Jahoda, K., McCammon, D. 1986, *Astrophys. J.* **302**, 432

2.44   Penzias, A.A., Wilson, R.W. 1965, *Astrophys. J.* **142**, 419

2.45   Reynolds, R.J. 1991, in *Proceedings IAU Symp.* **144**, *"The Interstellar Disk-Halo Connection in Galaxies"*, J.B.G.M. Bloemen, (ed.), Kluwer Acad. Pub., p. 67

2.46   Shuter, W.L.H. 1981, *Monthly Notices Roy. Astron. Soc.* **199**, 109

2.47   Shuter, W.L.H., Verschuur, G.L. 1964, *Monthly Notices Roy. Astron. Soc.* **127**, 387

2.48   Stacey, G.J., Geis, N., Genzel, R., Lugten, J.B., Poglitsch, A., Sternberg, A., Townes, C.H. 1991, *Astrophys. J.* **373**, 423

2.49   Stark, A.A., Gammie, C.F., Wilson, R.W., Bally, J., Linke, R.A., Heiles, C., Hurwitz, M. 1992, *Astrophys. J. Suppl. Ser.* **79**, 77

2.50   Stark, A.A., Bally, J., Knapp, G.R., Wilson, R.W. 1988, in *Molecular Clouds in the Milky Way and in External Galaxies*, R.L. Dickman, R.L. Snell, J.S. Young, (eds.), Springer Verlag, p. 303

2.51   Sullivan, W.T. 1984, *The Early Years of Radio Astronomy*, Cambridge University Press

2.52   van de Hulst, H.C. 1945, *Ned. Tijd. Natuurkunde*, **11**, 201

2.53   Wagoner, R.V. 1973, *Astrophys. J.* **179**, 343

2.54   Wilson, R.W., Jefferts, K.B., Penzias, A.A. 1970, *Astrophys. J.* **161**, L43

2.55   Weaver, H.F., Williams, D.R.W. 1973, *Astron. Astrophys. Suppl. Ser.* **8**, 1

3 Galactic Rotation and the Constants Specifying the Kinematic and Linear Scales of the Milky Way; Line-broadening Mechanisms Determining Profile Shapes at Low Latitudes

Abstract. Global distances in the Milky Way are commonly based on the acceptance of a galactic rotation curve as a known, well-behaved function. The determination of the rotation curve, $\Theta(R)$, is discussed in the context of the assumption of simple, circular motion. Also discussed is the observational background of the two galactic constants, R_0 and Θ_0, which determine the kinematic and linear scales entering morphological investigations. The assumption of circular rotation is known to be wrong, but only at the level of about two percent of the rotation motions. Kinematic irregularities can be separated into several distinct categories, including intrinsic broadening within a single feature; random motions of individual clouds with respect to each other; systematic flows, for example along spiral arms; and large-scale distortions from circularity of the Galaxy as a whole. The perceived kinematics of the interstellar medium dominate (for example over the effects of density variations) the appearance of low-latitude tracers of galactic morphology, through the behavior of the velocity-crowding parameter $|dv/dr|$. The dominance of this parameter, in the known presence of substantial kinematic activity, frustrates many mapping ambitions in our own system.

3.1 Introduction; Galactic Rotation

Our embedded perspective hinders study of some global properties of our own Galaxy, notably concerning details of its spiral structure and decomposition of its space motions, but facilitates others, notably of kinematic and physical details of the interstellar medium and of such comparative measures for the different constituents of the galactic gas layer as the radial and vertical scale lengths and the degree of flatness. Kinematic distances have proven indispensable for the derivation of many morphological properties of the Milky Way: it has long been clear that crucial assumptions, in particular of circular kinematic symmetry, are not valid in reality; it is important to judge the importance of the various assumptions to the validity of the derived morphology.

Kinematic distances are commonly based on the assumption that the orbits about the galactic center are circular and that the angular velocity of circular galactic rotation, $\Omega(R)$, is a well-behaved, monotonic, function only of distance from the center. The left-hand panel of Fig. 12 (from Burton 1988) illustrates the construction entering derivation of the fundamental equation for determination of kinematic distances. If the Milky Way rotates as assumed, then the radial velocity measured by an observer embedded in the equatorial plane of the rotating system, at a distance R_0 from the center and rotating about it with a linear velocity $\Theta_0 = \Omega_0 R_0$, is the difference between the component

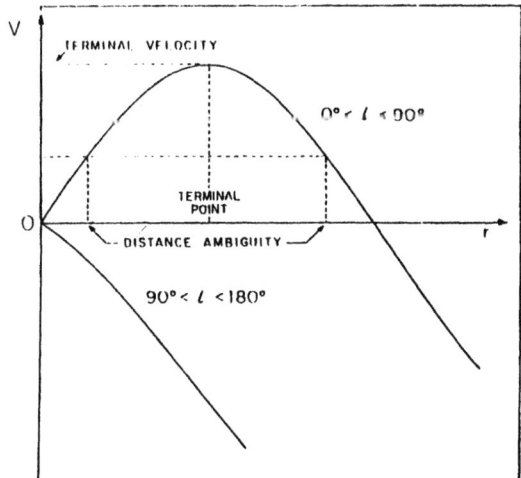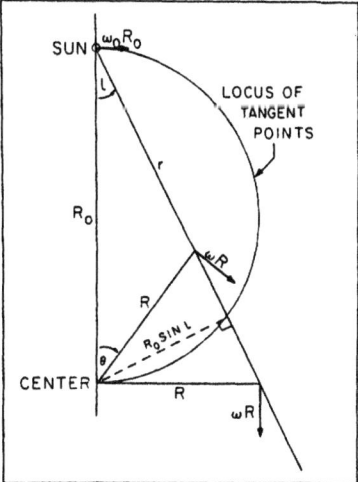

Fig. 12. Left-hand panel: Schematic indication of the radial-velocity variation which would be measured with respect to the Local Standard of Rest along a line of sight in the galactic equator traversing a representative longitude through the inner Galaxy (upper curve) and along a representative longitude traversing only the outer Galaxy (lower curve). The diagram illustrates the distance ambiguity which plagues analyses of inner-Galaxy material.

Right-hand panel: Construction used in the derivation of the fundamental equation $v(R, l) = R_0[\Omega(R) - \Omega_0]\sin l$, which allows distances to be attributed to measured velocities under the assumption of good kinematic behavior

of $\Omega(R)R$ Doppler-shifted along the line of sight minus the component of the observer's own motion, $\Omega_0 R_0$, along the line of sight. This difference is

$$
\begin{aligned}
v(R, l) &= \Omega(R)R\cos(90° - l - \theta) - \Omega_0 R_0\cos(90° - l) \\
&= \Omega(R)R(\sin\theta\cos l + \cos\theta\sin l) - \Omega_0 R_0\sin l,
\end{aligned}
\tag{7}
$$

where l is the galactic longitude (at $b = 0°$) of the line of sight in question and θ is galactocentric azimuth. The construction shows that $R\sin\theta = r\sin l$ and that $R_0 - R\cos\theta = r\cos l$, so that the radial velocity may be written

$$
v(R, l) = R_0[\Omega(R) - \Omega_0]\sin l.
\tag{8}
$$

If the assumptions inherent in this fundamental equation are valid, and if the galactic linear-velocity, axially-symmetric rotation curve $\Theta(R) = \Omega(R)R$ is known, together with the values of the galactic constants evaluated at the Sun, R_0 and Θ_0, then, in principle, distances along lines of sight through the Milky Way may be attributed to each measured radial velocity. The determination of $\Theta(R)$, R_0, and Θ_0 are discussed below, followed by some remarks on the validity of the assumption of circular rotation and some cautions regarding use of the fundamental equation.

3.2 The Distance of the Sun from the Galactic Center, R_0

The galactic constants R_0 and Θ_0 set the linear and kinematic scales of the Galaxy, and therefore are fundamental to most discussions of the morphology of our system. Commission 33 of the International Astronomical Union ("Structure and Dynamics of the Galactic System") has on several occasions coordinated work leading to specification and standard use of these and other galactic constants. Blaauw et al. (1960) introduced a series of five papers prepared on behalf of a working group of Commission 33 dealing with specification of the principal plane of the Galaxy, of the direction to its center, as well as of its poles; many details of these considerations have changed since that series was prepared, but the issues remain, and the series provides an instructive background discussion.

Kerr & Lynden-Bell (1986) summarize more modern investigations which led IAU Commission 33 to recommend standard usage for most purposes of the values $R_0 = 8.5\,\mathrm{kpc}$ and $\Theta_0 = 220\,\mathrm{km\,s^{-1}}$, replacing the values of $10\,\mathrm{kpc}$ and $250\,\mathrm{km\,s^{-1}}$ in wide prior use. The recommended standard values represent unweighted means of various determinations made using a wide range of techniques; the scatter of the values determined by different techniques indicates that systematic errors are more important than internal ones. Many galactic quantities scale simply by the appropriate power of R_0, but this scaling is not simple in all cases because of the vagaries of the velocity-to-distance transformation.

Several independent methods yield a reasonable consensus on the value of R_0. The determination of R_0 which involves the fewest intervening assumptions aims at direct measurement of the parallax of objects at the galactic center (see Reid & Moran 1988). VLBI observations of the statistical parallax of proper motions of H_2O maser features associated with the Sgr B2 molecular complex, which is probably located within a few hundred pc of the true galactic center, yield distances with accuracies approaching 10%. The internal accuracy of the proper motion method is currently limited by the number of maser spots which can be observed.

Geometrical and kinematic determinations are less direct than proper-motion determinations. Geometrical measurements of R_0 have been used since Shapley's classical determination of the scale of the Galaxy based on distances to globular clusters, found from the period-luminosity relationship for Cepheid variable stars. This type of approach assumes, of course, that the distribution centroid found coincides with the galactic barycenter, even though the total galactic mass is dominated by invisible matter distributed over a much larger volume and essentially inaccessible. The globular-cluster method is limited by the small number of clusters; these are, furthermore, only weakly concentrated toward the galactic center. Observations of Cepheids in the Magellanic Clouds show a very small internal dispersion in distances derived from the period-luminosity relationship. Although the intrinsic internal scatter of the Cepheid P-L relationship is small, calibration of the zero-point of the relationship remains problematic. A conceptually similar geometrical determination of R_0

involves finding the centroid of the distribution of field RR Lyrae variables. These variables can be identified in several windows of relatively low obscuration towards the galactic bulge. As with Cepheids, the intrinsic scatter in RR Lyrae absolute magnitudes is small, but the true RR Lyrae brightness and its possible dependence on metallicity and on R are uncertain.

Mira variables are strongly concentrated toward the galactic center. Miras with long periods commonly have dust shells radiating in the far infrared: their promise as distance indicators has been enhanced by the discovery of an infrared period-luminosity relation of small intrinsic dispersion. The large sample of variable IRAS sources detected in the galactic bulge holds promise for accurate determination of the shape and centroid of the bulge (see Habing 1987). Accurate photometry in the infrared would alleviate some of the problems of the large and variable interstellar extinction which plague work on Cepheid and RR Lyrae variables at optical wavelengths. Infrared variable stars in the bulge are also found in association with OH maser sources, and this association can provide another quite direct determination of R_0. A linear dimension across the circumstellar shell follows from the time lag between the maser responses of the near and far sides of the shell to variability in the central star; the angular dimension can be measured independently using radio interferometry, and the distance derived.

The restrictive assumptions required for kinematic determinations of R_0 make these generally less accurate than the more direct geometrical determinations. The most important restrictive assumptions require well-behaved kinematics and general galactic symmetry: these assumptions are not consistent with the kinematic and spatial irregularities discussed below. The principal kinematic determination of R_0 involves identifying objects of known distance whose radial velocity of zero $km\,s^{-1}$ indicates that they are located on the circle $R = R_0$. Stellar tracers used in this way include Cepheids and OB stars, and, as is the case with the use of optical tracers in geometric programs, absolute magnitudes and extinction corrections must be well known. HII regions with associated molecular clouds are particularly useful for kinematic measures of the distance scale; the distances follow spectrophotometrically from the optically identified HII regions which are selected on the basis of the radial velocities near zero of the associated molecular emission measured in the radio regime.

3.3 The Galactic Circular-Velocity Rotation Curve Evaluated at $R < R_0$; the Value of $\Theta(R)$ at R_0, Θ_0

The galactic circular-velocity rotation curve, $\Theta(R)$, is defined as the function describing the linear velocity of an object moving in a circle of radius R such that centrifugal force balances the gravitational attraction of the Galaxy. The constant Θ_0 is the circular velocity at R_0; it gives, by definition, the velocity of the Local Standard of Rest. As mentioned elsewhere in these notes, the Sun itself has a velocity of some $20\,km\,s^{-1}$ with respect to the LSR; the error on the peculiar velocity of the Sun with respect to the LSR remains difficult to specify. The radial component of the gravitational force of the Galaxy is given by $K_R = -\Theta^2/R$: knowledge of the rotation curve is thus fundamental to construction of models of the galactic mass distribution.

The value of the constant Θ_0 has been determined in several ways, none very direct (see references in Kerr & Lynden-Bell 1986). The motion of the LSR has been measured with respect to the ensemble of globular clusters in the galactic halo; this measurement then has to be corrected for the rotation of the globular-cluster system. The rather small degree of flattening of this system implies that its rotation is slow, but the details of the correction have proven evasive. RR Lyrae variables at large z-heights and halo field stars have been similarly used, and in both cases require additional information on the degree of rotation of the halo itself. The LSR motion has also been determined with respect to galaxies in the Local Group. The LSR motion with respect to HI at the distant edges of the Galactic gas layer in the first and second quadrants yields a value of $220\,\mathrm{km\,s^{-1}}$ (Knapp et al. 1978), but a similar study in the southern hemisphere data yields a substantially higher value; the difference casts doubt on the validity of the assumption of total circular symmetry.

The dependence of Θ on R (see Fig. 13) must be determined differently in the inner Galaxy than in the outer. Special considerations are necessary when deriving the rotation curve in the bulge region of the Galaxy; these considerations are discussed separately in Chap. 6 of these notes. Within the bulge region, deviations from circular rotation are of approximately the same amplitude as the rotational motions themselves, ruling out entirely the — in all cases precarious — assumption of solely circular motion. (Application of the terminal-velocity method under the associated assumption of circular rotation has resulted in a putative rotation curve for the gas layer in the bulge region which shows a rapid rise at small R and a definite peak between 1 and 2 kpc; such a curve is almost certainly not correct.)

HI observations have served as the best basis for derivation of the inner-Galaxy rotation curve, because 21-cm emission is widely spread throughout the Galaxy and is easily observed, largely transparent, and characterized by a low ($\sim 5\,\mathrm{km\,s^{-1}}$) intrinsic velocity dispersion. CO data is also very useful, especially because of its low intrinsic dispersion, but the lower volume filling factor introduces more scatter than characterizes the HI derivations.

If the Galaxy rotates as assumed, then the extreme velocity along a line of sight through the inner Galaxy is contributed by material at the point closest to the galactic center. Here, as Fig. 12 shows, the line of sight is tangent to a galactocentric circle and $R = R_{\min} = R_0|\sin l|$ is at its *minimum* value, i.e. at the subcentral point, for that line of sight, and $\Theta(R)$, and thus $|v(R_{\min}, l)|$, at their *maxima*. This "terminal velocity", $v_t(l)$, may be found from the suitably-defined kinematic cutoff value on HI spectra observed near $b = 0°$ in the inner Galaxy. (The requirement "suitably-defined" involves the manner in which the cutoff is measured; if the intrinsic dispersion of the well-behaved gas were zero, the cutoff would be abrupt; in fact, a correction is necessary to account for the blurring of the cutoff due to the intrinsic dispersion, which amounts to about $5\,\mathrm{km\,s^{-1}}$ for the case of HI data, and less for the CO data.) The inner-Galaxy rotation curve is then $\Theta(R_0|\sin l|) = |v_t(l)| + \Theta_0|\sin l|$. It is important

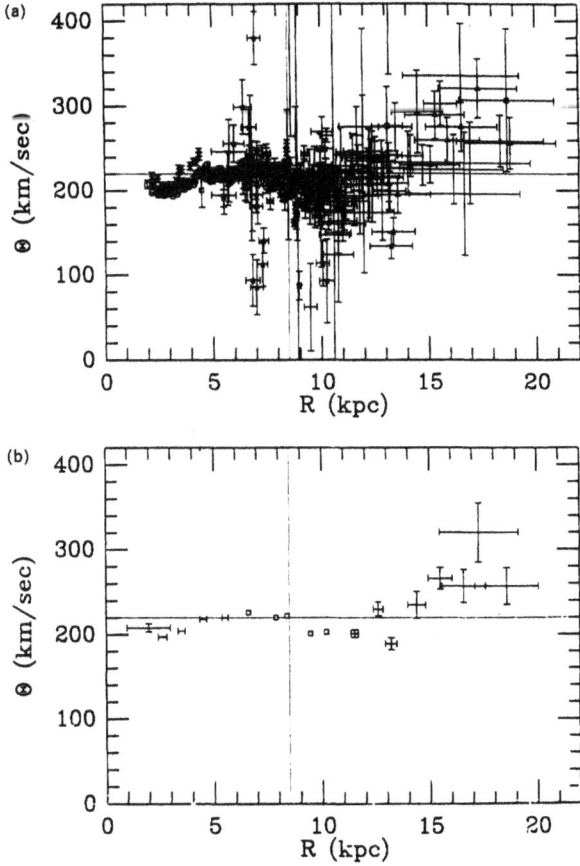

Fig. 13. Upper panel: Rotation curve data points representing a collation of material which includes CO and HI terminal-velocity measurements at $R < R_0$ and CO velocities toward optically identified HII regions at $R > R_0$. Data from the inner few kpc are excluded from the compilation.
Lower panel: Rotation curve of the Galaxy following from the data in the upper panel, suitably averaged. The scaling incorporates the galactic-constant values $R_0 = 8.5$ kpc and $\Theta_0 = 220$ km s^{-1}. (From Fich & Tremaine 1991)

to realize that the rotation curve derived from the terminal-velocity method formally represents only motions contributed from the locus of tangent points, i.e. the circle passing through the Sun and the galactic center. Viewed within the context of the assumption of axial symmetry, this constraint is obvious; but when the function so derived is viewed simply as a measure of galactic kinematics, it is well to explicitly recognize that the measure applies only to that locus.

The length of a galactic year, considered as the period of revolution of the LSR around the galactic center, is $2\pi R_0/\Theta_0 = 2.4 \times 10^8$ years, for the IAU-recommended values $R_0 = 8.5\,\mathrm{kpc}$ and $\Theta_0 = 220\,\mathrm{km\,s^{-1}}$. This period is about one or two percent of the age of the galactic disk. The period of galactic rotation scales approximately with R, because $\Theta(R)$ is quite constant over most of the Galaxy; how physical structures in the galactic disk can survive the shearing forces of such strong differential rotation constitutes the so-called "winding dilemma".

3.4 The Galactic Circular-Velocity Rotation Curve Evaluated at $R > R_0$

The terminal-velocity method cannot be used to derive the rotation curve in the outer Galaxy: because the amplitude of the radial velocity increases smoothly with increasing distance from the Sun along lines of sight at $R > R_0$, there is no uniquely specified velocity corresponding to a known distance, analogous to the correspondence of the terminal velocity to the subcentral-point distance.

The first indication that the rotation curve of the Galaxy is substantially flatter at $R > R_0$ than the Keplerian variation, for which $\Theta(R) \propto R^{-1/2}$, was given by Burton & Bania (1974). In that work, a compilation of radial velocities and optical spectroscopic distances of 668 supergiants, associations, and HII regions was analyzed for kinematic patterns. Then, the kinematic patterns were found which would, by themselves, in the manner described below in Sect. 3.6 and illustrated in Fig. 22, produce the observed form of HI data observed at $R > R_0$; the HI kinematics were derived in terms of a rotation curve whose flatness was specified by a free parameter. By adjusting this free rotation-curve parameter, the HI kinematic patterns were effectively slid inwards and outwards in R, until the maximum correlation between the HI, radio, patterns and the optically determined ones was reached. The correlation analysis required that a rotation curve be adopted that was substantially flatter than the Schmidt (1965), approximately Keplerian, curve if the gas and stellar kinematics were to match.

Determination of $\Theta(R > R_0)$ based on *directly* measured velocities and distances was introduced by the work done by Moffat et al. (1979) and by Jackson et al. (1979). The current basis of the outer-Galaxy rotation curve rests on work by Fich et al. (1989), by Brand (1986), and by others in papers cited by Fich & Tremaine (1991), which involves combining spectrophotometric distances to O- and B-type stars in optically visible HII regions, with radial velocities measured in the radio regime for the molecular lines which are invariably associated with HII regions. Other optical tracers which have been used to investigate the outer-Galaxy rotation curve include Cepheids (Welch 1988), carbon stars (Schechter et al. 1988), and planetary nebulae (Schneider & Terzian 1983); the results of these investigations are generally consistent with the flat, or slightly rising, rotation curve found from distant HII regions.

Unlike the inner-Galaxy situation based on the terminal-velocity method, derivation of the outer-Galaxy rotation curve is not restricted to a particular locus and can therefore in principal yield the generalized galactic velocity *field*, rather than the azimuthally smoothed rotation curve. The error bars shown in

Fig. 13 indicate that substantial uncertainties in the derived velocity field remain. The errors are dominated by the relatively small number of HII regions, or other suitable objects with distances and velocities independently determined. Optical obscuration limits the number of HII regions visible at large R, and, in any case, HII regions occur in relatively small numbers in the far outer Galaxy.

A new approach to calculating the outer-Galaxy rotation curve from HI data has recently been suggested by Merrifield (1992). This approach involves measuring the variation with longitude of the angular thickness of the gas layer, and then specifying the dependence of Θ on R which is required to maintain a constant layer thickness at a given R; this analysis also results in a measure of the R-variation of the HI layer thickness. The rotation curve which follows from Merrifield's analysis has a flattish, or slightly rising form (depending on the specific values of R_0 and Θ_0), consistent with the data plotted in Fig. 13. We note that the Fig. 13 outer-Galaxy rotation curve would rise more steeply if new information were to require a value of R_0 less than 8.5 kpc; the rise in Θ with R would decrease if a value of Θ_0 less than $220 \, \mathrm{km \, s^{-1}}$ were to be required.

Although additional work will no doubt reduce the size of the error bars in plots like the one shown in Fig. 13, it is well established that the rotation curve of our Galaxy, like those in comparable external systems, does not exhibit the steep Keplerian falloff which would be expected if the total mass distribution was similar to the distribution of visible stars. This implies the existence of mass which is only detected through its gravitational force. Fich & Tremaine (1991) review the observational constraints on the mass distribution in the outer Galaxy.

3.5 Line-Broadening Mechanisms Other than Simple Circular Galactic Rotation

There are several quite definite indications that the assumption of global circular symmetry is not generally valid in the Galaxy at large. For example, irregularities at the few percent level in the inner-Galaxy rotation curve derived from HI or CO terminal-velocity data in the first galactic quadrant are somewhat differently placed than those in the curve derived from fourth-quadrant data. Axial symmetry would require that the two curves agree in the mean. In fact, there is a systematic difference of about $7 \, \mathrm{km \, s^{-1}}$ between the two curves over the region $3 < R < 8 \, \mathrm{kpc}$; Fig. 14 indicates this difference in the run of $v_t(l)$ values. Another straightforward general indication of the failure of the symmetry assumption is given by the mean velocity of emission observed in the direction of the galactic center, as well as by that observed in the direction of the anticenter, which differ from $0 \, \mathrm{km \, s^{-1}}$ by about $7 \, \mathrm{km \, s^{-1}}$ (see also Blitz & Spergel 1991).

There are other kinematic properties of the Galaxy which cannot be reconciled with the assumption of circular rotation, or with that assumption qualified by the presence of localized streaming and an error in the determination of the

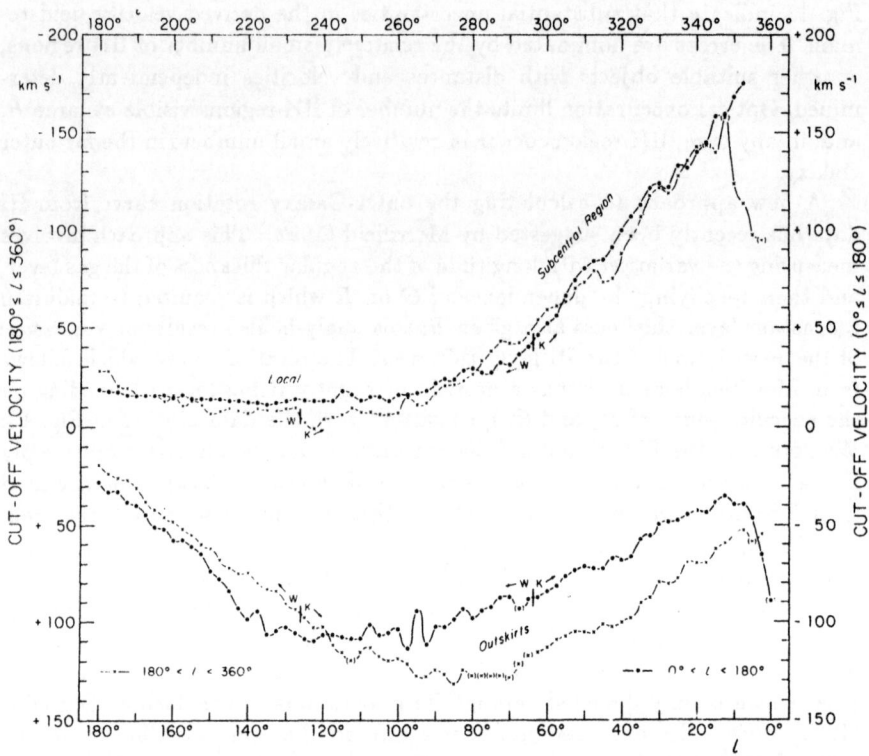

Fig. 14. Deviations from global axial symmetry illustrated by systematic differences in the cutoff velocities as measured in the longitude range $0° < l < 180°$ and in the complementary range $180° < l < 360°$. The cutoff velocities at $|l| < 90°$ refer to the terminal velocities which enter the inner-Galaxy rotation-curve determination; the cutoff velocities at $v < 0\,\mathrm{km\,s^{-1}}$ for the longitude range $0° < l < 180°$ and at $v > 0\,\mathrm{km\,s^{-1}}$ for the $180° < l < 360°$ range refer to the outer edge of the galactic gas layer. The data are from the $b = 0°$ observations of Westerhout (1969) and Kerr & Hindman (1970), as indicated; Blitz & Spergel (1991) give a similar plot for data taken at $|b| \leq 10°$ and projected onto the galactic plane

LSR. For example, Fig. 14 (from Burton 1973 and 1988) shows that the kinematic cutoff at $b = 0°$ corresponding to the far outskirts of the Galaxy occurs at velocities some $25\,\mathrm{km\,s^{-1}}$ more extreme in the fourth quadrant than in the first. Whether or not this lopsidedness is of kinematic or of structural origin, it indicates in either case violation of axial symmetry. Other indications of the failure of axial symmetry are rather clearly of a structural, rather than a kinematic, nature. One notes, for example, that the warped outer-Galaxy gas layer extends to higher angular distances in the northern-hemisphere data than at comparable distances from the line of nodes in the southern data.

It is instructive to identify deviations from strictly circular rotation in several categories, which are probably physically distinct:

1. line broadening intrinsic to a single interstellar structure, $\sigma_v \simeq 2-5\,\mathrm{km\,s^{-1}}$;
2. an evidently random component to the motions with respect to each other of the individual members of the ensemble of interstellar clouds, $\sigma_{c-c} \simeq 4\,\mathrm{km\,s^{-1}}$;
3. ordered, often substantial, motions of individual fragments of the interstellar ensemble, $|\Delta v|$ up to $50\,\mathrm{km\,s^{-1}}$ or more;
4. systematic streaming motions of the medium as a whole, incorporating length scales of several kpc, presumably induced by large-scale mass concentrations in the disk, $|\Delta v| \simeq 7\,\mathrm{km\,s^{-1}}$; and
5. possible deviations from circularity in the streamlines of the gas flows, $|\Delta v| \simeq 10\,\mathrm{km\,s^{-1}}$.

The first two of these five kinematic measures can best be made along the locus contributing the terminal velocities (although measurements at high z-heights provide detailed and important information on vertical motions in the local neighborhood), and thus are measured in the inner Galaxy; the third measure requires that individual features be separated from the blending which plagues low-latitude data, and so usually refers to material out of the standard galactic disk; the last two measures can be made in the outer- as well as in the inner-Galaxy.

Some additional remarks on each of these different kinematic measures are appropriate.

3.5.1 Line Broadening Intrinsic to a Single Interstellar Structure. Most galactic-morphology information comes from observations along low-latitude lines of sight where line blending and shadowing effectively prevent identification of individual structures. Line blending certainly overwhelms individual low-$|b|$ HI fragments, and substantially discriminates against ^{12}CO ones. An upper limit to the line broadening due to internal motions can be measured from the shape of the wings at the terminal velocities, where the line edges are determined by a kinematic cutoff. Even at the terminal-velocity cutoff, however, lengths of path longer than the characteristic length scale of single gas entities will contribute to the measured broadening; this implies that the effective dispersion measured at the terminal velocity will be an upper limit to the true, single-entity, internal motion. A measure of this broadening for the HI gas is given by the well-behaved nature of the terminal-velocity kinematic cutoff, such as characterizes the data in the longitude sector shown under high resolution in Fig. 9; this cutoff is fit by a Gaussian form with dispersion of about $5\,\mathrm{km\,s^{-1}}$ (Burton & Gordon 1978). Although the terminal-velocity measure spans galactocentric radii over the range $0.2 < R < R_0$ there is no indication of a variation in this quantity with R. (It is, of course, important to specify that the emitting material referred to in this regard is that HI which is encompassed by the density- and temperature-ranges whose emission dominates the single-dish data; we note in particular that the broad ($\sigma_v \simeq 25\,\mathrm{km\,s^{-1}}$), but weak, wings found on some high-$|b|$ HI profiles and attributed to material at temperatures of several thousands of degrees are submerged at low latitudes by ubiquitous, more intense emission.) Dickey et al. (1990) report observations of the nearly

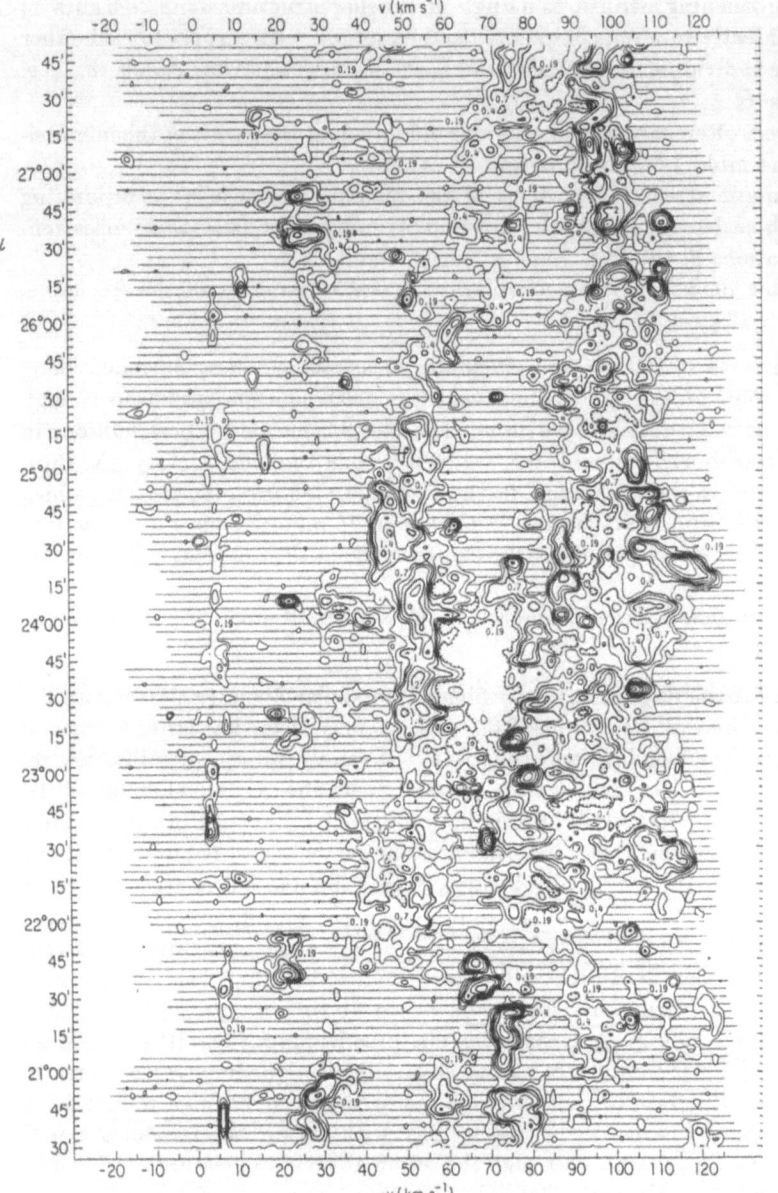

Fig. 15. Longitude, velocity arrangement of ^{13}CO emission observed over a sector of the galactic equator. The scatter of the measured terminal velocities about a mean corresponding, for example, to that derived from a data set like that shown in Fig. 9 for the more ubiquitously distributed HI tracer, yields a measure of the cloud-to-cloud velocity dispersion of members of the ensemble of molecular clouds. Generally ^{13}CO is more useful than ^{12}CO as a probe of the molecular-cloud ensemble: because the ^{13}CO features are narrower and more transparent than the ^{12}CO ones, the perceived nature of the cloud ensemble suffers less from blending. (From Liszt et al. 1981)

face-on external spiral NGC 1058, made using the VLA and sampling kinematic and spatial elements of sizes comparable to those sampled in the Milky Way morphology data. They found an upper limit to the characteristic dispersion of 6 km s^{-1} and very little variation of this dispersion over the gaseous disk; in both these regards the NGC 1058 situation is consistent with that measured at low-$|b|$ in our own system.

A value for the intrinsic broadening representative of the single-cloud velocity dispersion derived from ^{13}CO survey observations of the sort shown in Fig. 15 is lower than the HI value, namely about 2 or 3 km s^{-1}. The error on the CO linewidth measurement is probably smaller than is the case for the HI because the molecular-cloud ensemble is sampled at higher angular resolution, and because these clouds are less ubiquitously distributed, and thus suffer less blending, than is the case for HI gas. It is even more obvious for the molecular-cloud case than for the HI case that this broadening must be carefully specified, since high-resolution data show that kinematic and spatial structure appears on essentially all scales; the representative intrinsic dispersion value of 2.5 km s^{-1} derived for ^{12}CO data by Burton & Gordon (1978) refers to clouds populating the molecular annulus as they are sampled with one-arcminute resolution. In view of the fact that heavy blending generally precludes identification of individual features at low-$|b|$, at least in HI emission data, and in view of the fact that the intrinsic broadening is in any case less that other motions under consideration, it is not necessary to pursue this broadening further here.

3.5.2 Random Motions of the Individual Members of the Ensemble of Interstellar Clouds; the Molecular-Cloud "Size-Linewidth Relation".

The most explicit information on low-$|b|$ gas-tracer kinematics is provided by the terminal velocities, $v_t(l)$, contributed by material near the locus at $R = R_{min} = R_0 |\sin l|$. If material is always common along this locus, then the observed *scatter* in the $v_t(l)$ values is determined only by the bulk motions near the tangent points. Such ubiquity is the case for HI, whose distribution is adequately described for many low-$|b|$ purposes as smooth and of volume-filling-factor unity. The entities making up the ensemble of molecular clouds populating the molecular annulus typically have mean free paths of order 1 kpc or more, and in this case the discreteness of the emitting ensemble becomes important in determining the scatter of the terminal velocities. Random bulk motions may preclude the presence of clouds at the velocity corresponding to the tangent point, and, furthermore, spatial irregularities may leave the tangent point unpopulated. Both effects will lead to a scatter about a mean of the $v_t(l)$ measure. Burton & Gordon (1978) introduced a technique of using this scatter to derive both the spatial mean-free-path as well as the cloud-to-cloud velocity dispersion of the ^{12}CO molecular cloud ensemble; the technique has been applied to ^{13}CO data by Liszt et al. (1981, 1984) and by Liszt & Burton (1983), and to more modern ^{12}CO data by Clemens (1985). Figure 16 illustrates the observed situation.

The *scatter* of the measured terminal velocities about a local mean value is a measure — indeed probably the most accurate one available — of the macroscopic, cloud-to-cloud, motions of the individual entities of interstellar gas

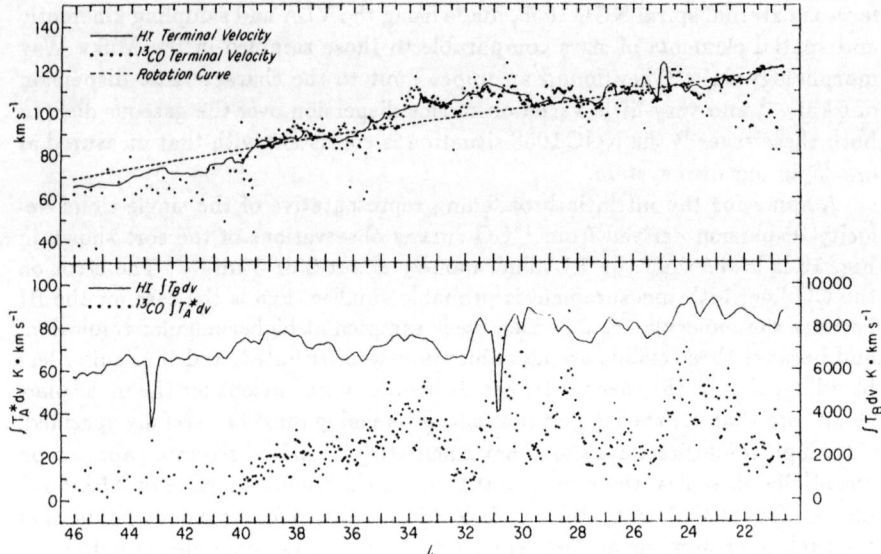

Fig. 16. Upper panel: Variation with longitude of the terminal velocities measured from ^{13}CO data (dots) as well as from HI data (full-drawn line). The dashed line shows the velocity contributed from the locus of tangent points, at $R = R_0 \sin l$, corresponding to the smoothed rotation curve given by Burton & Gordon (1978), for $R_0 = 10.0$ kpc and $\Theta_0 = 250$ km s^{-1}. **Lower panel:** Variation with longitude of the intensities integrated over positive velocities for ^{13}CO emission (dots) and for HI (line). The HI and the CO tracers share the same gradual variations in v_t when smoothed over several degrees of longitude, implying that both tracers respond similarly to the underlying gravitational potential; the *scatter* in the CO v_t values measures the cloud-to-cloud velocity dispersion as well as the cloud-to-cloud separation. There is no significant correlation of the HI and CO line integrals. (From Liszt et al. 1984)

with respect to each other. Figure 17 shows that the cloud-cloud kinematics can be characterized by a one-dimensional line-of-sight dispersion of $\sigma_{c-c} = 4.2 \pm 1.0$ km s^{-1}; the estimate of the accuracy of the σ_{c-c} determination is based on Monte Carlo simulations of the cloud ensemble (see e.g. Liszt et al. 1981).

Observations such as those shown in Fig. 15 have been subject to substantial efforts to derive properties of the individual entities making up the ensemble of galactic molecular clouds. Of particular potential importance is the relationship which has been derived (e.g. by Solomon et al. 1987; see their Fig. 7) which purports a tight correlation between the *size* of a molecular cloud and its internal *linewidth*, such that the FWHM, Δv, varies with the cloud characteristic dimension, D_c, as $\Delta v \propto D_c^{0.5 \pm 0.05}$. The size-linewidth relationship purported by Solomon et al. was based on visual inspection of data like that in Fig. 15 which attempted to isolate individual entities, for example by identification of regions of closed contours at a certain threshold temperature. It is, however, notoriously difficult to account for the effects of line blending and cloud shadowing when basing such an analysis on visual inspection of the (l, v)

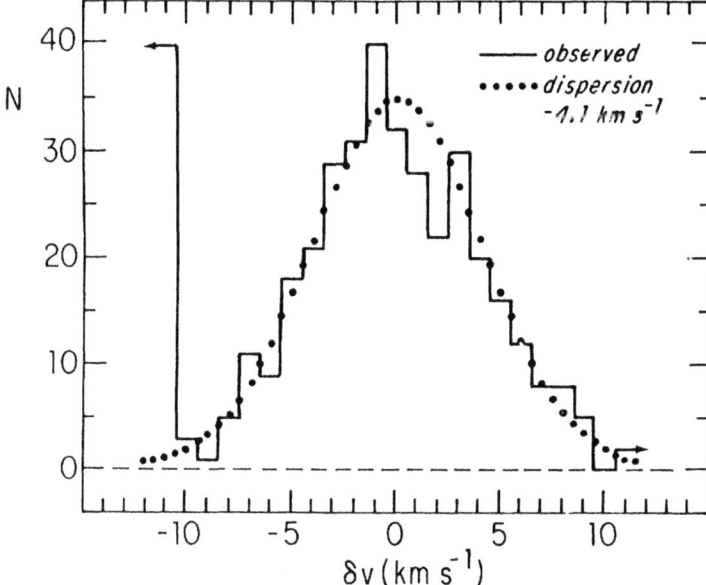

Fig. 17. Histogram of the observed differences $v_t(^{13}CO)$ -$v_t(HI)$ between the terminal velocities of the clumped molecular component and the more ubiquitously distributed atomic hydrogen component. This measure reflects the cloud-to-cloud velocity dispersion of the molecular-cloud ensemble; the dots show a Gaussian distribution characterized by a dispersion of $4.1 \, km \, s^{-1}$. Similar plots have been made for a Monte Carlo simulation of the ensemble of molecular clouds; that analysis shows that the statistical measure of the σ_{c-c} parameter is accurate to about $1 \, km \, s^{-1}$. (From Liszt et al. 1984)

maps. This point has been stressed by a number of authors, including Liszt & Burton (1981), and, specifically in the size-linewidth context, by MacLaren et al. (1988), Issa et al. (1990), and Combes (1991); see also Kegel (1989) and Scalo (1990).

Simulations of low-latitude data which take into account the vagaries of the velocity-to-distance transformation, of line blending, of cloud shadowing, etc., can be instructive and, occasionally, sobering. Figure 18 shows a simple Monte Carlo simulation of a collection of ^{12}CO molecular clouds sprinkled near the galactic equator and "observed" in a manner which mimics parameters such as antenna pattern, beam separation, velocity resolution, and sensitivity which enter the real ^{13}CO observations shown in Fig. 15. Rather comparable simulated (l, v) diagrams have been shown by Liszt & Burton (1981). Figure 18 represents, however, a sprinkling of molecular clouds all of which are the *same* size, and all of which have the *same* internal linewidth. It seems clear that a visual analysis of this figure, based on identification of the angular sizes and velocity widths of closed-contour patterns, would result in a relationship, at variance with the input, in which *larger* patterns were also *broader*. Such an analysis would then result in an overestimate of cloud masses and would ultimately lead to an exaggeration of the role played by "giant" molecular clouds as well as

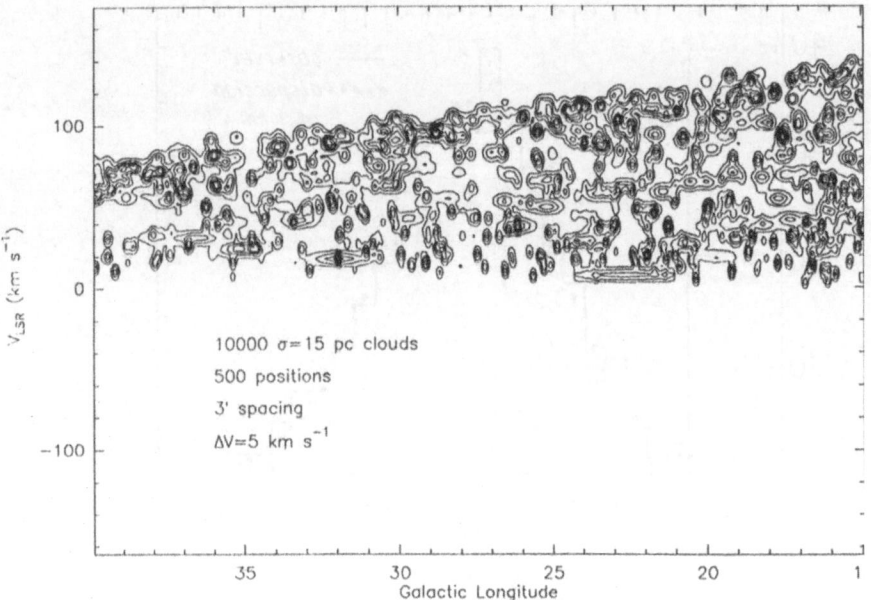

Fig. 18. Longitude, velocity diagram representing simulated observations the galactic equator at longitude intervals of 3′ for ^{12}CO emission from an ensemble of 10,000 molecular clouds, all of the *same size* (15 pc diameter), and all of the *same linewidth* (5 km s^{-1} FWHM). The problem of blending and shadowing inherent in low-latitude investigations of the molecular annulus is a severe one: a size-linewidth relation derived from this diagram in the same way as this relation has been derived from real observations (e.g. Solomon et al. 1987) would not reproduce the input values. (From Maloney & Burton, unpublished)

to an overestimate of the total mass of hydrogen in the Galaxy (see Issa et al. 1990). Additional work seems called for in order to rule out the possibility of contamination of the size-linewidth measure by blending effects. We do note, however, that the dependence of these quantities reported by Solomon et al. (1987) does fit into the more general relationship suggested by Larson (1981) to be valid over some six orders of magnitudes in the masses of a variety of interstellar phenomena.

3.5.3 Ordered, Often Substantial, Motions of Unusual Individual Fragments of the Interstellar Ensemble. The third kinematic characteristic mentioned above, namely the anomalous motions associated with certain individual structures such as the Heiles shell phenomenon (e.g. Heiles 1979) or infrared cirri (Deul & Burton 1990; Burton et al. 1992) require for their identification that the structure in question be seen quite isolated from surrounding emission, in velocity as well as in position. Such isolation is observed almost solely at latitudes more than about 10° from the galactic plane. Thus information on these motions is biased toward the local solar neighborhood. The line-of-sight velocities involved with these structures range up to 50 km s^{-1} or more; at the higher end of the range, the velocities embrace the class of objects called intermediate-velocity clouds. It is important to realize that the *vertical* component of the velocity dominates the kinematic information on these isolated structures, whereas it is the *horizontal* component which is measured along the terminal-velocity locus. It is curious, and not yet understood, why the well-behaved nature of the terminal velocities is not contaminated by contributions from the structures with more extreme kinematic peculiarities which are commonly seen at higher latitudes. It is interesting in this regard to compare the *well-behaved* terminal-velocity cutoff shown in the HI observations of Fig. 1, and, in more detail in the HI (l, v) map of Fig. 9, and which consistently characterizes HI data near $b = 0°$, with the frequent appearance at higher $|b|$ of highly anomalous velocities commonly found in association with infrared cirri (see Deul & Burton 1990; Burton et al. 1992).

3.5.4 Streaming Motions Systematic over Length Scales of Several kpc, Probably Induced by Large-Scale Gravitational Effects.

The last two kinematic measures mentioned above involve systematic streaming motions, coherent over length scales of a kpc or longer. Streaming motions induced gravitationally by major mass concentrations show up as bumps on the terminal-velocity rotation curve (Shane & Bieger-Smith 1966; Burton 1971). These bumps represent kinematic deviations of about $\pm 7 \, \mathrm{km \, s^{-1}}$ from a smoothly-drawn fit to the terminal velocities. Figure 41 shows the rotation curve for the part of the Galaxy interior to the solar orbit as given by a smooth-line fit to measurements of the terminal velocities in northern-hemisphere HI data. Although the data refer only to motions along the terminal-velocity locus, there is no reason to believe that the flows resulting in the bumps are not a general characteristic of our system. Similar motions are observed in association with spiral structure in external systems. The alternative possibility of interpreting the irregularities as due to lengthy regions in which the terminal-velocity region is relatively empty of HI gas, and where as a consequence the kinematic cutoff is contributed from material away from the subcentral point and thus at a lower velocity, was ruled out by Shane & Bieger-Smith (1966) by the argument that the terminal-velocity cutoff remains sharp at all longitudes: more or less empty subcentral regions would result in shallow wings at the highest velocities, and this situation is not observed. In view of the difficulties in mapping the spiral structure of the Milky Way, a case can be made that the firmest indication of the location of spiral arms in our Galaxy is that which follows from the locations of the rotation-curve bumps. We note that an increase in tangential velocity outwards across a spiral arm is observed commonly in external galaxies, where the perspective for such studies is much better than that afforded by our own Galaxy.

One of the most dramatic examples of unambiguous noncircular motion in the Galaxy is provided by the 3-kpc arm (see van Woerden et al. 1957; Rougoor & Oort 1960). This feature is observed spanning some 30° of longitude, and is observed in emission and absorption in HI as well as in molecular lines. The 3-kpc arm displays a net motion away from the galactic center of some $53 \, \mathrm{km \, s^{-1}}$. The total gas mass involved in this feature is about $10^8 \, M_\odot$. The mechanism responsible for the 3-kpc arm remains unknown. It is in particular not clear if it is a transient feature or if it represents a permanent flow of gas in closed orbits; the latter explanation is currently favored.

3.5.5 Possible Deviations from Circularity in the Streamlines of the Gas Flows. The final kinematic measure mentioned above refers to the failure of our system to show all of the cylindrical symmetries which are required for a system rotating with complete circular symmetry. It has been known for some time that the total profile extents measured at complementary northern and southern longitudes are not identical, nor are the placements of the bumps; furthermore, the maximum symmetry between the two hemispheres of the Galaxy is not reached along the line of nodes defined by the longitudes 0° and 180°. Several attempts have been directed to understanding this lopsidedness of the Milky Way, both in structural, as well as kinematic terms. Blitz & Spergel (1991) argue for ellipticity of the gas motions responding to a gravitational potential of triaxial form. The amplitude of the noncircular component of these motions is typically less than $10\,\mathrm{km\,s^{-1}}$.

3.6 Considerations Regarding the Velocity-to-Distance Transformation

The above discussion indicates that galactic kinematics considered over the entire range $3 \leq R \leq 20\,\mathrm{kpc}$ shows irregular as well as systematic deviations from a cylindrically symmetric, circular-motion idealization which amount to less than a few percent of the rotation velocity, which is itself about constant over this range. These deviations are not of sufficient amplitude to preclude derivation of the galactic rotation curve in a fairly direct manner, over all of the galactic disk beyond the bulge region. The kinematic irregularities do, of course, reflect astrophysical information, especially important for considerations of galactic dynamics. The deviations are also a hindrance to interpretations of galactic morphology which are largely based on kinematic determinations of distance.

There are many purposes for which it is useful to have a simple analytic expression for the galactic rotation curve. The expression

$$\Theta(R) = \Theta_0 \left[1.0074 \left(\frac{R}{R_0} \right)^{0.0382} + 0.00698 \right] \quad [\mathrm{km\,s^{-1}}] \qquad (9)$$

fits (with the values $R_0 = 8.5\,\mathrm{kpc}$ and $\Theta_0 = 220\,\mathrm{km\,s^{-1}}$) the smoothed outer-Galaxy rotation curve derived by Brand (1986) as well as the inner-Galaxy terminal-velocity curve (averaged over both northern and southern data) and is an adequate representation of the curve plotted in Fig. 12. This expression can inserted into the fundamental equation $v(R,l) = R_0[\Theta(R)/R - \Theta_0/R_0]\sin l$, and a simple geometrical conversion made to distance, r, along the line of sight for each (R,l) pair, to yield the velocities with respect to the Local Standard of Rest which follow for the idealization of circular rotation. Figure 19 (from Burton 1988) shows contours of radial velocity with respect to the Local Standard of

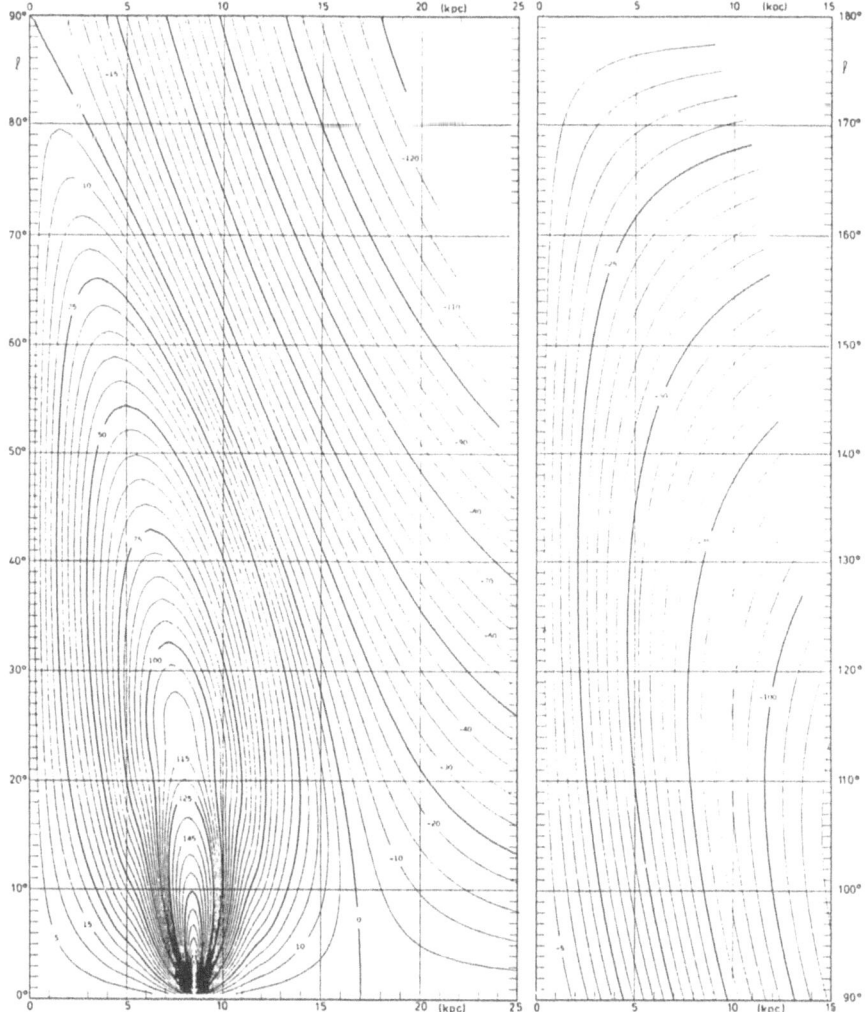

Fig. 19. Velocity-to-distance conversion chart. Contours of radial velocities with respect to the Local Standard of Rest are plotted against longitude (at $b = 0°$) and distance from the Sun. The conversion is based on the rotation curve given by the analytic expression mentioned in the text, for $R < R_0 = 8.5\,\mathrm{kpc}$, and on a flat rotation curve $\left(\Theta(R) = 220\,\mathrm{km\,s^{-1}}\right)$, for $R > R_0$

Rest plotted as a function of galactic longitude and distance from the Sun. Such a kinematic-distance conversion chart can be used to assign distances to low-$|b|$ objects which have measured velocities, assuming that the motions are governed only by circular rotation.

Under the simplifying assumption of circular motion, the velocity-to-distance transformation is conceptually very straightforward. But in practice, even

53

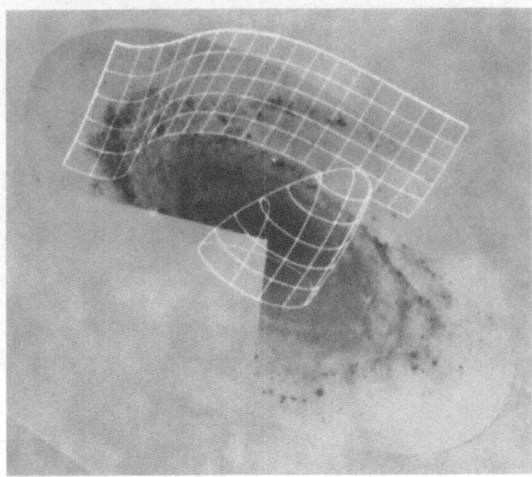

Fig. 20. Schematic indication of the transformation of velocity to distance for an observer embedded within a galactic system. The two surfaces correspond to locations of material perceived at a single radial velocity (about $50\,\mathrm{km\,s^{-1}}$ for a galaxy rotating as ours); in the outer reaches of the system, the velocity-to-distance transformation is single-valued, whereas in the inner portions it is double-valued. (From a diagram prepared by V. Icke in Burton 1988)

maintaining this assumption, the global transformation of the observed (l, b, v) data cube of measured intensities of HI, or of other spectral-line tracers, is not very simple. Figure 20 indicates that the transformation is in any case an interesting topographical exercise. The problem of displaying a large three-dimensional data cube in maps which can be printed on a piece of paper has led to the practice of viewing the data in "channel maps", in which emission at a particular constant velocity (or emission integrated over a small range of velocities) is plotted in (l, b) coordinates. (A catalog of such plots is given by Burton & te Lintel Hekkert 1985.) This practice is easier to justify when done for a small-scale spectral feature which has in some way or another been isolated from the global background. Figure 20 shows schematically how a constant-velocity slice through the three-dimensional data cube would be perceived by an observer embedded in the Milky Way. Two topological surfaces are indicated, neither of which sample the system in a particularly convenient manner; the inner-Galaxy surface indicates the double-valuedness of the velocity-to-distance transformation.

Even under the simplest assumptions the manner in which the radial velocity varies along a given line of sight is not linear with distance from the Sun, so that mapping is in fact more complicated than the direct conversion represented schematically in Fig. 20 would suggest. Along some lines of sight, the radial velocity varies slowly with distance, affording little resolution in the velocity-to-distance transformation; on other lines of sight, or on portions of

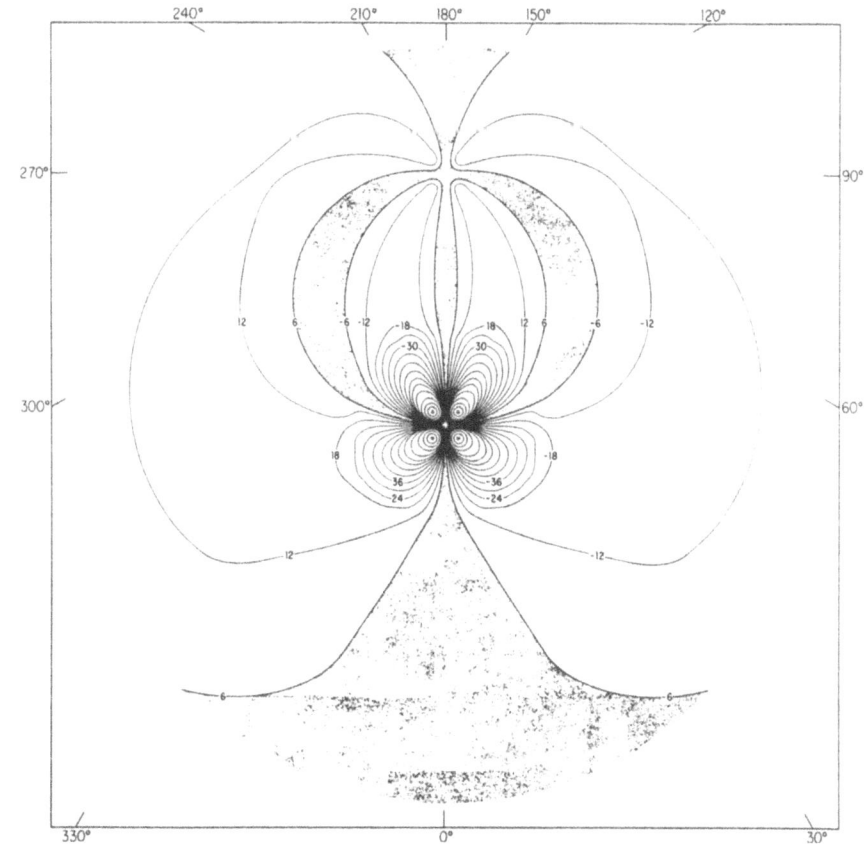

Fig. 21. Distribution over the face of the Galaxy of the kinematic "velocity-crowding" parameter $|dv/dr|$. The contours are labelled in units of km s^{-1} kpc^{-1}. Material lying in portions of the Galaxy where this parameter is small, say $|dv/dr| < 6$ km s^{-1} kpc^{-1}, is perceived with insufficient resolution in the velocity-to-distance transformation to allow distances to be determined; the compression of emission from a long path into a narrow velocity range will also lead to substantial optical depths. Portions of the Galaxy where $|dv/dr|$ is relatively large, say > 12 km s^{-1} kpc^{-1}, can be more easily mapped and the optical-depth effects are likely to be less important

them, the radial velocity varies substantially with changing distance, affording better resolution in the transformation.

Figure 21 (from Burton 1976) shows how the kinematic parameter $|dv/dr|$ varies across the face of the Galaxy under the idealized, circular-rotation case. The discussion in Chap. 2 stressed the importance of this parameter to determination of many aspects of the appearance and the degree of saturation of low-latitude 21-cm profiles; for the case of CO data, it also determines the degree to which members of the molecular-cloud ensemble shadow each other. Gas lying in regions where the value of $|dv/dr|$ is low will be particularly prone

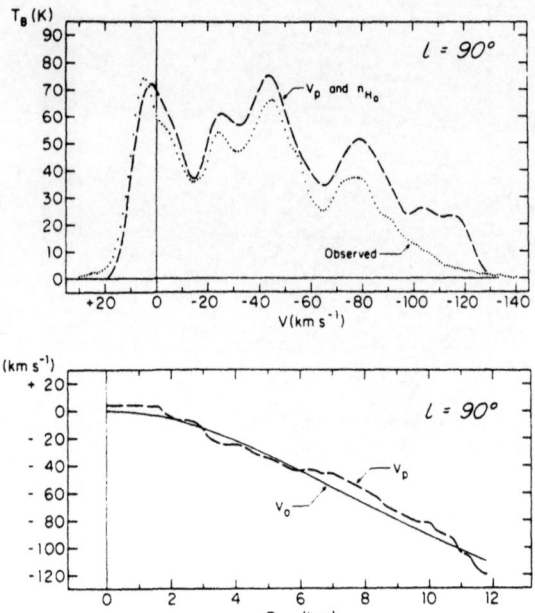

Fig. 22. Illustration of the sensitivity of hydrogen-line profiles to even small-amplitude kinematic variations from the basic circular rotation. The general shape of any observed HI profile in the galactic plane can be simulated by a model in which only the modelled velocity field is perturbed, while the gas density is held uniform. The maximum amplitude of the velocity irregularities necessary to reproduce the observed situation is about $10\,\mathrm{km\,s^{-1}}$, consistent with the amplitude of such irregularities which are in fact known to exist. Profile simulations based only on varying gas density, retaining circular motion, do not account for the dominance of the kinematic perturbations which are known to exist, can not reproduce "forbidden" velocities (such as the emission shown here for $l = 90°$ at $v > 0\,\mathrm{km\,s^{-1}}$), and require a density contrast at such a level that it can be ruled out on other grounds

to saturation and line-blending effects; the resolution in the velocity-to-distance transformation will be poor, and mapping difficult.

The arrangement of the kinematic parameter $|dv/dr|$ plotted in Fig. 21 corresponds to simple circular rotation. The importance of the mechanisms other than simple rotation discussed in Sect. 3.5 can be considered in terms of the behavior of the kinematic parameter. It is instructive, for example, to imagine the perceived velocity field plotted in Fig. 19 as printed on a rubber sheet and then to imagine it distorted in a way consistent with a realistic incorporation of the streaming motions, kinematic asymmetries, and other perturbations itemized above. Such perturbations would then show up on a plot like the one in Fig. 21 in which the arrangement of the $|dv/dr|$ parameter would be correspondingly more complex.

Just how important perturbations to the velocity field are to the appearance of galactic-structure spectra is illustrated by Fig. 22 (from Burton 1973). Essentially any low-latitude HI profile can be reproduced by a simulation in which the only free parameter is the line-of-sight variation of the velocity field (see

Burton 1972). The kinematic perturbations necessary to reproduce the profiles are of the same amplitude, sense, and spatial distribution as those which are in fact known to exist. It is not possible to similarly reproduce the observations by a simulation in which only the gas density is varied. This does not, of course, imply in any way that the gas density at low latitudes does not vary; it is simply the case that the *dominant determinant of profile shape is the kinematics*.

Thus although over much of the Galaxy the deviations from circular velocity are of the order of a few percent or less of the rotation velocity, these noncircular motions considerably complicate analysis of Galactic spectral-line data. A local error of $5\,\mathrm{km\,s^{-1}}$ (amounting only to about 2% of the rotation speed) in the determination of $\Theta(R)$ can cause a large error (amounting to 50% or more) in a kinematic distance, especially in regions where the radial velocity changes only slowly with distance from the Sun; such an error can also cause an ambiguity in the distance, if the velocity-to-distance transformation becomes locally multivalued. If the regions shaded in Fig. 21 are largely inaccessible to detailed mapping, it is easy to imagine that an analogous figure calculated in the presence of a more complicated, and more realistic, velocity field would have shaded regions sprinkled across it in a *much* more complicated way, frustrating general, detailed, low-latitude mapping.

The problems attributable to the vagaries of the $|dv/dr|$ parameter stem largely from our embedded perspective. Many questions originally posed for the Milky Way, including whether or not it displays a "grand design" of spiral structure, if the spiral arms are trailing or leading with respect to rotation, what pitch angle and spacing characterize the arms, what relative motions and spacings characterize the distributions of stars and gas, etc., can be more profitably pursued in external systems.

References

3.1 Blaauw, A., Gum, C.S., Pawsey, J.L., Westerhout, G. 1960, *Monthly Notices Roy. Astron. Soc.* **121**, 10
3.2 Blitz, L., Spergel, D.N. 1991, *Astrophys. J.* **370**, 205
3.3 Brand, J. 1986, Ph.D. Thesis, University of Leiden
3.4 Burton, W.B. 1971, *Astron. Astrophys.* **10**, 76
3.5 Burton, W.B. 1972, *Astron. Astrophys.* **19**, 51
3.6 Burton, W.B. 1973, *Publ. Astron. Soc. Pacific* **85**, 679
3.7 Burton, W.B. 1976, *Ann. Rev. Astron. Astrophys.* **14**, 275
3.8 Burton, W.B. 1988, in *Galactic and Extragalactic Radio Astronomy*, G.L. Verschuur, K.I. Kellermann, (eds.), Springer-Verlag, New York, p. 295
3.9 Burton, W.B., Bania, T.M. 1974, *Astron. Astrophys.* **33**, 425
3.10 Burton, W.B., Bania, T.M., Hartmann, D., Tang Yuan 1992, in *Proc. CTS Workshop No. 1. "Evolution of Interstellar Matter and Dynamics of Galaxies"*, J. Palouš, W.B. Burton, P.O. Lindblad, (eds.), Cambridge University Press, in press
3.11 Burton, W.B., Gordon, M.A. 1978, *Astron. Astrophys.* **63**, 7
3.12 Burton, W.B., te Lintel Hekkert, P. 1985, *Astron. Astrophys. Suppl. Ser.* **62**, 645
3.13 Clemens, D.P. 1985, *Astrophys. J.* **295**, 422
3.14 Combes, F. 1991, *Ann. Rev. Astron. Astrophys.* **29**, 195
3.15 Deul, E.R., Burton, W.B. 1990, *Astron. Astrophys.* **230**, 153

3.16 Dickey, J.M., Hanson, M.M., Helou, G. 1990, *Astrophys. J.* **352**, 522

3.17 Fich, M., Blitz, L., Stark, A.A. 1989, *Astrophys. J.* **342**, 272

3.18 Fich, M., Tremaine, S. 1991, *Ann. Rev. Astron. Astrophys.* **29**, 409

3.19 Habing, H.J. 1987, in *The Galaxy*, G. Gilmore, B. Carswell, (eds.), D. Reidel Pub. Co., p. 173

3.20 Heiles, C. 1979, *Astrophys. J.* **229**, 533

3.21 Issa, M., MacLaren, I., Wolfendale, A.W. 1990, *Astrophys. J.* **352**, 132

3.22 Jackson, P.D., FitzGerald, M.P., Moffat, A.F.J. 1979, in *Proc. IAU Symposium* **84**, "The Large-Scale Characteristics of the Galaxy", W.B. Burton, (ed.), D. Reidel Pub. Co., p.611

3.23 Kegel, W.H. 1989, *Astron. Astrophys.* **225**, 517

3.24 Kerr, F.J., Hindman, J.V. 1970, *Australian J. Phys. Astrophys. Suppl.* **18**, 1

3.25 Kerr, F.J., Lynden-Bell, D. 1986, *Monthly Notices Roy. Astron. Soc.* **221**, 1023

3.26 Knapp, G.R., Tremaine, S.D., Gunn, J.E. 1978, *Astron. J.* **83**, 1585

3.27 Larson, R.B. 1981, *Monthly Notices Roy. Astron. Soc.* **194**, 809

3.28 Liszt, H.S., Burton, W.B. 1983, in *Kinematics, Dynamics, and Structure of the Milky Way*, W.L.H. Shuter (ed.), D. Reidel Pub. Co., 135

3.29 Liszt, H.S., Burton, W.B., Xiang, D.-L. 1984, *Astron. Astrophys.* **140**, 303

3.30 Liszt, H.S., Xiang, D., Burton, W.B. 1981, *Astrophys. J.* **249**, 532

3.31 MacLaren, I., Richardson, K.M., Wolfendale, A.W. 1988, *Astrophys. J.* **333**, 821

3.32 Merrifield, M.R. 1992, preprint, Can. Inst. Theoret. Astrophys., Toronto

3.33 Moffat, A.F.J., FitzGerald, M.P., Jackson, P.D. 1979, *Astron. Astrophys. Suppl. Ser.* **38**, 197

3.34 Reid, M.J., Moran, J.M. 1988, in *Galactic and Extragalactic Radio Astronomy*, G.L. Verschuur, K.I. Kellermann, (eds.), Springer-Verlag, New York, p. 255

3.35 Scalo, J. 1990, in *Physical Processes in Fragmentation and Star Formation*, R. Capuzzo-Dolcetta, C. Chiosi, A. deFazio, (eds.), Kluwer Acad. Pub. Co., 151

3.36 Schechter, P., Aaronson, M., Cook, K.H., Blanco, V.M. 1988, in *The Outer Galaxy*, L. Blitz, F.J. Lockman, (eds.), Springer-Verlag, Berlin, p. 291

3.37 Schmidt, M. 1965, in *Stars and Stellar Systems* **V**, *"Galactic Structure"*, A. Blaauw, M. Schmidt, (eds.), University of Chicago Press, p. 606

3.38 Schneider, S.E., Terzian, Y. 1983, *Astrophys. J.* **274**, L61

3.39 Shane, W.W., Bieger-Smith, G.P. 1966, *Bull. Astron. Inst. Neth.* **18**, 263

3.40 Solomon, P.M., Rivolo, A.R., Barret, J., Yahil, A. 1987, *Astrophys. J.* **319**, 730

3.41 Welch, D.W. 1988, in *The Mass of the Galaxy*, M. Fich, (ed.), Can. Inst. Theoret. Astrophys., Toronto, p. 68

3.42 Westerhout, G. 1969, *Maryland-Green Bank Galactic 21-cm Line Survey*, 2nd edition, University of Maryland

4 Global Structural Distribution of Low-Latitude Tracers of the Inner Galaxy

Abstract. It has not proven possible to derive a detailed map of the spatial distribution of gas in the Milky Way which improves on the original Leiden-Sydney map in a way commensurate with the improved quantity and quality of data currently available. Many other aspects of the azimuthally-smoothed morphology may, however, be studied in detail. The longitudinal and radial distributions of integrated emission from tracers accessible along transgalactic paths reveal, for example, the degree of morphological confinement to the inner Galaxy. The galactic disk as defined by atomic hydrogen has a diameter about three times larger than that defined by the ionized and molecular states of hydrogen and by other tracers including the [CII]-line observed at 158 μm, the far-infrared emission from thermal dust emission, as well as the near-infrared emission observed at 2.4 μm following the distribution of stellar luminosity. Although the longitudinal distributions of most galactic tracers follow directly from the observations, interpretation of the fluctuations in these distributions is less straightforward. Molecular clouds with hotter cores are preferentially located on the inner side of the molecular annulus, where the mean cloud density is also higher than it is for clouds located further outwards. The shape of the inner-Galaxy annulus revealed by the observations of [CII] and by the near-infrared observations at 2.4 μm shows strong confinement of both of these tracers to the inner parts of the annulus defined by the ensemble of ^{12}CO molecular clouds, and a sharper drop of emissivity inward of the outer edge of the CO annulus. These morphological differences are among the indications that star formation does not occur in direct proportionality to ^{12}CO emissivity or to the number density of clouds constituting the galactic molecular-cloud ensemble. The thickness and degree of flatness of the different constituents of the gas and dust layer are other important aspects of galactic morphology which can be better studied in the Milky Way than in external systems. The inner-Galaxy gas and dust layer is flat to within substantially less than 1% of its total radial extent; its thickness amounts to about 1% of its radial extent.

4.1 Mapping Limitations Imposed by Our Embedded Perspective

The goal of the early Dutch and Australian 21-cm work was to derive as accurately as possible the distribution of neutral hydrogen in our Galaxy (see e.g. Schmidt 1957; Westerhout 1957; Oort et al. 1958; Kerr 1962). Work beginning in the late 1950's and continuing until the early 1970's led to re-definition of the galactic coordinate system on the basis of the location of the fundamental plane of the HI distribution, to derivation of the galactic rotation curve in the inner Galaxy and to recognition of the importance of the irregularities in it, to discovery of the warped and flaring nature of the HI gas layer in the outer reaches of the Milky Way, to discovery of kinematically anomalous structures of which the 3-kpc arm is the most impressive example, and to a measure of the midplane of the HI gas layer and of its vertical thickness. The Leiden-Sydney map (see Oort et al. 1958) purported to give the distribution of HI densities throughout the galactic equator; even though the quantity and quality of 21-cm data have improved enormously since the Leiden-Sydney map was derived, it has not proven possible to derive a convincing improvement on that map with correspondingly more detail.

The discussion in the preceding chapters indicates some of the difficulties which plague, and in fact largely prevent, mapping of the structural *details* of the low-latitude layer of interstellar material in our Galaxy. These difficulties are principally ones of analysis; the observations themselves are not in dispute. For the HI component, these difficulties are compounded by the sensitivity with which the 21-cm profile forms respond to the underlying kinematics. This sensitivity makes it difficult to specify the role which other parameters (for example, density, temperature, optical depth, etc.) might play in determining the appearance of a given feature. For the colder, more clumpy gas component traced by CO, many of the same problems hinder derivation of accurate distances. A much more general problem, discussed briefly below, concerns uncertainties in the conversion from CO intensities to molecular densities. These uncertainties involve not only the global calibration of the conversion to H_2 densities, but also galactic gradients in temperature, degree of line blending and optical depth, and perhaps in cloud composition, which make it difficult to determine, as in the HI situation, the locally dominant parameters. For the components of the interstellar layer which can only be observed in the continuum, distances are not available except through unfolding techniques which involve so much spatial averaging that structural details generally can not be mapped at low latitudes.

Because of the problems associated with distance determinations, many of which stem from our embedded perspective, it has proven difficult to determine, convincingly, such aspects of the structure of the Milky Way as whether or not it displays a "grand design" spiral pattern, whether spiral arms are trailing or winding, what pitch angles and spacings characterize the arms, how many spiral arms there are, what relative motions and relative locations characterize the different components associated with the spiral structure (and which might characterize a spiral arm seen in cross-section), and whether these relative motions and locations are those predicted by a linear, or nonlinear, application of the density-wave theory. These matters are of general importance to many aspects of galactic dynamics and to other astrophysical considerations.

Table 2 summarizes in a qualitative manner some of the questions posed in Milky Way studies, and my own subjective impression of the robustness of the answers achieved for these questions in our own Galaxy (at least in the parts at $R < R_0$), and in comparable external systems. The reader might find it an instructive exercise to form his or her own opinion on the degree of confidence which should be afforded the various answers to these questions. The review by Liszt (1985; see especially his Fig. 7), and the discussion printed following it, are both quite apropos to the spirit of the table.

Table 2. Subjective indication of the robustness of answers to some questions regarding the morphology of the gas layer in the inner Galaxy and in external systems

Questions posed:	Questions answered:	
	In Milky Way	in other galaxies
regarding structural aspects of spiral structure:		
– identification of spiral arms; dimensions, locations of elongated fragments	–	+
– leading or trailing	(–)	(+)
– arm inclinations	–	+
– number of arms	–	+
– arm-to-arm separation	–	+
– arm-interarm surface-density contrast	–	+
regarding kinematics of spiral structure:		
– azimuthally averaged rotation curve	+	+
– "bumps" in rotation curve	+	+
– identification of relative phases of density and velocity enhancements	–	(+)
regarding global structural properties:		
– radial surface density profile	+	+
– gas layer thickness	+	(–)
– behavior of midplane of gas layer; scalloping, corrugations	+	(–)
– radial gradients in thickness, temperature, etc.	+	(+)

4.2 Longitudinal Distributions of Low-$|b|$ Tracers

The longitudinal distribution of integrated emission from tracers accessible along transgalactic paths reveals the degree of morphological confinement to the inner Galaxy in manner which is, in principle, quite straightforward. Intervening assumptions regarding the kinematics may be largely ignored when the integrated emissivities are displayed against longitude; for example, the effects of kinematic distortions of the profiles are largely removed by the integration.

Until the mid-1970's, there were few compelling reasons to expect that the distribution of HI derived from 21-cm observations was not representative of the global morphology of the galactic interstellar medium in general. Until this time, spectral information on a galactic scale from other constituents of the interstellar medium was not available. Beginning in the mid-1970's, information from other constituents of the gas layer resulted in a quite different picture than had been presented by the 21-cm data. The HI had been shown to have a projected surface density which was more or less constant over the entire face of the Milky Way gas layer, with a radial extent continuing well beyond the Solar orbit; many of the parameters of the HI constituent had been shown to vary rather slowly across the Milky Way.

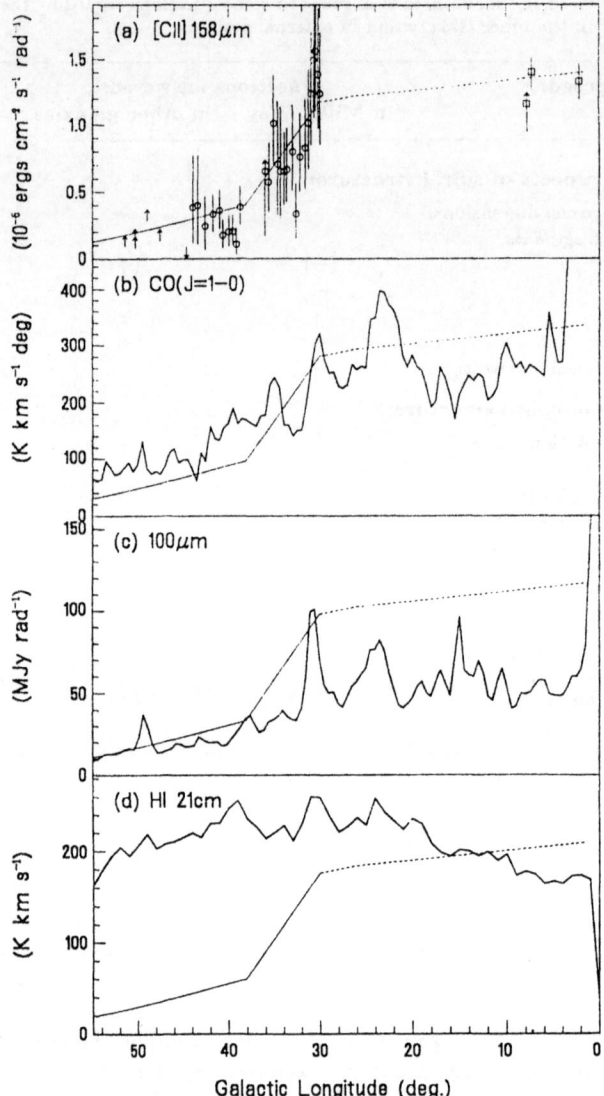

Fig. 23. Longitudinal distribution of the integrated emission measured in the galactic disk from the [CII] line at 158 μm, representing the diffuse gas probably defining the transition region between molecular clouds and the general medium; from the ^{12}CO line, representing the ensemble of molecular clouds; from the infrared continuum measured at 100 μm, representing interstellar dust illuminated by the ambient galactic radiation field; and from the HI line at 21 cm, representing the cool, diffuse component of the gas layer (from Shibai et al. 1991). The longitudinal distributions show that the galactic disk as defined by HI is substantially larger than the disk defined by the other tracers. The line-intensity ratio [CII]/^{12}CO measures the rate at which the reservoir of high-density material is in fact entering into stars

The morphological picture which followed from the first surveys of the CO molecule, by Scoville & Solomon (1975) and by Burton et al. (1975), was quite different. The CO data, serving as a general surrogate for the inaccessible, but largely dominant, molecule H_2, showed a distribution of clumps of high local density but of low volume filling factor, and a confinement to an annulus extending over the range $4 < R < 7\,\text{kpc}$ (for $R_0 = 10\,\text{kpc}$). As other global data became available, it became clear that the distribution of HI, instead of being typical, is in several ways unique. In particular, the galactic disk as defined by atomic hydrogen has a diameter about three times larger than that defined by the ionized and molecular states of hydrogen, and by all other molecules accessible along transgalactic paths. The morphological confinement to the inner Galaxy is also shared by γ-radiation, by synchrotron radiation, by pulsars and supernova remnants, by HII regions, and by radiation from dust, as well as, for that matter, the distribution of stellar luminosity.

The longitudinal variations of integrated emission from HI and from ^{13}CO are compared in Fig. 16; the longitudinal variation of infrared emission from dust is shown in Fig. 36. These plots show that molecular clouds, and dust emitting in the IRAS bands, are clearly much more confined to the inner Galaxy than is the distribution of the low density, quite diffuse, gas traced by HI. Figures 23 and 24 show longitude profiles for several other important, recently observed constituents of the Milky Way disk.

Figure 23 compares the HI, CO, and dust longitude profiles with that from singly-ionized carbon observed in the 158-μm infrared line by Shibai et al. (1991). The importance of observations of the [CII] line as a general indicator of star-formation activity has been reviewed by Stacey et al. (1991). The [CII] line tracers the warm ($T_k \geq 200\,\text{K}$) and dense ($n_H \geq 10^3\,\text{cm}^{-3}$) gas at the photodissociated interfaces between molecular clouds and the fully ionized medium beyond these interfaces; the line is correlated with the strength of the UV radiation field, largely contributed by recently formed OB stars. The 158-μm line is the main coolant of the cool interstellar medium at temperatures below a few hundred K, corresponding to those of the diffusely distributed material and of the cloud-interface material. The opacity of the earth's atmosphere in the far-infrared wavelengths requires that the [CII] be observed from balloon-borne telescopes (see Okuda 1981, 1991; Shibai et al. 1991), from the high-altitude Kuiper Airborne Observatory (see Stacey et al. 1991), or from future satellite missions.

One of the particularly significant aspects of [CII] observations concerns the measurement of the efficiency of star-formation activity. It is worth stressing that although the intensity of emission from CO and its isotopes may measure (under important assumptions) the total amount of molecular material forming the reservoir which might be, or become, available for star formation, the CO intensity does *not*, by itself, measure the amount of current star formation. It has been amply demonstrated, for example, that many molecular clouds have no embedded heat source, and thus evidently no current star formation. The line-intensity ratio [CII]/^{12}CO does serve as a measure of the rate at which the reservoir of high-density material is in fact entering into stars.

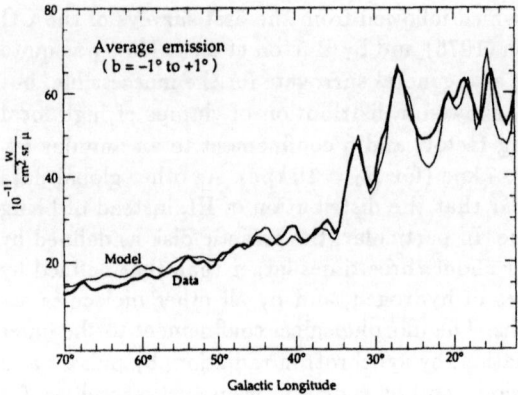

Fig. 24. Longitudinal distribution of the emission observed at 2.4 μm from the Infrared Telescope on board the Spacelab 2 mission (from Kent et al. 1991). This emission principally represents radiation from K and M giant stars. Interstellar extinction is a factor of 10 less at 2.4 μm than at visible wavelengths; contamination from thermal emission from dust either in the zodiacal cloud, or in interstellar space in general, may be ignored

Figure 24, from Kent et al. (1991), shows the longitude profile of emission measured in the near-infrared band at 2.4 μm using a telescope on board the Spacelab 2 mission. Earlier 2.4-μm material was gathered from balloon-borne telescopes, taking advantage of the narrow gap in the OH airglow emission centered at this wavelength (see e.g. Hayakawa et al. 1981; Oda et al. 1979; Hoffman et al. 1978). Near-infrared emission is dominated by radiation from stars: in this wavelength regime extinction by interstellar dust, although not negligible, is about an order of magnitude less than it is in the visible wavelength regime. Furthermore, the stellar emission at this wavelength is not contaminated by zodiacal light (such as does happen at the longer wavelengths of the IRAS bands discussed in Chap. 5), nor is it contaminated by thermal emission from interstellar dust. (Preparations for new, sensitive, all-sky surveys of 2-μm emission are currently being carried out; the efforts hold high promise for galactic studies.)

The longitude profiles shown in Figs. 23 and 24 (and in Figs. 16 and 36) follow quite directly from the observations. The general forms of these profiles indicate in a straightforward way the degree of confinement to the inner Galaxy; but interpretation of the *detailed* structure of these profiles is not so straightforward. It seems appropriate to mention a few of the complications.

In the case of the IRAS emission, the sharp peaks in the longitude profile are contributed by heat sources associated with localized regions of star formation; the discussion in Chap. 5 indicates how these localized effects are removed, via a rolling-wheel smoothing, before global conclusions are drawn. It is not at all obvious, however, that all of the emissivity enhancements may be accurately considered in terms of locally high dust densities. In particular, the generally enhanced emission at low-|l| values may reflect the enhanced radiation field associated with the high density of stars in the stellar bulge of the Galaxy;

similarly, it has not yet been convincingly demonstrated that the generally *low* level of 100-μm emissivity from the outer Galaxy is due to a dearth of dust particles rather than to a weakened illuminating flux.

In the case of the HI integrated emission, attempts have been made to associate the peaks, or steps, in the longitude profile with directions tangent to spiral arms. It is not clear that this is a generally valid way to locate structural features. If the HI within a spiral arm were to be compressed to higher optical depths, for example by the passage of a density wave, or if it were to be more commonly mixed *within* the arm with colder, denser, molecular material than would be the case *outside* of the spiral arm, then it is not implausible that the directions crossing a spiral arm feature would be associated not with a *peak* in the plots of $\int T_b(v)\,dv$ against l, but with a *trough*. Roberts & Burton (1977) discussed the complications of identifying structural features in HI and CO survey data in the case that large-scale density-wave shocks in the interstellar gas were to play an important role in determining the kinematics and relative distribution of various galactic tracers; in such a non-linear situation, the relative phases of the enhancements due to density and those due to kinematics are shifted with respect to each other, and both are shifted with respect to their locations in the linear density-wave case.

In the case of the CO integrated emission, some of the detailed structure in the longitude profile certainly has a local, rather than a global, origin. As is the case at $100\,\mu$m, local heating within a single molecular complex can produce a peak in the longitude variation, but, also as at $100\,\mu$m, there may be gradients of a global nature which influence the general shape of the CO longitude variation. Some remarks are made on these gradients in the following section. There may also be effects associated with major structures analogous to those mentioned for the HI case, and which have frustrated HI mapping; for example, the perceived integrated emission from molecular clouds occurring at higher spatial density may be systematically weakened due to enhanced cloud-to-cloud shadowing.

In the case of the 2.4-μm emission tracing the stellar distribution, the local peaks in the longitude variation probably do *not* indicate local enhancements of stellar density. Okuda (1981) pointed out that the near-infrared peaks are anticorrelated with the CO emission, and thus may be attributed to minima in the extinction rather than to maxima in the stellar surface density. The analysis of Kent et al. (1991) shown in Fig. 25 seems to verify this interpretation. Kent et al. first derived the vertical and radial scalelengths from an unfolding of the latitude and longitude distribution of the 2.4-μm emission. They simulated the emission in a model which involved a smooth distribution of near-infrared emission characterized by these scalelengths, with no account taken of extinction by interstellar material; they then compared the degree to which this simulation reproduced the observations with the success of a second simulation, in which the effects of extinction *were* included. The extinction was estimated using survey observations of CO (Dame et al. 1987) and of HI (Weaver & Williams 1973) to derive the total amount of gas, and converting this estimate, using a plausible gas-to-dust ratio, to one giving the total amount of dust and thus

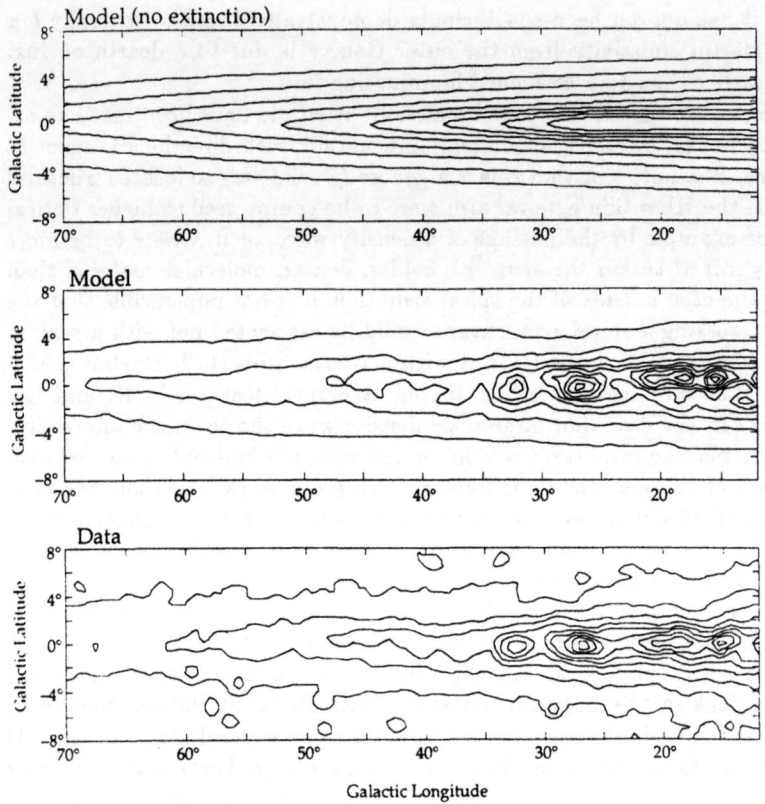

Fig. 25. Bottom panel: Observations of the Milky Way emission at 2.4 μm mapped by the IRT Spacelab mission and displayed in galactic coordinates.
Top panel: Model of the 2.4-μm emission incorporating the vertical and radial scalelengths following from the observations, but with no allowance made for interstellar extinction.
Center panel: Model of the 2.4-μm emission in which account is taken of interstellar extinction by dust, derived from the combined contributions from the galactic HI and CO constituents and using standard values of the gas-to-dust ratio. (From Kent et al. 1991)

of extinction. Kent et al. conclude that the luminosity fluctuations along the galactic plane are caused mainly by variations in the extinction along the long lines of sight through the galactic disk.

4.3 Radial Distributions Characterizing the Population I Layer

A global measurement of more direct physical relevance than the longitudinal distributions shown in the preceding section is given by the radial distribution of emissivities. For tracers with detailed spectral-line information, like HI, CO and other molecules, and HII regions, the radial distributions follow directly from the conversion of velocities to distances. Because distances from the galactic center are sought, rather than distances from the Sun, there is no distance ambiguity to be dealt with. For tracers measured in the continuum, as the 2.4-μm data, or with coarse spectral resolution, as pertains to the 158-μm [CII] line, the radial distribution follows from modelling based on a geometrical unfolding algorithm.

The kinematically-derived radial distributions of HI and CO are shown in Fig. 37, together with the geometrically-derived dust distributions. The radial distributions plotted in that figure are consistent with the conclusion which holds generally: all interstellar tracers accessible on a global scale, except HI, show a morphological confinement to the inner Galaxy. It has been known for some time (see e.g. the review by Burton 1976) that the radial distributions of all Population I tracers in the Milky Way share more or less the same confinement, with the exception of HI; this conclusion also holds (see e.g. the review by Young & Scoville 1991) for the molecular/atomic comparison in external galaxies.

About 65% of the ^{12}CO intensities emanate from the molecular annulus at $4 < R < 7$ kpc (for $R_0 = 10$ kpc), whereas only 36% of the HI distribution lies in this annulus. The total interstellar molecular mass in the Galaxy is about $3 \times 10^9 M_{\odot}$, and the total amount of HI in the Galaxy is about the same; all but a few percent of the molecular material is confined within the Solar orbit, whereas less than half of the total amount of HI is confined there. Essentially all external spiral galaxies show a comparable confinement of the molecular material to a region much smaller than the HI disk. Many galaxies, unlike our own, show a central hole with little HI, but still substantial CO; in these cases study of the molecular line becomes all the more important for determining the gas dynamics in the core. The dearth of molecular material in the central few kpc of the Galaxy is, however, a characteristic shared by only relatively few external systems.

Figure 26 shows the radial distributions of CO emissivities determined separately for the first and fourth longitude quadrants (from Bronfman et al. 1988; see also Robinson et al. 1988). The differences between the first- and fourth- quadrant radial distributions plotted in the figure are no doubt real, but in seeking details it is important to realize that both the kinematic and the geometric unfolding derivations of the radial distributions involve substantial azimuthal smoothing. Experiments under the controlled circumstances inherent in simulated spectra have indicated that the azimuthal smoothing dominates inaccuracies of the kinematic derivation attributable, for example, to irregularities in the adopted rotation curve. Details of the azimuthal structure are likely to remain out of reach in the Milky Way, although not in external galaxies.

Converting the longitudinal emissivity measure to a physically more meaningful one involves sources of uncertainties which are different for the different tracers.

For the HI situation, uncertainties in the derivation of number densities depend strongly on the optical-depth effects, which vary — as discussed in Chap. 3 — with location in the Galaxy. The vertical scale of the HI radial distribution indicated on the right-hand side of Fig. 37, and giving line-of-sight-averaged volume densities of emitting hydrogen atoms, depends on a cor-

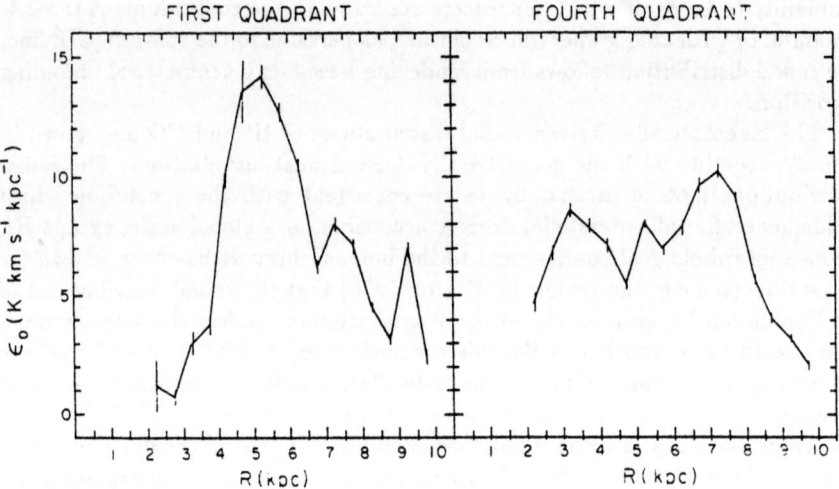

Fig. 26. Radial distributions of integrated emission determined separately from first- and fourth-quadrant ^{12}CO data (from Bronfman et al. 1988). The differences between the radial distributions derived for the two quadrants indicate deviations from cylindrical symmetry, but the global confinement of the molecular material to an annulus extending over the range $4 < R < 7$ kpc (for $R_0 = 10$ kpc) is evident in both sets of data. The radial distribution of CO emission is representative of the distribution of other tracers of the relatively compact, high-density regions of the interstellar medium as well as of tracers of the consequences or occurrence of star formation; but the CO emissivity itself is not simply proportional to the star-formation rate

rection for the global variation of the kinematic parameter $|dv/dr|$. The errors in this density scale which may occur as a consequence of temperature variations along the lines of sight, or from localized regions of self-absorption from cold HI residual in and near clumps of molecular gas, are difficult to ascertain quantitatively; these errors probably amount to at least 25%.

For the CO situation, conversion of measured intensities of the ^{12}CO surrogate to H_2 abundances involves major uncertainties. This conversion is important if the total number of interstellar nucleons is to be derived, but it remains a matter of controversy. Aspects of the CO-to-H_2 conversion have been discussed by Gordon & Burton (1976), Lequeux (1981), Liszt (1984), van Dishoeck & Black (1987), Polk et al. (1988), Bloemen (1989), Issa et al. (1990), and in the reviews by Combes (1991) and by Young & Scoville (1991) and in other references cited in these reviews. What is sought is the ratio $X \equiv N(H_2)/I(CO)$, where $I(CO) = \int T_b(CO) \, dv$. If this ratio were to be known with some confidence, and if it were to be taken as constant over the face of the Milky Way, then the number of hydrogen molecules in a particular direction would follow from the integrated ^{12}CO emission; dependent on additional assumptions, the total number of nucleons would then be dominated by $2N(H_2) + N(HI)$, with smaller contributions from helium and ionized hydrogen.

The numerical value of the conversion ratio of $N(H_2)/I(CO) = 3.0 \times 10^{20}$ (K km s^{-1})$^{-1}$ has been adopted by many authors. It has proven difficult to

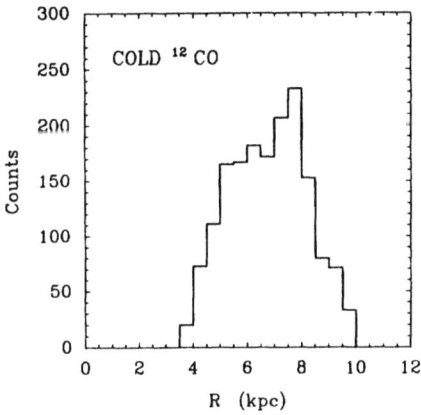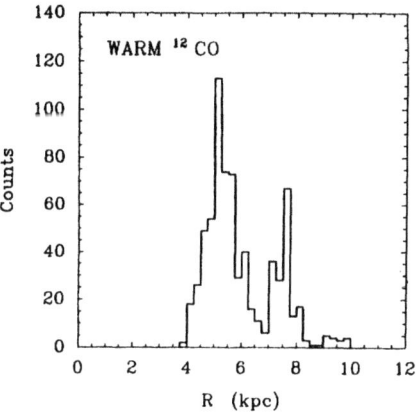

Fig. 27. Radial distribution of the number of *cold* ^{12}CO molecular clouds (left-hand panel) compared with the distribution of *warm* clouds (right-hand panel). Clouds with a core kinetic temperature $T_k < 10$ K are rather evenly sprinkled over the inner galactic disk; those with $T_k > 10$ K are clustered irregularly, follow the distribution of HII regions rather closely, and are predominantly found in the inner part of the molecular annulus. The data refer to the longitude sector $20° < l < 50°$; the radial scale corresponds to $R_0 = 10$ kpc. (From Solomon et al. 1985)

specify the error on this adopted value. One of the main problems in calibrating the conversion ratio stems from the high optical depth of the ^{12}CO line typical of most molecular clouds. Because of the high optical depths, only the outer envelopes of many clouds contribute to the observed emission; furthermore, shadowing of one cloud by another is likely to be important. Attempts have been made to calibrate the conversion ratio using the less abundant, but more transparent, isotopes ^{13}CO and C^{18}O; the scatters in the ratios $I(^{12}CO)/I(^{13}CO)$ and $I(^{12}CO)/I(C^{18}O)$ are large, and in any case there remain uncertainties in the isotopic abundance ratios. Other uncertainties include those concerning the C/H abundance ratio, the fraction of C contained in the CO molecule, and the fraction of CO is solid form. It is furthermore likely that the physical and chemical conditions vary from one molecular cloud to another, reflecting, for example, the differing star-formation history of the clouds. Dickman et al. (1986) interpret some aspects of the actually rather surprising degree of constancy of the conversion ratio X as due to the beam averaging which results for the CO, in a way analogous to the HI situation, in a statistically consistent measure of density and temperature. An independent approach to derivation of the conversion factor is based on observations of γ-radiation (see Bloemen 1989), which is produced by interactions of cosmic rays with interstellar nucleons.

In addition to the chemical and physical differences which may occur from cloud to cloud, there are also indications in the various radial distributions of *global* changes over the Galaxy. Figure 27 shows the radial distribution of the number of *cold* ^{12}CO molecular clouds (left-hand panel) compared with the distribution of *warm* clouds, following the distinction made by Solomon et al.

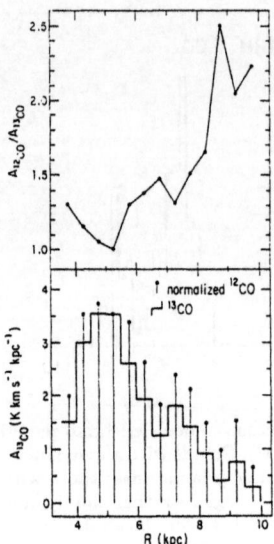

Fig. 28. Radial variation of the ^{12}CO and ^{13}CO integrated intensities (lower panel) and the radial variation of the ratio of the two emissivities (upper panel); the radial scale corresponds to $R_0 = 10$ kpc. The ratio of ^{12}CO/^{13}CO emissivities varies by more than a factor 2 across the galactic disk; the mean hydrogen column density is evidently higher near the inner edge of the molecular annulus than in the outer part. (From Liszt et al. 1984)

(1985) on the basis of the kinetic temperature observed in the direction of the cloud core. Although the statistics regarding the fraction of molecular clouds which contain evidence of current star formation are not yet firmly established, it is obvious that a large percentage of the members of the molecular-cloud ensemble do *not* show the enhanced-temperature cores which would be evidence of embedded heat sources. Those clouds which do show such hotter cores are preferentially located near the *inner edge* of the molecular-cloud annulus; the clouds with cooler cores are more generally spread throughout the radial extent $4 < R < 7$ kpc.

Another indication of global changes in the molecular-cloud properties over the inner Galaxy is shown in Fig. 28. The plots in this figure compare the radial variation of the ^{12}CO and ^{13}CO integrated intensities. The ratio of ^{12}CO/^{13}CO emissivities varies significantly across the galactic disk, in a manner indicating a higher mean molecular density for clouds located on the *inner edge* of the molecular-cloud annulus. Thus evidently the mean cloud column density changes over the galactic disk. Polk et al. (1988) have also investigated the ratio of ^{12}CO/^{13}CO emissivities, and found that this ratio is significantly lower in giant clouds than in average ones; thus the gas at lower optical depths makes an important contribution to total CO emission from the Milky Way. Application of a CO-to-H_2 conversion factor calibrated in *giant* molecular clouds would lead to an overestimate of the total molecular mass.

Investigations of external systems seek to establish the H_2/HI dependence on galaxy type and luminosity, as well as on radius and characteristics of lo-

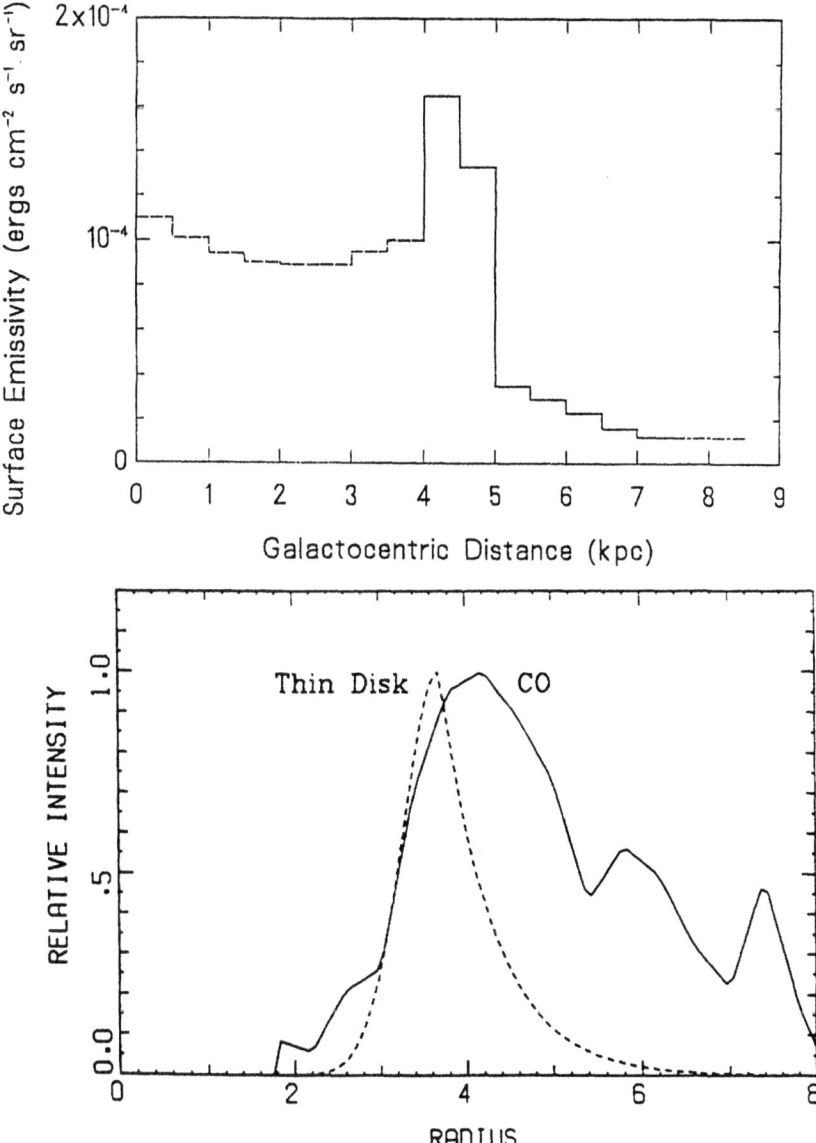

Fig. 29. Upper panel: Radial distribution of the surface emissivity of the galactic disk observed in the diffuse far-infrared [CII] line at 158 μm and attributed to extended regions surrounding molecular clouds (from Shibai et al. 1991). The radial scale is based on $R_0 = 8.5$ kpc.

Lower panel: Comparison of the radial distributions of CO integrated intensities and the emissivity observed in the near-infrared at 2.4 μm and attributed to the excess stellar luminosity of the Milky Way disk (from Kent et al. 1991). The CO radial profile corresponds to the northern-hemisphere data of Dame et al. (1987). The radial scale is based on $R_0 = 8.0$ kpc

cal environments (see Young & Scoville 1991, and references cited there). A wide range of the inferred ratio H_2/HI is observed, but there is little evidence supporting a radial variation in the CO-to-H_2 conversion factor.

The shape of the inner-Galaxy annulus revealed by the far-infrared observations of [CII] and by the near-infrared observations at $2.4\,\mu m$ are shown in Fig. 29. Both tracers show a preferential confinement to the inner parts of the annulus defined by the ensemble of ^{12}CO molecular clouds, and a sharper drop of emissivity at the outer edge of the annulus. The sharp drop displayed by the 158-μm data of Shibai et al. (1991) occurs at $R \simeq 5.8\,\mathrm{kpc}$, on the $R_0 = 10\,\mathrm{kpc}$ scale; at this radius the CO emission shown in the Fig. 26 remains quite intense. The radial profile of the 2.4-μm data discussed by Kent et al. (1991) also shows a narrower distribution, peaking at a radius ($R \simeq 4.6\,\mathrm{kpc}$ on the $R_0 = 10\,\mathrm{kpc}$ scale) on the inner edge of the CO molecular cloud distribution; the difference between the location of the molecular-annulus peak and the 2.4-μm peak is evidently significant.

Thus there are convincing indications that star formation does not occur in the Milky Way in direct proportionality to ^{12}CO emissivity, or to the number density of the clouds constituting the molecular Galaxy. The astrophysical implications stemming from this lack of direct proportionality are of substantial importance, and are the subject of much current work both regarding our own Galaxy and, particularly, regarding external systems.

The higher degree of confinement of the molecular material to the inner Galaxy compared to that shown by the HI gas has been interpreted by Wyse (1986) and by Wyse & Silk (1989) in terms of the radial variation of compression of the gas due to differential galactic rotation. Their approach modifies the proposal made by Schmidt (1959, 1963) that the star-formation rate would depend on a power, between 1 and 2, of the HI gas volume density. They explain the sharply peaked molecular distribution by proposing that a Schmidt law relates the atomic and molecular gas in a manner depending on the local amount of shearing due to differential galactic rotation. The resulting models account for several aspects of the radial dependence of star formation in the Milky Way and in a number of external systems.

4.4 Vertical Thickness and Degree of Flatness of the Inner-Galaxy Gas and Dust Layer

The thickness and degree of flatness of the galactic disk are important aspects of its morphology. These quantities cannot be measured accurately in external systems principally because of limitations imposed by telescope resolution, because of uncertainties about whether an external system is being seen exactly edge on, and because of the common occurrence of galactic warps. For tracers of our own Galaxy which present kinematic information, the linear thickness of the centroid of emission and the deviation of the location of this centroid from the equator $b = 0°$ can be studied at distances $R < R_0$ by measuring the latitude distribution of intensities contributed from near the terminal velocities, where there is no distance ambiguity. The value of the thickness derived from

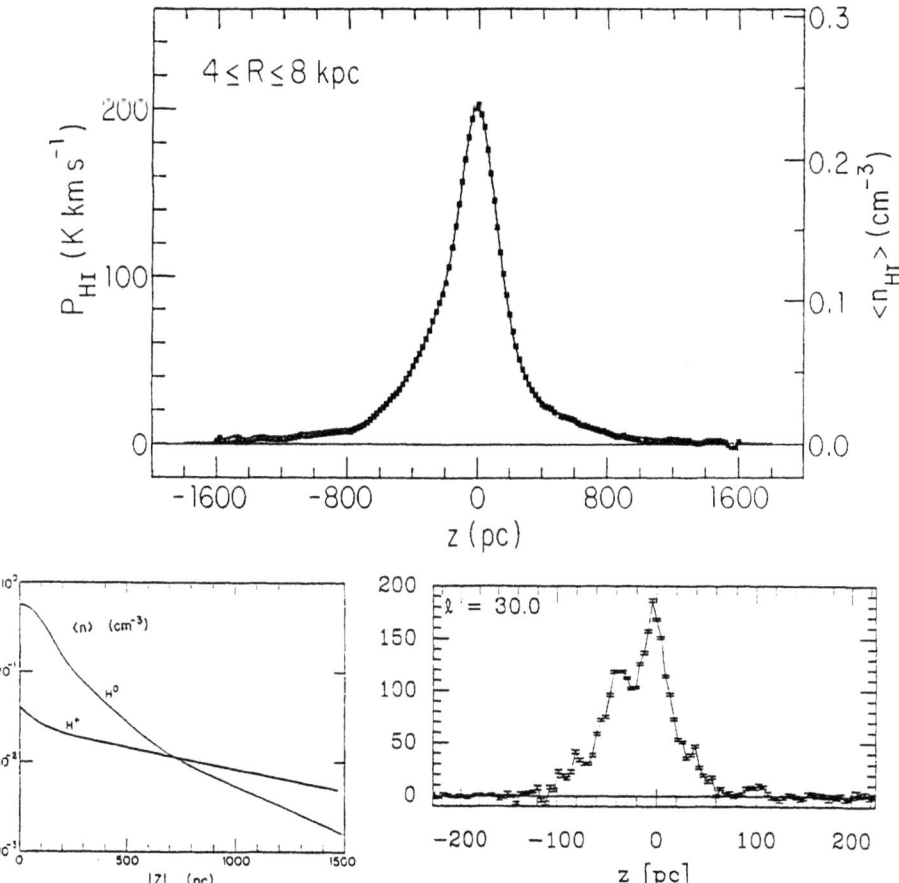

Fig. 30. Measures of the vertical thickness of the inner-Galaxy layer of interstellar material.
Upper panel: z-distribution of HI densities characterizing the region $4 < R < 8$ kpc (from Lockman 1984).
Lower-lefthand panel: Comparison of the vertical distribution of volume densities of the neutral and of the diffuse, ionized hydrogen components (from Reynolds 1991).
Lower-righthand panel: z-distribution of ^{12}CO emissivities from tangent-point velocities near $l = 30°$, corresponding approximately to the direction of the peak of the molecular annulus (from Knapp 1988)

the terminal-velocity locus will be an upper limit to the true value, because there may in fact be variations in the mean z-value of the emission centroid along the long lines of sight which contribute to each terminal-velocity measure.

Figure 30 shows that the typical thickness of the HI gas layer, measured to half-density points, is about 220 pc at $R < 9$ kpc (Lockman 1984). The distribution perpendicular to the galactic plane of densities in the HI gas layer is approximately Gaussian at the higher-intensity levels. But there is also a rather structureless, relatively weak, gas component extending to z-heights of several hundred pc. The low-intensity wings extending to higher z-distances cause the

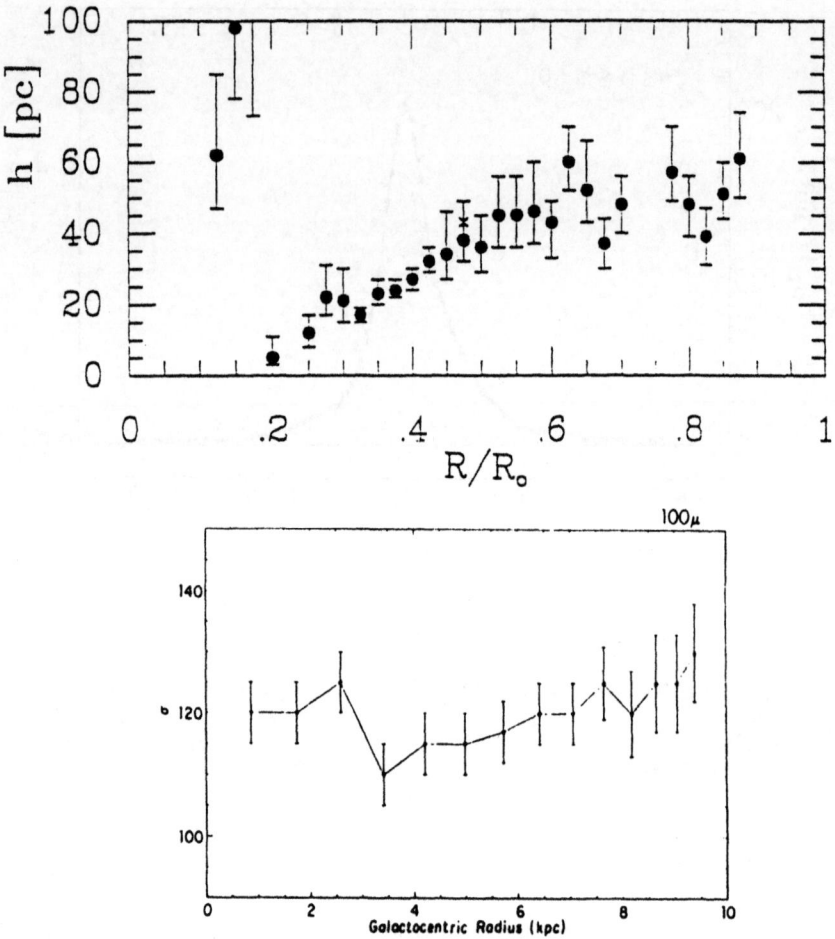

Fig. 31. Measures of the radial variation of vertical thickness of the inner-Galaxy layer of interstellar material.
Upper panel: Thickness (defined as the dispersion of a Gaussian) of the layer of ^{12}CO emission as a function of R/R_0 (from Knapp 1988).
Lower panel: Radial distribution of Gaussian half-widths of IRAS emissivities at $100\,\mu$m determined as discussed in Chap. 5.

overall HI vertical profile to deviate from the Gaussian form. Emission from the diffuse background evidently fills much of the volume within a kpc or so from the galactic equator; the velocity dispersion of this high z-height material indicates that it has a correspondingly high temperature. Reynolds (1991) has stressed the importance of a warm ($T_k \simeq 10^4$ K), diffuse, component of ionized hydrogen which may in fact be the dominant state of interstellar matter at about a kpc above the galactic equator. The comparative vertical distributions of the neutral and ionized hydrogen are plotted in the lower-lefthand panel of Fig. 30. If the ionizing photons responsible for the diffuse HII layer are contributed from hot stars at low z-heights, then the filling factor of the neutral HI layer must be less than unity.

The initial CO surveys of the Galaxy showed that the vertical thickness of the ensemble of molecular clouds is about half that of the HI layer (Burton & Gordon 1976; Cohen & Thaddeus 1977). This result has been confirmed in detail by the subsequent more complete material, and extended also to measurements of the location of the z-height of the mean plane (Sanders et al. 1984; Dame et al. 1987; Bronfman et al. 1988). The lower temperatures characterizing the molecular-cloud ensemble, and the smaller velocities dispersions, are consistent with the greater degree of vertical confinement shown by the CO data compared to the HI situation. The youngest stars, formed from the molecular clouds, are similarly tightly confined to a thin layer (see Blaauw 1965).

The layer thickness of tracers observed only in continuum radiation can be derived using unfolding techniques. The distribution of dust emitting at 60 μm and at 100 μm has a z-thickness comparable to that of the HI gas. Dust emission is generally well correlated with HI emission at $R < R_0$; the correlation is difficult to follow in the outer parts of the Galaxy because of the weakness of the dust emission.

Figure 31 shows that there is little variation of the thickness over the range $0.2R_0 < R < R_0$. The HI and CO layers are thinner in the part of the inner Galaxy encompassed by the bulge, and — as discussed in Chap. 7 — the gas layer traced by both HI and CO flares to much greater thickness beyond R_0. The constant vertical thickness over a large range of R leads to the expectation that the vertical motions would increase as R decreases in order to maintain the constant thickness against the total mass density. In fact, the vertical velocity dispersion is difficult to measure for the molecular-cloud ensemble, because of the low filling factor of the clouds seen at high $|b|$. Essentially all direct information on motions in the z-direction refer to the vicinity of the Sun; the viewing perspective of face-on external galaxies is more favorable in this regard. The horizontal line-of-sight velocity dispersion, measured both from HI and from CO data along the terminal-velocity locus, seems to remain constant over the distance range $0.2R_0 < R < R_0$.

The inner Galaxy layer of gas and dust is remarkable flat, from the location of the 3-kpc arm outwards to the vicinity of the Sun. Figure 32 shows that the layer of interstellar matter does show some systematic deviations from $b = 0°$, but these deviations amount to only about 30 pc for the HI gas layer and for the dust layer emitting at 100 μm; for the ensemble of molecular clouds, the deviations from flatness are typically even less. Thus the inner-Galaxy gas layer is flat to within much less than 1% of its total radial extent; its thickness amounts to only of order 1% of its radial extent.

Several different terms have been used to designate the confinement of the molecular-cloud ensemble to the inner Galaxy. The term "molecular annulus" seems more appropriate than the term "molecular ring", which suggests a much narrower band than is in fact observed. "Molecular annulus" also seems more appropriate than the term "molecular torus", which suggests a thicker distribu-

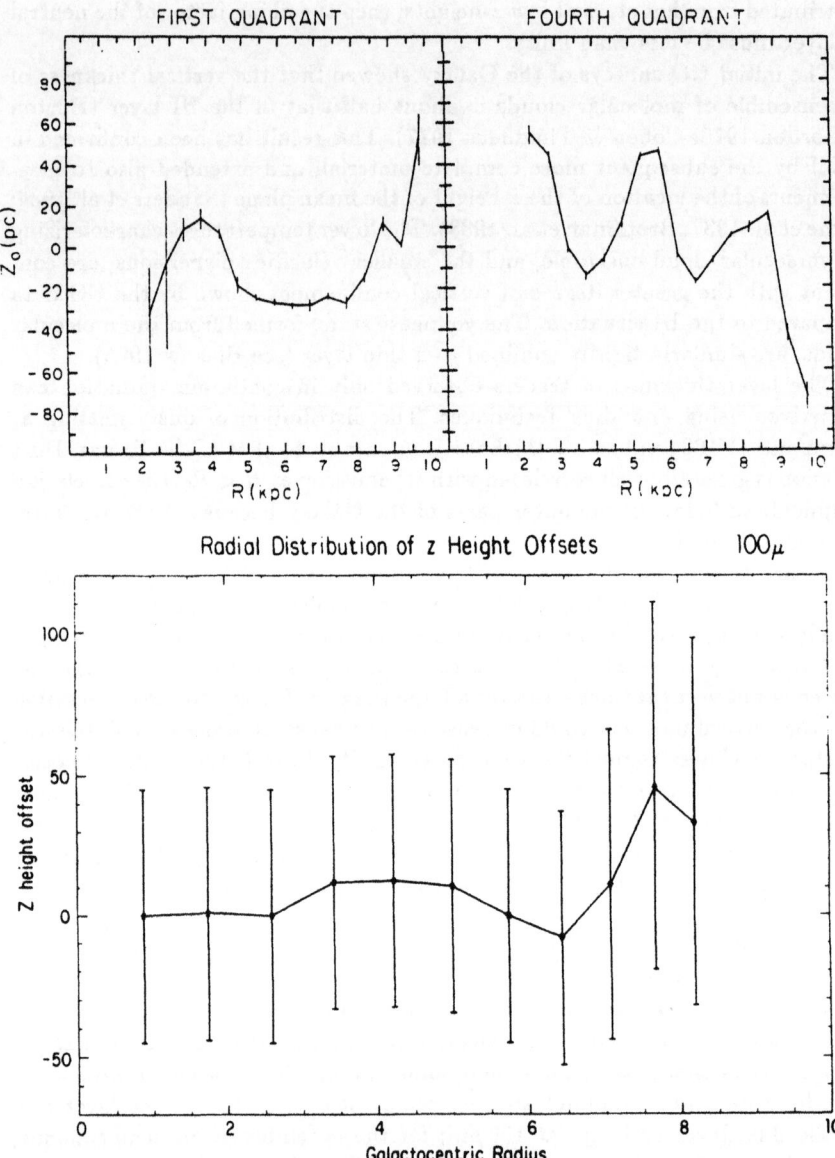

Fig. 32. Deviations from flatness of the midplane of the inner-Galaxy layer of interstellar material.

Upper panels: Vertical z-height of the centroid of ^{12}CO emission determined separately from first- and fourth-quadrant data (from Bronfman et al. 1988).

Lower panel: Vertical z-height of the centroid of IRAS 100-μm emission determined by the radial-unfolding procedure discussed in Chap. 5. The inner-Galaxy gas layer is flat to within less than 1% of its diameter, over the region $0.2R_0 < R < R_0$; the deviations from flatness in the outer Galaxy are discussed in Chap. 7, and are substantial

tion than is observed in the inner Galaxy. The layer shape in the outer Galaxy *is* clearly three-dimensional, and is discussed separately in Chap. 7.

References

4.1   Blaauw, A. 1965, in *Stars and Stellar Systems* **V**, *"Galactic Structure"*, A. Blaauw, M. Schmidt, (eds.), University of Chicago Press, p. 435
4.2   Bloemen, J.B.G.M. 1989, *Ann. Rev. Astron. Astrophys.* **27**, 469
4.3   Bronfman, L., Cohen, R.S., Alvarez, H., May, J., Thaddeus, P. 1988, *Astrophys. J.* **324**, 248
4.4   Burton, W.B. 1976, *Ann. Rev. Astron. Astrophys.* **14**, 275
4.5   Burton, W.B., Bania, T.M., Hartmann, D., Tang Yuan 1992, in *Proc. CTS Workshop No. 1. "Evolution of Interstellar Matter and Dynamics of Galaxies"*, J. Palouš, W.B. Burton, P.O. Lindblad, (eds.), Cambridge University Press, in press
4.6   Burton, W.B., Gordon, M.A. 1976, *Astrophys. J.* **207**, L189
4.7   Burton, W.B., Gordon, M.A., Bania, T.M., Lockman, F.J. 1975, *Astrophys. J.* **202**, 30
4.8   Cohen, R.S., Thaddeus, P. 1977, *Astrophys. J.* **217**, L155
4.9   Combes, F. 1991, *Ann. Rev. Astron. Astrophys.* **29**, 195
4.10   Dame, T., Ungerechts, H., Cohen, R.S., de Geus, E., Grenier, I., May, J., Murphy, D.C., Nyman, L.-Å., Thaddeus, P. 1987, *Astrophys. J.* **322**, 706
4.11   Dickman, R.L., Snell, R.L., Schloerb, F.P. 1986, *Astrophys. J.* **309**, 326
4.12   Gordon, M.A., Burton, W.B. 1976, *Astrophys. J.* **208**, 346
4.13   Hayakawa, S., Matsumoto, T., Murakami, H., Uyama, K., Thomas, J.A., Yamagami, T. 1981, *Astron. Astrophys.* **100**, 116
4.14   Hoffman,W., Lemke, D., Frey, A. 1978, *Astron. Astrophys.* **70**, 427
4.15   Issa, M., MacLaren, I., Wolfendale, A.W. 1990, *Astrophys. J.* **352**, 132
4.16   Kent, S.M., Dame, T.M., Fazio, G. 1991, *Astrophys. J.* **378**, 131
4.17   Kerr, F.J. 1962, *Monthly Notices Roy. Astron. Soc.* **123**, 327
4.18   Knapp, G.R. 1988, in *The Mass of the Galaxy*, M. Fich, (ed.), Can. Inst. Theoret. Astrophys., Toronto, p. 35
4.19   Lequeux, J. 1981, *Comments on Astrophysics* **9**, 117
4.20   Liszt, H.S. 1984, *Comments on Astrophysics* **10**, 137
4.21   Liszt, H.S. 1985, in *Proceedings IAU Symp.* **106**, *"The Milky Way Galaxy"*, H. van Woerden, R.J. Allen, W.B. Burton, (eds.), Reidel Pub. Co., p. 283
4.22   Liszt, H.S., Burton, W.B., Xiang, D.-L. 1984, *Astron. Astrophys.* **140**, 303
4.23   Lockman, F.J. 1984, *Astrophys. J.* **283**, 429
4.24   Oda, N., Maihara, T., Sugiyama, T., Okuda, H. 1979, *Astron. Astrophys.* **72**, 309
4.25   Okuda, H. 1981, in *Proceedings IAU Symp.* **96**, *"Infrared Astronomy"*, C.G. Wynn-Williams, D.P. Cruikshank, (eds.), Reidel Pub. Co., p. 247
4.26   Okuda, H. 1991, *Infrared Physics* **32**, 365
4.27   Oort, J.H., Kerr, F.J., Westerhout, G. 1958, *Monthly Notices Roy. Astron. Soc.* **118**, 379
4.28   Polk, K.S., Knapp, G.R., Stark, A.A., Wilson, R.W. 1988, *Astrophys. J.* **332**, 432
4.29   Reynolds, R.J. 1991, in *Proceedings IAU Symp.* **144**, *"The Interstellar Disk-Halo Connection in Galaxies"*, J.B.G.M. Bloemen, (ed.), Kluwer Acad. Pub., p. 67
4.30   Roberts, W.W., Burton, W.B. 1977, in *Topics in Interstellar Matter*, H. van Woerden, (ed.), Reidel Pub. Co., p. 195
4.31   Robinson, B.J., Manchester, R.N., Whiteoak, J.B., Otrucek, R.E., McCutcheon, W.H. 1988, *Astron. Astrophys.* **193**, 60
4.32   Sanders, D.B., Scoville, N.Z., Solomon, P.M. 1984, *Astrophys. J.* **276**, 182
4.33   Schmidt, M. 1957, *Bull. Astron. Inst. Neth.* **13**, 247
4.34   Schmidt, M. 1959, *Astrophys. J.* **129**, 243
4.35   Schmidt, M. 1963, *Astrophys. J.* **137**, 758
4.36   Scoville, N.Z., Solomon, P.M. 1975, *Astrophys. J.* **199**, L105
4.37   Shibai, H., Okuda, H., Nakagawa, T., Matsuhara, H., Maihara, T., Mizutani, K., Kobayashi, Y., Hiromoto, N., Nishimura, T., Low, F.J. 1991, *Astrophys. J.* **374**, 522

4.38 Solomon, P.M., Sanders, D.B., Rivolo, A.R. 1985, *Astrophys. J.* **292**, L19
4.39 Stacey, G.J., Geis, N., Genzel, R., Lugten, J.B., Poglitsch, A., Sternberg, A., Townes, C.H. 1991, *Astrophys. J.* **373**, 423
4.40 van Dishoeck, E.F., Black, J.H. 1987, *Astrophys. J.* **334**, 771
4.41 Weaver, H.F., Williams, D.R.W. 1973, *Astron. Astrophys. Suppl. Ser.* **8**, 1
4.42 Westerhout, G. 1957, *Bull. Astron. Inst. Neth.* **13**, 201
4.43 Wyse, R.F.G. 1986, *Astrophys. J.* **311**, L41
4.44 Wyse, R.F.G., Silk, J. 1989, *Astrophys. J.* **339**, 700
4.45 Young, J.S., Scoville, N.Z. 1991, *Ann. Rev. Astron. Astrophys.* **29**, 581

5 Comparative Global Properties of Interstellar Dust and Gas in the Galaxy

E.R. Deul & W.B. Burton

Abstract. The morphology of the galactic infrared dust emission observed with IRAS is compared here with that of the neutral gas component on scales ranging from kiloparsecs down to those corresponding to the resolution limits of the surveys. Correlations are particularly strong for the 100-μm emission at latitudes $|b| > 5°$. After subtraction of the contaminating zodiacal emission, the morphologies at 12 μm, 25 μm, and 60 μm also show close correspondence to that of the HI. For low latitudes, the correlation shows a clear non-linear behavior such that the HI emission is underluminous at the higher intensity levels. Three effects were found which influence the $I_{100\,\mu m}/N_{HI}$ intensity ratio: optical depths of HI at lower latitudes, increasing strength of the interstellar radiation field toward the galactic center, and contributions of dust associated with molecular clouds. Using a radial unfolding technique, we derive the variation of the far infrared emissivities with galactocentric distance at all IRAS wavelengths. Intensity ratio profiles show a rather constant 60 μm/100 μm ratio over the Galaxy. The 12 μm/100 μm ratio, however, shows a deficit of 12-μm flux at $R < 3$ kpc, indicating a significant difference between the bulge region and the Galaxy at large in the relative number of small dust grains and the standard, large grains in the rest of the Galaxy. The molecular annul· : of star-forming material is clearly visible in the infrared data at galactocentric distances between 4 kpc and 8 kpc. Comparison of the infrared profiles with kinematically derived radial profiles of HI and ^{12}CO shows that the dust morphology is generally similar to that of CO, but is characteristic of a distribution more confined to the inner Galaxy than is the case for HI. Assuming a constant dust-to-gas ratio, as well as a constant conversion factor between CO integrated intensities and column density of molecular hydrogen, and incorporating the radial behaviour of the Lyman-continuum flux, the infrared emissivities can be accounted for at $R > 4$ kpc. Interior to the molecular annulus, however, infrared emissivities are much higher than expected from an extrapolation of the situation beyond 4 kpc. Radiation from stars in the bulge of the Galaxy may be responsible for the enhanced far-infrared emission from the inner few kpc.

5.1 Introduction

Galactic studies using balloon-borne observations (e.g. Nishimura et al. 1980; Okuda 1981; Hauser et al. 1984; Caux & Serra 1986) showed that at far-infrared wavelengths the Galaxy displays a diffuse emission component, persisting along the entire galactic disk, superimposed on a large number of bright point sources, generally coincident with known HII regions or molecular cloud complexes. The limited spatial coverage of the earlier data has been greatly extended by the

IRAS survey, allowing aspects of the morphology of interstellar dust to be studied in detail, both at high latitudes and over the entire galactic disk.

Most of the far-infrared emission of the Galaxy originates from thermal radiation of interstellar dust grains that are heated by the general interstellar radiation field. Because a large fraction of the energy radiated by stars is absorbed and re-radiated at infrared wavelengths by these dust grains, the morphology of the infrared radiation contains information on the density distribution and physical properties of the dust grains as well as information on the stellar population.

Since the recognition of the diffuse emission from galactic dust, substantial effort has been put into correlating this aspect of the sky with observations at other wavelengths. A number of authors have reported linear relationships between the 100-μm intensities and either neutral atomic hydrogen or carbon monoxide integrated brightnesses. These studies, however, have been largely restricted to regions at high latitudes and to limited spatial extents where the interstellar radiation field responsible for heating the dust grains can be expected to be rather constant. There is a considerable spread in the published values of the coefficient for the dust emissivity and gas column-density correlation for different regions; most of the published values refer, furthermore, to the situation at high galactic latitudes where the interstellar radiation field is not expected to vary much (see Boulanger et al. 1985; Terebey & Fich 1986; Weiland et al. 1986; de Vries et al. 1987; Boulanger & Pérault 1988; Deul & Burton 1990). For this reason it is interesting to examine the global characteristics of the infrared dust emission by deriving the radial distribution of emissivities across the Galaxy and by comparing the behaviour of the ratio of dust-to-gas emissivities in different regions of the Galaxy. One might then obtain information on the physical conditions governing the correlation between the dust emissivities and gas column densities.

A generally strong correlation exists between high latitude features ($|b| >$ 10°) and structures observed in the 21-cm line of neutral atomic hydrogen (see e.g. Boulanger et al. 1985; Burton & Deul 1987; Deul & Burton 1990; Boulanger & Pérault 1988). These studies dealt with infrared cirrus features which can be individually identified at high latitudes. At low latitudes, both dust and gas structures are severely blended along long lengths of path through the interstellar medium.

Comparison of the dust and gas emission from the heavily blended galactic-plane lines of sight shows that significant deviations from a linear correlation exist between the 100-μm intensities and the HI column densities, and that the slope of the correlation factor depends on the position in the Galaxy. We show below that the determination of the dust-to-gas ratio is influenced by the radiation properties of both the dust and the gas.

At the higher infrared intensity levels, ^{12}CO emission associated with molecular-cloud ensemble is also correlated with the infrared emission. We remark below on the nature of this correlation on the low-latitude, blended paths. At high latitudes, relatively little dust emission is associated with isolated molecular clouds; although this association is interesting in other contexts, it is not important to this discussion.

In Sect. 5.2 we examine the general characteristics of the 100 μm-to-HI intensity relation by comparing the observed intensities on a point-by-point basis. Possible influences on the value of the correlation factor will be discussed.

To allow deconvolution of the observed intensities into the radial distribution of emissivities/densities the unfolding technique described by Deul (1988) is applied. Section 5.3 uses this unfolding technique to obtain radial profiles of the emissivities in the plane of the Milky Way. The dust distributions are compared with those obtained through kinematic unfolding of HI and CO observations of the diffuse and compressed neutral components of the interstellar gas.

We use the derived variation of the infrared emissivities with galactocentric distance to estimate the z-height distribution of the infrared emissivity. The z-height distribution, approximated with a Gaussian profile, can be determined by least-squares fitting of the calculated accumulated emissivities to the observed intensities.

The radial unfolding method is constrained by the several assumptions involved and by the small latitude range used in the determination of the radial profiles; it is also limited to the inner Galaxy. These considerations have to be taken into account when interpreting the results.

5.2 Correlation Between HI and 100-μm Intensities

Reliable subtraction of the contamination corresponding to the model of the zodiacal emission described by Deul & Wolstencroft (1988) yields maps at 12 μm, 25 μm, 60 μm, and 100 μm which may be compared directly with observations of other galactic tracers. The 100-μm infrared brightness map, shown in Fig. 33, was obtained by regridding and combining scans from the corrected Zodiacal Observations History File (see Deul & Walker 1989), onto a grid of galactic coordinates and smoothed to the resolution defined by the composite HI dataset. The total column-density map of neutral atomic hydrogen, shown in Fig. 34, was made from the composite dataset described by Deul & Burton (1988). The area shown in the figures contains contributions from all the HI surveys contained in that dataset.

Figures 33 and 34 reveal a general good correlation between the dust and gas intensities at the higher latitudes. Because at these latitudes the contribution of dust associated with molecular gas is small, a significant amount of infrared emission must be associated with neutral atomic hydrogen (see further Terebey & Fich 1986; Boulanger & Pérault 1988; Deul & Burton 1990). However, the distribution of 100-μm emission at the lower latitudes shows, for the inner Galaxy, greater confinement toward the galactic plane than is the case for the outer Galaxy. This situation contrasts with that pertaining for the HI column-density map. Similarly, the contrast of the 100-μm intensities between the inner Galaxy and the outer Galaxy is greater than that for the HI column density. If the dust and neutral gas are well mixed in a way that produces a constant ratio

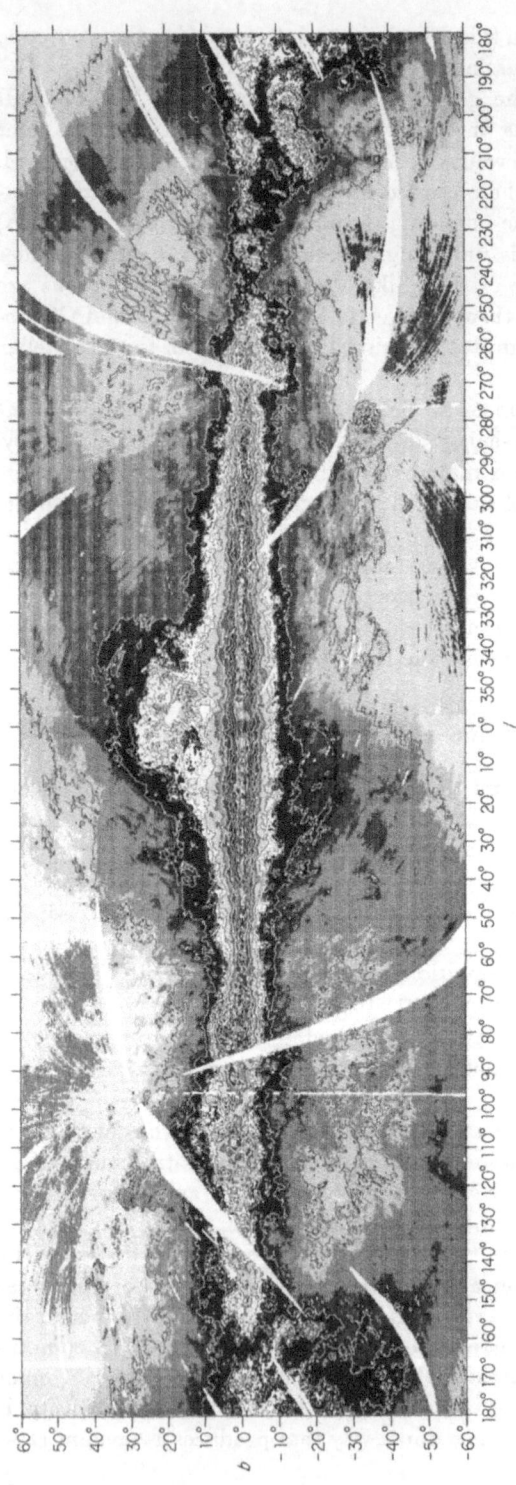

Fig. 33. Sky brightness distribution at 100 μm displayed in galactic coordinates over the range $-180° \leq l \leq 180°$ and $-60° \leq b \leq 60°$. The zodiacal emission, not as intense at 100 μm as at the other IRAS wavelengths but still significant, has been subtracted to obtain an all-sky map that primarily contains emission from the diffuse component of interstellar dust. The white lines indicate regions of missing data. At latitudes more than about 10° away from the galactic equator, individual dust cirrus features are recognized; at lower latitudes, blending along long lines of sight is severe

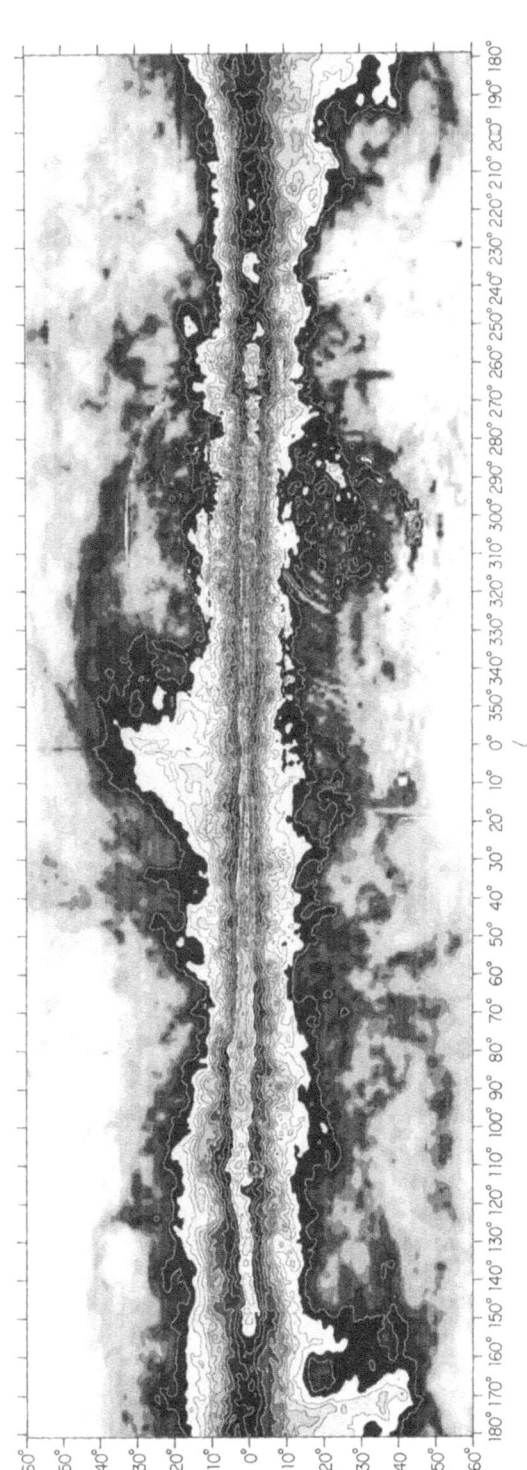

Fig. 34. Distribution of the integrated 21-cm line emission from HI displayed in galactic coordinates over the range $-180° \leq l \leq 180°$ and $-60° \leq b \leq 60°$. Under the assumption of optical thinness, which is generally valid for $|b| > 5°$, the integrated emission corresponds to the column density of neutral atomic hydrogen. The composite HI dataset entering this figure is described by Deul & Burton (1988)

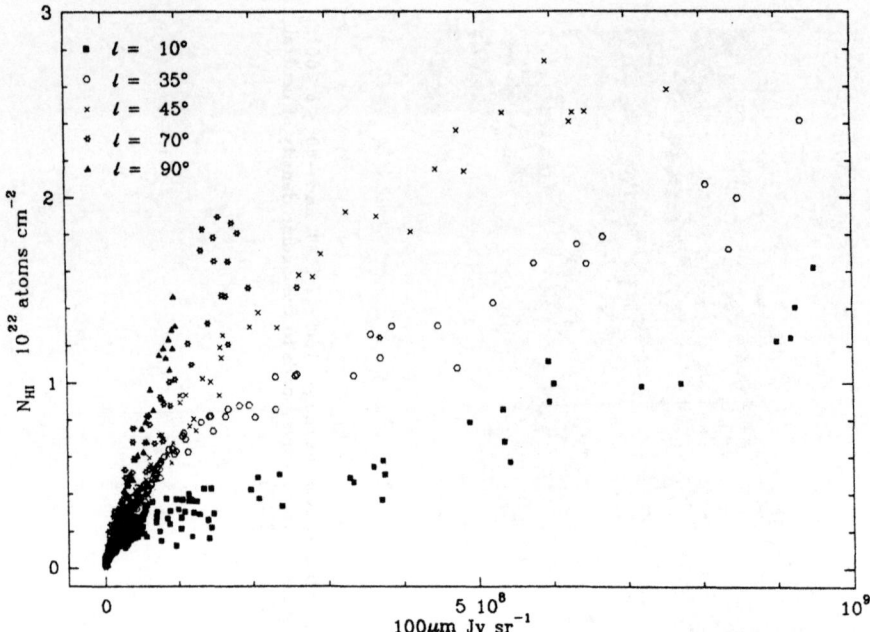

Fig. 35. Scatter diagram showing the correlation on a point-by-point basis between the intensities of dust emission at 100 μm and the total column density of HI. Strips at constant longitude, extending over the latitude range ($-60° < b < 60°$), and $0°.5$ wide in longitude, were taken to illustrate global variations in the correlation characteristics with position in the Galaxy

of dust-to-gas emission then comparison with the outer Galaxy shows that the infrared emission observed at the inner longitudes is overluminous with respect to the neutral gas content.

Figure 35 is a scatter diagram showing brightnesses at 100 μm plotted on a pixel-by-pixel basis against 21-cm integrated intensities. The correlations refer to $0°.25$ wide strips at the indicated constant longitudes and extending from $-60°$ to $+60°$ latitude. The nature of the correlation between the infrared brightnesses and the integrated neutral atomic hydrogen is evidently not constant throughout the Galaxy. Two aspects of the correlation are noticeable. First, there is a gradual flattening of the correlation at the higher intensity levels. This flattening is particularly evident for lines of sight through the inner part of the Galaxy at the lower latitudes ($b < |5|°$), where the infrared emission is particularly intense. Secondly, the strips at constant longitude are displaced with respect to each other in the sense that for lines sampling more of the outer Galaxy the $I_{100 \mu m}/N_{HI}$ ratio is generally lower than it is for lines sampling the inner Galaxy.

Three causes determine these characteristics of the dust/gas correlation. The first cause is to be found in the non-negligible optical depth of the HI line at some directions through the galactic gas layer. The gradual flattening of the correlation at high intensities is noticeable for lines of sight where the HI emission profiles show optical depths close to unity (Baker & Burton 1975). At 100 μm, the entire Galaxy is optically thin. Consequently the dust-to-gas brightness correlation will flatten for lines of sight traversing the inner galactic disk.

The second cause for the flattening of the correlation stems from the fact that the infrared intensities observed at 100 μm contain not only emission associated with neutral atomic hydrogen, but, as will be shown below, also contain a significant contribution from dust associated with molecular complexes. Molecular clouds predominantly reside at low latitudes; if the dust-to-gas ratio is approximately constant, then the correlation between HI and 100 μm will show, at the lower latitudes, that the infrared intensities are overluminous compared to the neutral material alone. This situation does in fact occur at the higher infrared intensities and is similar to that observed from local cirrus material by Weiland et al. (1986), de Vries et al. (1987), and Deul & Burton (1990).

A third cause for changes in the correlation can be attributed to a non-uniform interstellar radiation field. The number density of young O and B stars responsible for much of the heating of the interstellar dust increases sharply toward the galactic center. This increase implies a greater flux of blue light as well as a hardening of the optical spectrum both toward the galactic plane and toward the center (Mathis et al. 1983). Both the optical depth and the non-uniform interstellar radiation field become increasingly important toward the inner Galaxy, as well as toward lower galactic latitudes in general.

The line-of-sight integration inherent to the infrared continuum observations causes blending of the above three effects, particularly at the lower latitudes, where the line of sight traverses transgalactic paths. Therefore, we can only attribute the flattening of the correlation at higher intensities, and the decrease of the correlation coefficient with decrease in longitude, to the combined effects of HI optical depth, the presence of molecular material, and the changing interstellar radiation field. Scatter diagrams from other publications, which pertain to regions of limited extent (see Boulanger & Pérault 1988; Terebey & Fich 1986; Deul & Burton 1990) cover the area in our scatter diagram generally representing the most extreme cases. In cases where a known amount of molecular gas is present (de Vries et al. 1987; Weiland et al. 1986) the scatter points lie toward the regime of higher 100-μm intensities and of lower HI column densities in our plots. At 100 μm the IRAS observations generally trace the cool, diffuse component of the interstellar medium. The spikes of infrared emission contributed from known HII regions where removed by a rolling-wheel smoothing.

5.3 Radial Unfolding

A radial unfolding technique allows deconvolution of continuum data to derive galactic radial distributions. Techniques somewhat similar to the one described by Deul (1988) have been used by Strong (1975), Strong & Worrall (1976), and

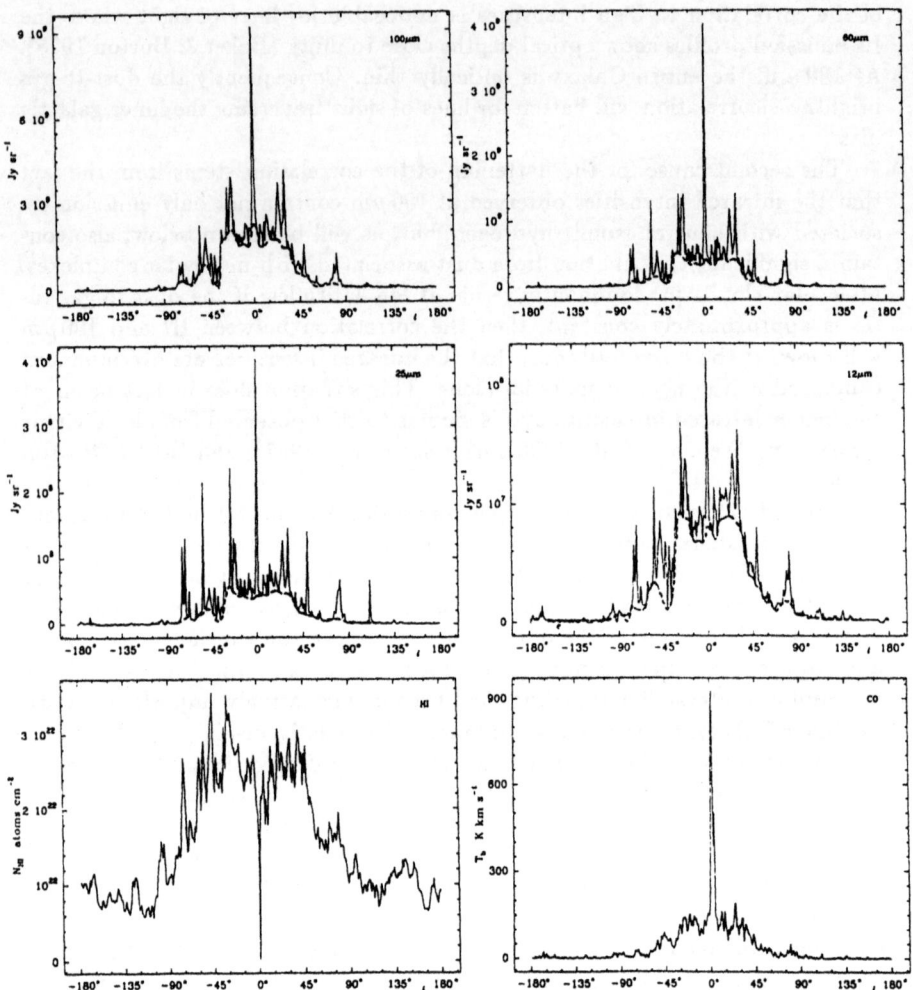

Fig. 36. Longitude profiles derived as described in the text from the dust infrared data, and from the composite CO and HI datasets of Dame et al. (1987) and Deul & Burton (1988), respectively. For the top four panels, displaying the profiles from the four IRAS passbands, the full-drawn lines represent the intensities with zodiacal emission subtracted. The dashed curves show the lower envelope, rolling-wheel, fit to these profiles. Note that, apart from those pertaining to the inner 10 degrees of longitude, the profiles look rather similar. The HI longitudinal profile contains considerable emission contributed from the outer Galaxy, but the inner-Galaxy contribution dominates

Caravane & Paul (1979) in the analysis of the SAS-II gamma-ray survey, by Kanbach & Beuermann (1979) and Phillips et al. (1981) for radio-continuum surveys, and by Caux et al. (1984) for infrared balloon surveys. Although the radio-continuum and earlier infrared surveys were unfolded using a (largely assumed) spiral-arm pattern, we restrict the emissivity distribution here to a purely radial dependence. The HI and CO radial distributions have also been derived using the cylindrical symmetry assumption. The radial unfolding technique uses the longitudinal intensity information only; its latitude range is limited to the bin width in that direction.

The infrared data used are those from the Zodiacal Observations History File (see IRAS Explanatory Supplement 1985). Contaminating zodiacal emission was subtracted from the four all-sky maps in the manner described by Deul & Wolstencroft (1988). The material was also corrected for the instrumental effects (substantial increase of responsivity of the 100-μm detectors for long periods of time after passing intense infrared sources and similar, but shorter lived, sensitivity changes for the 12-μm and 25-μm detectors) described by Deul & Walker (1989). The resulting maps should then contain emission from galactic or extragalactic origin only. The resolution of the maps at $12\,\mu$m, $25\,\mu$m, $60\,\mu$m, and $100\,\mu$m is $0°.5$ both in l and b. After applying the rolling-wheel method we obtain longitudinal profiles of the diffuse infrared emission (the dashed lines in Fig. 36). For comparison we show the longitudinal profiles of HI and CO in the bottom two panels of Fig. 36. These profiles are derived from the composite HI dataset described by Deul & Burton (1988) and from the CO survey described by Dame et al. (1987).

The unfolding process provided the radial profiles for the four IRAS passbands plotted in the upper four panels of Fig. 37. The full-drawn lines represent the radial distributions derived from the first quadrant longitude profile; the dashed lines, those from the fourth quadrant. Estimated error bars, based on an assumed constant error per longitude bin, are indicated at the top of each panel. Only the negative part of each error bar is shown. We point out that the increase of the emissivities toward the galactic center is uncertain in detail. Errors at $R < 2\,$kpc are large in any case, but in addition the strong intensities from the inner few degrees are broadened by the smoothing inherent in the data manipulation. To allow direct comparison with other data we have used $R_0 = 10\,$kpc. Conversion to $R_0 = 8.7\,$kpc can be performed by multiplying the radii by 0.87 and the emissivities by $1/0.87$. The lower two panels of Fig. 37 show the number density of neutral hydrogen atoms and the CO emissivities, respectively. The radial profiles from HI and CO observations were derived separately from the northern and southern hemisphere data.

5.4 Morphological Properties

5.4.1 Radial Infrared Properties. Infrared radial profiles plotted in Fig. 37 show similarities at the different wavelengths. This suggests that we have correctly subtracted the point sources from the longitude profiles to yield diffuse-emission contribution only. The emission at the shorter wavelengths would otherwise be dominated by local heat sources, causing overestimation of the emissivities in some rings and, consequently, underestimation for those neighboring rings just interior. This only happens for a few rings near $9\,$kpc. The general similarity of the radial profiles indicates that the interstellar dust responsible for the diffuse infrared emission has generally similar morphology at the four wavelengths.

Some of the differences may be accounted for by the wavelength-dependence of the radiating properties on dust-particle size. It is known that the interstellar

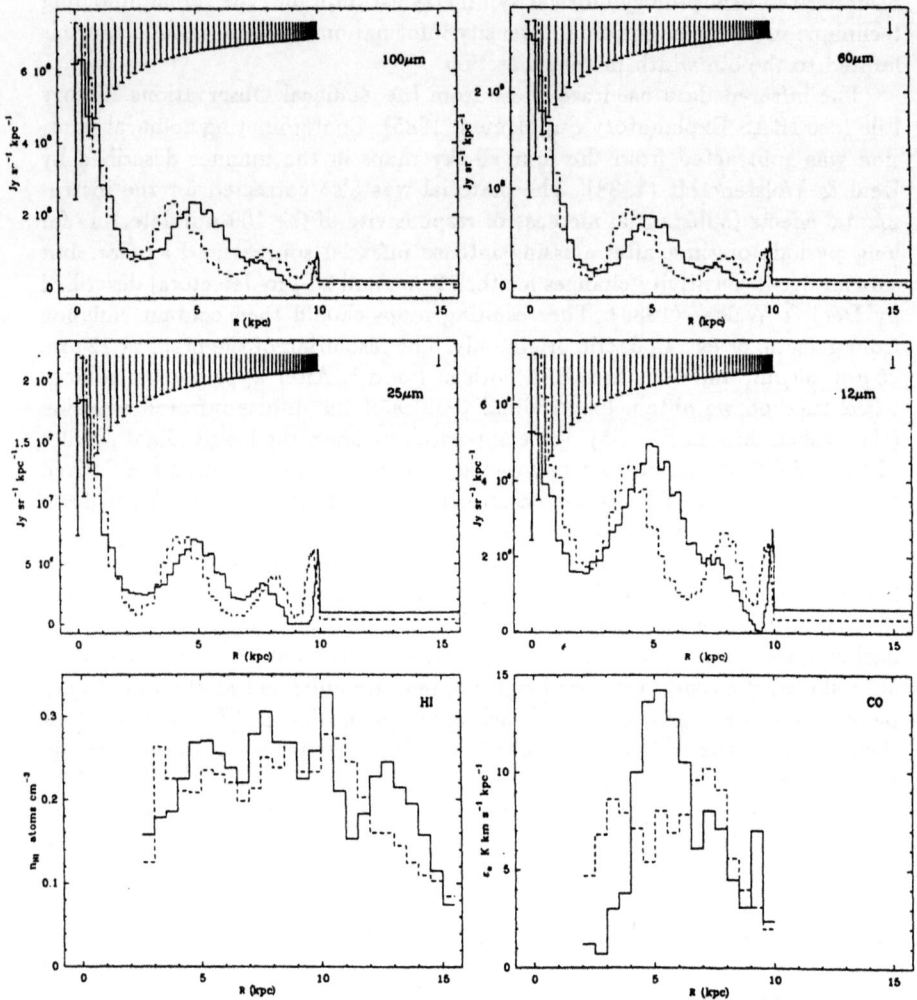

Fig. 37. Radial profiles of the dust distribution at $b = 0°$ derived from unfolding the data at the four IRAS wavelengths, together with those derived for the gas distribution from spectral-line observations of CO and HI. Full-drawn lines represent data from the first galactic quadrant; dashed lines, from the fourth. Only the negative part of the estimated errors for the infrared unfolding results have been indicated at the top of the relevant panels. Uncertainties in the unfolding method increase toward the galactic center. The interstellar gas profiles are based on CO data of Bronfman et al. (1988) and on the HI dataset compiled by Deul & Burton (1988), respectively

dust consists of an ensemble of particles with different sizes (Mathis et al. 1983; Draine & Anderson 1985; Puget et al. 1985). There are indications that the emission at $12\,\mu$m and $25\,\mu$m, and a significant fraction of the emission at $60\,\mu$m, is dominated by non-equilibrium emission from small ($a < 0.005\,\mu$m) particles (Draine & Anderson 1985; PAH's: Puget et al. 1985); the 100-μm

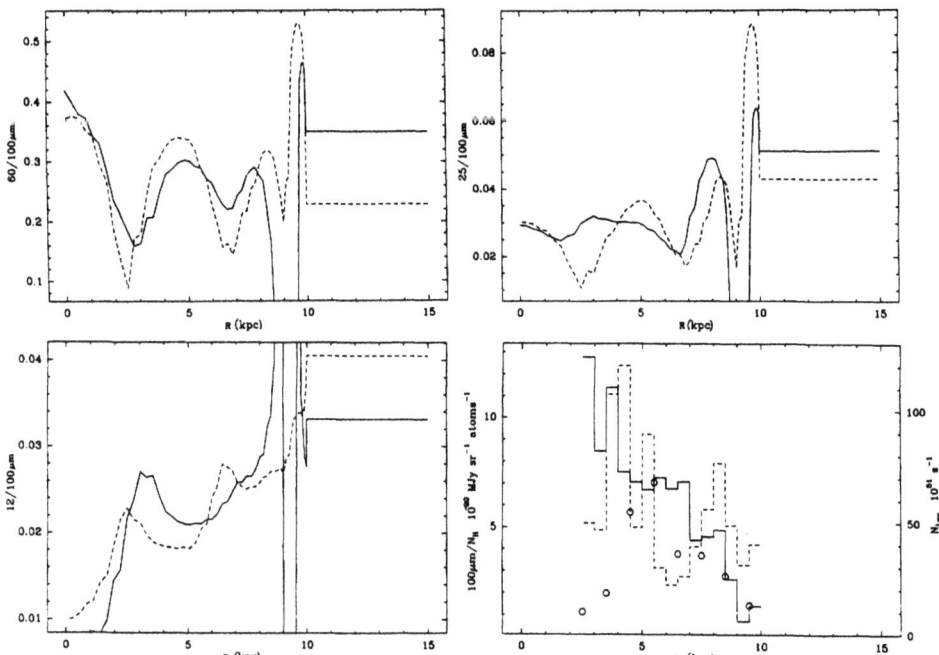

Fig. 38. Derived radial dependences of infrared emissivity ratios are indicated in three of these panels. The bottom right-hand one shows the radial dependence of the dust-to-gas emissivity ratio. The full-drawn lines give the ratio dependence for the first quadrant; the dashed lines for the fourth quadrant. The circles indicate the number of Lyman continuum photons per ring element (Güsten & Mezger 1983). The $12 \, \mu m / 100 \, \mu m$ emissivity ratio suggests a general lack of small particles in the inner Galaxy; the $25 \, \mu m / 100 \, \mu m$ ratio is rather constant. The $60 \, \mu m / 100 \, \mu m$ ratio, which may represent the temperature of the dust, is also rather constant, although there are regions of higher temperature correlated with the inner part of the molecular annulus. The dust-to-gas emissivity ratio shows a steep increase toward smaller galactic radii

emission is not dominated by the small particle emission. The *general* similarity of the radial profiles at the different wavelengths suggests, however, that the small particles are well mixed with the larger ones on scales ranging from a few hundred parsecs to kiloparsecs.

That there may be some dependence on R is illustrated by the ratio plots of Fig. 38. Although the estimated errors ($\geq 20\%$) are rather large, the $12 \, \mu m / 100 \, \mu m$ emissivity ratio does decrease significantly with decreasing galactocentric distance, particularly for $R < 3 \, \text{kpc}$. The $25 \, \mu m / 100 \, \mu m$ emissivity ratio, which shows structure reminiscent of the radial emissivity profile, is, within the errors, constant throughout the Galaxy. The difference between the behaviour of the $12 \, \mu m / 100 \, \mu m$ and $25 \, \mu m / 100 \, \mu m$ ratios may be explained by the fact that emission in the 12-μm passband is probably dominated by emission lines from polycyclic aromatic hydrocarbon (PAH) molecules, while at $25 \, \mu m$ the majority of the emission is due to small grains. Draine & Anderson (1985) show that the maximum non-equilibrium temperature for small grains decreases with increasing grain size, so that in their model the $25\,\mu\mathrm{m}/100\,\mu\mathrm{m}$ ratio decreases less dramatically with increasing minimum grain size than the $12\,\mu\mathrm{m}/100\,\mu\mathrm{m}$ ratio. Because the influence of the interstellar radiation field on these small grains is rather small, the behaviour of the $12\,\mu\mathrm{m}/100\,\mu\mathrm{m}$ and $25\,\mu\mathrm{m}/100\,\mu\mathrm{m}$ ratio with galactocentric radius suggests that the number of small ($a < 0.005\,\mu\mathrm{m}$) grains, that are well mixed with the large ($a > 0.01\,\mu\mathrm{m}$) grains, decreases sharply inside $R < 3\,\mathrm{kpc}$. The increase of temperature of the large grains due to the more intense radiation field in the galactic center region is not enough (following the Draine & Anderson model) to explain the $12\,\mu\mathrm{m}/100\,\mu\mathrm{m}$ behavior. The deficiency of very small particles for the center region — where the interstellar radiation field is very intense — may be caused by the relative ease, compared to the case for large grains, with which small ones are evaporated (see Puget 1985). Similar effects are observed around 5 kpc, where most young stars reside; the $60\,\mu\mathrm{m}/100\,\mu\mathrm{m}$ emissivity ratio shows a minimum at the position where the $12\,\mu\mathrm{m}/100\,\mu\mathrm{m}$ ratio peaks.

Although the radial profiles derived from the four IRAS bands show general agreement, substantial differences characterize the profiles derived from the northern and southern data. These differences indicate deviations from azimuthal symmetry, and may be identified in a rough way with global aspects of the spiral structure of the Galaxy. The presence of a known complex of emission at $l = 280°$, often identified with the Carina spiral arm, causes an enhancement near $R = 9.5\,\mathrm{kpc}$ of the emissivities calculated from the fourth quadrant data. A similar enhancement in the first quadrant results from the Cygnus feature. Near $7 - 8\,\mathrm{kpc}$ there is an enhancement in the radial profiles of the fourth quadrant which may be attributed to the enhanced emission in the longitudinal profile that is associated with what has been called the Norma spiral arm. The first quadrant extension of this feature causes the enhancement near 5 kpc. An enhancement in the fourth quadrant near 4 kpc may be associated with the Scutum-Crux feature observable at $l = 330°$. Finally, we note the presence near the galactic center of infrared emission intense at all wavelengths.

5.4.2 Radial Dependence of the Dust-to-Gas Ratio ($R < R_0$).

The radial infrared morphology is generally similar to that of the cold, compressed interstellar gas traced by CO, but differs especially in its degree of confinement to the inner Galaxy from the morphology of the diffusely distributed interstellar gas traced by HI. Some similarity between the dust and molecular-gas morphologies is expected in any case, because cold dust, well-mixed with gas within molecular clouds, contributes substantial infrared emission.

Some assumptions are required before discussing the dust-to-gas ratio derived from the emissivities of the IRAS passbands and the integrated spectral lines of CO and HI. These assumptions must be borne in mind when considering the dust-to-gas ratio plotted in Fig. 38.

First, we assume that the dust seen by IRAS represents both in fraction, as well as in morphology, the total dust content on the line of sight. This is correct for the higher latitudes, because the characteristic optical extinction at $|b| > 5°$ is less than about $3A_V$ (Burstein & Heiles 1984) causing the dust to be optically thin to the ultraviolet part of the interstellar radiation field (Mathis et al. 1983). Apart from a few exceptions of dark molecular clouds, the 100-μm intensities thus represent fully the total dust mass away from the galactic plane (de Vries et al. 1987; Boulanger & Pérault 1988; Deul & Burton 1990). Lower-latitude molecular clouds are typically denser than higher-latitude ones, with optical depths corresponding to extinctions exceeding 3 magnitudes; this fact, together with the recognition of the biased sensitivity of the IRAS detectors to the warmer ($T > 18$ K) dust, leads to underestimates of the total amount of dust. Another relevant effect concerns the range of temperature components which will generally be sampled along each long line of sight. The derivation of total dust content using the average temperature will then provide a lower limit. Thus the observed 100-μm intensities may not represent, by themselves, the total amount of interstellar dust on lines of sight close to the galactic plane (Young et al. 1986; Pajot et al. 1986; Sodroski et al. 1987).

Second, we assume that the molecular gas content of the Galaxy is proportional to the observed column density of ^{12}CO. This assumption has been discussed widely (e.g. Scoville & Solomon 1975; Gordon & Burton 1976; Bloemen et al. 1986, 1990). The rather paradoxical situation of an approximately constant ratio of observed intensities in the ^{13}CO line with respect to the thick ^{12}CO line has been explained in terms of the rarity of shadowing of individual clouds on the same line of sight. The ^{12}CO line is evidently an effective tracer of the *number* of clouds, and in that sense of the total mass of the ensemble of galactic clouds.

Third, we assume that the atomic gas content of the Galaxy is adequately given by the integrated HI intensities, corrected at the lowest latitudes for non-negligible saturation. The 21-cm spectral line is optically thin over most of the Galaxy. For some lines of sight, however, the optical depth hovers around unity. Although the true optical depths are still poorly known, a first-order correction can be applied, assuming a uniform spin temperature of 125 K. The corrections are significant in regions of the galactic plane ($10° < |l| < 70°$ at $|b| < 0°.5$) and for regions of exceptional velocity crowding ($l = 0° \pm 10°$, $l = 180° \pm 10°$, and $70° < |l| < 80°$ at $|b| < 3°$). The corrections, which are typically $10\% - 20\%$, are calculated such that the derived densities can adequately reproduce the observed situation in modelling experiments.

The total masses of the molecular and atomic hydrogen components of the interstellar medium are about equal at $R < R_0$ (Bloemen et al. 1986). Because the radial CO emissivity profile peaks near 5 kpc, whereas the radial HI density distribution is more or less flat, the total amount of molecular gas is relatively high in that region; the shape of the radial total hydrogen gas abundance thus closely resembles that of the CO profile at $R < R_0$. At 5 kpc the contribution of neutral atomic hydrogen to the total gas content is of order 20%. In addition Lockman (1984) finds evidence, based on the velocity structure of high-z HI material, for an enhanced number of heat sources for the inner molecular annulus. Enhanced star formation for smaller galactocentric radii in the molecular annulus is also supported by Liszt et al. (1981), who find the more energetic molecular clouds on the inner portion of the molecular annulus.

Therefore, both the gas density and the interstellar radiation field increase, consequently causing an increase in the total intensity of the infrared emission from the inner molecular annulus.

The radial dependence of the dust-to-gas ratio of 100-μm emissivity to total hydrogen column density has been plotted in Fig. 38. A value of 2×10^{20} molecules $(\mathrm{K\,km\,s^{-1}})^{-1}$ for the conversion ratio $W_{\mathrm{CO}}/N_{\mathrm{H_2}}$ (Bloemen et al. 1986, 1990) was adopted. Also plotted are values for the number of Lyman-continuum photons, N_{Lyc}, per ring (Güsten & Mezger 1983). The derivation of the N_{Lyc} photon counts was based on first quadrant data only; therefore the circles should be compared with the full-drawn line only in the figure. The dust-to-gas emissivity ratio appears to increase with decreasing galactocentric distance. Some of the radial structure of the $I_{100\,\mu\mathrm{m}}/N_{\mathrm{H}}$ ratio can be explained by the increase in N_{Lyc}. The increase in N_{Lyc} represents an increase in the UV component of the interstellar radiation field and, consequently, an increase in the heating of the interstellar dust grains. That this situation does not result in a corresponding increase of the $60\,\mu\mathrm{m}/100\,\mu\mathrm{m}$ intensity ratio may be understood in terms of the Draine & Anderson (1985) model. Their Fig. 4 shows that the $60\,\mu\mathrm{m}/100\,\mu\mathrm{m}$ intensity ratio is rather insensitive to the strength of the interstellar radiation field in case of higher (~ 0.3) ratio values. At $R \leq 3$ or $4\,\mathrm{kpc}$ the number of Lyman-continuum photons is significantly less, with respect to the dust-to-gas emissivity ratio, than at larger radii. This suggests that the intense blue light of the bulge plays an important role in heating the dust particles. In any case, details of the distribution of HI are poorly known at $R \leq 2\,\mathrm{kpc}$ because the galactic rotation curve is poorly known there and because the optical depth effects are important (see Burton 1988), but we predict that the $I_{100\,\mu\mathrm{m}}/N_{\mathrm{H}}$ ratio would show a dramatic increase toward the galactic center, both because the radial profile at $100\,\mu\mathrm{m}$ increases toward the center and because the Galaxy shows a relative deficiency of hydrogen gas inside the molecular annulus. A steep increase in the dust-to-gas ratio is also observed in M31 (Walterbos & Schwering 1987), but in that galaxy this increase may be largely due to the relative lack of HI and H_2 in the bulge region. An exponential-law fit to the galactic $I_{100\,\mu\mathrm{m}}/N_{\mathrm{H}}$ profile for $2.5\,\mathrm{kpc} < R < 10.0\,\mathrm{kpc}$, results in

$$\frac{I_{100\,\mu\mathrm{m}}}{N_{\mathrm{H}}} = 6.0 \pm 0.1 \times 10^{-20} \exp\left(-\frac{R-5}{2.9 \pm 0.3}\right) \quad [\mathrm{MJy\,sr^{-1}\,atoms^{-1}}]. \quad (10)$$

5.4.3 Dust-to-Gas Mass Ratio. Assuming a constant dust-to-gas mass ratio and a constant dust grain size, the radial behaviour of the $I_{100\,\mu\mathrm{m}}/N_{\mathrm{H}}$ ratio is dominated by the distribution of the emissivity properties of individual dust particles. If heating of interstellar dust grains is caused by photons from the blue part of the visual spectrum, then the distribution of blue light could yield information on the radial behaviour of the emissivity per dust particle. Based on the local distribution of the blue surface brightness, de Vaucouleurs & Pence (1978) found a scale length of 3.5 kpc for the stellar disk component. More global information on the galactic distribution of stellar light yields an exponential scale length of 5.5 kpc (van der Kruit 1986). Both these lengths are greater than the scale length of 2.9 ± 0.3 kpc characterizing the $I_{100\,\mu m}/N_H$ dust-to-gas ratio. Evidently either the dust-to-gas mass ratio must change with galactocentric distance, or the heating of dust grains is caused by more than only the ambient blue-light background illumination. The second option seems the more likely, because the extinction curve increases considerably with inverse wavelength, making the absorptivity of a typical dust grain greater for shorter wavelengths (more energetic photons) so that ultra-violet light will also contribute to the heating. On the other hand it is not expected that the dust-to-gas ratio varies much with galactocentric distance because the metal abundance has a very shallow gradient, with a scale length of 12.5 ± 0.25 kpc (Shaver et al. 1983). The metal abundance gradient for M31 is also very shallow, and does not account for the $3.6 - 4.8$ kpc scale length found by Walterbos & Schwering (1987).

In our further discussion of the dust-to-gas mass ratio we assume that the optical depth at $100\,\mu m$, determined from the radial emissivity profile and the $60\,\mu m/100\,\mu m$ radial temperature profile, is representative of the total dust content in the Galaxy, and that the total gas content is proportional to the velocity-integrated sum of ^{12}CO and HI intensities. Following Hauser et al. (1984), the dust-to-gas mass ratio may then be given by

$$\frac{M_d}{M_H} = 4.30 \times 10^{23} \frac{\tau(\lambda)}{N_H \kappa(\lambda)}, \tag{11}$$

where $\kappa(\lambda)$ is the absorption coefficient per mass element of the dust and $\tau(\lambda)$ is the optical depth in the infrared wavelength λ. For $\kappa(100\,\mu m)$ we use a value of $41\,cm^2\,g^{-1}$ derived by Hauser et al. (1984). The derived values for the mass ratio lie in the 3×10^{-3} range for $R > 4$ kpc, which is considerably less than the values of ~ 0.01 obtained by Savage & Mathis (1979) for the local galactic surroundings. At $R < 4$ kpc the mass ratio increases sharply, but the errors grow likewise.

The values derived above are quite similar to those found by Sodroski et al. (1987), indicating that it is of little significance that the background-removal technique employed by Sodroski et al. differed from that employed here. Extrapolation of our radial profiles to beyond the solar distance leads to a lower mass-ratio value than for the inner Galaxy, indicative of the general weakness of diffuse dust emission from the outer Galaxy.

We determined the mass ratio between dust and gas using the MRN (Mathis, Rumpl, & Nordsieck; 1977) grain model of interstellar graphite and silicate. Derivation of the temperature may be influenced by a contribution from non-equilibrium emission from small grains (Draine & Anderson 1985) or to a lesser extent by emission from polycyclic aromatic hydrocarbon (PAH) molecules (Léger & Puget 1984; Allamandola et al. 1985). The temperatures derived, however, are in agreement with the values given by Hauser et al. (1984), who used observations at longer wavelengths $(150 - 350\,\mu m)$, which are less influenced by small-grain emission or by PAH's and which are better suited than the IRAS observations to temperature determinations of the very cold dust (Pajot et al. 1986). A significant amount of non-equilibrium emission at

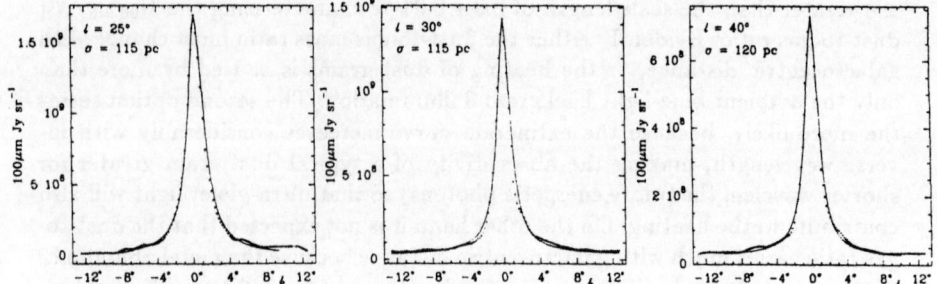

Fig. 39. Model fits to the 100-μm data in latitude cuts, through the galactic equator at the indicated longitudes. The full-drawn line represents the model fit; the crosses, the observations. The use of the derived radial emissivity distributions is essential in determining the width of the Gaussian z-height distribution for the dust. The indicated widths of the galactic dust layer are greater than the half-intensity width for CO emission, and more closely resemble those derived from the HI

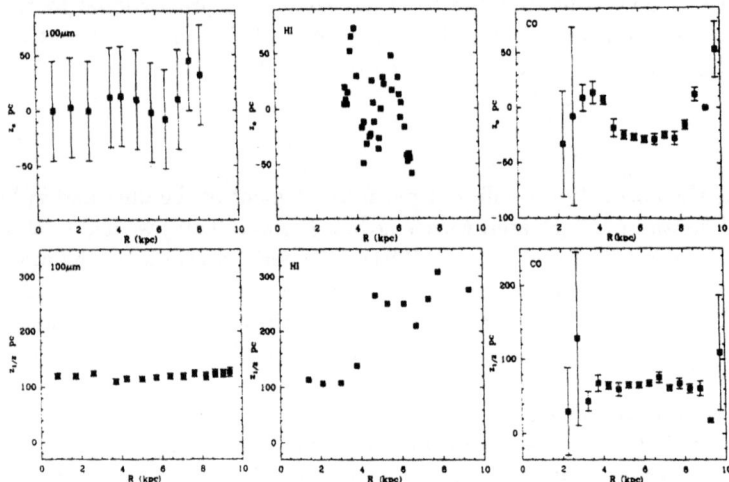

Fig. 40. Compilation of the derived mid-plane positions and the z-height widths for the first-quadrant data at 100 μm. The values for the CO and HI data are taken from Bronfman et al. (1988), and from Jackson & Kellman (1974) and Celnik et al. (1979), respectively. Note the similarities in the radial trends for the 100-μm and CO data

60 μm, or multiple temperature components along the line of sight, would lead to an overestimate of the temperature and consequently to an underestimate of the mass ratio. To obtain a global mass ratio similar to the local value (Savage & Mathis 1979), an average temperature of 17 K would be required. The fraction of non-equilibrium emission contributing at 60 μm to raise the ratio 60 μm/100 μm from 17 K to the average 23 K is almost 200%. The Draine & Anderson (1985) extrapolation of the MRN grain size distribution is only capable of obtaining such a large contribution to the 60-μm intensity by incorporating and enhancing the relative numbers of grains down to 3×10^{-4} μm

sizes. The model of Draine & Anderson assumes that almost 19% of the carbon abundance is in graphite grains smaller than $5 \times 10^{-3}\,\mu$m, which is in contradiction to the average extinction curve for the local interstellar medium. Alternatively, if small grain contributions are not so important, IRAS must have detected only a fraction ($< 30\%$) of the total dust mass in the Galaxy. If IRAS has only detected some $10\% - 15\%$ of the total dust mass, as suggested by Pajot et al. (1986), then our estimate for the dust-to-gas mass ratio is in agreement with values derived for the solar neighborhood.

Bloemen et al. (1990) found from a decomposition of the HI and CO profiles that the dust associated with the neutral atomic hydrogen accounts for about 70% of the total diffuse infrared emission. This factor is in agreement with our analysis of the radial profiles.

5.4.4 Gas-Layer z-Height Distribution.
Derivation of the mid-plane location of the centroid and the layer thickness of the interstellar dust at $100\,\mu$m is illustrated by the latitude profile fits shown in Fig. 39. We restricted this part of the analysis to the 100-μm data, because (within the uncertainties of the method used) all IRAS passbands exhibit the same scale-height characteristics. The scale height of the dust density was derived using the radial dependence of the dust emissivity. Latitude cross-cuts are themselves not sufficient for comparative purposes because such cuts do not incorporate the radial behaviour of the emitting material, which is quite different among the 100-μm, HI, and CO tracers. Therefore, any conclusion solely based on the comparison between latitude profiles must be viewed cautiously.

The radial profiles of the dust emissivity derived are used to calculate the line-of-sight contribution for a given ring, attenuated by a Gaussian z-distribution. Figure 39 shows that the latitude profiles could be accurately fit in this way. The average scale height for the interstellar dust distribution, $120\,$pc, is about twice that of the cold, compressed molecular gas, and is similar to the HI layer thickness. The conclusion that the vertical morphology of the dust is similar to that of the HI, but different from that of the CO, contrasts with the conclusion that the radial distribution of the interstellar dust is dominated by the molecular gas distribution.

Mid-plane offsets and scale heights are shown in Fig. 40. The parameters for the CO and HI emission pertain to first-quadrant data from Bronfman et al. (1988), and from Jackson & Kellman (1974) and Celnik et al. (1979), respectively. The lack of kinematic distance information results in larger error bars for the continuum infrared data than for the CO and HI data. The CO emission shows greater confinement to the galactic plane than the HI emission.

The mid-plane offsets of the dust emission at $100\,\mu$m show other trends similar to those of the molecular gas traced by CO. An increase of the mid-plane offsets around $4\,$kpc and beyond $8\,$kpc for both $100\,\mu$m and CO is evident from the top panels of Fig. 40. This suggests that the 100-μm intensities near the plane contain a large contribution from emission from dust associated with the cold, compressed molecular gas.

The variation of the 100-μm z-height distribution with galactocentric distance is rather constant and does not show the increase with distance known to exist for HI (Lockman 1984). The derived values for the 100-μm z-height are high compared to those for the CO and more closely resemble those of the inner HI disk distribution. Therefore, the 100-μm intensities at $|b| > 2°$ predominantly contain emission from dust associated with neutral atomic hydrogen.

The infrared z-height distribution of the Galaxy thus shows the combined effects of those from the molecular and atomic components. Because the molecular gas is confined more to the galactic plane than the atomic gas, the infrared z-height distribution reveals characteristics associated with the molecular component at low latitudes. For $|b| > 2°$, however, the characteristics of infrared z-height distribution resemble those of the atomic component. These findings agree with the results of Sodroski et al. (1987), who also state that the wings of the profiles show similarities with the neutral gas distribution. The cosecant z-height distribution used by Boulanger & Pérault (1988) also indicates the association of the higher-latitude infrared emission with the neutral gas component. Our results show that the scale heights of the infrared emission and the HI are similar in the inner- and local-Galaxy; in the outer Galaxy, the infrared emissivities are so weak that comparisons with the HI thickness are difficult to make.

5.5 Concluding Remarks

The comparative large-scale distribution of interstellar dust and gas have been considered here. Because of interest in the global properties the area of the sky used was limited to a narrow latitude range around the galactic equator and to a small number of latitude strips at longitudes corresponding to the inner Galaxy. The method employed to derive the radial characteristics of the infrared emissivity assumes that the global properties of the infrared emission are well represented by the limited galactic coverage used.

The radial emissivity distributions at the four IRAS wavelengths obtained from unfolding the longitude profiles closely resemble each other. Assuming an emissivity law proportional to λ^{-2}, we derived a nearly constant temperature of 23 K for the dust with variations in temperature of less than 10%. Because the dust is heated by the general interstellar radiation field, where flux decreases rapidly with increasing galactocentric distance (Mathis et al. 1983; Cox et al. 1986), we would expect the dust temperature to change too (radiation energy $\propto T^6$). Assuming, however, standard large grains only, the factor of ~ 5 change suggested by Mathis et al. (1983) for the decrease of the interstellar radiation field between the inner and outer Galaxy would be too high compared to the value that may be derived from the $60\,\mu$m/$100\,\mu$m ratio ($1.1^6 \approx 1.8$). This can be understood in terms of the Draine & Anderson (1985) model, whereby a considerable amount of the 60-μm flux is due to non-equilibrium emission from small grains, causing the $60\,\mu$m/$100\,\mu$m intensity ratio to become less dependent on the strength of the interstellar radiation field.

The $I_{100\,\mu m}/N_H$ dust-to-gas ratio plotted in Fig. 38 steeply decreases from 2.5 kpc to 10 kpc. An exponential fit has a scale length of 2.9 ± 0.3 kpc which is considerably smaller than similar derivations for the blue light. If the heating of interstellar grains is dominated by the UV part of the interstellar radiation field, then the number density of Lyman-continuum photons, representative of star-formation activity, should show a tighter correlation with the $I_{100\,\mu m}/N_H$ ratio. This is indeed the case for $R > 4$ kpc. Converting the units of the dust-to-gas ratio to $L_\odot/M_\odot$, assuming that the dust emits at the derived line-of-sight averaged temperature with a λ^{-2} emissivity law, we obtain values ranging from $20 L_\odot/M_\odot$ near 2.5 kpc to $1 L_\odot/M_\odot$ at 10 kpc. The average value for the Galaxy over the range $4 < R < 10$ kpc is $6.4 L_\odot/M_\odot$. Integrating these luminosities over the surface of the Galaxy yields a value of $1.5 \times 10^{10} L_\odot$, which is in agreement with the estimates of Hauser et al. (1984), Cox et al. (1986), and Boulanger et al. (1988).

With the N_{Lyc} and blue light distributions in mind the steep increase of the infrared emissivities at galactocentric distances less than 3 kpc is not consistent with a constant dust-to-gas ratio. The magnitude of this effect is subject to rather large errors; determination of the location at which the increase starts is likewise subject to large errors. The effective radius of the bulge is 2.7 kpc (de Vaucouleurs 1983), which is similar, within the errors of determination, to the galactocentric distance below which the ratio plots of Fig. 38 show different infrared properties. The light from the galactic bulge component evidently influences the infrared properties considerably. A similar situation has been noted in M31 by Walterbos & Schwering (1987).

Several conclusions may be summarized:

1. The well-defined correlation between HI emission and infrared radiation at high ($|b| > 2°$) latitudes weakens nearer the galactic plane, where the HI emission suffers from optical depth effects and where infrared emission association with molecular material becomes relatively more important.

2. The molecular annulus is clearly visible in the infrared data, suggesting that the dust associated with molecular material plays an important role along lines of sight traversing the inner regime of the galactic equator.

3. The $60\,\mu m/100\,\mu m$ emissivity ratio remains rather constant throughout the Galaxy, while the $12\,\mu m/100\,\mu m$ ratio shows a decrease for the inner Galaxy, consistent with a decrease in the number of small particles relative to the number of large ones.

4. Using the radial distribution of HI, CO, and Lyman continuum emissivities, the infrared profiles show the expected behavior for $R > 4$ kpc, but the intense infrared emission from the bulge region of the Galaxy requires an additional heating source, probably to be identified with the high stellar density in that region.

References

5.1   Allamandola, L.J., Thielens, A.G.G.M., Barker, J.R. 1985, *Astrophys. J.* **290**, L25
5.2   Baker, P.L., Burton, W.B. 1975, *Astrophys. J.* **198**, 281
5.3   Bloemen, J.B.G.M., Strong, A.W., Blitz, L., Cohen, R.S., Dame, T.M., Grabelsky, D.A., Hermsen, W., Lebrun, F., Mayer-Hasselwander, H.A., Thaddeus, P. 1986, *Astron. Astrophys.* **154**, 25
5.4   Bloemen, J.B.G.M., Deul, E.R., Thaddeus ,P. 1990, *Astron. Astrophys.* **233**, 437
5.5   Boulanger, F., Baud, B., van Albada, G.D. 1985, *Astron. Astrophys.* **144**, L9
5.6   Boulanger, F., Pérault, M. 1988, *Astrophys. J.* **330**, 964
5.7   Boulanger, F., Beichman, C., Désert, F.X., Helou, G., Pérault, M., Ryter. C. 1988, *Astrophys. J.* **332**, 328
5.8   Bronfman, L., Cohen, R.S., Alvarez, H., May, J., Thaddeus, P. 1988, *Astrophys. J.* **324**, 248
5.9   Burstein, D., Heiles, C. 1984, *Astrophys. J. Suppl. Ser.* **54**, 33
5.10   Burton, W.B. 1988, in *Galactic and Extragalactic Radio Astronomy*, G.L. Verschuur, K.I. Kellermann, eds., Springer-Verlag, New York, p. 295
5.11   Burton, W.B., Deul, E.R. 1987, in *The Galaxy*, G. Gilmore, R. Carswell, eds., Reidel, Dordrecht, p. 141
5.12   Caravane, P.A., Paul, J.A. 1979, *Astron. Astrophys.* **75**, 340
5.13   Caux, E., Puget, J.-L., Serra, G., Gispert, R., Ryter, C. 1984, *Astron. Astrophys.* **144**, 37
5.14   Caux, E., Serra, G. 1986, *Astron. Astrophys.* **165**, L5
5.15   Celnik, W., Rohlfs, K., Braunsfurth, E. 1979, *Astron. Astrophys.* **76**, 24
5.16   Cox, P., Krügel, E., Mezger, P.G. 1986, *Astron. Astrophys.* **155**, 380
5.17   Dame, T.M., Ungerechts, H., Cohen, R.S., de Geus, E.J., Grenier, I.A., May, J., Murphy, D.C., Nyman, L.-Å., Thaddeus, P. 1987, *Astrophys. J.* **322**, 706
5.18   Deul, E.R. 1988, Ph.D. Thesis, University of Leiden
5.19   Deul, E.R., Burton, W.B. 1988, in *Proceedings of IAU Symp.* **133** *"Mapping the Sky: Past Heritage and Future Directions"*, S. Debarat, J.A. Eddy, H.K. Eichhorn, A.R. Upgren, eds., Kluwer Academic Pub., Dordrecht
5.20   Deul, E.R., Burton, W.B. 1990, *Astron. Astrophys.* **230**, 153
5.21   Deul, E.R., Walker, H.J. 1989, *Astron. Astrophys. Suppl. Ser.* **81**, 207
5.22   Deul, E.R., Wolstencroft, R.W. 1988, *Astron. Astrophys.* **196**, 277
5.23   de Vaucouleurs, G. 1983, *Astrophys. J.* **268**, 451
5.24   de Vaucouleurs, G., Pence, W.D. 1978, *Astron. J.* **83**, 1163
5.25   de Vries, H.W., Heithausen, A., Thaddeus, P. 1987, *Astrophys. J.* **319**, 723
5.26   Draine, B.T., Anderson, N. 1985, *Astrophys. J.* **292**, 494
5.27   Gordon, M.A., Burton, W.B. 1976, *Astrophys. J.* **208**, 346
5.28   Güsten, R., Mezger, P.G. 1983, *Vistas in Astronomy* **26**, 159
5.29   Hauser, M.G., Silverberg, R.F., Stier, M.T., Kelsall, T., Gezari, D.Y., Dwek E., Walser, D., Mather, J.C. 1984, *Astrophys. J.* **285**, 74
5.30   IRAS Explanatory Supplement 1985, C.A. Beichman, G. Neugebauer, H.J. Habing, P.E. Clegg, T.J. Chester, eds., Jet Propulsion Laboratories
5.31   Jackson, P.D., Kellman, S. A. 1974, *Astrophys. J.* **190**, 53
5.32   Kanbach, G., Beuermann, K. 1979, in *Proceedings of the XVI International Conference on Cosmic Rays* **1**, 75
5.33   Léger, A., Puget, J.-L. 1984, *Astron. Astrophys.* **137**, L5
5.34   Liszt, H.S., Xiang, D., Burton, W.B. 1981, *Astrophys. J.* **249**, 532
5.35   Lockman, F.J. 1984, *Astrophys. J.* **283**, 90
5.36   Mathis, J.S., Mezger, P.G., Panagia, N. 1983, *Astron. Astrophys.* **128**, 212
5.37   Mathis, J.S., Rumpl, W., Nordsieck, K.H. 1977, *Astrophys. J.* **217**, 425
5.38   Nishimura, T., Low, F.J., Kurtz, R.F. 1980, *Astrophys. J.* **239**, L101
5.39   Okuda, H. 1981, in *Proceedings of IAU Symp.* **96** *"Infrared Astronomy"*, C.G. Wynn-Williams, D.P. Cruikshank, eds., Dordrecht, Reidel, p. 247
5.40   Pajot, F., Boissé, P., Gispert, R., Lamarre, J.M., Puget, J.-L., Serra, G. 1986, *Astron. Astrophys.* **157**, 393

5.41 Phillips, S., Kearsey, S., Osborne, J.L., Haslam, C.G.T., Stoffel, H. 1981, *Astron. Astrophys.* **98**, 286

5.42 Puget, J.-L. 1985, in *Birth and Infancy of Stars*, Proceedings Les Houches Summer School (1983), A. Omont, R. Lucas, eds., North Holland Publ. Co.

5.43 Puget, J. L., Léger, A., Boulanger, F. 1985, *Astron. Astrophys.* **142**, L19

5.44 Savage, B.D., Mathis, J.S. 1979, *Ann. Rev. Astron. Astrophys.* **17**, 73

5.45 Scoville, N.Z., Solomon, P.M. 1975, *Astrophys. J.* **199**, L105

5.46 Shaver, P.A., McGee, R.X., Newton, L.M., Danks, A.C., Pottasch, S.R. 1983, *Monthly Notices Roy. Astron. Soc.* **204**, 53

5.47 Sodroski, T.J., Dwek, E., Hauser, M.G., Kerr, F.J. 1987, *Astrophys. J.* **322**, 101

5.48 Strong, A.W. 1975, *J. Phys. A. Math. Gen.* **8**, 617

5.49 Strong, A.W., Worrall, D.M. 1976, *J. Phys. A. Math. Gen.* **9**, 823

5.50 Terebey, S., Fich, M. 1986, *Astrophys. J.* **309**, L73

5.51 van der Kruit, P.C. 1986, *Astron. Astrophys.* **157**, 230

5.52 Walterbos, R.A.M., Schwering, P.B. 1987, *Astron. Astrophys.* **180**, 27

5.53 Weiland, J.L., Blitz, L., Dwek, E., Hauser, M.G., Magnani, L., Rickard, L.J 1986, *Astrophys. J.* **306**, L101

5.54 Young, J.S., Schloerb, F.P., Kenney, J.D., Lord, S.D. 1986, *Astrophys. J.* **304**, 443

6 Kinematics and Distribution of Neutral Interstellar Gas in the Galactic Bulge Region

W.B. Burton & H.S. Liszt

Abstract. The distribution and motions of the neutral gas layer in the inner few kpc of the Galaxy, corresponding roughly to the radial extent encompassing the bulge, show properties which distinguish the gas layer in this region from that at larger radii. The inner region is pervaded by a combination of circular and noncircular motions of approximately the same amplitude. This combination is recognized most easily near $b = 0°$ when adequate account is taken of contamination of the HI data by absorption against continuum radiation associated with the Sagittarius source complex in the galactic core. The abrupt kinematic shift in the HI velocities centered on $l = 0°$ is a consequence of distortion of the HI pattern by effects of absorption against sources of continuum radiation confined to the galactic nucleus. This is demonstrated by considering tracers largely unaffected by absorption, like CO, and by considering tracers which follow the absorption directly, like OH and H_2CO. The envelope of HI emission around the galactic core is thus not a reliable guide to the run of rotation velocity within the bulge. The noncircular component disqualifies use of the terminal-velocity method to derive the circular-velocity rotation curve for gas lying within the bulge region. The inner-Galaxy gas layer is both tilted, or warped, and flared. The fundamental plane of kinematic symmetry is tilted some 20° with respect to the plane defined by $b = 0°$. This tilt is revealed by the atomic and by the molecular gas, as well as by the distribution of infrared emission from stars. Lines of sight through the tilted and flared gas layer sample a much more complicated structural and kinematic geometry than prevails in the Galaxy at large. This complicated sampling geometry can, however, be held responsible for a number of characteristics of the observations, including the large line widths observed for the spectral-line tracers and the isolated, but only apparently anomalous, structures occurring in position, velocity maps of the region.

6.1 Use of the Terminal-Velocity Measure to Determine the Inner-Galaxy Rotation Curve

It is convenient to discuss many global properties separately for three distinct radial regimes of the galactic gas layer, namely that contained within the bulge (at, roughly, $R < 0.3R_0$), that defining the star-forming, molecule-rich annulus (at $0.3R_0 < R < R_0$), and that (at $R > R_0$) tracing the warped, flaring, outer reaches of the Milky Way[1]. This distinction is not arbitrary, because the form of the dominant gravitational potential, the nature of the radiation environment,

---

[1]   In order to facilitate comparisons made with earlier papers, we use in this chapter the galactic-constant values $\Theta_0 = 250\,\mathrm{km\,s^{-1}}$ and $R_0 = 10\,\mathrm{kpc}$.

the content of the interstellar medium, and the stage of evolution, differ strongly in these three regions. Attention is directed here to the kinematic circumstances prevailing in the inner few kpc of the gas layer.

Neutral atomic hydrogen can be observed easily throughout the galactic gas layer, from the innermost nucleus to the far outer reaches of the Milky Way. Over the entire Galaxy, such observations yield important information on physical properties of the interstellar gas, such as density, temperature, and motions. Information on densities and temperatures is extracted from the observations in often indirect ways and under assumptions whose verification has proven difficult; information on the *kinematics* of the gas, on the other hand, is presented by the HI 21-cm spectra in a way which is, generally, much more straightforward.

A kinematic measure of particular importance involves determination of the rotation curve of the Galaxy. Detailed knowledge of the rotation characteristics of a galaxy is prerequisite to essentially all discussions of galactic dynamics. The rotation curve has been determined from the terminal-velocity method along the locus of subcentral points for $R < R_0$. HI data have been the prime tracer of galactic kinematics for the entire gaseous disk within the solar orbit. Analyses of tracers other than HI, in particular of CO, are consistent with the kinematic results from HI over the region $0.3R_0 < R < R_0$; in the region of the galactic gas layer encompassed by the bulge, other gas tracers serve special purposes but are not so easily traced as HI and CO because the emission levels are very low (except from within the innermost few hundred pc).

The most stringent assumption required for the validity of the terminal-velocity method is that of circular rotation. It was argued in Chap. 3 that this assumption seems to be valid, to within a few % of the rotation speed, over the entire disk of the Galaxy *outside* of the bulge region. The deviations from a completely well-behaved kinematic situation are, of course, interesting in themselves, but ignoring these deviations does not invalidate the simplest derivations of galactic morphology; these deviations amount to less than about $10\,\mathrm{km\,s^{-1}}$ in the Galaxy at large, compared to the rotation speed of about $250\,\mathrm{km\,s^{-1}}$. In any case, the deviations from circular motion are well enough understood outside of the bulge region that the terminal-velocity analysis can be carried out such to avoid their influence. Over the portion of the galactic gaseous disk contained *within* the bulge region, on the other hand, noncircular motions are quite pervasive and of an amplitude comparable to the circular ones. We stress in the following that these motions *do* largely disqualify the terminal-velocity method in the region of the inner-galactic gas layer encompassed by the bulge.

Figure 41 illustrates that derivations of the rotation curve in the inner few kpc which depend on the validity of the assumption of circular rotation yield a curve with a sharp rise from the galactic nucleus to a distinct peak of very rapid rotation at $R = 500\,\mathrm{pc}$, then a decrease in rotation speed persisting until about $R = 2.5\,\mathrm{kpc}$. The presence of such a peak in the rotation velocity has important consequences on dynamic models of the Milky Way and to the derivation of the galactic mass distribution (see, e.g., Bahcall et al. 1982; Schmidt 1985). Sufficiently many rotation curves of external galaxies are now known that one

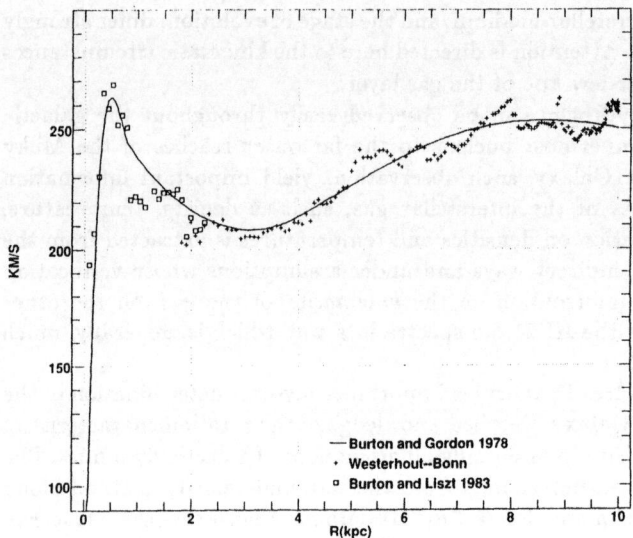

Fig. 41. Putative rotation curve for the inner Galaxy as given by a smooth-line fit to measurements of the terminal velocities in northern-hemisphere HI data (Burton & Gordon 1978; Burton & Liszt 1983; Westerhout, private communication), and normalized with $R_0 = 10\,\mathrm{kpc}$ and $\Theta_0 = 250\,\mathrm{km\,s^{-1}}$. The assumption of circular motion is probably correct to within a few percent at $R \geq 3\,\mathrm{kpc}$, but fails at smaller radii. The form of the rotation curve in the inner few kpc depicted here, involving an abrupt rise to a high rotation level and subsequent dip, is not characteristic of other galaxies and is probably incorrect. (Other references to the inner-Galaxy rotation curve so derived include: Rougoor & Oort 1960; Sanders & Wrixon 1973; Simonson & Mader 1973; Sinha 1978; and Clemens 1985)

may note that an inner peak and subsequent dip such as purported for the inner Milky Way rarely, if ever, occur in other systems. Rubin (1983), Burstein & Rubin (1985), and Casertano & van Gorkom (1991) have compiled a large number of external-galaxy rotation curves: compared to all of these, the curve shown in Fig. 41 is an oddity. We question here the reality of this peak, by questioning the validity of the terminal-velocity analysis in terms of circular rotation.

Figure 42 shows the sort of HI observations involved in the rotation-curve determination. It has been argued, largely for one of two reasons, that the terminal-velocity method could be applied to the gaseous disk in the inner few kpc, despite the evidence for structural and kinematic noncircularity. Firstly, the noncircular motions commonly were viewed as motions due to discrete features, superimposed upon a general kinematic field of pure rotation, whose kinematic structure did not interfere with determination of terminal velocities. The 3-kpc arm is the most obvious specific example of such a feature. Secondly, the HI data displayed in a longitude, velocity map at $b = 0°$ show, at negative velocities, the abrupt shift of the emission pattern across the $l = 0°$ plane; this shift, if confirmed as accurately reflecting the distribution of gas, is the undeniable signature of pure rotation. That a complementary shift is not ob-

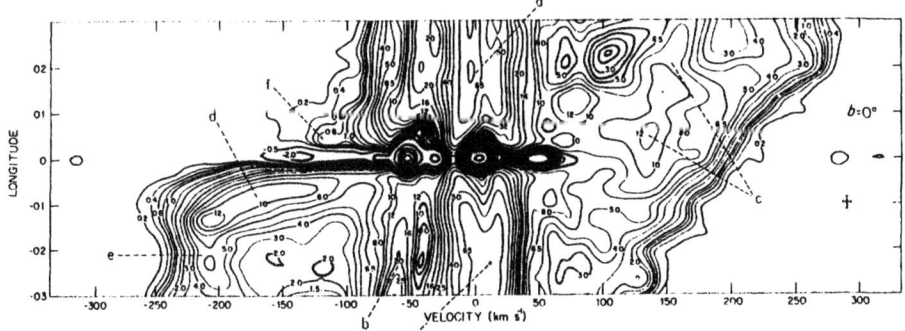

Fig. 42. Longitude, velocity distribution of HI observed at $b = 0°$ between $l = -3°$ and $l = +3°$ (from Sanders & Wrixon 1973). Cylindrical structural and kinematic symmetry requires that this map be symmetric when flipped around the $l = 0°$ and $v = 0\,\mathrm{km\,s^{-1}}$ axes. In particular, purely circular motions require that the quadrants $l < 0°$, $v > 0\,\mathrm{km\,s^{-1}}$ and $l > 0°$, $v < 0\,\mathrm{km\,s^{-1}}$ be essentially empty of emission. The letters give the Sanders-Wrixon labelling of different features: e and d label the "rotating nuclear disk"; other features have long been recognized as indicative of noncircular motion, including the "3-kpc arm" (feature b) and the "expanding arm at $+135\,\mathrm{km\,s^{-1}}$" (feature c)

served at *positive* velocities was viewed as due to contamination from discrete features superimposed on a field of general rotation; the superposition must, furthermore, be such that the Sun is placed in a special position. It is important to address this abrupt shift explicitly. Models which invoke a pervasive non-circular velocity field have failed to explain this aspect of the HI data, which is "allowed" by rotational kinematics, just as models based only on nonradial motions have failed to account for the emission in the "forbidden" quadrant at $l < 0°$, $v > 0\,\mathrm{km\,s^{-1}}$.

The presence of peculiar motions over substantial scales in the inner few kpc of the Galaxy have, of course, been recognized since the first surveys of 21-cm emission were made (see, especially, van Woerden et al. 1957, and Oort & Rougoor 1960). The 3-kpc arm is the largest feature in the core region bearing an unambiguous signature of expansion. The influence of the 3-kpc arm on the terminal-velocity determination of the rotation curve may be discounted in this discussion, however, because this feature is not found near the terminal velocities at the longitudes in question. Various other features with clearly noncircular motions have been recognized in maps of HI emission which violate the assumption of general circular motion. As is the 3-kpc arm, several such features are also particularly easy to recognize in the HI longitude, velocity map at $b = 0°$ in Fig. 42 within a few degrees of $l = 0°$; prominent among these features is the so-called "expanding arm at $+135\,\mathrm{km\,s^{-1}}$". (We note that other, largely interferometric, spectral-line data show apparently localized anomalous-velocity features in the innermost nucleus of the Galaxy and confined to short angular scales of minutes of arc or less, and to linear scales of a few tens of pc or less. Such features are too small-scale to have influenced the determination of the rotation curve in the bulge region; for this region, observations are involved

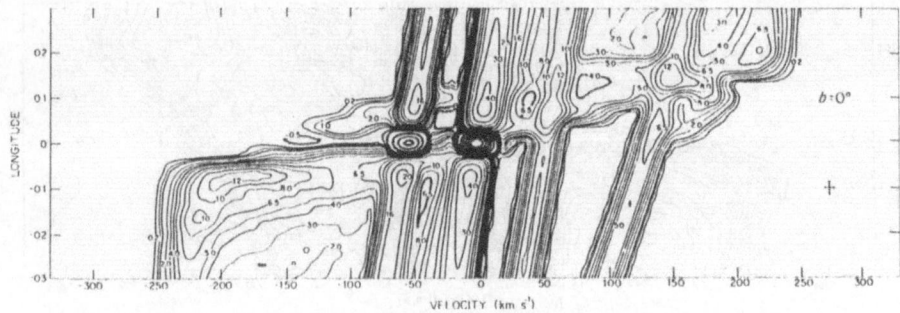

Fig. 43. Longitude, velocity distribution of HI as modelled by Sanders & Wrixon (1973). The model superposes a number of discrete features moving radially outward (in order to account for the emission in the $l < 0°$, $v > 0\,\mathrm{km\,s^{-1}}$ quadrant) on a half-disk of gas moving on purely circular orbits (in order to account for the sharp drop in intensities at $l = 0°$, $v < 0\,\mathrm{km\,s^{-1}}$)

which are typically made with angular resolutions exceeding 10 minutes of arc and which cover some 20° of longitude.)

But whatever structural interpretation is given to the gas contributing the emission, the very fact that the emission patterns can be followed easily from the (l, v) quadrant at $l > 0°$, $v > 0\,\mathrm{km\,s^{-1}}$, *allowed* in terms of circular rotation, to the *forbidden* quadrant at $l < 0°$, $v > 0\,\mathrm{km\,s^{-1}}$, disqualify the assumption of purely circular motion: the emission in the quadrant at $l < 0°$, $v > 0\,\mathrm{km\,s^{-1}}$, unambiguously forbidden, and the continuity of the emission pattern from that quadrant through to $l > 0°$, $v > 0\,\mathrm{km\,s^{-1}}$, implies that the terminal velocities in the allowed quadrant at positive velocities are themselves contaminated by noncircular motions.

The situation in the allowed quadrant at negative velocities, that is at $l < 0°$, appears different, because important 21-cm emission patterns (other than the 3-kpc arm) do not cross $l = 0°$ at negative velocities and because of the general paucity of emission at $l > 0°$, $v < 0\,\mathrm{km\,s^{-1}}$. These characteristics of the HI data have provided the principal justification for the circular-rotation assumptions which have led to the peaked rotation curve shown in Fig. 41. These characteristics, in particular the paucity of emission at $l > 0°$, $v < 0\,\mathrm{km\,s^{-1}}$, have similarly been held incompatible with the presence of expanding features (other than the 3-kpc arm) at longitudes southward of the direction to the galactic center. The models incorporating discrete expanding features have thus shown a bias with respect to the Sun-center line, with the expanding features harbored in a half-disk, oriented favorably with respect to the position of the observer, lying at $l > 0°$; similarly, models based on circular rotation have also only been defensible for a favorably-oriented half-disk lying southward of the Sun-center line. This dichotomous situation is illustrated in the schematic representations shown by Oort (1977; see his Fig. 9) and by Sanders & Wrixon (1973; see their Fig. 10). Figure 43 shows the longitude, velocity distribution

104

which results from this type of model, involving two half-disks of differing dominant kinematics.

There are, of course, other analyses of the gas-layer motions in the inner few kpc of the Galaxy which do not involve circular motions. References to specific models, each with a different dynamical foundation but each giving kinematic predictions, include, among others, those of Shane (1972), Simonson & Mader (1973), Peters (1975), Simonson (1976), Roberts (1979), Yuan (1984), van Albada (1985), Mulder & Liem (1986), Gerhard & Vietri (1986), and Binney et al. (1991). These models invoke nonradial motions, generally along closed streamlines, to explain, in particular, the observational situation at $v > 0\,\mathrm{km\,s^{-1}}$. The troublesomely-oriented dichotomy shown by the HI remains for these models, too, however, because they also predict a general, flipped-velocity, symmetry with respect to $l = 0°$, and in that regard fail to account for lack of this symmetry shown in Fig. 42.

In what follows, we argue that the weakness of HI emission in the $l > 0°$, $v < 0\,\mathrm{km\,s^{-1}}$ quadrant is a consequence of absorption of HI intensities against the continuum radiation from the direction of the Sagittarius source complex, and that if account is taken of this absorption, the $l > 0°$ forbidden-velocity quadrant appears complementary to the $l < 0°$ one. This situation invalidates application of the terminal-velocity derivation of circular motion, but at the same time allows discussion of noncircular motions in general terms.

6.2 The Role of Absorption in Determining the Paucity of HI Emission in the $l > 0°$, $v < 0\,\mathrm{km\,s^{-1}}$ Quadrant at $b \simeq 0°$

At the specific direction of the galactic nucleus, the 21-cm profile shows evidence of absorption over the entire relevant range of negative velocities. Such evidence is shown by the indentation near $l = 0°$ in the Fig. 44 map of HI emission at $b = 0°$ over a wider range of longitudes than presented in Fig. 41. The *vertical* cut through the inner-Galaxy HI layer shown in Fig. 45 as a latitude, velocity diagram at $l = 0°$ illustrates the same point. No plausible spatial and kinematic distribution of gas could directly account for these indentations.

The high-resolution VLA absorption spectrum observed toward Sgr A* published by Liszt et al. (1985) shows absorption over the entire range of negative velocities, to about $-160\,\mathrm{km\,s^{-1}}$. It is clear that the kinematics of the absorbing gas represents motions in the sense of expansion outwards from the nucleus. The continuum radiation from the direction of Sgr A* is, of course, very intense. But the radio continuum emission is *generally* intense within a few degrees of the direction of the nucleus; in addition to the extended sources Sgr A, B, C, D, and E (see Fig. 8.1 of Liszt 1988) there is intense *diffuse* continuum emission from the inner few degrees (see Fig. 4 of Reich et al. 1984). The distribution of the continuum radiation suggests asking if a larger region in the negative-velocity quadrants might not be contaminated by absorption against the continuum radiation in the core over several degrees of longitude and within a half degree or so of $b = 0°$.

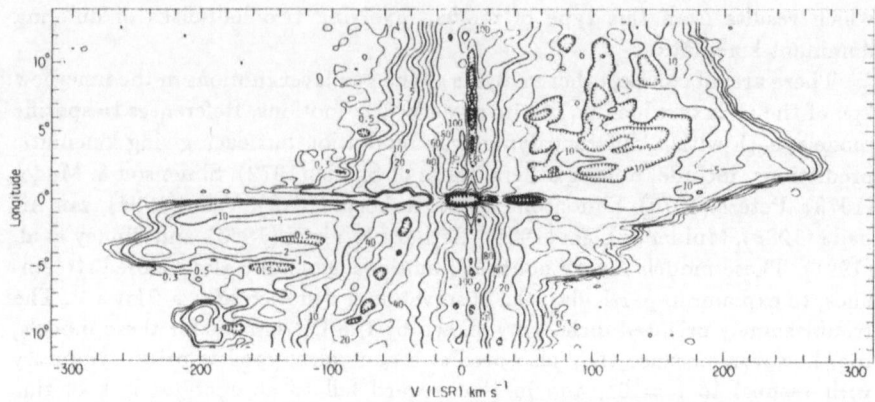

Fig. 44. Longitude, velocity distribution of HI intensities at $b = 0°$. The diagram, made using data of Burton & Liszt (1983), shows intensities labelled in units of brightness temperatures. Note the continuity of the emission crossing $l = 0°$ at positive velocities, contrary to the situation at the negative-velocity crossing. The consequences of absorption at the specific direction $l = 0°$ are obvious; we argue here that the negative-velocity regime is *generally* contaminated by absorption against a distribution of continuum sources in the galactic core

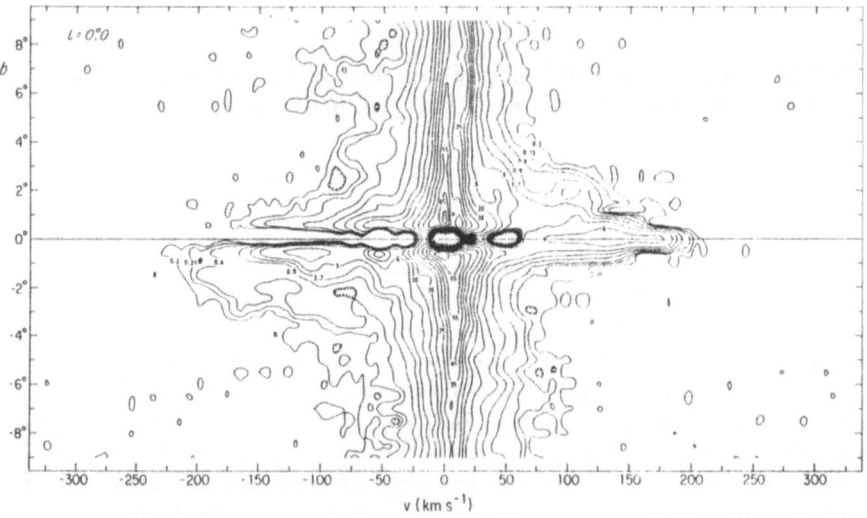

Fig. 45. Latitude, velocity distribution of HI intensities at $l = 0°$. The diagram (from Burton & Liszt 1983) shows intensities labelled in units of antenna temperature. This vertical cut through the gas layer in the direction of the galactic center shows that the low-latitude negative velocities contribute emission generally, except where absorption prevails. That the emission is relatively stronger in the opposed quadrants $b < 0°$, $v < 0\,\mathrm{km\,s^{-1}}$ and $b > 0°$, $v > 0\,\mathrm{km\,s^{-1}}$ is evidence of the tilted nature of the gas layer

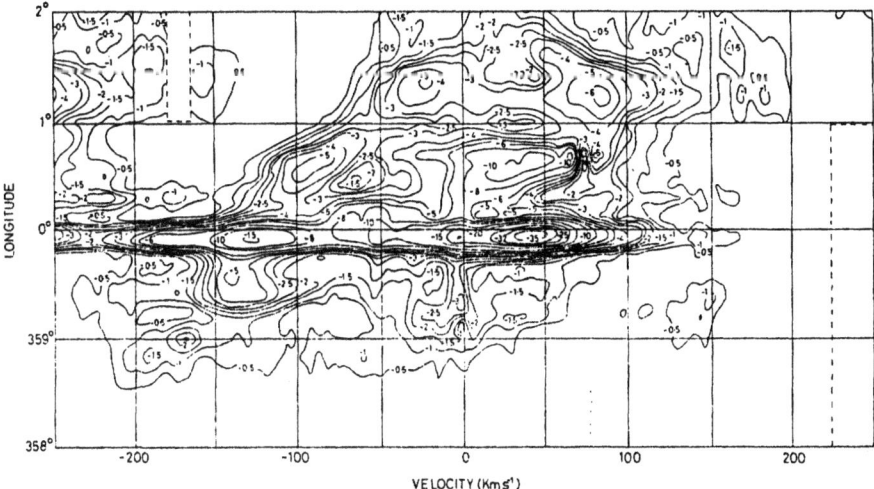

Fig. 46. Longitude, velocity map of OH absorption in the 1665-MHz (18-cm) line at $b = 0°$ (from Cohen & Few 1976). (Contours at the extreme negative velocities are contamination from the 1667-MHz line)

It is notoriously difficult to separate the effects of emission, absorption, and self-absorption in low-latitude single-dish galactic 21-cm observations. Therefore it seems reasonable to consider especially the crucial forbidden-velocity quadrants in spectral lines which trace *solely* absorption, and in lines which trace solely emission, but not the *combined* effects as is the case with HI.

The tracers OH and H_2CO, observed in absorption, are both widely-enough distributed throughout the inner-Galaxy gas layer to serve as general tracers of the absorbing gas. Figure 46 reproduces the Cohen & Few (1976) (l, v) map of OH absorption observed along $b = 0°$; Fig. 47 reproduces the Cohen & Few (1981) (l, v) map of H_2CO absorption over approximately the same region. In both maps, the forbidden quadrant at negative velocities and positive longitudes — which is so sparsely populated by HI emission — *is* filled, and it is, furthermore, filled in a way complementary to the HI situation in the positive-velocity, negative-longitude forbidden quadrant. It seems obvious that the skewness of the Cohen & Few absorption patterns indicates a rotation component of the inner-Galaxy gas, and that the confinement of the absorption to largely negative velocities indicates an expansion component. It seems plausible that the general paucity of absorption in the positive-velocity, negative-longitude quadrant is not due to the absence of gas in that regime, but to its location beyond the sources of enhanced continuum radiation. Continuum maps show that these sources are confined to within a few hundred pc from the galactic nucleus; this situation is evidently the absorption analogue to the paucity of HI emission in the complementary forbidden quadrant.

The gas distribution which is revealed by the absorption tracers is confirmed by spectral-line observations of tracers which are — unlike those of HI — gen-

107

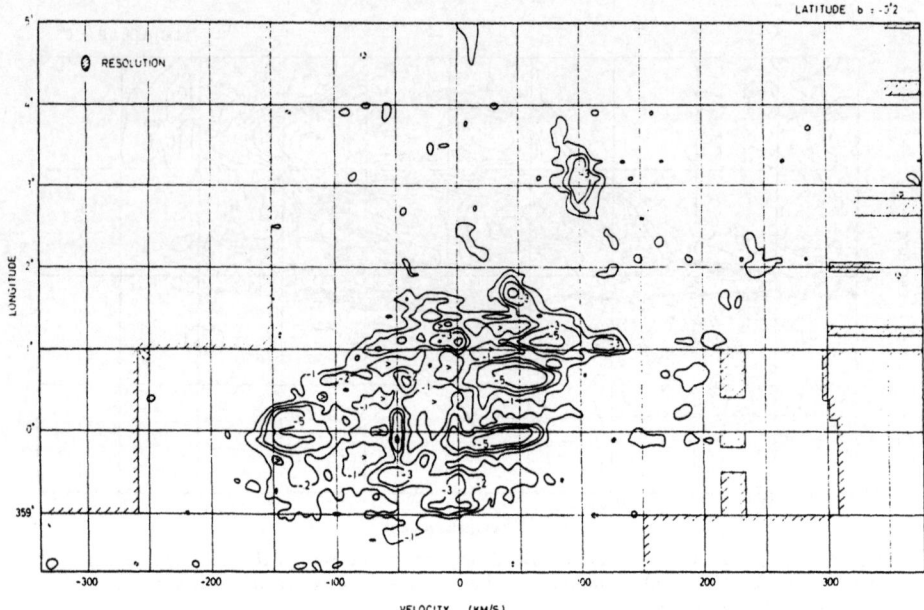

LATITUDE b : -3.2

Fig. 47. Longitude, velocity map of 6-cm H₂CO absorption at $b = -0°.2$ (from Cohen & Few 1981). As is the case with the hydroxyl gas, absorption against the enhanced continuum radiation confined to the inner few degrees of the galactic core by formaldehyde fills the forbidden quadrant at $l > 0°$. The (l, v) structure in this quadrant is quite similar to the structure in the complementary $l < 0°$, $v > 0$ km s⁻¹ forbidden quadrant as shown by *emission* observations of HI and CO

erally *insensitive* to absorption. Absorption-insensitive tracers provide, in addition, more complete information than the absorption-dominated ones, because they provide information from gas lying *beyond* the sources of continuum radiation in addition to information on gas lying between the continuum region and the observer. Furthermore, such tracers can follow the gas distribution away from the equator more directly than observations of OH or H₂CO, which depend on the continuum background which is itself dominated by radiation from a flat layer in the inner few hundred pc from the nucleus. The background continuum radiation is so weak at mm-wavelengths that absorption against such background sources in tracers like CO is negligible. (Self-absorption due to cold foreground material occurs in galactic CO data under some circumstances, but is not important in the present context, largely because the ensemble of molecular clouds in the Galaxy at large, at, say, $R > 3$ kpc, contributes negligibly to the forbidden-velocity quadrants.)

Figure 48 shows ¹³CO data (from Liszt 1988) taken along the galactic equator. Emission fills the same pattern in the forbidden $l > 0°$ portion of (l, v) space at $b = 0°$ as is filled there by OH and H₂CO absorption, but which is evidently depleted in HI emission by absorption contamination. We note the symmetry of the CO emission pattern when it is flipped around the $l = 0°$

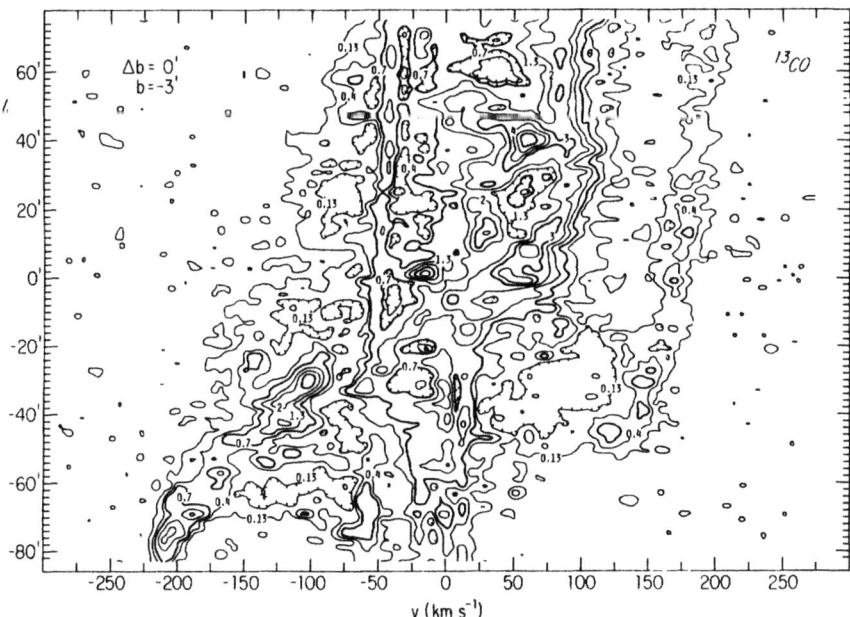

Fig. 48. Longitude, velocity map of ^{13}CO data taken along the inner few degrees of the galactic equator (from Liszt 1988). Absorption effects are negligible in these data because of the weakness of the background continuum radiation at mm-wavelengths; self-absorption effects can also be ignored because the kinematics of the ensemble of molecular clouds in the Galaxy at large is restricted in these directions to within about $10\,\mathrm{km\,s^{-1}}$ of $v = 0\,\mathrm{km\,s^{-1}}$. Emission is pervasive in the forbidden, negative-velocity, quadrant at $l > 0°$. The (l, v) structure in this quadrant is quite similar to that shown by the *absorption* observations of OH and H_2CO. The structure in the complementary $l < 0°$, $v > 0\,\mathrm{km\,s^{-1}}$ forbidden quadrant is similar to that shown by observations of HI. (Observations of the ^{12}CO tracer — see e.g. Fig. 2 of Liszt & Burton 1978 — confirm the same kinematics as shown here for ^{13}CO)

and $v = 0\,\mathrm{km\,s^{-1}}$ axes. The lack of this symmetry between the complementary forbidden-velocity quadrants in the HI data, which has constrained its interpretation, can be understood in terms of absorption contamination.

It has been considered a puzzling aspect of the CO distribution in the inner few kpc, especially that of ^{12}CO, that the molecular emission is apparently so diffusely distributed in position, velocity space, unlike the much more clumpy, cloudy distribution presented by molecular material in the molecular annulus at larger R. We argue below that, just as we attribute the large velocity widths observed to the large range of velocities which are sampled along each line of sight through the inner-Galaxy gas layer, we may attribute the pervasiveness of the emission to the line-of-sight sampling properties. It could be the case that discrete molecular clouds characterize the bulge region, but, if so, they appear heavily blended in the data because of the sampling properties inherent in observations through this region.

The observations of the CII line at $157.7\,\mu$m made from balloons by Okuda et al. (1989), Shibai et al. (1991), and Mizutani et al. (1991) confirm that future

109

space-borne CII missions will be an especially rich source of information on the physical, kinematic, and structural aspects of the interstellar gas in the bulge region. Carbon has a lower ionization potential (11.3 eV) than hydrogen and is thus easily ionized by ambient low energy UV photons; furthermore the CII line is easily excited by collisions by virtue of its low excitation temperature (92 K). The data of Okuda et al. and Mizutani et al. show that CII emission is spread diffusely throughout the inner few degrees.

Emission observations of absorption-insensitive tracers potentially provide more complete information than observations dominated by (or sensitive to) absorption, because they provide information from gas lying in or near the galactic equator but *beyond* the sources of continuum radiation in addition to information on gas lying *between* the continuum sources and the observer. Furthermore, observations of such tracers follow the gas distribution away from the galactic equator and, in general, away from the region on the sky to which the continuum radiation is confined. The confinement of the continuum radiation presents an important restriction to the use of OH or H_2CO data in this regard. The Fig. 8.1 plot of Liszt (1988), and the Fig. 4 plot of Reich et al. (1984), show that the relevant continuum radiation is largely confined to $|l| < 1°.2$, and $|b| < 0°.2$, occupying a thin layer of a few hundred pc radial extent and a few tens of pc thick. It is clear from the observations that both the HI and the CO distributions are much more extensive than the distribution of continuum radiation, both horizontally as well as vertically. It is natural to ask if the HI emission situation is clarified by examining its structure *outside* of the region of absorption contamination. In this regard it is important to take account of the fact that both the HI and the CO distributions within the general bulge region are not centered on the plane $b = 0°$, but show spatial and kinematic symmetry about a plane which is tilted with respect to the equator.

Before discussing, in the following section, the evidence for the tilt of the inner-Galaxy gas layer, and some of the consequences of this tilt, it is interesting to look at the (l, v) distribution in an HI dataset from which the spectra most likely contaminated by absorption have been eliminated, and in which the higher-latitude data have been projected onto a single plane. Such an (l, v) map is shown in Fig. 49. In producing this figure spectra taken within 1° of the galactic center were eliminated, and then the remaining spectra at $|b| \le 6°$ were projected onto $b = 0°$. Figure 49, therefore, represents the simplest possible attempt to negate the effects of absorption and of the tilt.

The (l, v) projection shown in Fig. 49 shows a complementary symmetry between the two forbidden-velocity quadrants. The abrupt shift of the emission pattern as the line of sight crosses $l = 0°$, which characterizes the $b = 0°$ data shown in Fig. 44, is not present in this projection. This dichotomous shift provided the principal justification of the assumption of the dominance of circular rotation and of the validity of application of the terminal-velocity method which led to the sharply rising and peaked rotation curve plotted in Fig. 41. The Fig. 49 HI emission pattern is consistent with the absorption-dominated patterns such as displayed in the Cohen & Few data reproduced in Figs. 46 and 47, as well as with the absorption-insensitive CO data shown in Fig. 48.

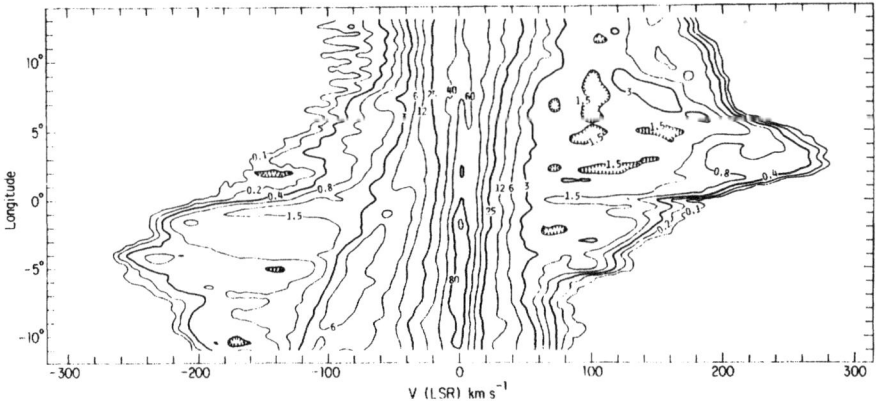

Fig. 49. Longitude, velocity map of HI intensities, as in Fig. 44, but from a dataset in which spectra taken within 1° of the center were eliminated (in order to suppress the effects of absorption and the continuum emission concentrated in the inner few hundred pc) and in which spectra were averaged over $|b| \leq 6°$ (in order to suppress the effects of the inner-Galaxy tilt). This projection largely restores complementary symmetry in the two forbidden-velocity quadrants

It seems plausible to conclude that noncircular motions are in fact pervasive in *both* forbidden-velocity quadrants of the HI tracer and that consequently straightforward derivation of the rotation curve using the terminal-velocity method is not justified in *either* quadrant. Thus the Fig. 41 curve is not a correct description of the rotation in the inner few kpc of the Galaxy.

6.3 The Tilted Nature of the Inner-Galaxy Gas Distribution

The gas layer in the Galaxy at large, that is in the region extending from the outer boundary of the bulge at 2 or 3 kpc to near the outer edge of the optical disk near about 12 kpc, is remarkable flat. Deviations of the center of mass of the gas layer from the equator $b = 0°$ are less than 30 pc, whether this centroid is measured for the atomic component in the HI line or for the molecular component in a CO line. The dust layer is evidently similarly flat over this region. The deviations from flatness amount to much less than 1% of the radial extent.

The situation in the inner few kpc of the Galaxy is very different. Most of the gas between a few hundred pc from the nucleus and a few kpc, roughly the radial extent of the galactic bulge, lies in a disk which is tilted some 20° with respect to the plane $b = 0°$. The tilted nature of the layer is well established both for the atomic gas (see, e.g., Kerr 1967; Cohen 1975; Burton & Liszt 1978; Sinha 1979; Liszt & Burton 1980; Burton & Liszt 1983) and for the molecular gas (e.g. Liszt & Burton 1978; Sanders et al. 1984; Burton & Liszt 1992). (We comment below on the work of Blitz & Spergel (1991), who showed that the distribution of stars detected at 2.4 μm by Matusumoto et al. (1982) is likewise tilted.)

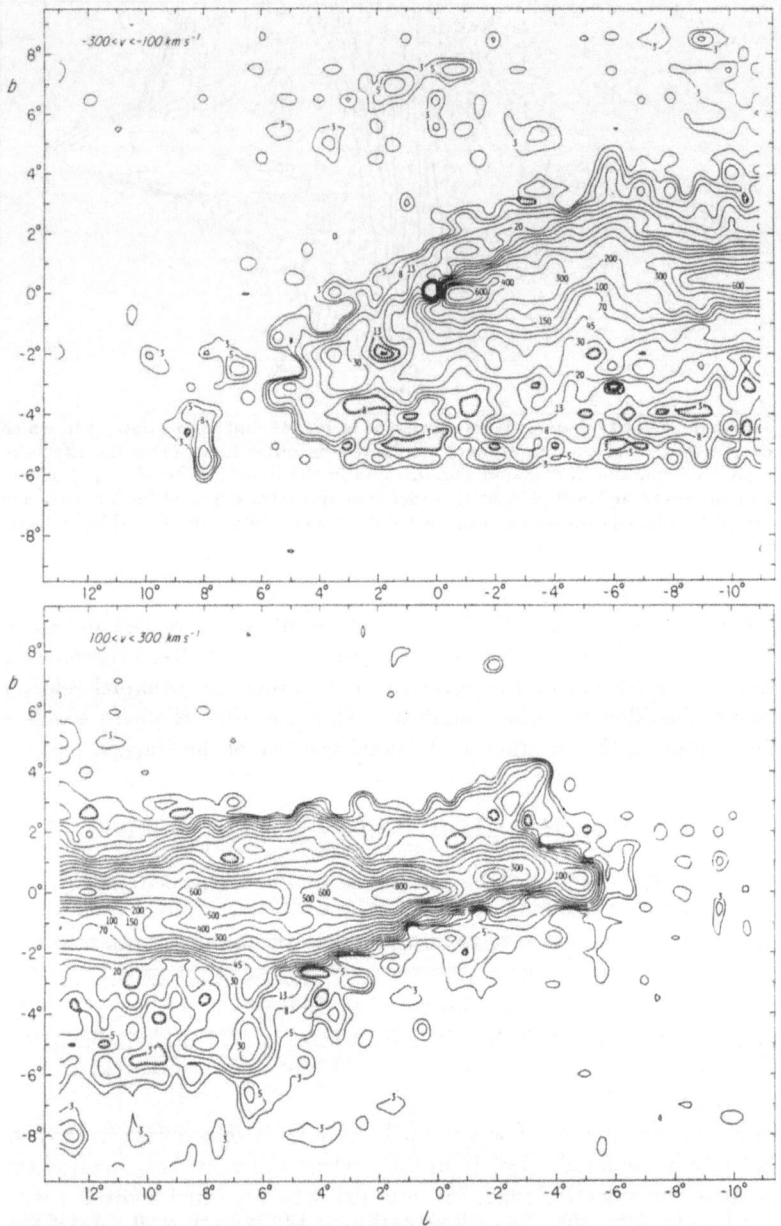

Fig. 50. Demonstration of the tilt of the inner-Galaxy HI gas layer shown by two spatial maps (from Burton & Liszt 1983) of HI intensities integrated over the indicated, wide velocity ranges. Material in the velocity range $-300 < v < -100\,\mathrm{km\,s^{-1}}$ is *forbidden*, in terms of circular rotation, at $l > 0°$, but *allowed* at $l < 0°$. Material in the velocity range $+100 < v < +300\,\mathrm{km\,s^{-1}}$ is *forbidden* at $l < 0°$, but *allowed* at $l > 0°$. The integrated emission departs coherently from $b = 0°$ on both sides of the meridian plane $l = 0°$, irrespective of the forbidden or permitted nature of the motions

The most straightforward measure of the shape of the emission centroid of the inner-Galaxy gas layer is provided by maps of integrated intensities. By suitably choosing the range of integration when making these first-moment maps, gas associated with the core region can be separated from other line-of-sight material. The possibility of such kinematic discrimination represents, of course, a general and major advantage of spectral-line data over data from continuum tracers. The two plots shown in Fig. 50 portray integrated intensities over velocity ranges which exclude emission at $|v| < -100\,\mathrm{km\,s^{-1}}$. Almost all contamination from HI gas in the Galaxy at large at $R > 3\,\mathrm{km\,s^{-1}}$ (where the gas layer is kinematically well behaved) is avoided by this choice of range of integration.

The upper panel of Fig. 50 shows contours of HI intensities integrated over the range $-300 < v < -100\,\mathrm{km\,s^{-1}}$, where no emission is predicted for the case of purely circular motion. That forbidden-velocity emission occurs at all indicates radial motion, and that this emission is absorbed in front of the Sagittarius continuum sources indicates that the radial motions are in the sense of expansion. The forbidden-velocity radiation lies in a coherent distribution, tilted with respect to the equatorial plane $b = 0°$ and crossing the meridian plane $l = 0°$ at negative latitude. A complementary situation is shown for the positive-velocity gas in the lower panel of Fig. 50; here essentially all the gas in the forbidden quadrant at $l < 0°$ also lies in a tilted distribution, crossing the meridian $l = 0°$ at positive latitude. At $l < 0°$ in the upper panel, and at $l > 0°$ in the lower, the integrated radiation is contributed by a combination of both forbidden- and allowed-velocity gas. It seems justified to conclude that this combined material largely describes the same coherent distribution as shown by the unambiguously-forbidden material, especially if one realizes that at $|l| > 6°$, approximately, emission from portions of the lines of sight through the Galaxy at large is not fully isolated from the moment maps.

Kinematic data pertaining to the larger-scale properties of the Galaxy are commonly displayed either in (l, v) maps taken parallel to the plane $b = 0°$, or in (b, v) maps taken parallel to the plane $l = 0°$. For gas distributed as in either of the models shown in Fig. 51, such cuts would not be the most physically significant.

It is possible to determine the spatial orientation of the tilted distribution. Two angles are necessary to specify the tilt of the circular disk shown on the left-hand side of Fig. 51, namely the angle in the plane of the sky between the rotation axis of the disk and the Galaxy at large; a third angle, that specifying the orientation of the semimajor axis of the elliptical streamlines, must be fixed for the bar-shaped structure on the right-hand side of Fig. 51. These angles can be fixed rather confidently because of the sensitivity of simulated spectra to perceived kinematics. The observations also show clearly which side of the tilted structure is nearer the Sun, and which side is further away. The panel on the right-hand side of Fig. 51 illustrates that the nearer side will appear substantially larger because of projection effects. This difference is observed, and shows that the near side of the distribution is that located in the first longitude quadrant (see Liszt & Burton 1980).

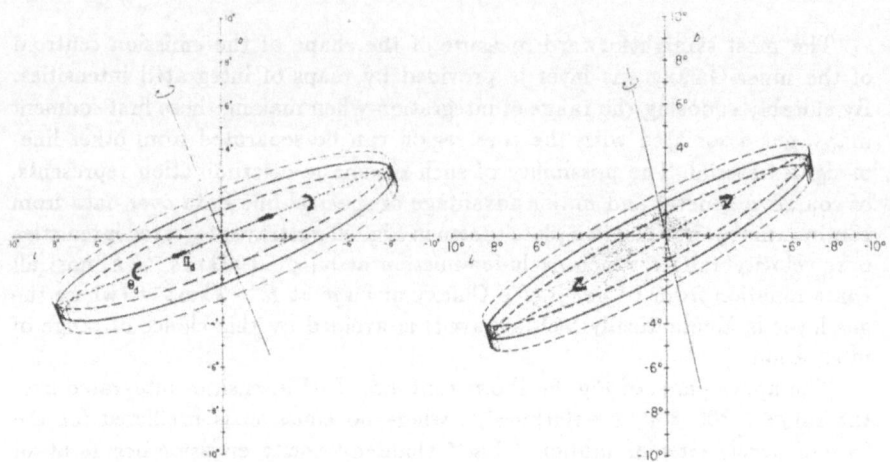

Fig. 51. Appearance of model tilted gas distribution as projected onto the plane of the sky in angular coordinates. The left-hand panel (from Burton & Liszt 1978) shows a model in which the gas layer is held confined to a layer of scale height 0.1 kpc in a circular disk of diameter 3 kpc which is tilted 22° with respect to the plane $b = 0°$ and 78° with respect to the plane of the sky. Within this disk the kinematics involve rotation and expansion of approximately equal amplitude. The right-hand panel (from Liszt & Burton 1980) shows the projection of a tilted gas distribution of elliptical cross section. The motions in this bar-like case are along closed elliptical paths and thereby involve no net flow of mass out of the inner Galaxy. Because the projections of the kinematics are quite similar for both models, both lead to rather similar predictions when confronted with the observations

The *fundamental plane of kinematic symmetry* as well as of gas-density symmetry is inclined some 20° from the galactic equator. Figure 52 shows that substantial kinematic symmetry is displayed if the HI data are sampled in a position, velocity cut through the bulge region along the line $b = -l \tan 22°$, the locus corresponding to the observed centroid of the HI column depths at velocities associated with the inner Galaxy.

The method of deriving the Fig. 41 peaked, circular-velocity rotation curve was justified by the lack of HI emission, observed at $b = 0°$, in the negative-velocity quadrant at $l > 0°$. Sampled along the locus of the tilt, this quadrant is occupied in much the same way as the forbidden-velocity quadrant at $l < 0°$. The abrupt discontinuity as the line of sight crosses $l = 0°$ (which is the required signature of purely circular rotation) is missing when the HI data are sampled along the fundamental, but tilted, plane.

6.4 Modelling Some Aspects of the Kinematics and Distribution of the Gas Layer in the Bulge

6.4.1 Rotation and Expansion Components of the Motions in the Inner Few kpc.
The discussion above indicates that the gas layer in the bulge region does show substantial kinematic regularity if account is taken of the tilt of the fundamental plane of the gas layer and, for the case of the HI, if account is taken

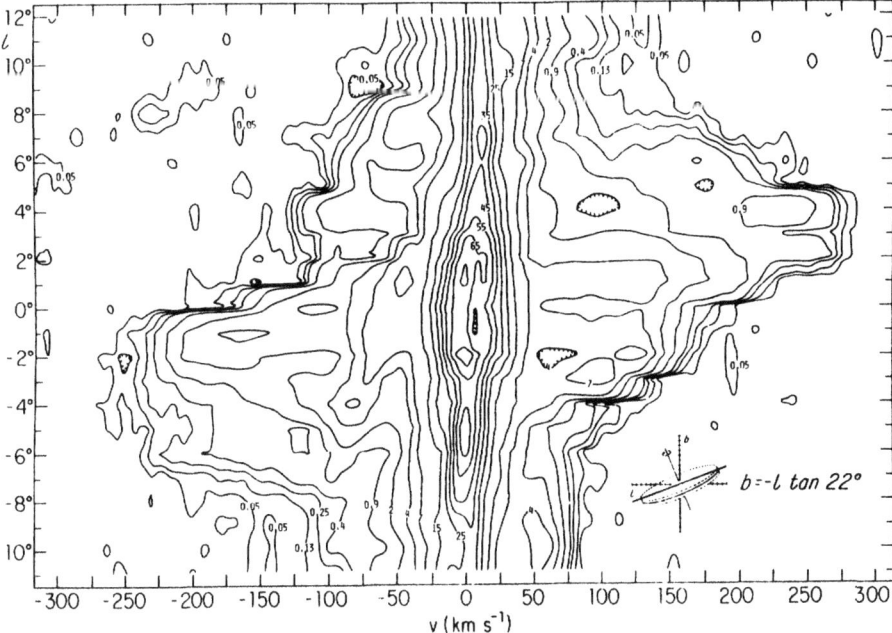

Fig. 52. Longitude, velocity arrangement of HI emission observed from the direction $b = -l \tan 22°$ (from Burton & Liszt 1978). This cut through the inner Galaxy reveals the high degree of kinematic symmetry characterizing the tilted gas distribution; that the complementary forbidden quadrants $l > 0°$, $v < 0\,\mathrm{km\,s^{-1}}$ and $l < 0°$, $v > 0\,\mathrm{km\,s^{-1}}$ are populated in this cut indicates that noncircular motions are important. The profile at $l = 0°$, $b = 0°$ is distorted by absorption and has been excluded from this map. The *skewness* of the emission pattern reflects the rotation component of the motions; the *breadth* of the pattern reflects the expansion component

of the distortions caused by absorption against the sources of continuum radiation distributed throughout the central few hundred pc of the Galaxy. The (l, v) pattern displayed by the HI data as projected onto the plane $b = 0°$ in Fig. 49, or as sampled along the fundamental plane of symmetry in Fig. 52, indicates the rotation component through its *skewness* and the expansion component through its *breadth* near $l = 0°$.

It is possible to separate the rotation component from the expansion component by kinematic modelling. Such a procedure leads to the motions which then, subsequently, must be incorporated into a dynamical interpretation. We have described such model fitting in several papers, confronting observations of HI as well as of CO (e.g. for HI: Burton & Liszt 1978; Liszt & Burton 1980, and for CO: Liszt & Burton 1978; and Burton & Liszt 1992). Figure 53 shows the decomposition of the rotation and expansion components based on modelling of ^{12}CO spectra taken at $|l| < 1°$ by Burton & Liszt (1992). Although the longitude range of the CO data is not large, the distribution of the gas sampled is such that material from the entire bulge region, spread over lines of sight several kpc long, enter the modelling. There is abundant evidence found in de-

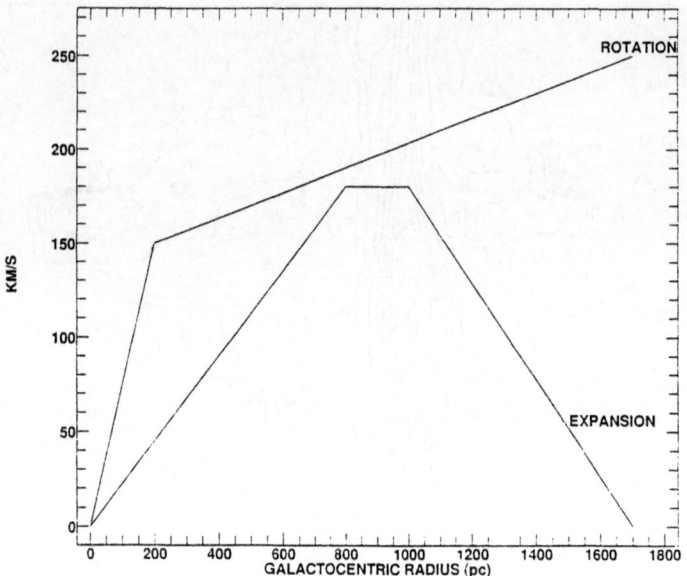

Fig. 53. Decomposition of the rotation and expansion components of motion in the tilted, or warped, molecular gas layer pervading the region encompassed by the galactic bulge (from Burton & Liszt 1992). Although this decomposition followed from modelling of ^{12}CO observations, simulations of HI data based on these motions account for many aspects of that data. The rotation plotted here rises more slowly with distance from the galactic center then would be the case if noncircular motions were excluded. The motions plotted are derived from line-of-sight perceptions: it is not unlikely that these perceived motions in fact represent a decomposition of motions along closed elliptical orbits, involving no net flux of mass out of the inner regions. Essentially all the neutral gas in the gas layer lying between a few hundred pc and a few kpc from the center partakes in these motions

tailed comparison of HI spectra with their CO counterparts that the atomic gas and the molecular gas are globally well-mixed. Thus it is not surprising that the kinematic decomposition shown in Fig. 53 leads to simulated HI profiles which mimic many aspects of the observations.

Figure 54 shows two simulated (l, v) maps of HI emission, calculated in the presence of the velocity field shown in Fig. 53. Modelling requires that, in addition to the kinematics, a number of physical parameters be specified. The profiles entering Fig. 54 utilized parameters which have proven useful in simulating many aspects of the HI distribution, including mean midplane density of the smoothly-filled gas layer of $0.3\,\mathrm{cm}^{-3}$, spin temperature of $135\,\mathrm{K}$, vertical scale height of the gas layer of $50\,\mathrm{pc}$, and internal gas dispersion of $8\,\mathrm{km\,s}^{-1}$. We note here that essentially *all* of the structures which can be identified in the modelled spectra show widths which are *much* greater than the input value of the gas dispersion; this is a consequence of the large range of velocities sampled along each line of sight.

The upper panel of Fig. 54 shows the simulated HI (l, v) diagram resulting from a straightforward solution of radiative transfer in the absence of any

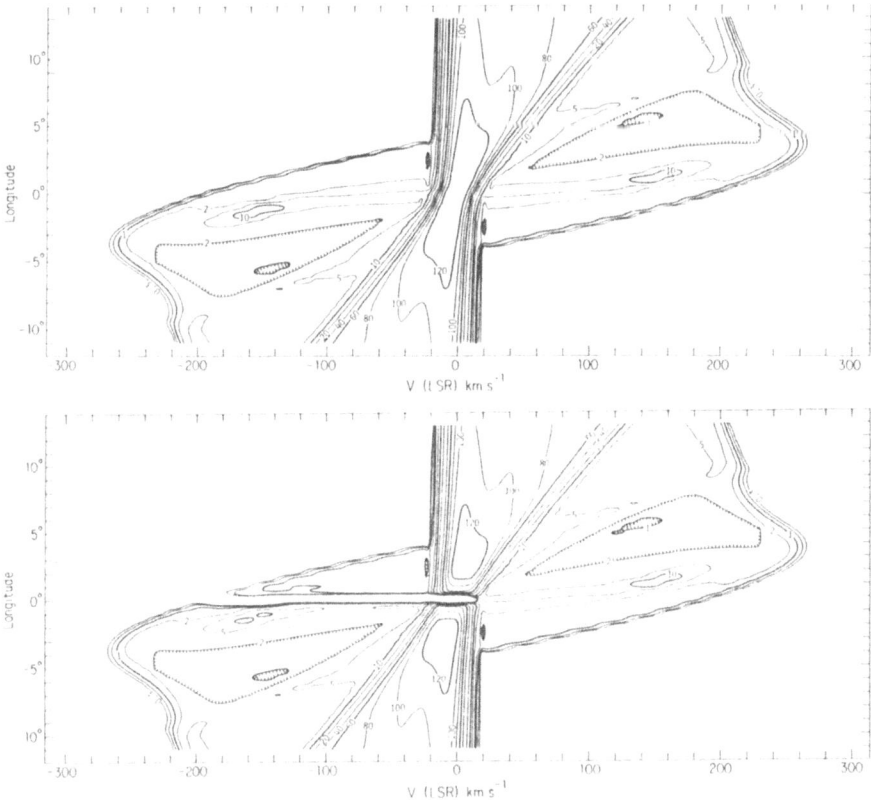

Fig. 54. Two longitude, velocity maps showing simulated HI intensities at $b = 0°$. Both maps were constructed from synthetic data modelled with the kinematics shown in Fig. 53 (and the simple gas parameters described in the text.) The upper map was calculated ignoring continuum radiation. The lower map was calculated in the presence of a simulated contribution of continuum radiation confined in the core of the modelled system. The upper map may be compared with the projected map shown in Fig. 49, in which the effects of absorption have been suppressed, and with the map in Fig. 52, containing data along the fundamental plane of the inner-Galaxy gas. The lower map may be compared with the observed situation, contaminated by absorption effects, shown in Fig. 44

continuum background. This modelled situation may be compared with the observed one shown in Fig. 49, in which the effects following from absorption against the continuum background have been suppressed, and with that shown in Fig. 52, which sampled the HI gas layer on its tilted fundamental plane of symmetry. The lower panel of Fig. 54 shows HI simulated spectra generated in the presence of a continuum background. The radio continuum was represented by the actual excess system temperature which occurred on the NRAO 140-foot telescope while the observations shown in Fig. 44 were being made. The lower map may be compared with the Fig. 44 observed one. It is, in particular, interesting to compare the shape of the terminal-velocity envelope encompassing the gas emission *observed* with the *simulated* situation shown in Fig. 54. At positive velocities, the continuity of the envelope is provided by the strong contribution of noncircular motion; at negative velocities, the behavior of the emission pattern within a degree or two of $l = 0°$ must be attributed to the presence the complex of continuum sources dominated by Sgr A and B. The filling of the two forbidden-velocity longitude quadrants in a symmetric way is proof of noncircular motion; the angular extent of the forbidden-velocity pattern indicates that noncircular motions are important over several kpc of the inner Galaxy. The rotation perceived determines the skewness of the emission pattern; the nonradial motion perceived determines the breadth of the pattern. The regularity of the pattern shows that essentially all of the inner-Galaxy gas lying between several hundred pc and several kpc partakes of the combined radial and circular motions: essentially all of this gas is confined to the tilted distribution.

Thus if account is taken of the tilt of the fundamental midplane of the inner-Galaxy neutral gas lying between several hundred pc and about $0.3R_0$, and of the role of absorption against continuum sources in the core which dominates the appearance certainly of the OH and H_2CO tracers, but also in an important way of HI, then the evidence for a pervasive noncircular component, in addition to a circular one, becomes convincing. The decomposition of the motions shown in Fig. 53 shows that the noncircular component and the circular one are of approximately the same amplitude over several kpc. The circular component increases much more slowly with R than is the case for the putative rotation curve plotted in Fig. 41. It seems clear that that curve, with its rapid rise at small R and definite peak between 1 and 2 kpc, as derived under the assumption of circular motion and with far-reaching consequences for models of the galactic mass distribution and dynamics, cannot be correct.

6.4.2 Comments on the Ejection v. Field Hypothesis for Interpreting Discrete Spectral Features; Comments on the Large Apparent Widths of These Features. HI observations toward the galactic bulge region have shown a large number of apparently anomalous spectral features which have been interpreted as evidence for ejecta produced by violent activity in the galactic nucleus (e.g. van der Kruit 1970; Sanders & Wrixon 1972; Cohen 1975). Oort (1977) reviewed the data on these apparently anomalous structures, and focussed on the necessity of deciding between an explanation for these structures in terms of violent ejection or in terms of their response to the characteristics of the gravitational field holding sway in the core. If the *ejection* hypothesis were to be correct, then the peculiar spectral features would be provided by isolated clouds moving at high velocities on ballistic orbits after ejection from the nucleus, and which are observed superimposed on a background distribution of gas in pure rotation. (Initial speeds of $\simeq 750\,\mathrm{km\,s^{-1}}$ would be required for the ejecta to reach the z-heights observed.) If, on the other hand, the *field* hypothesis were to be correct a plausible combination of motions would have to be found which satisfy the peculiarities in the observations.

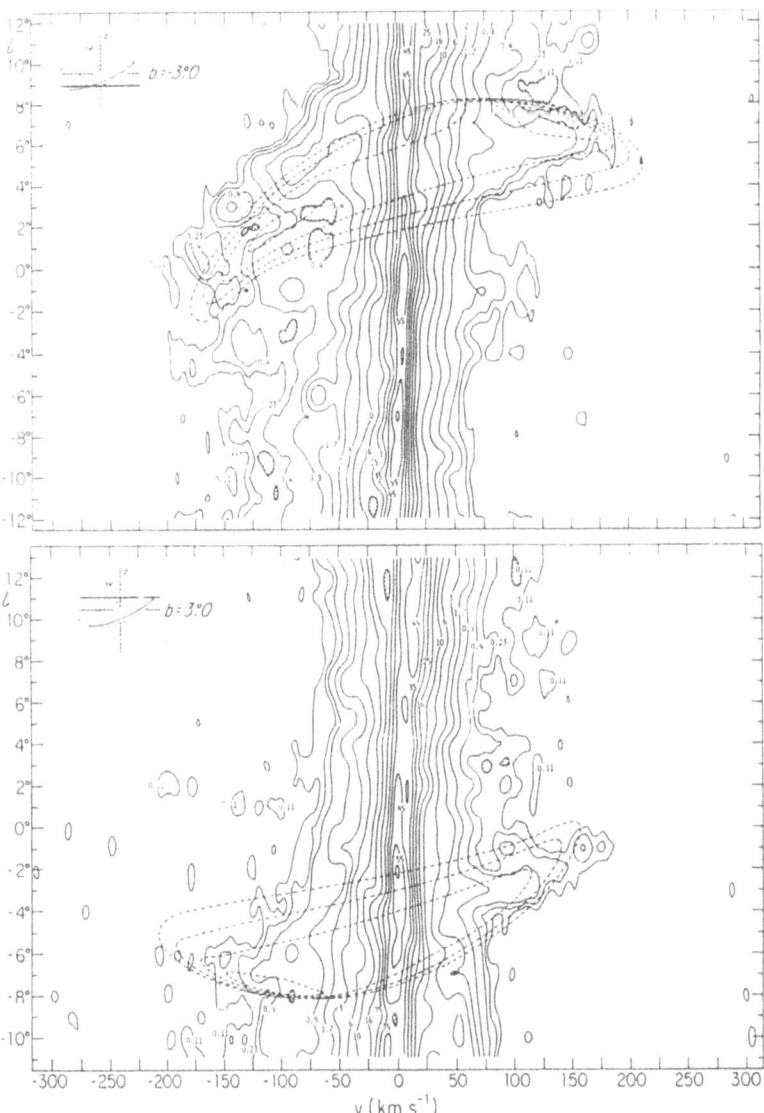

Fig. 55. Emission in the (l, v) plane at two representative latitudes, comparing the observations with simulated spectra calculated for the indicated cuts through a smooth, tilted gas distribution whose structure is indicated by the left-hand diagram of Fig. 51 and whose motions are (approximately) those shown in Fig. 53 (from Burton & Liszt 1978). The structural and kinematic description of the tilted disk is very simple, but the manner in which the gaseous disk is perceived in simulated observations results in apparently anomalous spectral features: these features have observed counterparts. It seems that the observed features are not spatially or kinematically discrete, and do not indicate ejection from the nucleus. (Only emission from the tilted gas distribution is including in the modelled spectra; emission from the Galaxy at large dominates the observed situation at $|v| < 100\,\mathrm{km\,s^{-1}}$, and can be removed in the mind's eye)

119

The characteristics of the anomalous-velocity HI features located between a few hundred pc and a few kpc from the galactic center can be summarized as follows:

1. The anomalous-velocity features, and the core HI and CO gas in general, lie in a tilted distribution.
2. The anomalous-velocity features, and the core gas in general, are confined within a definite kinematic envelope.
3. The anomalous-velocity features show substantial continuity in position and velocity.
4. The principal anomalous-velocity features occur in pairs at opposite velocities in the same (l, b) quadrant.
5. The anomalous-velocity features, and the core HI and CO gas in general, show exceptionally large velocity widths.

These characteristics of the observations are largely accounted for by the kinematic and structural vagaries along lines of sight sampling a pervasive gas layer tilted as in Fig. 51 and moving as in Fig. 53 (see e.g. Burton & Liszt 1978, 1980, and 1992; Liszt & Burton 1978, 1980). Figure 55 shows examples of two (l, v) cuts through the HI data, both of which contain apparently anomalous spectral features. Superposed on the observations are simulated HI spectra sampling the tilted distribution, and which seem to adequately reproduce the observed situation. Similar modelling, done with the same structural and kinematic parameters represented in Fig. 53, can account for most of the apparently anomalous features. Because of the wealth of evidence for the gas-layer structure and kinematics, and because of the success of predictions based on these gas-layer properties, it seems reasonable to favor the field hypothesis over the ejection one. Thus it does not seem necessary to invoke a large number of unrelated violent events, each separately collimated, to account for the data.

We comment separately on the large total velocity widths commonly shown by all of the gas tracers of the inner region, including the HI, CO, OH, and H_2CO observations portrayed in the figures of this chapter. The spectral features of these tracers are commonly found to extend over some $50 \, \mathrm{km \, s^{-1}}$ or even $100 \, \mathrm{km \, s^{-1}}$ in those cases where apparently isolated spectral features can be identified. Understanding the physical state of the core gas requires deciding if these measured widths are *intrinsic* to spatially separate features (see Spergel & Blitz 1992). We mentioned above the evidence that the perceived gas distribution in the core region is quite pervasive, and that apparently spatially separate features are generally artifacts of the sampling vagaries. We have argued in the papers cited in the preceding paragraph that the large velocity widths (in CO as well as in HI) are also artifacts of the sampling, and thus are *not* intrinsic indicators of physical properties of gas clumps. Several points may be given in support of this argument. First, we note that the most valid measure of the true gas dispersion is given by the breadth of the wing measured along the terminal-velocity locus. This locus is determined by a kinematic turnover, and is thus quite unambiguous as a measure of dispersion. Fitting a curve of Gaussian shape to this wing almost anywhere in the inner-Galaxy material returns a dispersion of order 5 to $8\,\mathrm{km\,s^{-1}}$, both for the HI and for the CO data. In other words, the spectral features of some 50 or $100\,\mathrm{km\,s^{-1}}$ total width always have a very non-Gaussian shape: they are indeed broad, but their edges do not show the shallow wings which would be consistent with large intrinsic dispersions. Second, we note that the run of terminal-velocities with position is very smooth and well behaved. We would expect that individual, high-dispersion clumps whose central velocity happened to lie within, say, 10 or $20\,\mathrm{km\,s^{-1}}$ of the terminal-velocity envelope would cast their high-dispersion, shallow, wings to velocities beyond this envelope. This situation is not observed. The envelope is smooth, regular, and otherwise well behaved. Finally, we note that the modelling of the tilted distribution which accounts for so many of the *positional* characteristics of the gas also accounts for many of the observed *kinematic* characteristics, including the large widths, but non-Gaussian shapes, the narrow edges at the kinematic cutoffs, and the smoothness of the terminal-velocity locus.

6.4.3 Questioning the Concept of a "Rotation Curve" for the Not-Only-Rotating, Inner-Galaxy Gas. In the above we argued that the rotational component of motion of the inner-Galaxy gas may be isolated, even if there is no gas moving in closed circular orbits. The behavior, in particular the slope, of the terminal-velocity envelope, as determined in our Galaxy as well as in other galaxies, can be a useful comparative kinematic measure, but it does not provide what is usually thought of as a "rotation curve" except under certain circumstances. The kinematic decomposition of perceived motions is useful primarily as an indicator of the response of the gas to the underlying potential, but the circular component which results from such a decomposition cannot be considered as a "rotation curve".

In this regard we remark that the circular component in the kinematic decomposition also cannot be expected to coincide with the equilibrium circular velocity which would follow a radial unfolding of the distribution of light from the stellar population (see Binney et al. 1991). The slow rise of the circular component plotted in Fig. 53 is thus not consistent with the very sharply peaked radial distribution of the near-infrared emission at $2\,\mu\mathrm{m}$ from the galactic core (Sanders & Lowinger 1972; Allen et al. 1983).

It is in any case not clear if there is a single variation of the kinematic components which holds for all gaseous constituents of the gas layer. Different families of closed orbits are imaginable; presumably these would be constrained so that stream-lines would not cross as far as the gas constituents are concerned, but a variety of stellar kinematics could co-exist with the gaseous. It also seems likely that a single variation of the kinematics need not hold on all scales. The kinematics of the gas layer in the Galaxy at large clearly reflects a different dynamical situation than prevails in the bulge region; similarly, the kinematics within the innermost several hundred pc cannot be predicted by an extrapolation inward of the situation at larger radii.

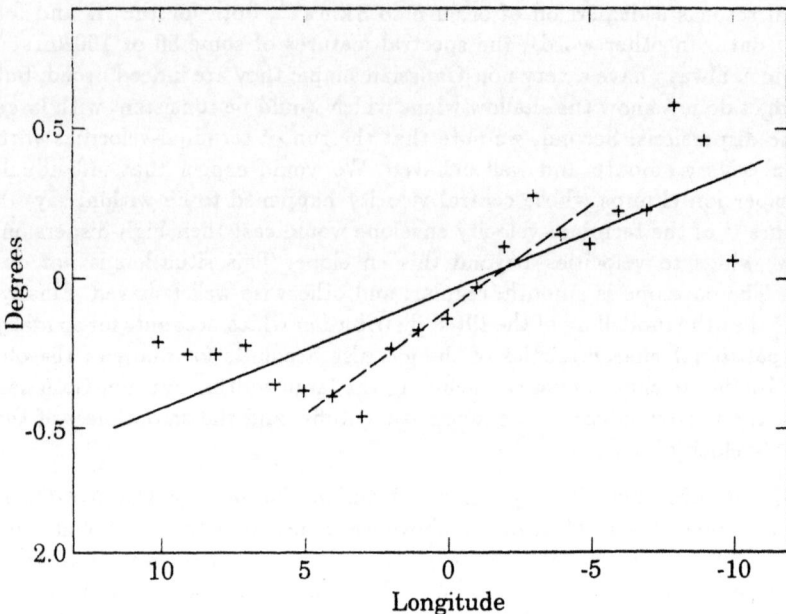

Fig. 56. Location on the plane of the sky of the centroid of the surface brightness of emissivity measured around the galactic center at 2.4 μm from balloons by Matsumoto et al. (1982) and attributed to the old stellar population (from Blitz & Spergel 1991). The dashed line refers to data at $|l| < 5°$; the full-drawn line, to $|l| < 10°$

6.5 Some Brief Remarks on the Form of the Potential Dominating the Motions in the Bulge Region, and on the Situation in M31

It seems plausible that the motions in the region dominated by the potential of the bulge might enjoy a degree of dynamic independence from the motions elsewhere in the Galaxy, for example in the region dominated by the probably rather spherical potential of the outer Galaxy. Substantial effort has gone into predicting the response of the gas to a bar potential (see references in Sect. 6.1, above). Some of the modelling has demonstrated that the observed kinematics are consistent with an assumed bar, but direct evidence for such a potential shape has only recently been offered. Blitz & Spergel (1991) have interpreted the observations at 2.4 μm made by Matsumoto et al. (1982), and identified with the old stellar population of the bulge, in terms of a stellar bar. The orientation of the bar found by Blitz & Spergel is such that the near side is located at positive longitudes. Figure 56 shows that the entire bar structure is tilted with respect to the plane $b = 0°$. Both regarding its orientation and its tilt the bar shows similarity to what has been found from the combined structural and kinematic information in the gas tracers. Weinberg (1991) analyzed IRAS data on AGB stars, and found evidence for a stellar bar with a semi-major axis of 5 kpc length. Because Weinberg's analysis found it necessary to discriminate against IRAS sources more than 3° from the equator, it is not clear if that large structure refers specifically to the bulge.

The class of OH/IR objects associated with the bulge region holds substantial promise for ascertaining not only the shape of the stellar potential, but also aspects of its kinematics (see review by de Zeeuw 1992). Stars seen in the bulge region show a large velocity dispersion, of order $100 \, \mathrm{km \, s^{-1}}$; this blurs much of the structural detail. The large intrinsic velocity dispersion of the stars makes it more difficult to interpret the breadth of the (l, v) distribution in terms of systematic nonradial motions then was the case for the low-dispersion gas tracers, but it remains possible to interpret the general *skewness* of the OH/IR (l, v) pattern in terms of the rotation component. It is interesting to note that the skewness of the (l, v) plot of OH/IR stars observed in the inner degree of the Galaxy by Lindqvist et al. (1992) shows a rise of the inferred rotation component consistent with the gradual one observed for the gas tracers.

It appears that misalignment of the bulge kinematics with that of the parent galaxy is not uncommon. Bertola et al. (1991) describe the bulge and disk alignments of 32 spirals and state that most of the bulges show not only misalignment between disk and bulge apparent major axes, but also internal isophote twisting which they consider indicative of a triaxial shape. Rubin et al. (1977) describe the particularly striking misalignment in the inner kpc of the galaxy NGC 3672.

It is often tempting to seek comparisons between the gas layer properties of the Milky Way with those observed in M31 and other nearby systems. The morphological similarities of the outer HI layers of M31 and the Milky Way have been commented on frequently: the outer layers of both galaxies are warped, and by about the same amount, away from the equator; both layers flare to increasing thickness with increasing R; and both show no precession of the line of nodes of the warp. There also appear to be similarities in the morphologies of the inner regimes. The stellar light from the bulge in M31 has long been known to be oriented differently from isophotes at larger distances. Light et al. (1974) found that the inner core of M31 shows a flattened nucleus which is tilted relative to the major axis of the main system. They concluded, as had Johnson (1961) earlier on the basis of the form of isophotes at somewhat larger radii, that the bulge of M31 is a distinct structure, to be distinguished dynamically from the rest of that galaxy.

Recently Ciardullo et al. (1988) described the distribution and kinematics of the gas confined within the bulge region of M31. The gas in the bulge of M31 lies in a disk whose plane is tipped relative to that of the general stellar disk of M31. In addition, the inner gas disk is warped such that the gas south of the nucleus is viewed closer to face-on than the gas in the northern half of the galaxy. Furthermore, they show that the velocities along the near side of M31's minor axis are not zero, but negative, indicating the presence of noncircular expansion motions. It seems plausible to interpret the motions in the bulge region of M31 in terms of motions in closed orbits because of the systematic nature of the motions and because of the large net flux which would be required by true expansion. For the case of the gas layer in the bulge region of the Milky

Way, a net flux of about $4\,M_\odot$ per year would be required if the perceived nonradial motions were to be interpreted as due to true expansion. Although it seems more plausible to interpret the perceived motions in terms of closed orbits, we do note, however, that the fate of matter ejected from the bulge stars in M31, or in the Milky Way, and subsequently exposed to a galactic wind from the stars packed into the bulge at high space density, is not yet clear.

References

6.1 Allen, D.A., Hyland, A.R., Jones, T.W. 1983, *Monthly Notices Roy. Astron. Soc.* **204**, 1145

6.2 Bahcall, J.N., Schmidt, M., Soneira, R. 1982, *Astrophys. J.* **258**, L23

6.3 Bertola, F., Vietri, M., Zeilinger, W.W. 1991, *Astrophys. J.* **374**, L13

6.4 Binney, J.J., Gerhard, O.E., Stark, A.A., Bally, J., Uchida, K.I. 1991, *Monthly Notices Roy. Astron. Soc.* **252**, 210

6.5 Blitz, L., Spergel, D.N. 1991, *Astrophys. J.* **379**, 631

6.6 Burstein, D., Rubin, V.C. 1985, *Astrophys. J.* **297**, 423

6.7 Burton, W.B. 1988, in *Galactic and Extragalactic Radio Astronomy*, G.L. Verschuur, K.I. Kellermann, eds., Springer-Verlag, New York, p. 295

6.8 Burton, W.B., Gordon, M.A. 1978, *Astron. Astrophys.* **63**, 7

6.9 Burton, W.B., Liszt, H.S. 1978, *Astrophys. J.* **225**, 815

6.10 Burton, W.B., Liszt, H.S. 1983, *Astron. Astrophys. Suppl. Ser.* **52**, 63

6.11 Burton, W.B., Liszt, H.S. 1992, *Astron. Astrophys. Suppl. Ser.* in press

6.12 Casertano, S., van Gorkom, J.H. 1991, *Astron. J.* **101**, 1234

6.13 Ciardullo, R., Rubin, V.C., Jacoby, G.H., Ford, H.C., Ford Jr., W.K. 1988, *Astron. J.* **95**, 438

6.14 Clemens, D.P. 1985, *Astrophys. J.* **295**, 422

6.15 Cohen, R.J. 1975, *Monthly Notices Roy. Astron. Soc.* **171**, 659

6.16 Cohen, R.J., Few, R.W. 1976, *Monthly Notices Roy. Astron. Soc.* **176**, 495

6.17 Cohen, R.J., Few, R.W. 1981, *Monthly Notices Roy. Astron. Soc.* **194**, 711

6.18 de Zeeuw, P.T. 1992, in *Proceedings of IAU Symp. 149 "The Stellar Populations of Galaxies"*, in press

6.19 Gerhard, O.E., Vietri, M. 1986, *Monthly Notices Roy. Astron. Soc.* **223**, 337

6.20 Johnson, H.M. 1961, *Astrophys. J.* **133**, 309

6.21 Kerr, F.J. 1967, in *Proceedings of IAU Symp. 31 "Radio Astronomy and the Galactic System"*, H. van Woerden, (ed.), Academic Press, London, p. 239

6.22 Light, E.S., Danielson, R.E., Schwarzschild, M. 1974, *Astrophys. J.* **194**, 257

6.23 Lindqvist, M., Winnberg, A., Habing, H.J., Matthews, H.E. 1992, preprint. (See also Lindqvist's Ph.D. Thesis, Chalmers Institute of Technology 1992)

6.24 Liszt, H.S. 1988, in *Galactic and Extragalactic Radio Astronomy*, G.L. Verschuur, K.I. Kellermann, eds., Springer-Verlag, New York, p. 359

6.25 Liszt, H.S., Burton, W.B. 1978, *Astrophys. J.* **226**, 790

6.26 Liszt, H.S., Burton, W.B. 1980, *Astrophys. J.* **236**, 779

6.27 Liszt, H.S., Burton, W.B., van der Hulst, J.M. 1985, *Astron. Astrophys.* **142**, 245

6.28 Matusumoto, T., Hayakawa, S., Koizumi, H., Murakawa, H. 1982, in *AIP Conf. No. 83 "The Galactic Center"*, G. Riegler, R. Blandford, (eds.), American Institute of Physics, New York, p. 48

6.29 Mizutani, K., et al. 1991, preprint

6.30 Mulder, W.A., Liem, B.T. 1986, *Astron. Astrophys.* **157**, 148

6.31 Okuda, H., Shibai, H., Nakagawa, T., Matsuhara, T., Maihara, T., Mizutani, K., Kobayashi, Y., Hiromoto, N., Low, F.J., Nishimura, T. 1989, in *Proceedings of IAU Symp. 136 "The Center of the Galaxy"*, M. Morris, (ed.), Kluwer Academic Pub. Co., Dordrecht, p. 145

6.32 Oort, J.H. 1977, *Ann. Rev. Astron. Astrophys.* **15**, 295

6.33 Peters, W.L. 1975, *Astrophys. J.* **196**, 617

6.34 Reich, W., Fürst, E., Steffen, P., Reif, K., Haslem, C.G.T. 1984, *Astron. Astrophys. Suppl. Ser.* **58**, 197

6.35 Roberts, W.W. 1979, in *Proceedings of IAU Symp.* **84** *"The Large-Scale Characteristics of the Galaxy"*, W.B. Burton, (ed.), Reidel Pub. Co., Dordrecht, p. 175

6.36 Rougoor, G.W., Oort, J.H. 1960, *Proc. Nat. Acad. Sci., U.S.A.* **46**, 1

6.37 Rubin, V.C. 1983, in *Kinematics, Structure, and Dynamics of the Milky Way*, W.L.H. Shuter, (ed.), Reidel Pub. Co., Dordrecht, p. 379

6.38 Rubin, V.C., Thonnard, N., Ford, W.K., Jr. 1977, *Astrophys. J.* **217**, L1

6.39 Sanders, D.B., Solomon, P.M., Scoville, N.Z. 1984, *Astrophys. J.* **276**, 182

6.40 Sanders, R.H., Lowinger, T. 1972, *Astron. J.* **77**, 292

6.41 Sanders, R.H., Wrixon, G.T. 1972, *Astron. Astrophys.* **18**, 467

6.42 Sanders, R.H., Wrixon, G.T. 1973, *Astron. Astrophys.* **26**, 365

6.43 Schmidt, M. 1985, in *Proceedings of IAU Symp.* **106** *"The Milky Way Galaxy"*, H. van Woerden, R.J. Allen, W.B. Burton, (eds.), Reidel Pub. Co., Dordrecht, p. 75

6.44 Shane, W.W. 1972, *Astron. Astrophys.* **16**, 118

6.45 Shibai, H., et al. 1991, *Astrophys. J.* **374**, 522

6.46 Simonson III, S.C. 1976, *Astron. Astrophys.* **46**, 261

6.47 Simonson III, S.C., Mader, G.L. 1973, *Astron. Astrophys.* **27**, 337

6.48 Sinha, R.P. 1978, *Astron. Astrophys.* **69**, 227

6.49 Sinha, R.P. 1979, in *Proceedings of IAU Symp.* **84** *"The Large-Scale Characteristics of the Galaxy"*, W.B. Burton, (ed.), Reidel Pub. Co., Dordrecht, p. 341

6.50 Spergel, D.N., Blitz, L. 1992, *Nature*, submitted

6.51 van Albada, G.D. 1985, in *Proceedings of IAU Symp.* **106** *"The Milky Way Galaxy"*, H. van Woerden, R.J. Allen, W.B. Burton, (eds.), Reidel Pub. Co., Dordrecht, p. 547

6.52 van der Kruit, P.C. 1970, *Astron. Astrophys.* **4**, 462

6.53 van Woerden, H., Rougoor, G.W., Oort, J.H. 1957, *Compt. Rend. Acad. Sci., Paris* **244** 1961

6.54 Weinberg, M.D. 1991, *Astrophys. J.* **384**, 81

6.55 Yuan, C. 1984, *Astrophys. J.* **281**, 600

7 The Warped and Flaring Layer of Atomic and Molecular Gas in the Outer Galaxy

Abstract. The Milky Way serves in several regards as a prototype of a warped galactic system. Our embedded perspective is, in fact, ideal for studying the morphology of the outer Galaxy. The improved sensitivity and positional- and velocity-coverage of recent HI data, together with improved confidence in the velocity field at large R, led to an improved description of the HI warp, although the essentials of the description dating from the late 1950's remain unchanged. In the outer Galaxy, the v-to-R transformation is single valued, so the heliocentric (l, b, v) data cube can be transformed to the more directly relevant galactocentric (R, θ, z) coordinates. HI can be traced to distances greater than $3R_0$, and molecular clouds can be traced to more than $2R_0$. Both tracers yield a quantitative description of the deviation of the density midplane from the equator plane $b = 0°$, the flare to increasing thickness at larger R, the location of the line of nodes of the warp, and the radial variation of the projected surface densities. The line of nodes of our Galaxy's warp is quite straight, showing no evidence of precession or of winding under rotational shear. The molecular clouds found with embedded IRAS heat sources at large R indicate that star formation occurs well beyond the conventionally-defined optical disk. The ensemble of molecular clouds shows the same warped shape and flaring thickness as shown by the outer-Galaxy HI layer. Preliminary evidence suggests that the chemical and physical circumstances in the warped outer reaches of the Galaxy differ from those in the inner regions. The parameters specifying the global, flaring warp in M31 are remarkably similar to those specifying the warp in our own Galaxy. Restrictive viewing requirements suggest that even substantial, regular external warps might be difficult to recognize from the projected shapes alone, without kinematic information. It seems difficult to argue against the proposition that *all* galactic disks are globally warped.

7.1 Introduction; the Shape of the Outer Galactic Disk

The early Leiden and Sydney surveys of the galactic 21-cm radiation led to the discovery in the 1950's that the gas layer in the outer parts of the Milky Way systematically deviates from the plane $b = 0°$, which had itself been *defined* as the flat centroid of the HI emission in the inner parts of our Galaxy. The work by Burke (1957), Kerr (1957), Westerhout (1957), Oort et al. (1958), and Gum et al. (1960) led to a description of the shape of the outer HI layer which recognized that the gas layer is warped above the plane $b = 0°$ in the part accessible from the Northern hemisphere, and below it in the part accessible from the Southern hemisphere; that the velocity extent of the layer is less in the northern data than in the southern; that the amplitude of the warp carries it to higher angular distances from the equator in the northern data than in the southern; and that the thickness of the gas layer increases regularly toward larger galactocentric radii. More modern data have led to a much more detailed description of the outer gas layer, but the essentials of the original morphological description have not changed.

Recent investigations of the global warp have been motivated by several considerations. It has become clear that most of the galactic mass is invisible to direct observation, and that this unobserved material dominating the galactic potential can be traced most effectively by study of the form of the outer gas layer and of its motions. It has also become evident that the warped outer layer is not a curiosity found only in our local system, but is an intrinsic property of the dynamics governing many, indeed probably most, if not all, spirals. Additional motivation is provided by the realization that the physical circumstances, including metallicity, differ for the general interstellar environments characterizing the outer reaches compared to those found in the inner, more commonly star-forming, regions.

Although many Milky Way studies are hindered by our embedded location, this embedded perspective is ideal for study of the morphology of our own warped layer. In general, galactic warps are most easily identified and studied in systems which are viewed edge-on, and with the line of nodes of the warp oriented along the line of sight. External systems will, statistically, not commonly present such viewing geometry; but it remains in all cases difficult to establish the inclination and location of the line of nodes as accurately for external systems as can be done in the Milky Way.

Several recent advances facilitate derivation of the morphology of the warped outer Milky Way layer. The first advance concerns the accuracy with which the rotation curve is known at large distances, and thus the accuracy with which distances may be determined. Kulkarni et al. (1982) reanalyzed the Weaver & Williams (1973) Northern-hemisphere HI data using the slightly-rising rotation curve which had been found by Blitz et al. (1980) from velocities of molecular clouds associated with HII regions at optically-determined distances. The earlier descriptions of the warp had, of course, been based on the distance scale appropriate to an outer rotation curve falling rapidly, in an approximately Keplerian fashion. Use of the new rotation curve led to a substantial upward revision of the scale of the outer Galaxy (see review by Blitz et al. 1983).

A second advance concerns the extension of observational material covering the parts of the Milky Way only visible from the Southern hemisphere. The HI data take with the Parkes 18-m telescope by Kerr et al. (1986) complemented the Northern-hemisphere data of the Weaver & Williams (1973) survey to the extent that the full 360° sweep of longitude was represented within the latitude range $-10° \leq b \leq +10°$. The analysis by Henderson et al. (1982) made use of the combined northern and southern $|b| \leq 10°$ data, and a flat rotation curve $\left(\Theta_0(R>R_0=10\,\mathrm{kpc}) = 250\,\mathrm{km\,s^{-1}}\right)$ to produce the first detailed global description of the warped outer-Galaxy HI layer.

A third advance lies in the current availability of more sensitive HI data, with better kinematic and spatial coverage. The galactic warp is sufficiently severe that its signature extends to latitudes well beyond the $\pm 10°$ limits of the Weaver & Williams (1973) and Kerr et al. (1986) surveys. (The kinematic coverage of the Heiles & Habing (1974) survey, which does extend to higher latitudes, does not encompass the velocities of the warp signature.) The Leiden-Green Bank survey of Burton (1985) covers northern latitudes to $b \simeq 33°$, and thus samples most of the more extreme portions of the warp. Burton & te Lintel Hekkert (1986) combined the Leiden-Green Bank data with the Parkes data and with other low-latitude northern data, and presented cuts through the composite data cube which allow a quantitative description of the parameters describing the warped layer, its flare to increasing thickness at larger R, and the location of the line of nodes of the warp. Diplas & Savage (1991) derived the warp parameters using the Stark et al. (1992) HI material. The Stark et al. survey reliably probes lower intensity levels than the composite dataset used by Burton & te Lintel Hekkert, although its kinematic and spatial coverages are less detailed.

A fourth advance concerns the new possibility to probe to large R using tracers other than HI. Some results are mentioned below which follow from recent observations of molecular clouds found to distances as large as $R \simeq 2R_0$, most of which show evidence of star formation. The molecular and infrared results extend those on the global properties found in HI data to more detailed information pertaining to the physical conditions of the interstellar environment in the far outer reaches of the Galaxy.

A useful general reference to investigations of the outer parts of the Milky Way is the Kerr-symposium volume "The Outer Galaxy", edited by Blitz & Lockman (1988). The proceedings of the Pittsburgh workshop on "Warped Disks and Inclined Rings around Galaxies", edited by Casertano et al. (1991), deals with warps as a common characteristic of galaxies and with the dynamics involved.

7.2 The Shape of the Warped and Flaring HI Layer Extending to the Far Outer Galaxy

The most straightforward way to show that the outer-Galaxy HI layer is warped and flared involves plots following directly from the observations, after conversion from the heliocentric (l, b, v) data cube to one representing galactocentric cylindrical coordinates. This conversion is possible for outer-Galaxy material because radial velocities measured there correspond to galactocentric distances in a single-valued manner for plausibly well-behaved galactic kinematics. After the v-to-R conversion, the observed intensities are contained in cylindrical coordinates, (R, θ, z), representing galactocentric radius and azimuth, and vertical distance from the plane $b = 0°$. Cuts through the (R, θ, z) data cube (or rather data "cylinder") are intuitively more accessible than cuts through the directly observed (l, b, v) dataset (see, for example, Fig. 20 in Chap. 3 of these notes).

Several figures here display observations from the Leiden-Green Bank survey combined with other material in the composite used by Burton & te Lintel Hekkert (1986). Burton & te Lintel Hekkert give their results alternatively for the case of a flat rotation curve, as well as for a slightly rising one. The differences between the two sets of cuts are not entirely trivial, because of the sensitivity of the analysis to the perceived kinematics. (This point is also illustrated by Lockman (1988; see his Fig. 2), who compares the radial dependence of projected HI surface densities derived using a flat rotation curve with the substantially-different dependence derived from the same data using a curve which rises slightly, but which differs from the flat curve by at most 14%.) The galactocentric, cylindrical-coordinate (R, θ, z) dataset of Burton & te Lintel Hekkert (1986) contains HI volume-density information at spacings of 1° in θ, and at intervals of 250 pc in R and of 100 pc in z. (Unless mentioned otherwise, the remarks below correspond to the galactic-constant values $\Theta_0 = 250\,\mathrm{km\,s^{-1}}$ and $R_0 = 10\,\mathrm{kpc}$, and to a flat rotation curve, $\Theta(R > R_0) = \Theta_0$.)

7.2.1 Cylindrical Cuts Through the HI (R, θ, z) Dataset at Constant Radii.

The HI warp is most easily described pictorially. Figure 57 shows cuts through the HI (R, θ, z) dataset at progressively larger values of R. The galactic warp is revealed by the behavior of the density-centroid pattern plotted in (θ, z) coordinates. The HI gas layer remains quite flat until about $R = 12\,\mathrm{kpc}$ from the galactic center. On the cylinder at $R = 13\,\mathrm{kpc}$ (shown in the upper panel of Fig. 57a) the band of HI emission deviates perceptibly, but not dramatically, from $z = 0\,\mathrm{pc}$. Until $R \simeq 16\,\mathrm{kpc}$ from the center the amplitude of the warp increases approximately linearly with increasing R, and approximately equally in the hemisphere centered on azimuth $\theta \simeq 80°$ as in the hemisphere centered 180° in phase away, at $\theta \simeq 260°$. In the cylinder at $R = 16\,\mathrm{kpc}$ (shown in the lower panel of Fig. 57a) the deviations from flatness are substantial. The azimuthal variation of the z-height of the centroid of the variation is approximately sinusoidal in form, with the deviations *above* the galactic equator in the regions accessible to Northern-hemisphere of approximately the same amplitude as the deviations *below* the equator found in the southern data.

At larger R, the amplitude of the warp continues to increase in the Northern-hemisphere data. It can be followed in the Burton & te Lintel Hekkert analysis to z-heights of more than 4 kpc at $R \simeq 25\,\mathrm{kpc}$, where the sensitivity limits of the data are reached; in the Diplas & Savage (1991) analysis based on the more sensitive Stark et al. (1992) data, the warp can be followed to higher z-heights and to larger radii. The behavior of the warp in the Southern-hemisphere data is not symmetric with respect to the northern data at large R. The maximum excursion of the gas density centroid below the galactic equator is reached at about $R \simeq 18\,\mathrm{kpc}$; at larger R, the gas layer folds back towards the equator.

This asymmetric behavior is illustrated by the two cylindrical cuts, corresponding to larger R, plotted in Fig. 57b. At $R = 19\,\mathrm{kpc}$, the warp carries the HI density centroid to a maximum amplitude above the equator of about 2 kpc near azimuth $\theta \simeq 80°$; the corresponding maximum depression below the equator occurs again about 180° in azimuthal phase away, but amounts to a displacement of only about 1 kpc. At $R = 21\,\mathrm{kpc}$, the Northern-hemisphere data shows z-heights of some 4 kpc; the southern-data portion of the gas layer has flopped back to z-heights of only a few hundred pc. Similar floppy warps are

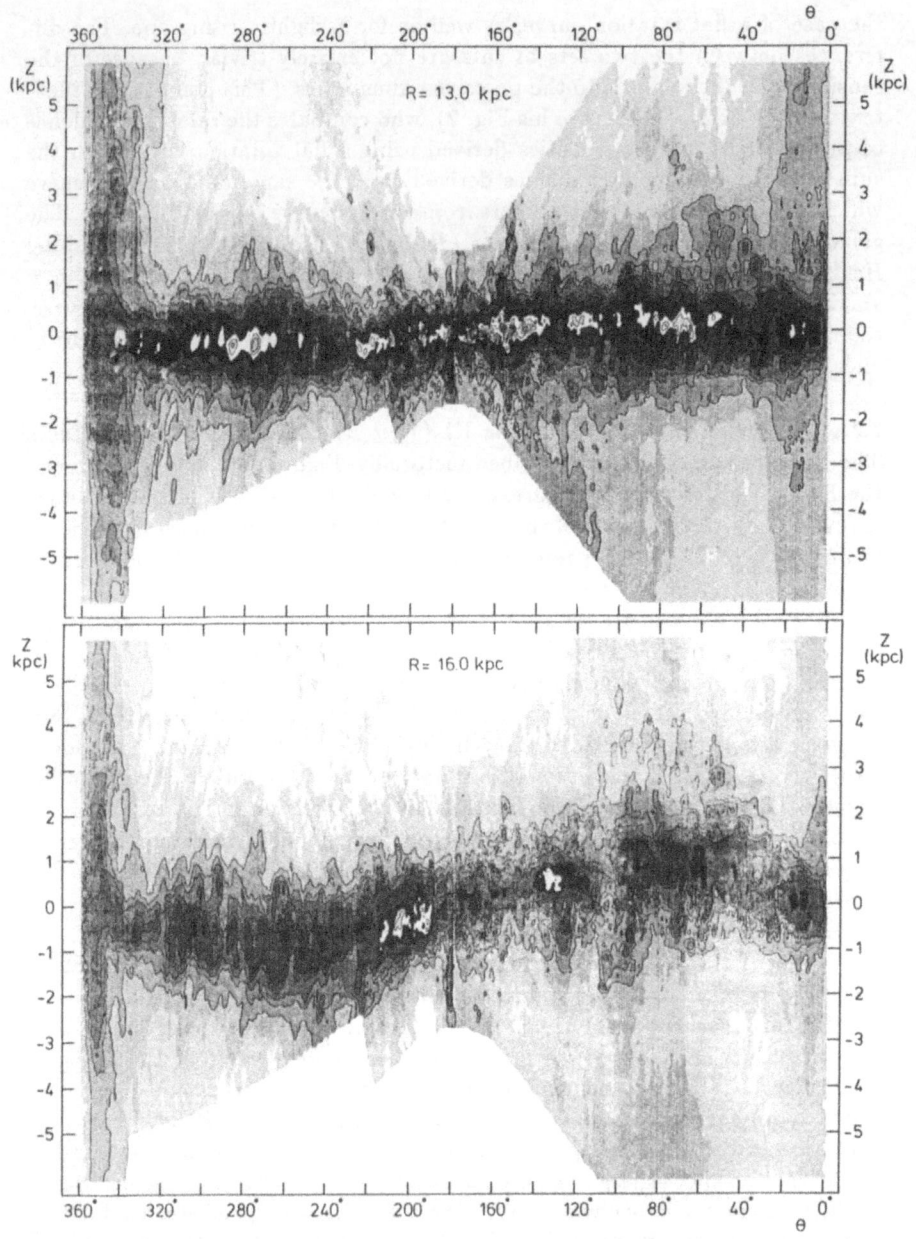

Fig. 57a. Arrangement of HI mean volume densities in cylinders with walls 500 pc thick centered at the galactocentric distances $R = 13.0$ and $R = 16.0$ kpc (from Burton & te Lintel Hekkert 1986). Grey-scale divisions occur at 0.001, 0.004, 0.01, 0.02, 0.03, 0.05, 0.07, 0.16, 0.30, 0.45, and 0.60 atoms cm^{-3}. The intensities have been corrected for global optical-depth effects. The cylindrical-coordinate (R, θ, z) dataset was sampled at $1°$ intervals in θ and 100-pc intervals in z

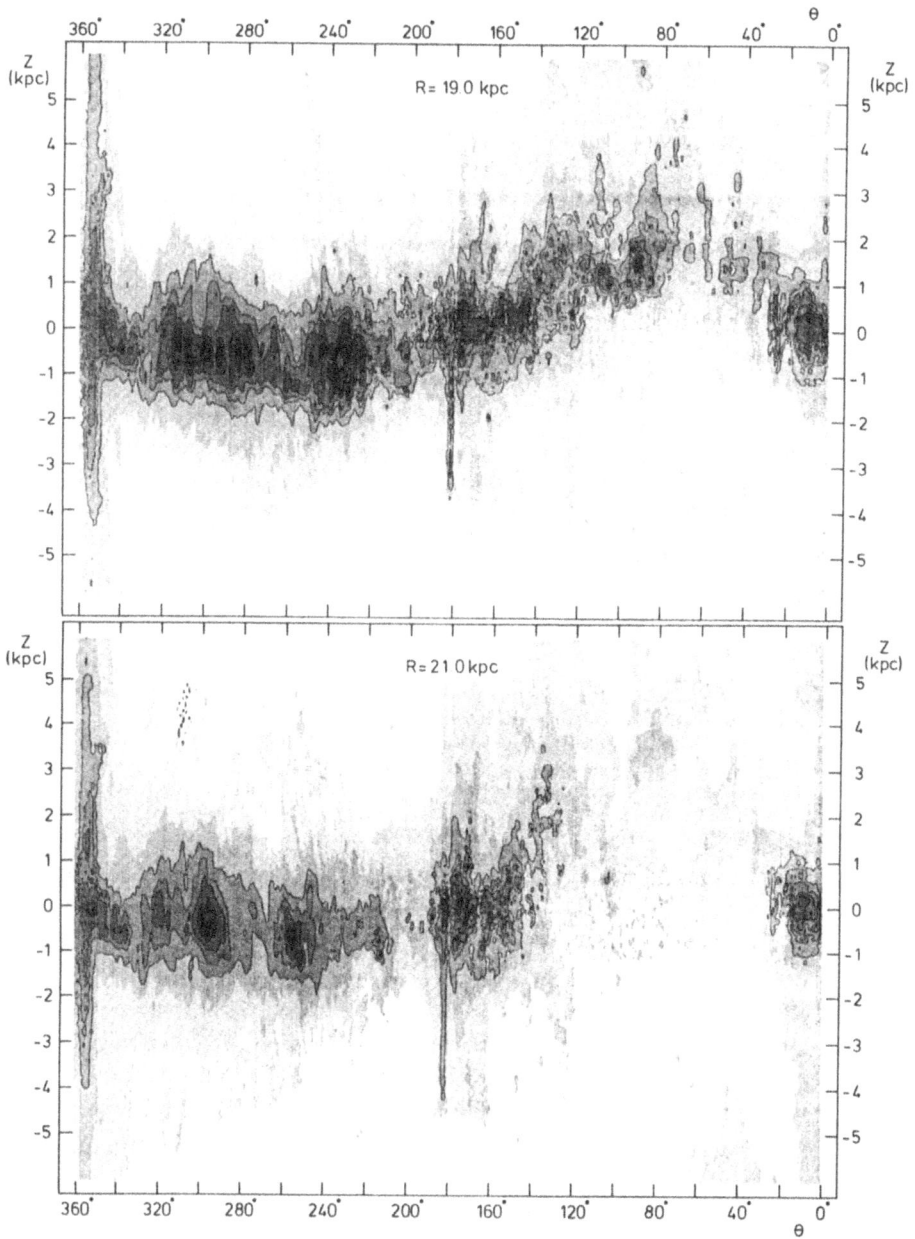

Fig. 57b. Arrangement of HI mean volume densities in cylinders with walls 500 pc thick centered at the galactocentric distances $R = 19.0$ and $R = 21.0$ kpc (from Burton & te Lintel Hekkert 1986). The intensity scaling and sampling intervals are as in Fig. 57a. The v-to-R transformation was based on a flat rotation curve, $\Theta_0(R > R_0 = 10\,\text{kpc}) = 250\,\text{km s}^{-1}$. This transformation is not robust within about 15° of $\theta = 0°$ and 180°; the apparent anomalies at these azimuthal values may be ignored

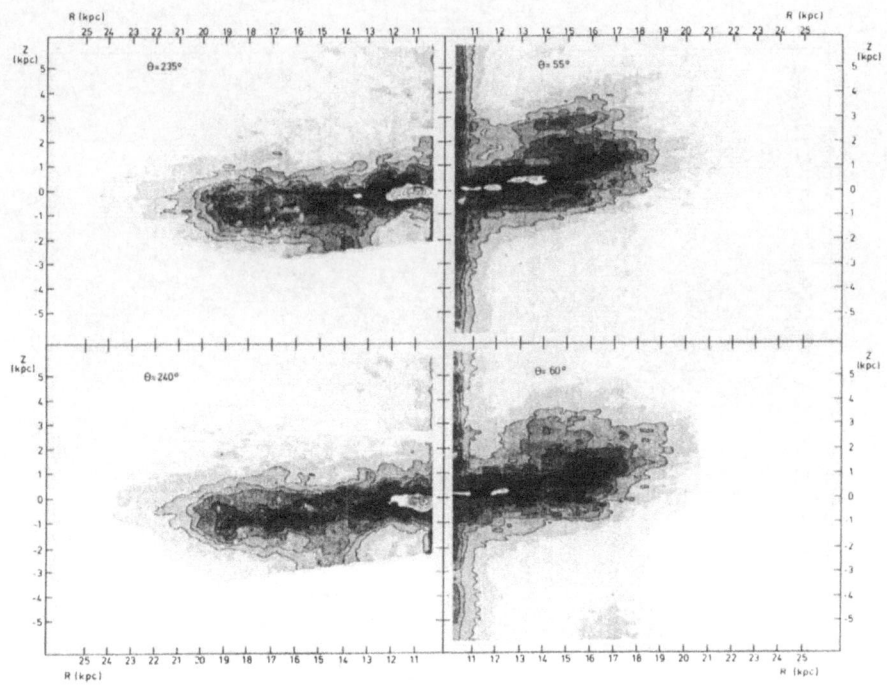

Fig. 58. Arrangement of HI mean volume densities in two representative perpendicular planes cutting the galactic equator at the opposed-azimuth pairs, $\theta = 55°$ and $235°$, and $\theta = 60°$ and $240°$ (from Burton & te Lintel Hekkert 1986). The scaling of the HI volume densities is the same as in Fig. 57. The cylindrical-coordinate (R, θ, z) dataset was sampled at 100-pc intervals in z and 250-pc intervals in R, over wedges of θ, $5°$ wide, centered at the indicated azimuths. The anomalous bands at $R < 11$ kpc are artifacts of the finite velocity dispersion of the gas, and may be ignored. The approximate 2π-symmetry of the galactic warp justifies pairing of maps separated by $180°$ of azimuth, so that a single plane cuts an entire galactic diameter

also found in external systems; an example of a galaxy which exhibits a floppy warp is given by the HI observations of NGC 3718 made by Schwarz (1985), who showed that the inclination angle of the warp of NGC 3718 flattens off at large radii after rising linearly at smaller radii.

7.2.2 Vertical Cuts Through the HI (R, θ, z) Dataset at Constant Azimuth.
The outer-Galaxy warped HI layer can also be demonstrated by cuts through the dataset which sample the material in sheets passing through the galactic center and perpendicular to the galactic equator. Two such vertical cuts are shown in Fig. 58. The approximate 2π-symmetry of the galactic warp justifies pairing of maps separated by $180°$ of azimuth. These cuts illustrate, as did the cylindrical surfaces represented in Fig. 57, that the warp amplitude reaches greater z-heights in the Northern-hemisphere data than in the Southern.

Diplas & Savage (1991) studied the distribution of HI in the parts of the outer Galaxy represented in the Bell Laboratories survey of Stark et al. (1992). That survey was carried out using the Bell Labs 18-m horn antenna, which

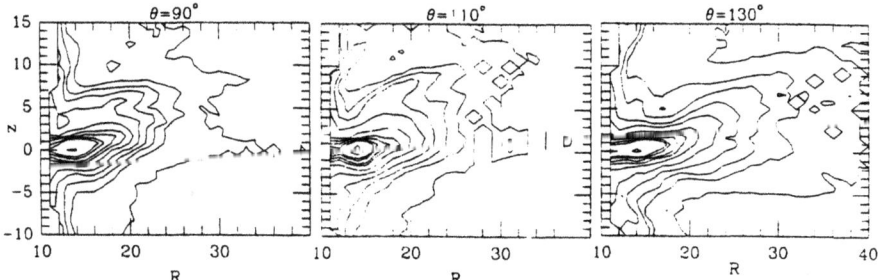

Fig. 59. Arrangement of HI mean volume densities in perpendicular planes cutting the galactic equator at the indicated azimuths, from the Diplas & Savage (1991) analysis of the Stark et al. (1992) HI survey made with low sidelobe contamination. In addition to the contour levels represented in Fig. 58, the more sensitive data entering this figure allowed contours at the levels of 0.0008 and 0.0005 atoms cm^{-3}. The Diplas & Savage analysis used the same galactic-constant values, $\Theta_0 = 250\,km\,s^{-1}$ and $R_0 = 10\,kpc$, as used in the Burton & te Lintel Hekkert analysis. The sensitivity of the Stark et al. data allows the Milky Way warp to be followed to galactocentric radii as large as $\sim 30\,kpc$

is well suited to study of weak emission because of its low response to stray-radiation contamination. Lockman et al. (1986), in particular, have stressed the possible influence of stray radiation entering the sidelobes of conventional single-dish antennæ. The level of this contamination can reach intensities up to $T_b \simeq 0.5\,K$. Such intensity levels are within the reach of modern survey data, and, although insignificantly low for the purposes of some investigations, these levels could well be important in studies of the outer-Galaxy warp. Therefore it is interesting to compare the results of Diplas & Savage, who used coarsely sampled but sensitive data, with those of Burton & te Lintel Hekkert, who used more complete, but less sensitive, data.

Figure 59 shows the arrangement of HI mean volume densities in perpendicular planes cutting the galactic equator at the indicated azimuths, as determined by Diplas & Savage (1991) from the Stark et al. (1992) data. These, and the other cuts through the dataset presented by Diplas & Savage, may be compared with those shown for the detailed, composite dataset by Burton & te Lintel Hekkert; the analyses used the same values of the constants Θ_0 and R_0, and the same, flat, rotation curve. The Diplas & Savage investigation largely confirmed the earlier work, but did extend information on the warp, and on the flaring z-thickness of the galactic gas layer, to larger galactocentric distances. The comparisons between the two analyses indicate that stray radiation is evidently not a significant problem in the composite dataset. This may be attributed to the fact that the kinematic signature of the warp lies predominantly at velocities more extreme than about $50\,km\,s^{-1}$; most directions on the sky contain negligible or only quite weak emission at $|v| > 50\,km\,s^{-1}$, and so most directions will contribute negligible sidelobe contamination at velocities of relevance to investigations of the warp. Some 10% of the sky, largely at latitudes $|b| < 10°$, do, however, emit intensely at the velocities characterizing the warp. One of the principal motivations for carrying out the new Leiden-Dwingeloo

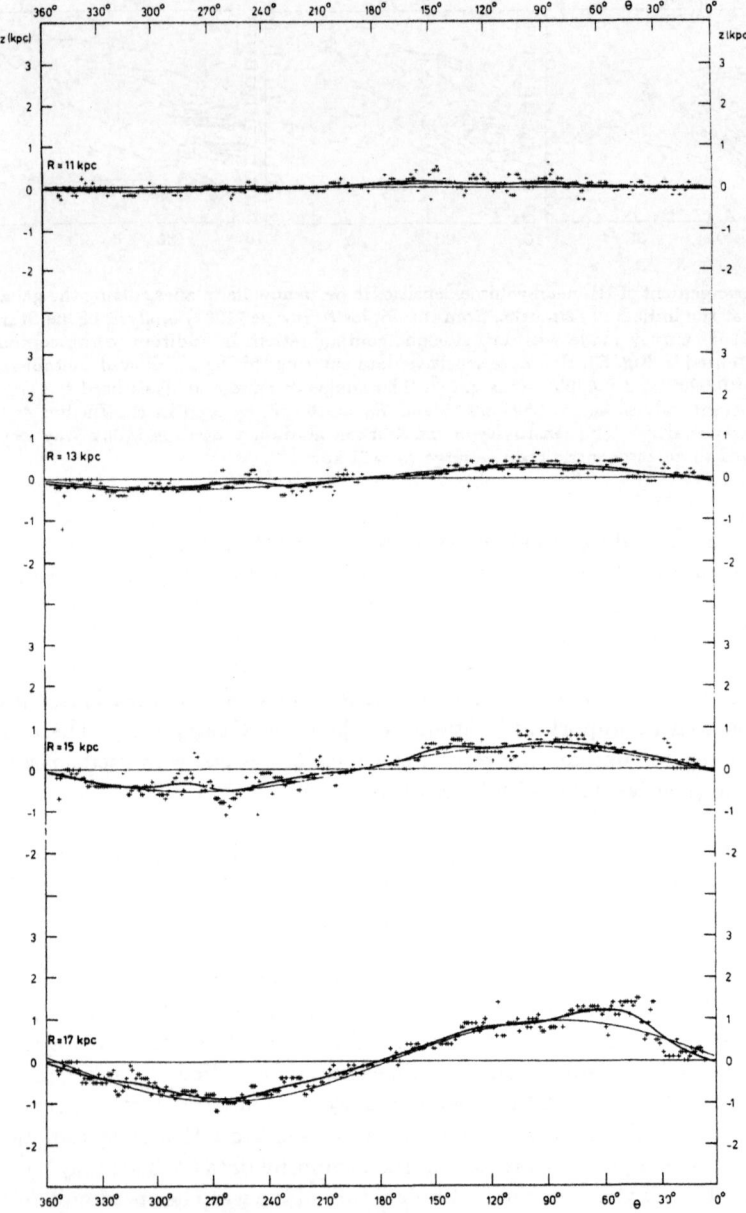

Fig. 60a. Vertical z-distance from the plane $b = 0°$ of the maximum HI density at the indicated galactocentric distances, derived from plots like those in Fig. 57 (from Burton 1988a). The more smoothly-varying lines represent sine curves, least-squares fit to the data points; the less regular curves represent fits (made by K.K. Kwee) using a cubic-spline algorithm. The location of the line of nodes of the warp may be specified by locating the galactic azimuths at which the gas layer crosses the equator, $z = 0$ pc

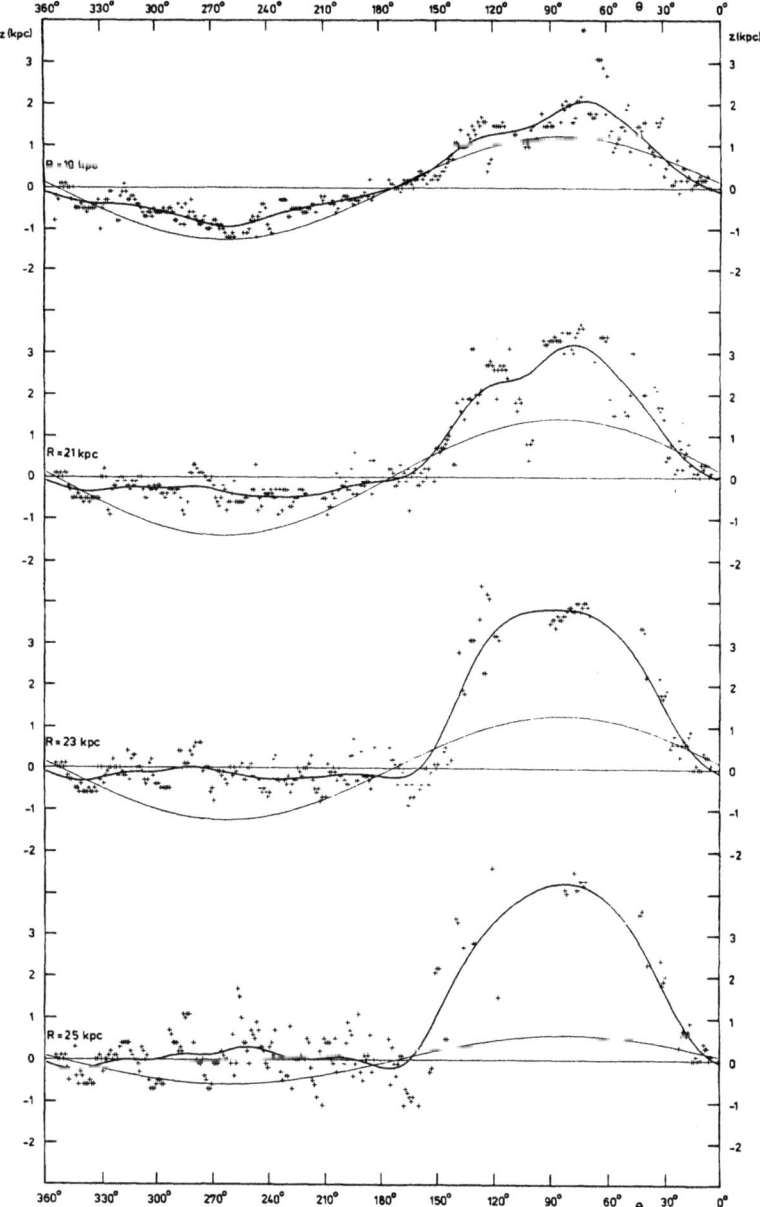

Fig. 60b. Vertical z-distance from the plane $b = 0°$ of the maximum HI density at the indicated galactocentric distances in the outer reaches of the Galaxy, derived from plots like those in Fig. 57 (from Burton 1988a). The symbols and lines have the same meanings as in Fig. 58a. The excursion of the mean gas layer above the galactic equator grows until a z-height of about 4 kpc is reached at the largest distances; the excursion below the equator increases with increasing R until about 16 kpc, beyond which it folds back to almost reach the equator again at the largest distances

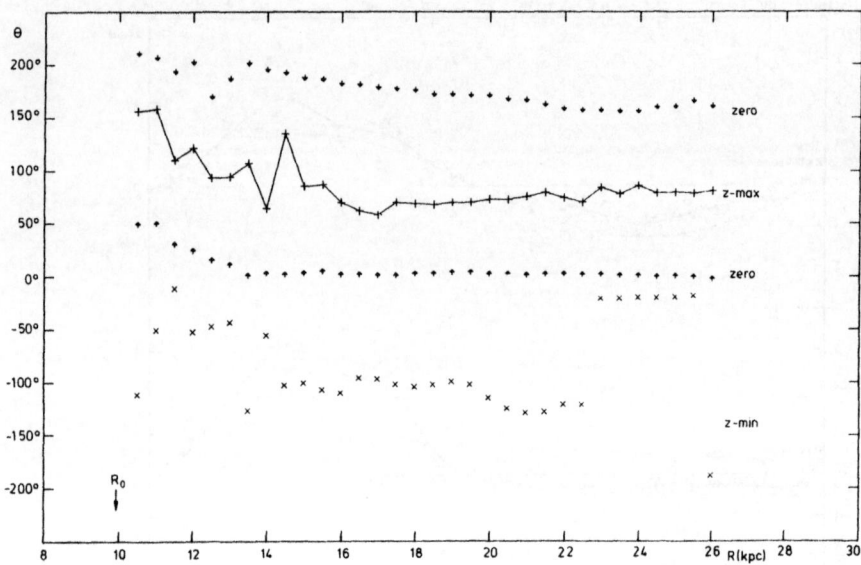

Fig. 61. Radial dependence on azimuth of the location of the line of nodes of the galactic warp. Indicated in the figure are the locations where the warped HI layer reaches its maximum height above the plane $z = 0$ pc (+ symbols), as well as the positions where the layer crosses $z = 0$ pc (↑ and ↓ symbols). These zero-crossings give the position of the line of nodes of the galactic warp. The location of the line of nodes varies but little with galactocentric radius. (From Burton 1988a,b)

HI survey is provided by the intention to describe the warp with detailed as well as sensitive data; these new data will be corrected for stray-radiation.

7.2.3 Location of the Line of Nodes of the Galactic Warp. The azimuthal directions at which the gas layer crosses the plane $b = 0°$ define the line of nodes of the galactic warp. Figure 60 shows, separately for cuts at different R through the cylindrical dataset, how the z-height of the maximum HI density varies with galactocentric azimuth. The location of the line of nodes of the warp, as well as the warp amplitude, and the dependences of these quantities on changing galactocentric radius, may be determined from such plots.

The series of plots shown in Fig. 60 indicates that the galactic warp becomes significant at $R \simeq 12$ kpc; that it grows in amplitude linearly, and approximately equally in the two galactic hemispheres, until $R \simeq 16$ kpc; that at larger radii the warp amplitude continues to grow in the Northern-hemisphere data up to a z-height of at least 4 kpc, but that the southern-data gas layer folds back toward the equator after reaching a maximum z-height of about 1 kpc at $R \simeq 18$ kpc. The series of plots also shows that the location of the plane crossings does not change with changing R. The line of nodes of our Galaxy's warp is evidently quite straight.

Figure 61 shows the radial dependence on azimuth of the positions in the Galaxy where the outer HI layer crosses the galactic-equator plane $z = 0$ pc, as well as the locations where the layer reaches its maximum excursions above and

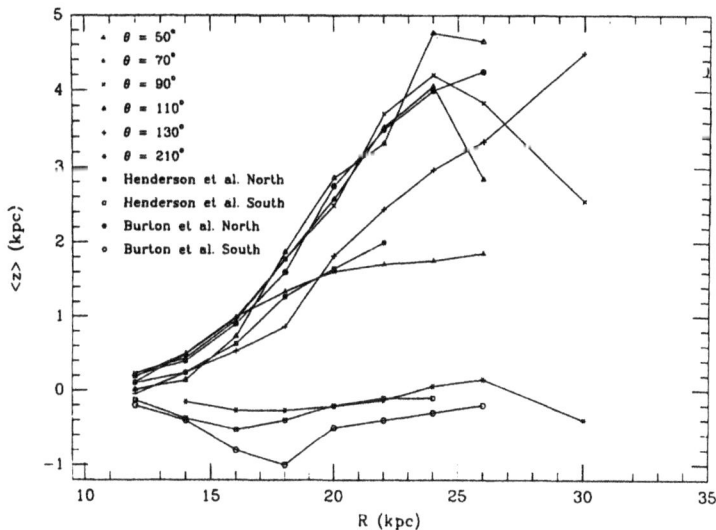

Fig. 62. Variation with galactocentric radius of the mean z-height of the HI layer measured at the indicated azimuths, and as measured — after azimuthal smoothing — separately in the galactic hemispheres accessible from the North and from the South (from Diplas & Savage 1991). The HI layer becomes significantly warped outward of $R \simeq 12.5$ kpc. The amplitude of the warp grows equally in the two hemispheres until $R \simeq 16$ kpc; at larger R the warp continues to increase in amplitude in the hemisphere accessible from the North, but flops back towards the equator in the southern data

below the plane. The azimuthal locations of the zero-crossings define the line of nodes. (There is no reason to think that the fact that the line of nodes lies within a few degrees of the Sun-center line is anything else than a coincidence; that is, there is no reason to think that it is due to an error in the analysis.) The straightness of the line of nodes indicates that the shape of the warp is not deformed by precession under the influence of differential galactic rotation. Note that a test particle on the inner part of the warp, at, say, $R = 12$ kpc, would rotate twice around the galactic center during the time a test particle deeper in the warp, at $R = 24$ kpc, would require for a single rotation, if both particles moved in accordance only with a flat rotation curve. The warp evidently is not deformed by such shearing, nor by precession.

The regularity of the warp displayed by the straight line of nodes is also indicated by the consistent location of the maximum upward and downward excursions at a separation of about 90° from the location of the zero-plane crossings. A straight line of nodes is a common aspect of warps seen in many external galaxies (see e.g. Briggs 1990).

7.2.4 The Galactic Warp Specified by the z-Height of the Midplane of the HI Layer.
Figure 62 shows the radial variation of the z-height of the mean HI layer, plotted separately for several of the azimuthal wedges investigated by Diplas & Savage (1991). The figure also includes the radial variation of the warp as determined by Henderson et al. (1982) and by Burton & te Lintel

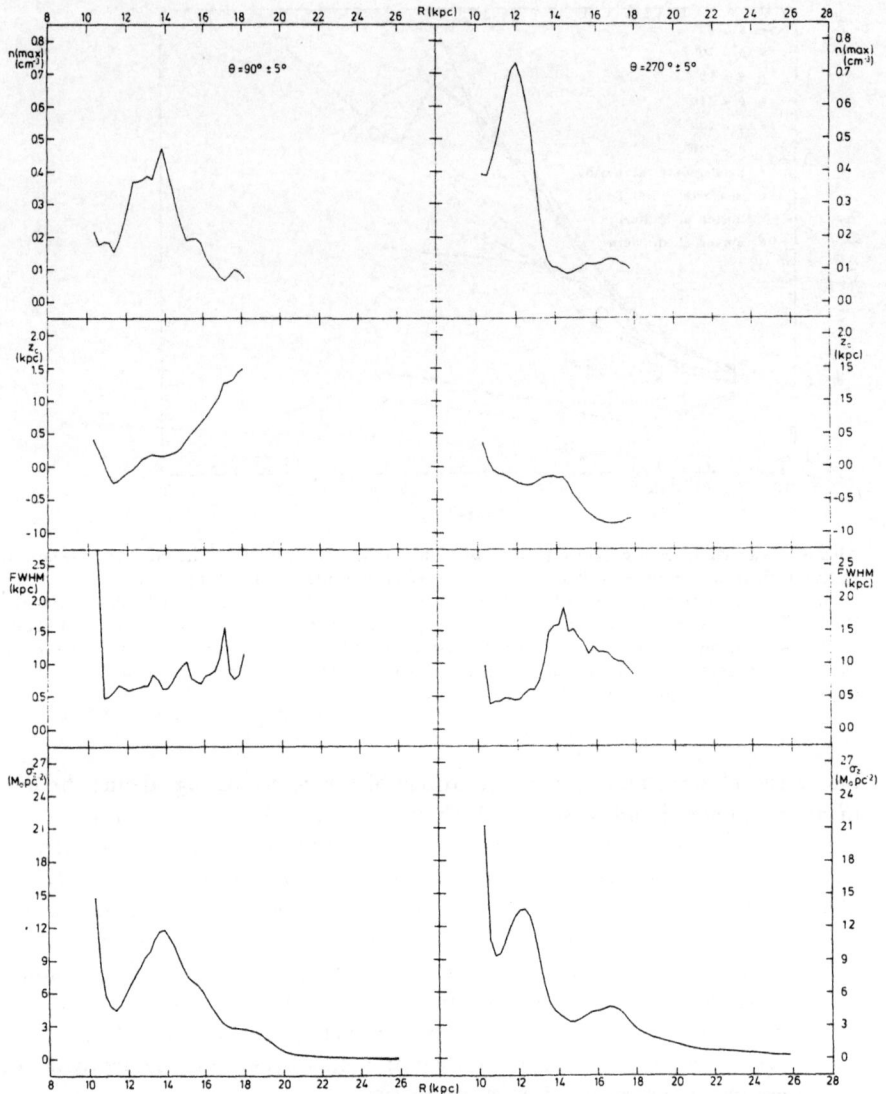

Fig. 63. Representative parameters of the outer-Galaxy HI layer as sampled along a perpendicular slice, 10° wide in azimuth, cutting the galactic equator at the opposed azimuths $\theta = 90°$ and 270°. The slice samples the outer Galaxy on a vertical plane approximately perpendicular to the line of nodes, where the out-of-plane excursions of the galactic warp are relatively extreme. The quantities plotted were found from analysis of cylindrical cuts through the Burton & te Lintel Hekkert (1986) composite dataset (as in Fig. 57) and include the maximum value of the HI volume density found in the warped swath of emission, the z-height of the centroid of this swath, its full-width to half-maximum thickness, and the total HI surface density projected through it

Hekkert (1986); both of these determinations involved substantial azimuthal smoothing.

It will be particularly interesting to confirm if the fold-over of the warp in the Northern-hemisphere data, hinted at by the data at the largest R in several of the azimuth wedges shown in Fig. 02, is in fact a property of the Milky Way warp; several external galaxies have indicated such a property. Such confirmation will require detailed observations made with high sensitivity, and, in addition, with sufficient reach in latitude to sample the full extent of the warp signature; in some directions in the first and second quadrants, the warp spans $\Delta b \simeq 40°$.

7.2.5 The Flaring Thickness of the Warped Galactic HI Layer. Figure 63 shows the variation with galactocentric radius of several morphological parameters of the warped gas layer, including the peak value of the HI volume density found in the layer, the z-height of the centroid of the gas layer, its full-width to half-maximum thickness, and the total HI surface density projected through the gas layer.

The vertical thickness of the HI layer increases approximately linearly from the onset of the warp, near $R \simeq 11\,\mathrm{kpc}$, where the FWHM thickness is about $320\,\mathrm{pc}$, until the outermost regions where the thickness is a full order of magnitude larger. We repeat that the signature of the warp commonly extends to latitudes beyond $25°$ or $30°$, so that values of the thickness derived from survey material confined to $|b| < 10°$ will likely underestimate the true thickness of the gas layer; this point has been stressed by Diplas & Savage (1991; see their Fig. 7b).

The flare of the outer gas layer to larger vertical thicknesses at larger galactocentric distances is a property which our Galaxy shares with other warped systems.

Measuring the flaring vertical thickness of the galactic warp is important to general dynamical considerations. The thickness of the gas layer presumably represents a balance between the gravitational force toward the mean layer and the restoring forces due to turbulent motions in the gas, to thermal pressure, and to pressures associated with cosmic rays and magnetic fields. Although the *vertical* velocity dispersion of the gas in our Galaxy cannot be measured at large R, observations of face-on external galaxies show little or no variation of the z-dispersion with R (see e.g. Dickey et al. 1990, and Bosma 1991). Measures of the *horizontal* velocity dispersion can be made, and there is no indication that the line-of-sight velocity dispersion changes systematically with increasing R. The flaring nature of the disk, together with the evidence for constant velocity dispersion (at least for the line-of-sight component of σ_v) and the generally flat rotation curve imply that the dark matter in the outer Milky Way cannot be confined strongly to the galactic equator.

An interesting use of the gas-layer thickness has recently been introduced by Merrifield (1992). Merrifield found the galactic scaling quantities $\Theta(R)$, Θ_0, and R_0 which led to a derived thickness of the HI layer which is constant over each galactocentric ring. (The assumption that the thickness of the HI layer

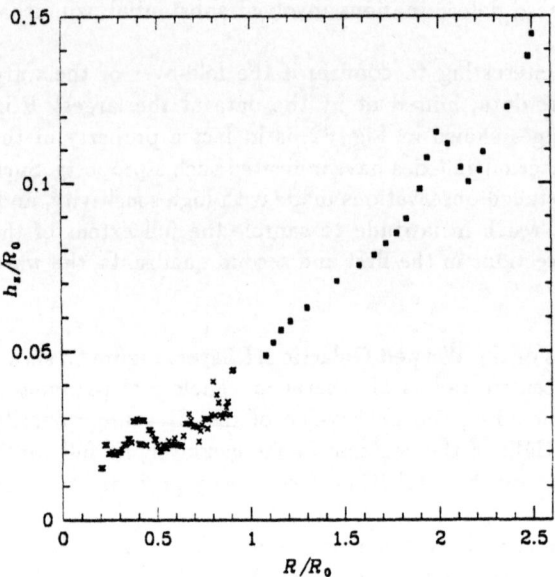

Fig. 64. Variation with galactocentric radius of the (azimuthally-smoothed) thickness of the HI layer in the outer Galaxy, found by an analysis which determines the galactic scaling quantities by requiring that the gas-layer thickness be constant in a given galactocentric ring (from Merrifield 1992)

in a ring at a given radius is constant is based on the observational results of Henderson et al. 1982.) This method also leads to the independent measure of the layer thickness which is shown in Fig. 64.

7.3 Comments Regarding the Apparent Lopsidedness of the Outer Galaxy

We briefly commented in Chap. 3 of these notes on the apparent lopsidedness of the Milky Way. Figure 14 shows the lack of cylindrical symmetry which pertains when the Galaxy is sliced along the equator plane $z = 0$ pc: the kinematic cutoff at $z = 0$ pc corresponding to the far outskirts of the Galaxy occurs at velocities some $25\,\mathrm{km\,s^{-1}}$ more extreme in the Southern-hemisphere data than in the Northern. The above discussion on the galactic warp shows, however, that measures of the far outer edge of the Milky Way should not be taken solely in the $z = 0$ plane (even though deviations from symmetry in the equator are valid indications of asymmetry, warp or no warp). It is clear, for example, from the plots shown in Fig. 58 of HI density in perpendicular, opposed-azimuth cuts through the galactic equator, that especially the northern-data cutoff would be identified at greater R, and thus at more extreme velocities, if measured at the z-value of the centroid of the gas layer. The comparative-cutoff plot given by Blitz & Spergel (1991) is an improvement on Fig. 14 in that it measures the cutoff-velocities for data taken at $|b| \leq 10°$, projected onto the galactic plane, and not just for data at $b = 0°$. A further improvement would be given if the data range were extended beyond $|b| = 10°$, because the signature of the galactic warp extends beyond this limit. Inspection of Fig. 58 shows that, although the amplitude of the cutoff-differences measured at complementary azimuths would decrease in that case, the evidence for asymmetry would remain.

The evidence for bilateral asymmetry of the outer-Galaxy edge has been accepted for some time, but it remains a matter of discussion whether this lop-sidedness requires a primarily kinematic interpretation, or primarily a spatial one. For example, Kerr (1962) recognized that interpretation of the early Sydney HI data using the same v-to-R transformation which had been used for the Leiden data led to implausible results. Kerr could reconcile the results by introducing an outward velocity component of the Local Standard of Rest; he also considered kinematic information which suggested *general* nonradial motions. Kerr (1983) later showed that some bilateral symmetry is restored if motions in the galactic HI layer were described in terms of elliptical, rather than circular, orbits. (Recent dynamical interpretations of gaseous as well as stellar orbits in galaxies, especially in terms of closed orbits governed by a triaxial potential, have provided well-founded alternatives to the conventionally-adopted circular-orbit situation.) Shuter (1981), and others, have considered the consequences of the quite plausible possibility that the motion of the Local Standard of Rest is substantially (i.e., $\sim 10\,\mathrm{km\,s^{-1}}$) in error; these consequences are far-reaching, of course, because of the dominant role which kinematics play in determining the characteristics of galactic HI and other observations.

Blitz & Spergel (1991) have reanalyzed the evidence for asymmetries found in the galactic HI (and molecular) data in terms of a triaxial spheroid with motions driven by a quadrupole term in the gravitational potential. Their interpretation involves an outward radial motion amounting to $14\,\mathrm{km\,s^{-1}}$ in the Solar vicinity, whose ellipticity decreases with increasing R. The influence of the oval distortion in the Blitz & Spergel model is stronger in the inner Galaxy than in the outer; gas inwards of the Sun moves on orbits of substantial ellipticity, but gas outwards of the Sun moves on orbits of decreasing ellipticity. The outermost gas in the model moves on essentially circular orbits. (The observational consequences of changing ellipticity have also been considered by Abramenko 1978.) Blitz & Spergel show that the kinematic predictions of their model are in accordance with several of the observed indicators of apparent asymmetry in the Milky Way; when account is taken of the motion of the LSR predicted by their model, the shape of the outer galactic disk is quite round and the shape of the dark-matter halo therefore quite axisymmetric.

Kuijken (1991), on the other hand, chooses for the spatial paradigm rather than the kinematic one to explain the outer-Galaxy asymmetries in the total-velocity extent, and shows a consistent dynamical context which can incorporate spatial lopsidedness. In Kuijken's view, the Milky Way gas layer would show, from an external perspective, the same sort of lopsidedness as shown by the galaxies discussed by Baldwin et al. (1980). Baldwin et al. showed that an $m = 1$ distortion (i.e., spatial lopsidedness, as opposed to an $m = 2$, quadrupole distortion) can persist for many galactic revolutions.

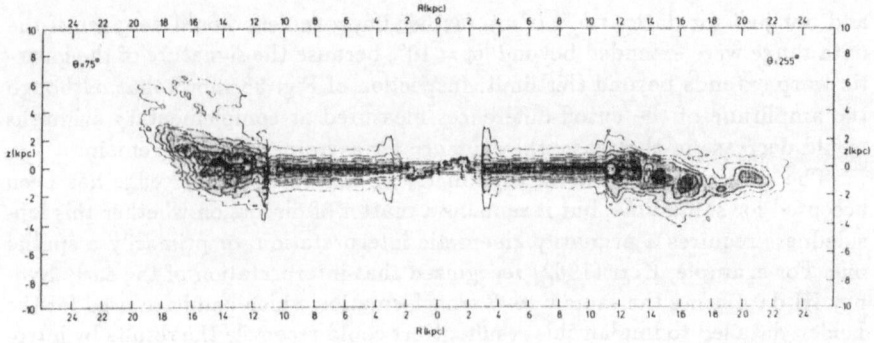

Fig. 65. Schematic vertical cross-section through the HI layer of our Galaxy at the indicated azimuths (from Burton 1988a,b). In the outer Galaxy, the location of this cross-section corresponds approximately to a position angle some 90° from the line of nodes, and thus samples the most extreme out-of-plane excursions of the galactic warp. In the inner Galaxy outside of the bulge region, the layer is indicated according to the parameters specified by Lockman (1984); in the region encompassed by the galactic bulge, the gas layer is tilted as described in Chap. 6 of these notes

Irrespective of the context chosen to explain the shape of the total-Galaxy cross-section, there remain two important indications of spatial asymmetry; namely, that the amplitude of the warp viewed in the Northern-hemisphere data extends to substantially larger z-heights than in the portion viewed from the southern data, and that the third- and fourth-quadrant warp folds back to almost reach the equator again at the largest distances, whereas the warp seen in the first and second quadrants grows more or less linearly in z with increasing R. In these regards, the warp in our Galaxy deviates from what one might consider as a prototype of a grand-design, integral-sign warp seen in some external systems.

Several of the properties of the warped, flaring outer-Galaxy HI are shown schematically in Fig. 65. The vertical cross-section shown corresponds to a cut approximately at right angles to the line of nodes.

7.4 The Shape of the Molecular-Cloud Ensemble in the Outer Galaxy

Although HI observations by themselves serve well in establishing many global structural properties of the Galaxy, it is natural to seek complementary information from other tracers in order, in particular, to learn more about the physical environment at large R.

The low optical surface brightness of the outer Galaxy has made it difficult to establish if the *stellar* distribution follows the geometry of the gas layer. Miyamoto et al. (1988) showed that O and B stars extending a few kpc beyond the Sun's position follow the nearby part of the warp defined from HI data. Carney & Seitzer (1992) have described the preliminary results of their observational program in which suitable "windows" are identified in the parameters

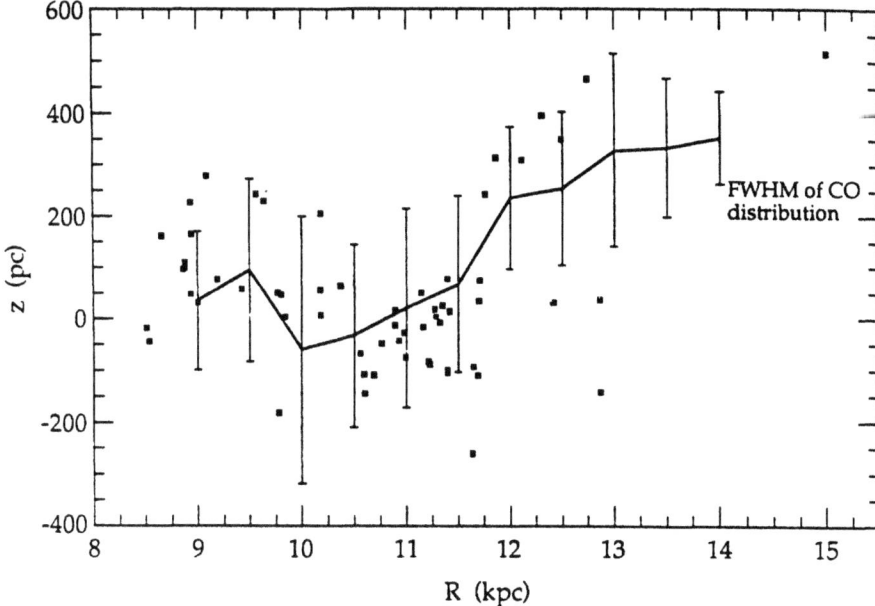

Fig. 66. Vertical z-distance of the molecular midplane determined from ^{12}CO data systematically sampled over the longitude range $65° < l < 116°$ by Digel (1991). The error bars indicate the FWHM of the CO distribution as measured in Digel's longitude segment. The squares indicate the location of optically-visible HII regions cataloged by Fich & Blitz (1984), which generally have molecular-cloud counterparts. The linear scale corresponds to $R_0 = 8.5\,kpc$

found describing the HI warp and subsequently searched for red-giant stars using deep CCD images; the mean chemical abundances of these stars are being determined with a view to comparing the chemical evolution at the largest R with that followed locally.

Djorgovski & Sosin (1989) sought evidence for a warp in the galactic stellar disk using a sample of some 90,000 IRAS point sources, selected on the basis of infrared colors similar to those expected from evolved, dust-shell stars. A plot of the mean b-height against l of the Djorgovski & Sosin data shows a systematic, approximately sinusoidal, variation; the stellar variation is consistent with that which would follow from the HI data after integrating away the kinematic information. (A comparison of Fig. 1 of Djorgovski & Sosin (1989) with the velocity-integrated HI data shown in Fig. 6 in Chap. 2 of these notes is instructive in this regard.) That the variation of the mean IRAS-source b-height against l is sinusoidal, and does not show the north/south asymmetries characteristic of the HI warp at the largest R, probably does not mean that the stellar disk is warped differently than the gaseous one; instead, it probably means that the IRAS sample is dominated by objects at $R < 16\,kpc$, i.e., at radii within the regime where the warp is still largely symmetric.

143

A few dozen HII regions have been optically identified at galactocentric radii up to about $2\,R_0$ and these also preferentially follow the HI warp. In general, however, optically-visible tracers are found in too small numbers at large R to reveal their spatial morphology in detail.

7.4.1 Outer-Galaxy Molecular Clouds Found Sampling on Regular Grids.

Molecular clouds are more sparsely distributed in the outer Galaxy than in the inner. The projected molecular-cloud surface density is low in any case, and the thickening of the cloud layer tends to further decrease the fraction of sky containing a cloud; in addition, the CO luminosity characteristic of outer-Galaxy clouds is lower than that found in the molecular-annulus clouds. Until recently, CO surveys sampled on a dense, uniform grid have not been able to provide the large-scale coverage and adequate sensitivity necessary for general investigations.

The CO observations made by Dame et al. (1987), May et al. (1988), Mead (1988), and Mead & Kutner (1988), among others (see review by Combes 1991), have recently been augmented by the systematically-sampled observations made (and compiled) by Grabelsky et al. (1987) and by Digel (1991). The Grabelsky et al. material covers a large sector in the third and fourth quadrants; Digel's data cover a large sector in the first and second quadrants.

Figure 66 shows the vertical z-distance of the molecular midplane determined from Digel's ^{12}CO data, systematically sampled in the first and second galactic quadrants. The outer-Galaxy molecular-cloud ensemble evidently follows the warp established by the HI data at least to $R \simeq 15\,\text{kpc}$ (on the $R_0 = 10\,\text{kpc}$ scale). The thickness of the cloud ensemble also flares as the HI layer does.

Figure 67 shows the volume density of H_2, derived by Grabelsky et al. from Southern-hemisphere ^{12}CO data, as a function of z-distance from the galactic equator.

7.4.2 Outer-Galaxy Molecular Clouds Found Using Infrared Heat-Source Finding Charts.

The description of the outer-Galaxy molecular-cloud layer given by Wouterloot et al. (1990) was based on an attempt to circumvent the time-consuming requirements of unbiased, regular-grid CO surveying. Wouterloot et al. (1988) had shown that IRAS sources selected from the *Point Source Catalog* on the basis of their infrared colors commonly had associated molecular emission which could deliver kinematic information from large distances. Wouterloot & Brand (1989) selected some 1300 outer-Galaxy IRAS candidates, observed them for CO emission, and thus determined radial velocities for more than 1000 clouds. Wouterloot et al. (1990) then used the kinematic distances determined for this sample to follow the shape of the molecular-cloud ensemble to $R \simeq 20\,\text{kpc}$ and compared the molecular morphological results with those derived from HI data.

Figure 68 shows the location of CO clouds with embedded heat sources as arranged in cylindrical sheets comparable to those displaying the HI data in Fig. 57.

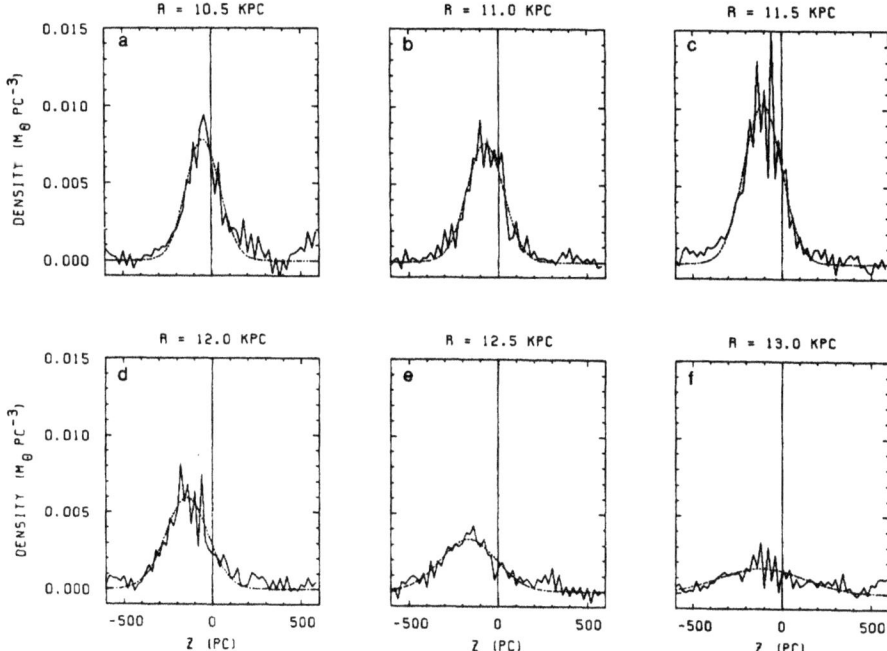

Fig. 67. Volume density of H_2 as a function of z-distance from the galactic equator, calculated from ^{12}CO data systematically sampled over the longitude range $270° < l < 330°$ and azimuthally smoothed in 500-pc rings centered at the indicated galactocentric distances (from Grabelsky et al. 1987). The linear scale corresponds to $R_0 = 10.0$ kpc. The half-thickness of the CO layer increases with increasing R, flaring as the HI layer does in the outer Galaxy

Figure 69 shows the variation with R of the thickness of the molecular-cloud layer, and compares this variation with the one representing the flare of the HI layer. The clouds in the sample used by Wouterloot et al. could be followed to $R \simeq 24$ kpc (on the $R_0 = 10.0$ kpc scale; note that Figs. 68 and 69 incorporate the $R_0 = 8.5$ kpc scale), and allowed the first global description of the far outer layer in a tracer other than HI. The fact that the cloud sample was selected on the basis of associated heat sources implies that star formation occurs in the far outer reaches of the Milky Way. The distribution of mean z-height of the cloud layer shows the same warped form as that of the HI diffuse gas layer. The cloud ensemble also shows a flaring thickness, with the thickness of the disk increasing at $R > 12$ kpc at the same rate for the CO as for the HI tracers. Comparison of the radial scalelengths of the projected surface densities of the HI and of the molecular clouds suggests that the HI gas layer terminates less abruptly than the cloud ensemble.

It is not yet known what fraction of molecular clouds is forming stars. A variety of evidence suggests that the answer to this question will have to sought separately in the outer Galaxy, at high $|z|$-distances, in the bulge region, and in the inner-Galaxy molecular annulus.

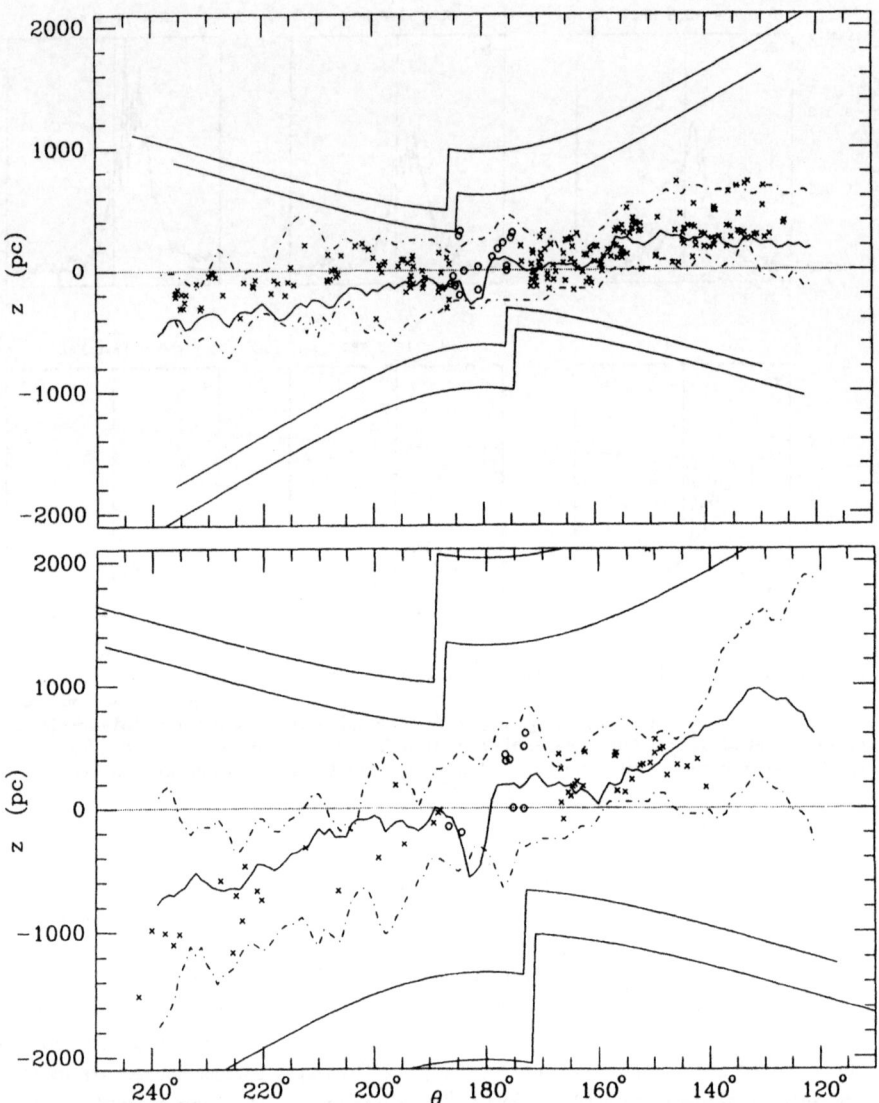

Fig. 68. Distribution with galactocentric azimuth of the z-heights of molecular clouds identified by their association with IRAS heat sources (from Wouterloot et al. 1990). The upper panel gives the shape of the molecular-cloud layer for CO sources with kinematic distances in the range $12.0 < R < 14.0\,\mathrm{kpc}$; the lower panel gives the shape of the molecular-cloud layer in the range $16.0 < R < 20.0\,\mathrm{kpc}$. The linear scale corresponds to $R_0 = 8.5\,\mathrm{kpc}$. The smoothly-drawn lines indicate the boundaries of the region searched in the IRAS point-source catalog for molecular-cloud candidates. The irregular line gives the mean centroid of the HI emission in the appropriate distance interval; the dashed lines indicate the z-heights at which the mean emission has fallen to half of its peak value

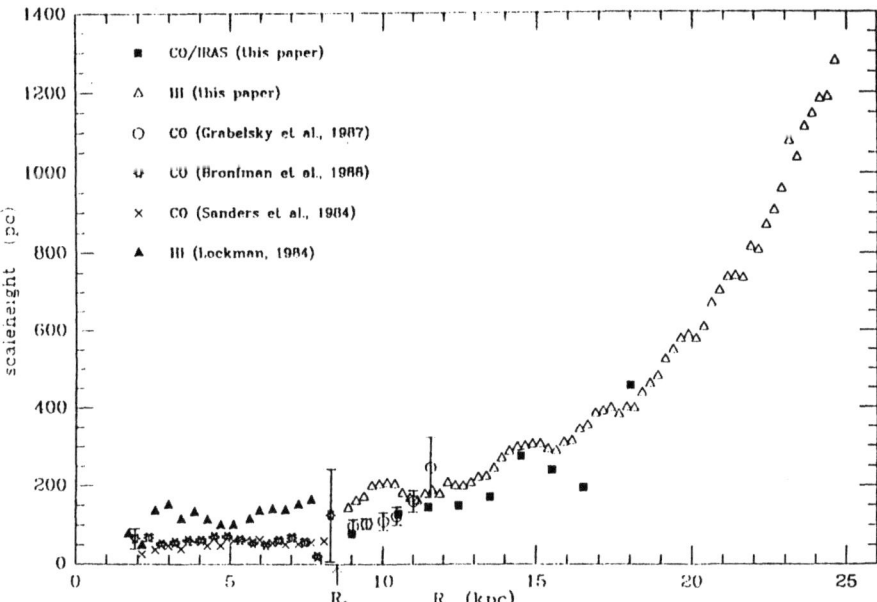

Fig. 69. Variation with galactocentric distance of the thickness of the HI gas layer and of the ensemble of molecular clouds in the outer Galaxy (from Wouterloot et al. 1990). The HI thickness refers to the half width to half maximum of the HI emissivity in cylindrical intervals 250 pc wide; the molecular-cloud thickness refers to the half width to half maximum of the number of clouds found in cylindrical intervals 1 kpc wide. The inner-Galaxy results (from the indicated references, and re-scaled to $R_0 = 8.5$ kpc) are included for comparison with the situation at large R. For both the HI and the CO tracers of the galactic gas layer, the warped shape as well as the flaring thickness become well-defined a kpc or two beyond R_0

7.4.3 Do the Molecular Clouds in the Outer Reaches of the Galaxy Differ from Those Populating the Inner-Galaxy Molecular Annulus?

The results on the global shape of the ensemble of molecular clouds with embedded heat sources seem firm irrespective of the completeness of the Wouterloot et al. sample (cf. Digel 1991). But it is interesting to compare the analysis of clouds with associated star formation with analyses based on more general CO mapping. It has long been known from inner-Galaxy work that not all molecular clouds show evidence for current star formation; it is therefore also interesting to ask, more generally, if the clouds in the far reaches of the Galaxy differ from those populating the molecular annulus.

There are, of course, important differences between the interstellar situation at large R and that pertaining inwards of R_0 (see Brand & Wouterloot 1991). The gradient of the gravitational potential flattens as the gas layer thickens. The ambient radiation field is weaker at large R than it is near the Sun or at $R < R_0$ (although the opacity of the medium to the ultraviolet flux may also be less in the outer reaches). The cosmic-ray flux is also weaker (see Bloemen et al. 1984), as is the metallicity (see Shaver et al. 1983).

147

The interstellar medium at large R is subject to relatively little global compression. The compression due to the passage of a density wave will be less important near the co-rotation radius than inwards of this radius. The galactic distribution of pulsars reveals that supernovæ occur relatively infrequently at large R, so compression by these violent events will be relatively unimportant there.

Compression due to cloud-cloud collisions will be substantially less in the outer Galaxy than within the molecular annulus. Regarding the collision cross section, we note that the outer-Galaxy clouds are more widely spaced than those in the molecular annulus. Because of the flaring thickness, the mean volume density of CO (as well as of HI) decreases even more rapidly with increasing R in the outer Galaxy than the surface density. There is no measure of the cloud-cloud velocity dispersion at $R > R_0$ comparable to the scatter of terminal velocities used in the inner Galaxy (see Fig. 16), but the *intrinsic* velocity broadening of outer-Galaxy molecular clouds evidently does not differ from that typical of clouds in the Galaxy at large.

Collisions of molecular clouds have been held responsible for initiating star formation (see e.g. Scoville et al. 1986), as well as for increasing the dispersion of stellar velocities. The role of cloud interactions in initiating star formation will no doubt be critically considered as more information at large R becomes available. It appears now that, despite the low collision cross section, star formation proceeds in the outer Galaxy as elsewhere in our system, although it is not yet clear if star formation is initiated more, or less, efficiently in the inner Galaxy than in the outer. Carpenter et al. (1990) studied the stellar content and physical properties of a representative sample of outer-Galaxy molecular clouds, and found no difference in the ratio of far-infrared luminosity to mass between the large-R clouds and those in the inner Galaxy; they concluded that collisions do not enhance the star-formation rate per unit mass within a cloud.

Analyses of mapping of some outer-Galaxy molecular clouds by Digel (1991) and by Digel et al. (1990) have shown that the largest outer-Galaxy clouds are comparable in size and in mass to the giant ones in the molecular annulus (cf. Mead et al. 1990), but that CO luminosities of the outer-Galaxy clouds are some 4 times lower than the comparable inner-Galaxy values. This difference has been interpreted in terms of a lower metallicity, or a lower temperature, for the clouds at large R. Mead & Kutner (1988) reported lower kinetic temperatures on the surfaces of outer-Galaxy clouds than characteristic of clouds in the molecular annulus.

One particular aspect of the above considerations which is of importance to the morphological subject of these lecture notes concerns the conversion factor $X = N(\mathrm{H_2})/I(\mathrm{CO})$, which, although found to be quite constant over most of the inner Galaxy, may increase with increasing R in the outer regions (see Digel et al. 1990; Elmegreen 1989). Assuming a constant conversion factor, Wouterloot et al. (1990) derived the mass of $\mathrm{H_2}$ clouds in the outer Galaxy (at $R > R_0 = 8.5\,\mathrm{kpc}$) to be $5.8 \times 10^8 \, M_0$, and the corresponding HI mass to be $5.3 \times 10^9 \, M_0$.

7.5 Warps in the Two Nearby Galaxies M31 and M33

Although large-scale warps are now known to be a common aspect of the morphology of the gas layers in the outer parts of spiral galaxies, few distant external systems reveal their shape with the clarity found for the Milky Way. The two nearest large spirals are both warped: the shape of the HI layer in M31 closely resembles that of our own Galaxy (see e.g. Henderson 1979; Brinks & Burton 1984), while the warp in M33 is much more severe (Rogstad et al. 1976).

Brinks & Burton (1984) have modelled the form and kinematics of the outer HI layer in M31 using a thin, flat gas disk extending over the region at $R < 16$ kpc, and a flaring, warped gas distribution at larger radii. Although some more or less edge-on galaxies (M31 is tilted only about 13° from the line of sight) reveal their warp by a characteristic integral-sign shape of the projected HI surface densities, under some circumstances more detailed information is available from the HI kinematics. The viewing geometry of M31 is sketched in Fig. 70. The warped gas layer folds back onto the lines of sight so that M31 is typically traversed *twice*, allowing — as shown in Fig. 70 — separate identification of the velocities contributed by the flat inner disk and by the flaring warp. The model by Brinks & Burton utilized this kinematic distinction to derive the velocity field and HI distribution separately in the inner and outer parts of M31; the modelling procedure involved comparing simulated HI spectra with those observed.

The parameters specifying the global, flaring warp in M31 are remarkably similar to those specifying the warp in our own Galaxy. The deviation of the warped part of M31 from the extended plane of its flat inner disk increases approximately linearly with increasing distance from the center of that galaxy. The maximum deviation of the warp from the plane of the extended disk is about 3.8 kpc. The thickness of the HI disk flares, reaching a maximum thickness at the outer edge of the warp of about 1.7 kpc.

As in our Galaxy, the flaring nature of the layer together with the quite flat rotation curve require that the mass distribution in M31 resides predominantly in a halo. Also as in our Galaxy, the approximate 2π-symmetry and the general regularity of the warp in M31 argue against it being a distortion induced by the passage of a small companion (e.g. the LMC in the case of the Milky Way, or M32 in the case of the Andromeda galaxy), or by a merger. The large degree of azimuthal symmetry and the general regularity of the Milky Way warp suggest that it is a persistent, relaxed, intrinsic global phenomenon, and it is along these lines that most dynamical interpretations for the warp phenomenon are being sought. We note in this regard that warped, but largely isolated, galaxies have also been observed.

Figure 71 shows schematic representations of the warped HI layer in the Milky Way and in M31, drawn to the same scale, and viewed under comparable perspectives. These representations suggest that even substantial, regular warps might be difficult to recognize from an external perspective from the projected shapes alone, without kinematic information.

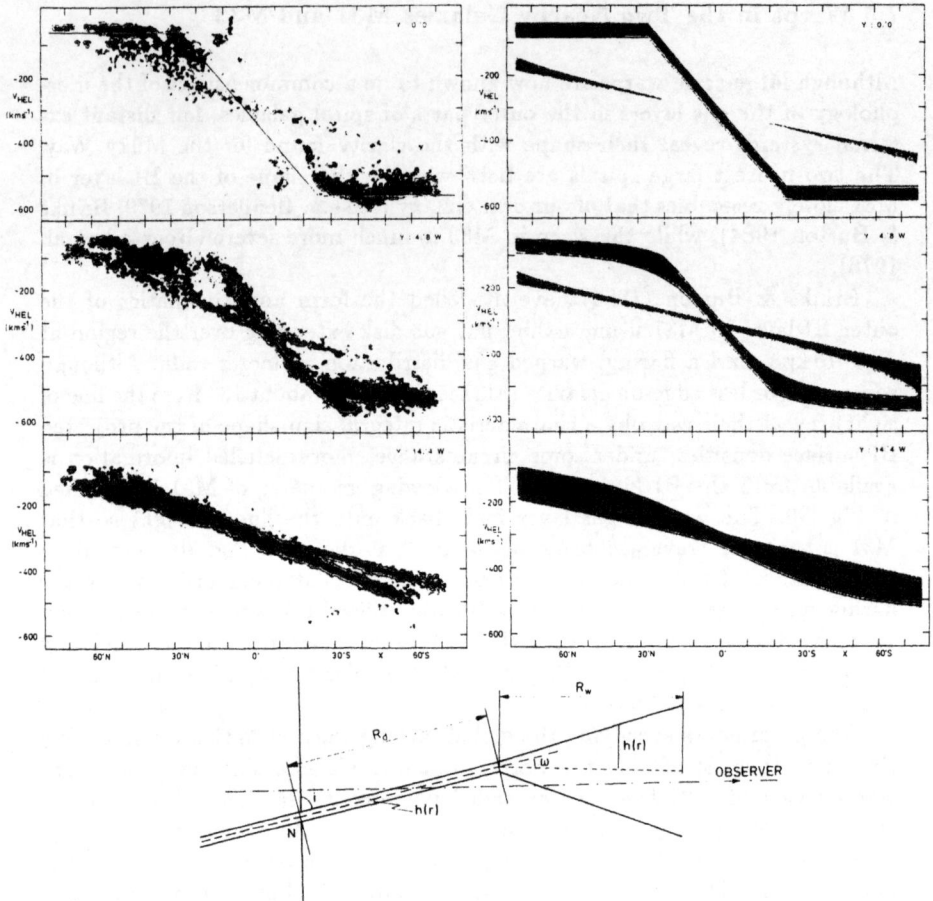

Fig. 70. Demonstration that the HI layer in M31 is globally warped and flared in its outer parts in a manner quite similar to that pertaining for the Milky Way (from Brinks & Burton 1984).
Left-hand panels: Observed position-velocity maps of HI emission from M31 made along and parallel to the major axis (from the data of Brinks & Shane 1984).
Right-hand panels: Simulated position-velocity maps along and parallel to the major axis corresponding to gas layer which is flat at radii within M31 $R < 16$ kpc, but which is warped and flaring at larger radii.
Bottom schematic: Sketch of the geometry presented by the flaring warp of the HI layer in M31, whereby lines of sight separately cross the flat inner, and the warped outer, parts of the galaxy, revealing different kinematics. The sketch is drawn to scale, but note that the thickness of the HI layer is indicated at ±3 times the scale height

   The third large spiral in the Local Group is M33, and this galaxy is also warped. There is some indication of a warp in the optical surface brightness (Sandage & Humphreys 1980) and in the integrated HI surface density (Wright et al. 1972), both of which can be rectified to a regular pattern by taking a warp into account. But as with the Milky Way and M31, the kinematic signature of

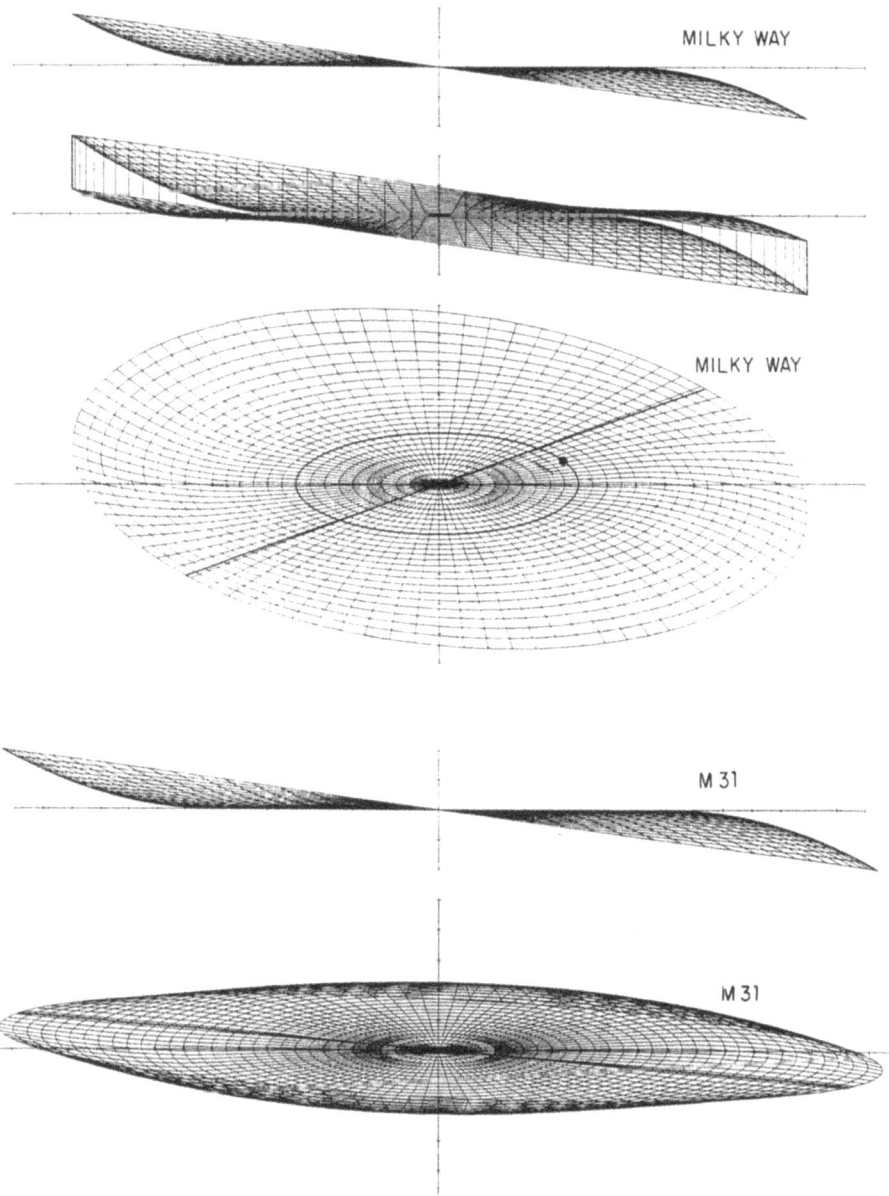

Fig. 71. Schematic representation of the warped HI layer in the Milky Way and in M31, drawn to the same scale (from Burton 1988a). The upper plot shows the midplane of the gas layer for the Galaxy viewed edge on, looking along the line of nodes of the warp; the second plot shows the surfaces at ±1 scale height, under the same orientation. The third plot shows the midplane of our Galaxy under the orientation it would present to a viewer in M31; the position of the Sun is indicated by the dot, the line of nodes of the warp is shown by the heavy line, and galactocentric rings are drawn at 1-kpc intervals. The upper plot in the M31 pair shows the warped midplane of M31, as viewed edge-on perpendicular to the line of nodes; the bottom plot shows the M31 midplane as viewed from our Galaxy

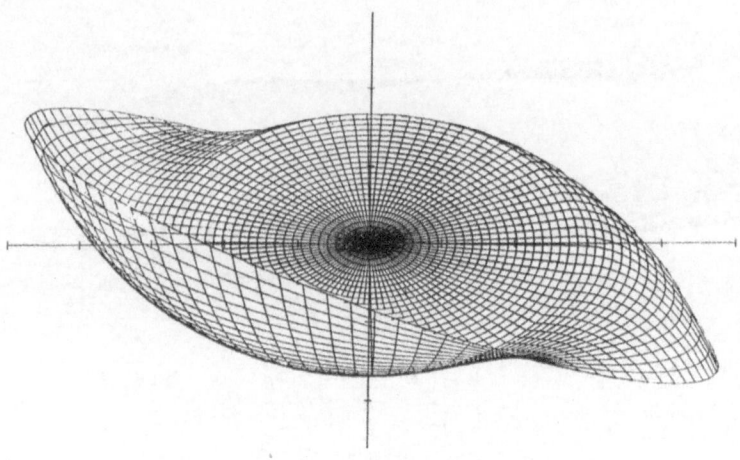

Fig. 72. Schematic model of the warped HI layer in M33, drawn to approximately 2.5 the scale used in Fig. 71 for M31 and the Milky Way (from Burton & Deul, unpublished). The parameters of the severely warped M33 HI layer were determined by fitting the modelled kinematics to those observed in the Westerbork observations of Deul & van der Hulst (1986)

the M33 warp is much clearer than the surface-density signature. Rogstad et al. (1986) showed that in M33, as in M31, the kinematics are more complicated than they would be for a single traverse of the lines of sight through a flat gas layer. Figure 72 shows the shape of the warped gas layer of M33 which leads to simulated HI profiles which approximately fit the kinematic patterns in the detailed Westerbork survey of Deul & van der Hulst (1986).

Thus all three of the local large spirals have warped outer gas layers. In the context of comparing the Milky Way morphology with that of M31, it is also interesting to note that, in addition to sharing the warped, flaring outer regions dominated by the presumably quite spheroidal potential distribution, they both show a tilted plane of symmetry in the inner few kpc — see Ciardullo et al. (1988) for the stellar situation in the inner disk of M31, and Braun (1991) for the HI situation — where the quite spheroidal bulge subsystem dominates.

It would of course be interesting to know what percentage of all galaxies with a regular gaseous disk are in fact warped. The statistics are not yet certain. Figure 71 suggests that the circumstances under which a warped galaxy would show the expected characteristic integral-sign shape are quite restrictive. The system would have to be viewed rather close to edge on, and under an orientation with the line of nodes pointing rather directly at the observer; the line of nodes would also have to remain quite straight. Even under those circumstances, the tendency might be to overestimate the gas-layer thickness, instead of to identify a warp. We note that it is evidently common that the HI layer starts to bend at radii where the stellar luminosity is terminating; this further complicates identification of warps from the shape of the optical surface brightness pattern.

Just as for our own Galaxy, and for M31 and M33, kinematic information is essential for a detailed description of the warped-layer morphology. Bosma

(1981a,b) pioneered the identification of warps in distant galaxies on the basis of their twisted kinematic signatures. A quite different, independent kinematic indication that warps are common is suggested by comparing the measures of HI line-widths made in *face-on* systems by Lewis (1987), using the Arecibo telescope whose beam, albeit small, is typically larger than the galaxies sampled, with high-resolution measures made with the VLA (e.g. Dickey et al. 1990) which resolve much structure in the face-on disks. The global velocity dispersions found by Lewis are typically an order of magnitude larger than the dispersions found in the resolved disks, suggesting that the larger beams are sampling entire distorted disks and are giving kinematic indications of warps.

Bosma (1991) estimates that the fraction of warped HI disks is of order 50% at least, based on statistics of edge-on, integral-sign systems as well as of more face-on systems revealing their warps via their kinematics. Battaner et al. (1991) suggest a frequency of warps higher than 80%, largely on the basis of optical cross sections. It is, of course, easier to say that a given system which happens to present a clear signature to that effect *is* warped, than to say that a system without such a signature is *not* warped; it seems that it would be very difficult to argue against the proposition that *all* galactic disks are globally warped.

References

7.1   Abramenko, B. 1978, *Astrophys. and Space Science* **54**, 323
7.2   Baldwin, J.E., Lynden-Bell, D., Sancisi, R. 1980, *Monthly Notices Roy. Astron. Soc.* **193**, 313
7.3   Battaner, E., Florido, E., Sanchez-Saavedra, M.-L., Prieto, M. 1991, in *Warped Disks and Inclined Rings around Galaxies*, S. Casertano, P.D. Sackett, F.H. Briggs, (eds.), Cambridge University Press, p. 200
7.4   Blitz, L., Fich, M., Kulkarni, S. 1983, *Science* **220**, 1233
7.5   Blitz, L., Fich, M., Stark, A.A. 1980, in *Interstellar Molecules*, B.H. Andrew, (ed.), Reidel Pub. Co., p. 213
7.6   Blitz, L., Lockman, F.J., (eds.) 1988, *The Outer Galaxy*, Springer-Verlag, Berlin
7.7   Blitz, L., Spergel, D.N. 1991, *Astrophys. J.* **370**, 205
7.8   Bloemen, J.B.G.M., Strong, A.W., Blitz, L., Cohen, R.S., Dame, T.M., Grabelsky, D.A., Hermsen, W., Lebrun, F., Mayer-Hasselwander, H.A., Thaddeus, P. 1984, *Astron. Astrophys.* **154**, 25
7.9   Bosma, A. 1981a, *Astron. J.* **86**, 1791
7.10 Bosma, A. 1981b, *Astron. J.* **86**, 1825
7.11 Bosma, A. 1991, in *Warped Disks and Inclined Rings around Galaxies*, S. Casertano, P.D. Sackett, F.H. Briggs, (eds.), Cambridge University Press, p. 181
7.12 Brand, J., Wouterloot, J.G.A. 1991, in *Proceedings IAU Symp. 144, "The Interstellar Disk-Halo Connection in Galaxies"*, J.B.G.M. Bloemen, (ed.), Kluwer Acad. Pub., p. 121
7.13 Braun, R. 1991, *Astrophys. J.* **372**, 54
7.14 Briggs, F.H. 1990, *Astrophys. J.* **352**, 15
7.15 Brinks, E., Burton, W.B. 1984, *Astron. Astrophys.* **141**, 195
7.16 Brinks, E., Shane, W.W. 1984, *Astron. Astrophys. Suppl. Ser.* **55**, 179
7.17 Burke, B.F. 1957, *Astron. J.* **62**, 90
7.18 Burton, W.B. 1985, *Astron. Astrophys. Suppl. Ser.* **62**, 365

7.19 Burton, W.B. 1988a, in *The Outer Galaxy*, L. Blitz, F.J. Lockman, (eds.), Springer-Verlag, Berlin, p. 94

7.20 Burton, W.B. 1988b, in *Galactic and Extragalactic Radio Astronomy*, G.L. Verschuur, K.I. Kellermann, (eds.), Springer-Verlag, New York, p. 295

7.21 Burton, W.B., te Lintel Hekkert, P. 1986, *Astron. Astrophys. Suppl. Ser.* **65**, 427

7.22 Carney, B.W., Seitzer, P. 1992, preprint, *IAU Symp.* **149**

7.23 Carpenter, J.M., Snell, R.L., Schloerb, F.P. 1990, *Astrophys. J.* **362**, 147

7.24 Casertano, S., Sackett, P.D., Briggs, F.H., (eds.) 1991, *Warped Disks and Inclined Rings around Galaxies*, Cambridge University Press

7.25 Ciardullo, R., Rubin, V.C., Jacoby, G.H., Ford, H.C., Ford Jr., W.K. 1988, *Astron. J.* **95**, 438

7.26 Combes, F. 1991, *Ann. Rev. Astron. Astrophys.* **29**, 195

7.27 Dame, T., Ungerechts, H., Cohen, R.S., de Geus, E., Grenier, I., May, J., Murphy, D.C., Nyman, L.-Å., Thaddeus, P. 1987, *Astrophys. J.* **322**, 706

7.28 Deul, E.R., van der Hulst, J.M. 1986, *Astron. Astrophys. Suppl. Ser.* **67**, 509

7.29 Dickey, J.M., Hanson, M.M., Helou, G. 1990, *Astrophys. J.* **352**, 522

7.30 Digel, S.W. 1991, Ph.D. Thesis, Harvard University

7.31 Digel, S.W., Bally, J., Thaddeus, P. 1990, *Astrophys. J.* **357**, L29

7.32 Diplas, A., Savage, B.D. 1991, *Astrophys. J.* **377**, 126

7.33 Djorgovski, S., Sosin, C. 1989, *Astrophys. J.* **341**, L13

7.34 Elmegreen, B.G. 1989, *Astrophys. J.* **338**, 178

7.35 Fich, M., Blitz, L. 1984, *Astrophys. J.* **279**, 125

7.36 Grabelsky, D.A., Cohen, R.S., Bronfman, L., Thaddeus, P. 1987, *Astrophys. J.* **315**, 122

7.37 Gum, C.S., Kerr, F.J., Westerhout, G. 1960, *Monthly Notices Roy. Astron. Soc.* **121**, 132

7.38 Heiles, C., Habing, H.J. 1974, *Astron. Astrophys. Suppl. Ser.* **14**, 1

7.39 Henderson, A.P. 1979, *Astron. Astrophys.* **75**, 311

7.40 Henderson, A.P., Jackson, P.D., Kerr, F.J. 1982, *Astrophys. J.* **263**, 116

7.41 Kerr, F.J. 1957, *Astron. J.* **62**, 93

7.42 Kerr, F.J. 1962, *Monthly Notices Roy. Astron. Soc.* **123**, 327

7.43 Kerr, F.J. 1983, in *Surveys of the Southern Galaxy*, W.B. Burton, F.P. Israel, (eds.), Reidel Pub. Co., p. 113

7.44 Kerr, F.J., Bowers, P.F. Jackson, P.D., Kerr, M. 1986, *Astron. Astrophys. Suppl. Ser.* **66**, 373

7.45 Kuijken, K. 1991, in *Warped Disks and Inclined Rings around Galaxies*, S. Casertano, P.D. Sackett, F.H. Briggs, (eds.), Cambridge University Press, p. 159

7.46 Kulkarni, S.R., Blitz, L., Heiles, C. 1982, *Astrophys. J.* **259**, L63

7.47 Lewis, B.M. 1987, *Astrophys. J. Suppl. Ser.* **63**, 515

7.48 Lockman, F.J. 1984, *Astrophys. J.* **283**, 429

7.49 Lockman, F.J. 1988, in *The Outer Galaxy*, L. Blitz, F.J. Lockman, (eds.), Springer-Verlag, Berlin, p. 79

7.50 Lockman, F.J., Jahoda, K., McCammon, D. 1986, *Astrophys. J.* **302**, 432

7.51 May, J., Murphy, D.C., Thaddeus, P. 1988, *Astron. Astrophys. Suppl. Ser.* **73**, 51

7.52 Mead, K.N. 1988, *Astrophys. J. Suppl. Ser.* **67**, 149

7.53 Mead, K.N., Kutner, M.L. 1988, *Astrophys. J.* **330**, 399

7.54 Mead, K.N., Kutner, M.L., Evans, N.J. 1990, *Astrophys. J.* **354**, 492

7.55 Merrifield, M.R. 1992, preprint, Can. Inst. Theoret. Astrophys., Toronto

7.56 Miyamoto, M., Yoshizawa, M., Suzuki, S. 1988, *Astron. Astrophys.* **194**, 107

7.57 Oort, J.H., Kerr, F.J., Westerhout, G. 1958, *Monthly Notices Roy. Astron. Soc.* **118**, 379

7.58 Rogstad, D.H., Wright, M.C.H., Lockhart, I.A. 1976, *Astrophys. J.* **204**, 703

7.59 Sandage, A., Humphreys, R.M. 1980, *Astrophys. J.* **236**, L1

7.60 Schwarz, U. 1985, *Astron. Astrophys.* **142**, 273

7.61 Scoville, N.Z., Sanders, D.B., Clemens, D.P. 1986, *Astrophys. J.* **310**, L77

7.62 Shaver, P.A., McGee, R.X., Newton, L.M., Danks, A.C., Pottasch, S.R. 1983, *Monthly Notices Roy. Astron. Soc.* **204**, 53

7.63 Shuter, W.L.H. 1981, *Monthly Notices Roy. Astron. Soc.* **199**, 109

7.64 Stark, A.A., Gammie, C.F., Wilson, R.W., Bally, J., Linke, R.A., Heiles, C., Hurwitz, M. 1992, *Astrophys. J. Suppl. Ser.* **79**, 77

7.65   Weaver, H.F., Williams, D.R.W. 1973, *Astron. Astrophys. Suppl. Ser.* **8**, 1

7.66   Westerhout, G. 1957, *Bull. Astron. Inst. Neth.* **13**, 201

7.67   Wouterloot, J.G.A., Brand, J. 1989, *Astron. Astrophys. Suppl. Ser.* **80**, 149

7.68   Wouterloot, J.G.A., Brand, J., Burton, W.B., Kwee, K.K. 1990, *Astron. Astrophys.* **230**, 21

7.69   Wouterloot, J.G.A., Brand, J., Henkel, C. 1988, *Astron. Astrophys.* **203**, 367

7.70   Wright, M.C.H., Warner, P.J., Baldwin, J.E. 1972, *Monthly Notices Roy. Astron. Soc.* **155**, 337

Large Scale Dynamics
of the Interstellar Medium

Bruce G. Elmegreen

IBM Research Division, T.J. Watson Research Center,
P.O. Box 218, Yorktown Heights, NY 10598, USA

1 Introduction

The interstellar medium (ISM) is like the ocean of a galaxy, composed largely of a cool fluid confined by gravity to a thin layer, and serving as a reservoir for all of the material in stars and planets that will ever form, evolve, and disperse. It was discovered about 60 years ago when observations of distant star clusters showed evidence for reddening and absorption by intervening dust [1]. The corresponding gas was observed soon after this in the form of narrow absorption lines [2], which eventually demonstrated that a large fraction of the matter is concentrated into clouds [3 – 5].

Before the early 1970's, there was little effort to consider this cloudy structure in theoretical studies of supernova remnants, HII regions, spiral density wave flows, gravitational instabilities, and so on. The terms "heating" and "cooling" referred to gas temperatures, shock fronts were imagined to be disturbances in uniform media, and large scale gas flows or instabilities were taken to be isothermal or adiabatic. Magnetic fields were not generally included in discussions of interstellar gas processes either (see review in [6]).

Now observations show clouds and structure everywhere with an intricate texture on a wide range of scales. The relative velocities between the cloudy parts of the fluid greatly exceed their internal thermal speeds, and usually exceed their internal Alfvén and gravitational escape speeds as well, so collisions between them are likely to be dissipative and sometimes destructive. There is also a pervasive magnetic field in the gas with an average uniform pressure slightly less than the turbulent or kinematic pressure, and with a random component in approximate energy equilibrium with these other motions.

These recent observations have led to a new and more complex picture of interstellar structure and dynamics, in which the concepts of heating and cooling apply not only to gas temperatures but also to the stirring and dissipation of supersonic cloud motions. Shock fronts in clumpy media are beginning to be studied, and large-scale gas flows are now modelled with the inclusion of various sources and sinks of turbulent energy, so that deviations from an adiabatic equation of state can be recognized.

The following lectures review these and other recent developments. The second chapter gives an observational overview of the structure of the interstellar medium, and the next four chapters discuss various theoretical topics: heating and cooling, explosions, spiral waves, and instabilities. The last chapter returns to observations of the global properties of the ISM and star formation.

References

1.1 Trumpler, R.J. 1930, *Lick Observatory Bulletin*, **14**, no. 420, 154
1.2 Dunham, T. Jr. 1937, *Publ. Astron. Soc. Pacific* **49**, 26
1.3 Adams, W.S. 1949, *Astrophys. J.* **109**, 354
1.4 Clark, B.G. 1965, *Astrophys. J.* **142**, 1398
1.5 Hobbs, L.M. 1969, *Astrophys. J.* **157**, 135
1.6 Kaplan, S.A., Pikel'ner, S.B. 1974, *Large Scale Dynamics of the Interstellar Medium*, *Ann. Rev. Astron. Astrophys.* **12**, 113

2 Basic State of the Interstellar Gas

2.1 Giant Clouds

High resolution observations of HI and CO in other galaxies, and surveys of the gas distribution in our Galaxy, show that most of the mass in the ISM is in the form of giant cloud complexes containing $10^5 M_\odot$ to $10^7 M_\odot$. These complexes often form one or more star clusters and OB associations, and they produce most of the new stars in a galaxy (see reviews in Efremov et al. [1,2]). They are highly confined to spiral arms, with generally smaller clouds between the arms [3 − 7]. There are also relatively smooth components of warm HI [8] and ionized gas [9] that extend for about a kiloparsec into the halo, and another component of hot gas [10] that may go further.

Figure 1 shows a diagram of the star clusters, HI clouds, and CO clouds in a spiral arm of M31 [11], where the CO emission tends to be concentrated in the densest parts of the giant HI clouds and in regions immediately adjacent to these clouds (see also [12 − 14]). This HI envelope/CO core structure is also typical for our Galaxy [15,16], the LMC [17] and similar irregular galaxies [18,19], and in some parts of NGC 6946 [7,20,21] and M101 [22]. Several of the giant HI clouds in NGC 6946 can be seen in Fig. 2, which plots HI emission at the local circular velocity.

In other galaxies, such as M51 [23 − 26], M33 [27], and M83 [28,29], the HI and CO emissions are separated by several hundred parsecs, with the HI in a dissociation region in the midst of intense star formation downstream from the CO ridge. In these cases, the whole ISM is highly molecular, and the gas becomes atomic only where the radiation field is large. The spiral confinement of the largest CO clouds in M51 is shown in Fig. 3.

Sometimes most of the CO is inside what is believed to be a galactic shock (e.g., M51 [25] and NGC 1068 [30]), and sometimes the molecular clouds are downstream from this shock (e.g., M81 [31,32] and M83 [29,33]), which appears primarily in the form of a dust lane or HI ridge with density-wave streaming motions (cf. Sect. 5.2.1). This relative position of the dense molecular clouds and shock may depend on the spiral arm strength (Sect. 5.2.1).

Giant HI and CO clouds appear in the spiral arms of the Milky Way, just as they do in other spiral galaxies. Figure 4 shows four (l, b) maps of HI emission from the first quadrant of our Galaxy [15], with each map at a different velocity, from $10 \, \text{km s}^{-1}$ at the bottom to $40 \, \text{km s}^{-1}$ at the top, with a $1 \, \text{km s}^{-1}$ interval around these velocities, plotted as a grayscale for HI brightness temperature. The longitude range is $10°$ to $70°$ and the latitude range is $-10°$ to $+10°$. The gas is clumped into $10^6 M_\odot$ to $10^7 M_\odot$ clouds, many of which are associated with known regions of star formation such as the one surrounding M17 at $V = 20 \, \text{km s}^{-1}$ and $l = 15°$ (the little black dot in the middle of this cloud is self-absorption from the M17 molecular cloud [34]). This type of cloudy structure dominates the neutral hydrogen distribution in the first Galactic quadrant and it is probably characteristic of gas in our Galaxy. Clouds like these were

159

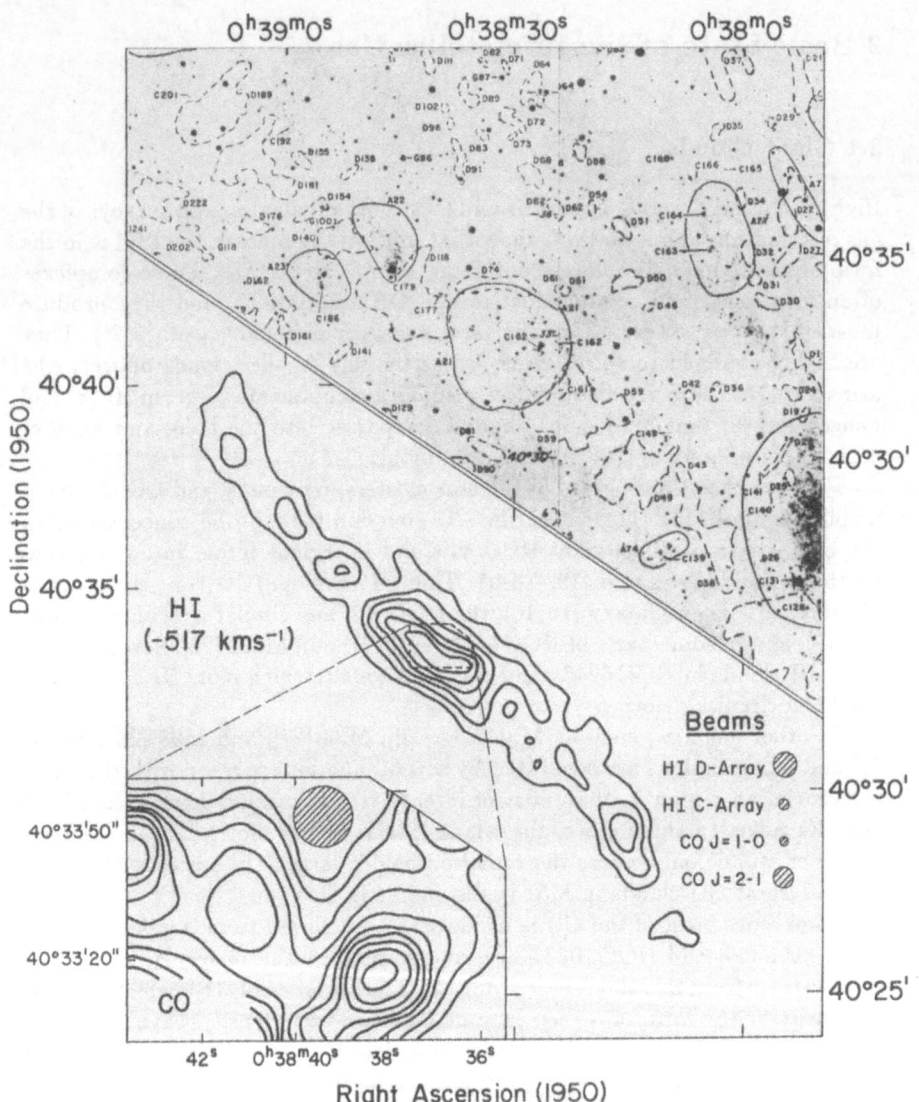

Fig. 1. Stars with HI and CO clouds in a spiral arm of M31 [11]

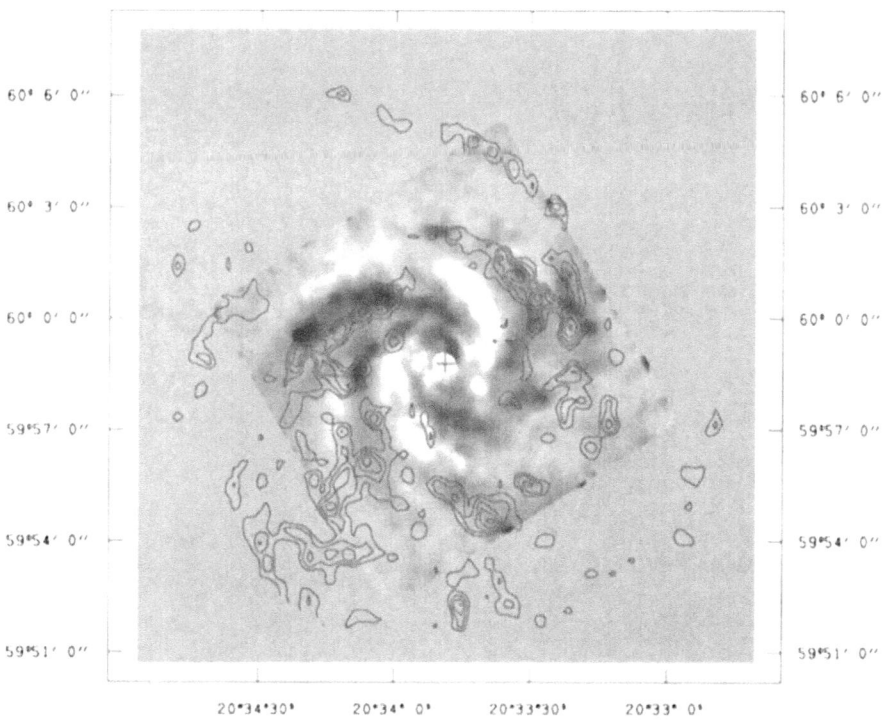

Fig. 2. HI in NGC 6946 [21] enhanced to show features at the local circular velocity

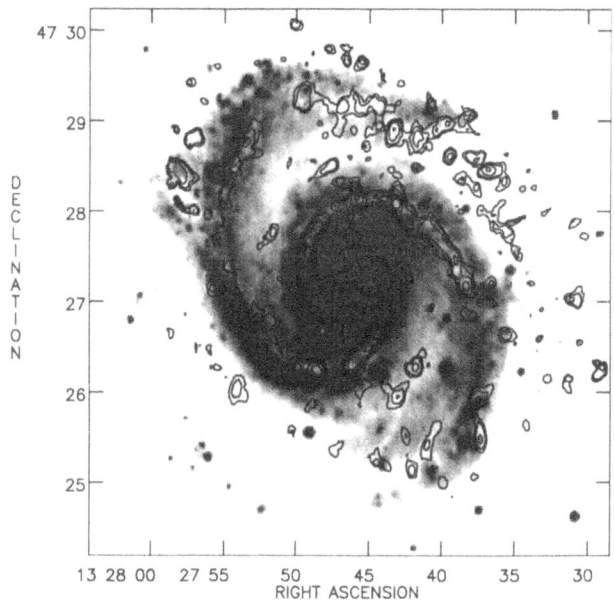

Fig. 3. CO in M51 [25] showing large cloud complexes in the spiral arms

161

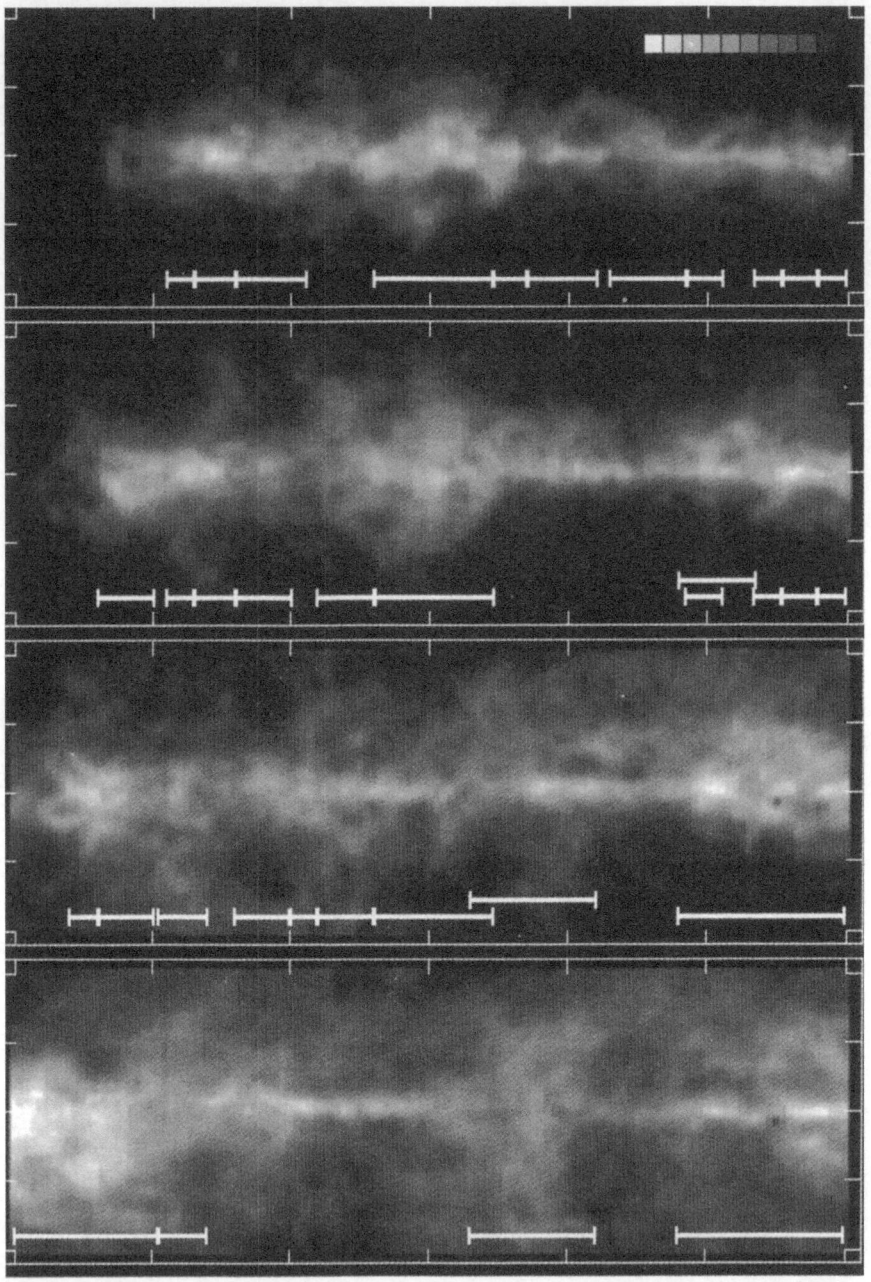

Fig. 4. HI in the inner Milky Way at four velocities, 10 (bottom), 20, 30 and $40 \, \text{km s}^{-1}$, plotted in galactic coordinates from $l = 10°$ to $l = 70°$ [15]. Large cloud complexes extend above and below the galactic plane

originally discovered by McGee & Milton [35], who displayed contours of HI emission from the *outer* Galaxy on an (l, b) map with the velocity a function of longitude, such that each longitude corresponded to a different distance, tracing out the spiral structure. The corresponding CO structure in the outer Galaxy was found by Grabelsky et al. [16].

The identification of giant HI clouds in the inner Galaxy (Fig. 4) is more ambiguous than it was for McGee & Milton [35] in the outer Galaxy because each velocity corresponds to two distances in the inner Galaxy and because velocity crowding is more severe there [36, 37]. Thus some of the giant HI clouds in Fig. 4, particularly on the left in each diagram, are illusions resulting from the superposition of several independent clouds along the line of sight. But many other clouds, especially those far from this "terminal velocity" region, appear to be real physical objects because they have well defined borders in (l, b, v) space, they are associated with known CO clouds and HII regions at the same distance, and their internal velocity dispersions are approximately equal to the virial theorem values for the observed masses and sizes [15]. They also resemble in mass, size, density, and mean separation the giant HI clouds seen in other galaxies like our own, so their presence is not surprising. Moreover, they have a density in excess of the limit for tidal binding, so they are unlikely to be disrupted by background tidal forces in their current positions.

For the 25 HI clouds in the inner Galaxy [15] that appear to be free from obvious blending or other ambiguities, the typical mass is $10^7 M_\odot$, the density is $10 \, \text{cm}^{-3}$, the radius is 100 pc, and the velocity dispersion along the line of sight is $5 \, \text{km s}^{-1}$. These are the largest clouds in our Galaxy. They trace the same spiral arms as the largest CO clouds and HII regions, which is reasonable considering the distribution of similar clouds in other galaxies like ours. Their spiral confinement also has a practical application, i.e., that there is no distance ambiguity for most of the largest clouds because very few lines of sight have them at both the near and far kinematic distances (which are on the same circle around the center of the Galaxy). A giant HI cloud at the near kinematic distance, for example, would be in a spiral arm, but then the far kinematic distance will usually be outside an arm and contain no similar HI cloud.

A remarkable property of the largest HI and CO clouds in our Galaxy and other galaxies is that they usually do not appear in the interarm regions. This implies that the clouds form in the arms and then disperse before (or when) they leave (the arms are presumably waves that flow through the gas and stars — see Chap. 5). The remarkable thing about this observation is that the time scale for the flow to go from an arm to the interarm region increases with galactocentric distance, up to the corotation radius. This implies that star formation inside the clouds, which probably has a fixed time scale, may not be the primary source of cloud destruction because then there would be more interarm superclouds or giant molecular associations in the inner part of a galaxy than in the outer part. More likely, the interarm transit itself destroys or loosens up the largest clouds (with help from star formation) because of the larger tidal forces in the interarm regions compared to the arms (cf. Chap. 5). Alternatively, the largest clouds in some galaxies could be unbound conglomerates [37, 38] and disperse in the interarm region because of the generally lower density there, regardless of tidal forces.

The dense molecular clouds that appear inside of these larger complexes are not so easily destroyed by tidal forces. Internal star formation probably destroys molecular clouds, which explains why there are more interarm molecular clouds in the inner Galaxy than in the outer Galaxy. For example, Stark [3] and others found small CO clouds between the arms in the inner Galaxy, but Dame et al. [4] showed that there are no comparable CO clouds in the interarm region between the solar neighborhood and the Perseus arm, which is in the outer Galaxy. Interarm molecular clouds could be the former cores of larger, lower-density clouds in the arms. These low-density clouds dispersed when the gas entered the interarm region because of large tidal forces, but the molecular clouds did not disperse this way because of their higher density. This suggests that the *unbound* molecular associations in the interarm regions of M51 [25] may be tidally disrupted remnants of formerly bound molecular associations in the arms.

2.2 Giant Holes

Cool interstellar gas on all scales can often be characterized as consisting of clumps and filaments in the high density regions, and holes and tunnels in the low density regions. This structure is evident even on galactic scales, so in addition to the giant spiral arm clouds discussed in the previous subsection, there are also giant holes in the gas, seen as voids in the emission from HI and various molecules. Such regions have been studied for a long time in the Large Magellanic Cloud [39, 40] and in our Galaxy [41 − 45], where they appear as radio continuum, Hα, dust, or HI shells or partial shells surrounding low density or X-ray-emitting [46] gas. Reviews of giant holes and expanding cavities may be found in McCray & Snow [47], Ikeuchi [48], and Tenorio-Tagle & Bodenheimer [49].

Giant holes and shells are usually observed in other galaxies using HI aperture synthesis techniques [21, 50, 51]. According to a survey of M31 by Brinks & Bajaja [50], the average diameter of a hole in M31 is $\sim$ 200 pc, the average density outside the hole is $\sim$ 0.5 cm^{-3}, the average expansion velocity is $\sim$ 10 km s^{-1}, the average age is 10^7 y, the average missing mass is $10^5 M_{\odot}$, and the average kinetic energy required to make the hole is 10^{50} erg. A clear example in M31 of a hollow shell with star formation on the perimeter was discussed by Brinks et al. [52]. The sizes and energies of most of the holes and shells in the Milky Way, M31, and other similar galaxies are consistent with their formation around OB associations by the combined action of ionization, stellar winds, and supernovae [53]. The largest holes could be the result of an impact between high velocity clouds and the galactic plane [54].

Extremely large shells have recently been found in M101 [55], where there is a shell 1.5 kpc in diameter and expanding at 50 km s^{-1}, in NGC 1620 [56], where there is a partial shell 3 kpc in radius, and in NGC 3079 [57], where there are several kpc-size shells that open up into the halo, presumably allowing some of

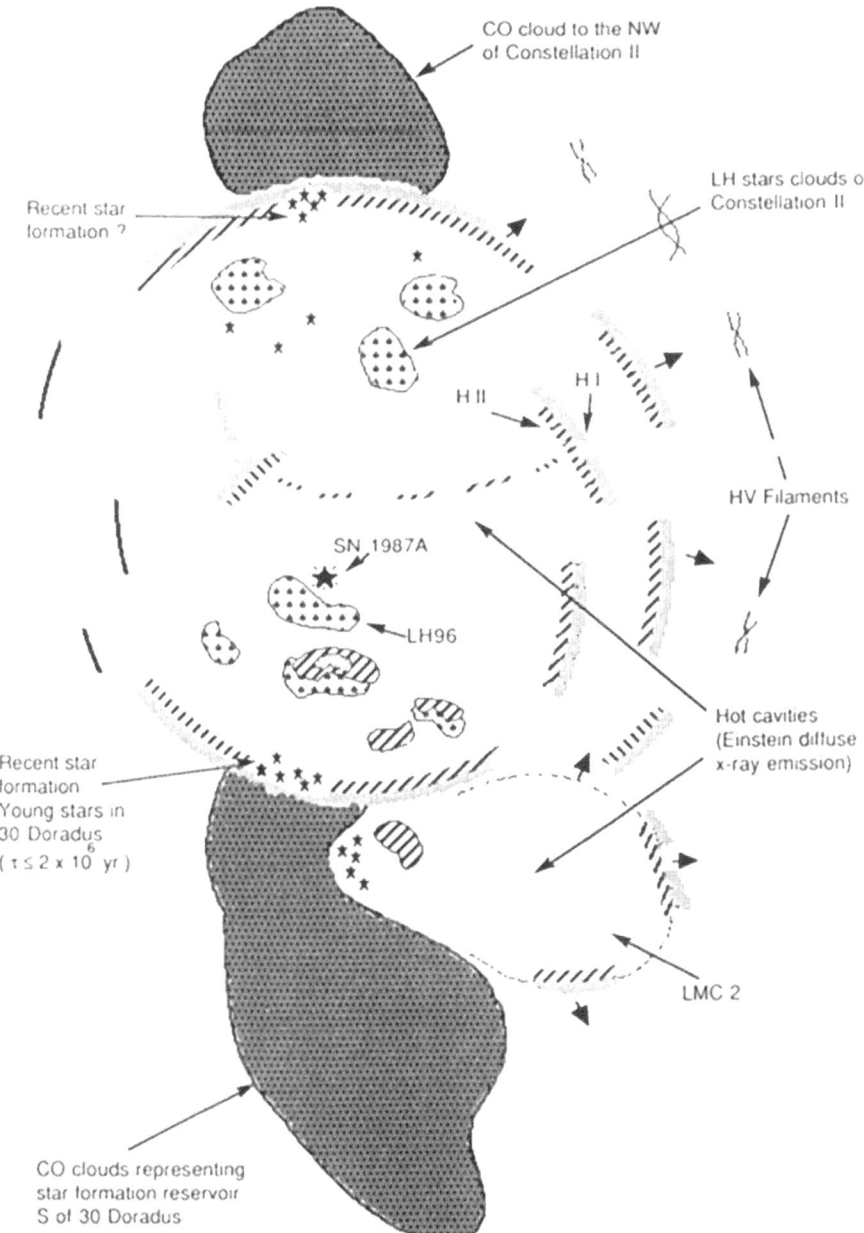

CO cloud to the NW
of Constellation II

LH stars clouds of
Constellation II

Recent star
formation ?

H II

H I

HV Filaments

SN 1987A

LH96

Hot cavities
(Einstein diffuse
x-ray emission)

Recent star
formation
Young stars in
30 Doradus
($\tau \leq 2 \times 10^{6}$ yr)

LMC 2

CO clouds representing
star formation reservoir
S of 30 Doradus

Fig. 5. Giant cavity near 30 Doradus in the Large Magellanic Cloud [65]

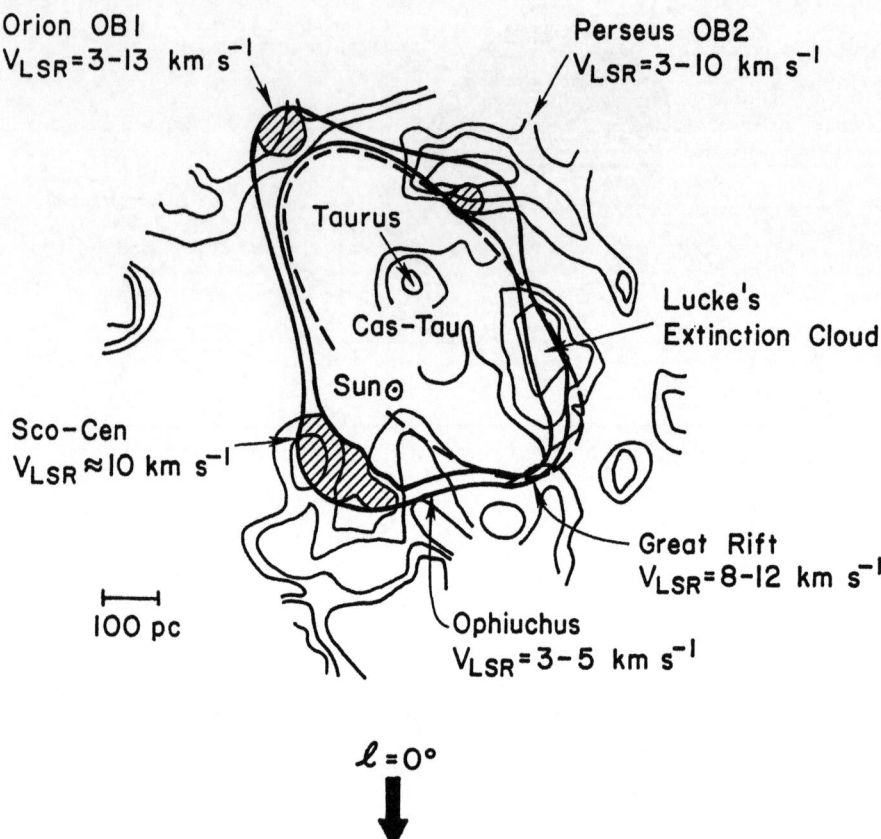

Fig. 6. Giant cavity around the Sun, with the Orion, Perseus and Sco-Cen OB associations on the periphery, from [74]

the hot gas to escape. Such large regions do not yet have a detailed theoretical explanation (cf. Sect. 4.5).

The best studied holes and shells are in the Large Magellanic Cloud [40, 58–61] and other irregular galaxies [62, 63], including the Small Magellanic Cloud [64]. A diagram of the cavity surrounding SN 1987A in the LMC and much of the recent star formation in this region is reproduced in Fig. 5, from Bruhweiler et al. [65]. The expansion from this cavity may have triggered the current epoch of star formation in the 30 Doradus region and elsewhere, and it apparently also created numerous HII filaments.

The solar neighborhood appears to be inside a large expanding cavity, and possibly inside two cavities. The largest is apparently defined by the HI ring around the Sun found by Lindblad et al. [66] and is centered on, and possibly driven by, the old Cas-Tau association [67, 68], which is $\approx 4 \times 10^7$ years old. Two slightly different models for the structure of this ring are in [69] and [70]. Reviews of these models and other features of Lindblad's ring are in Pöppel et al. [71, 72]. A recent study of Gould's Belt, which is the large local region of stars $\sim 7 \times 10^7$ years old that surrounds Lindblad's ring, is in Comerón & Torra [73].

Figure 6 shows a schematic diagram of the gas in Lindblad's ring (from [74]). The Orion, Perseus and Sco-Cen OB associations appear to be the main condensations in the ring (see also [75]). They could have formed by the gravitational collapse of a more uniform distribution of ring material, or they could have existed previously inside the cavity and been moved along with the general expansion. The first possibility seems preferred because the condensations appear to be at about the same radii as the lower density parts of the ring, and if the dense clouds were swept up, they should have lagged behind these lower density parts. In any case, these three associations apparently represent a second generation of star formation, triggered by the first generation Cas-Tau association. Star formation in the Ophiuchus cluster, which is part of the Sco-Cen cloud, and in Taurus, which may be a piece of the ring ejected inward from the perimeter [76], are probably in the third generation. One of these expanding regions that made the third generation (Ophiuchus) apparently also produced a second shell around the solar neighborhood, expanding away from the Sco-Cen association [77]. Other loops and bubbles connected with this local gas [78, 79], and with the Orion association [80, 81] illustrate the intense level of activity in regions of massive star formation.

A different origin for the expanding ring was proposed by Franco et al. [82], who suggested that a high latitude cloud hit the disk and made the ring as a splash. The formation of the individual clouds in the ring, such as Orion, could still be from gravitational instabilities as in the previous model, but then these clouds would be the first generation and the Cas-Tau association would be unrelated, or it could have been triggered immediately by the same cloud collision. Support for this model comes from observations by Alfaro et al. [83] who suggest that all of the largest complexes of star formation in the solar neighborhood and nearby spiral arms [2] are slightly offset from the galactic midplane, as if many of them were triggered by high-velocity cloud impacts.

The ubiquity of holes and shells has led to the suggestion that gas in the galactic midplane is continuously tossed up into the halo, perhaps in winds or chimney-like structures around each large OB association [84, 85] (cf. Sect. 4.5). Eventually some or all of this gas cools and falls back down.

2.3 Hierarchical Structure

The interstellar medium has a hierarchy of structures spanning a factor of at least 10^4 in length, from 0.1 pc to 1 kpc. This structure is evident on most maps that contain a sufficiently large number of pixels to resolve several stages in the hierarchy. Over small subsets of this range the structure is self-similar and fractal, both for gravitating and non-gravitating clouds, but it is not self-similar on all scales because galactic shear and spiral density waves are important on the largest scales. Thus, the large clouds and holes discussed above look different from the clumps and holes inside clouds. Reviews of hierarchical and clumpy

167

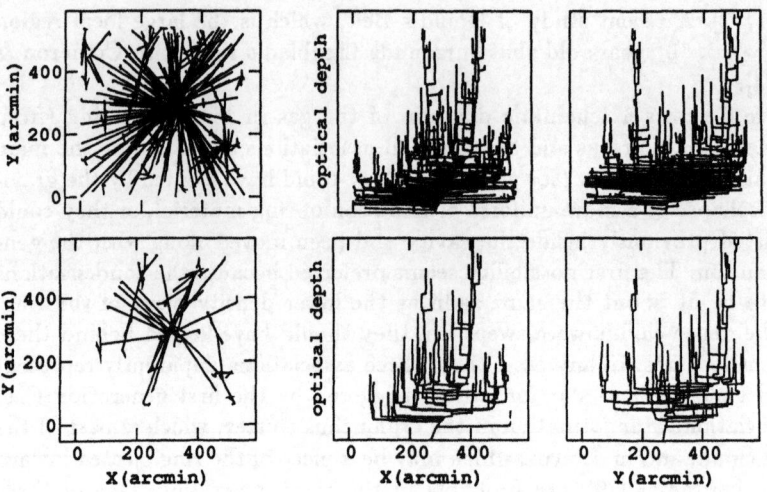

Fig. 7. Structure tree for a cloud in Taurus, showing branching from hierarchical structure [91]

structure are in Scalo [86, 87], Stutzki et al. [88], and Falgarone & Phillips [89]. A discussion of the cloud mass spectrum is in Dickey & Garwood [90].

Hierarchical structure implies that clouds contain clumps and these clumps contain even smaller subclumps, and so on. But in fact discrete clumps with sharp boundaries are difficult to find in high resolution studies — most of the gas is rather tenuous — and so stages in the hierarchy are poorly defined. Houlahan & Scalo [91] analyze hierarchical structure in terms of structure trees in which branches represent clumps inside the main cloud. Their structure tree for an extinction map of the Taurus region is shown in Fig. 7. The bottom trees have been pruned so that column density fluctuations less than a factor of 0.1 are not included. The hierarchical structure shows up as a branching of the tree toward higher opacity.

Another way to characterize the structure of a cloud is to determine the way in which the cloud's perimeter increases with area. If the cloud's perimeter is convoluted over a wide range of scales, with arcs and wiggles at all magnifications, then the total perimeter can increase faster than the $1/2$ power of the area. One way to measure the amount of convolution on the perimeter of an object is with the logarithmic derivative [92]

$$D_{\mathrm{H}} = 2\,\frac{d\log P}{d\log A} \qquad (1)$$

for perimeter P enclosing a cloud of area A. A sequence of nested regular polygons has $D_{\mathrm{H}} = 1$, but both atmospheric and interstellar clouds have $D_{\mathrm{H}} \approx 1.36$ [87, 92 − 96], which implies that the perimeter increases faster than the square root of the area. When this derivative is uniform over a large range of scales, the structure is fractal [97], and when it varies, the structure is multifractal. Such properties are studied for interstellar clouds because turbulence has a

168

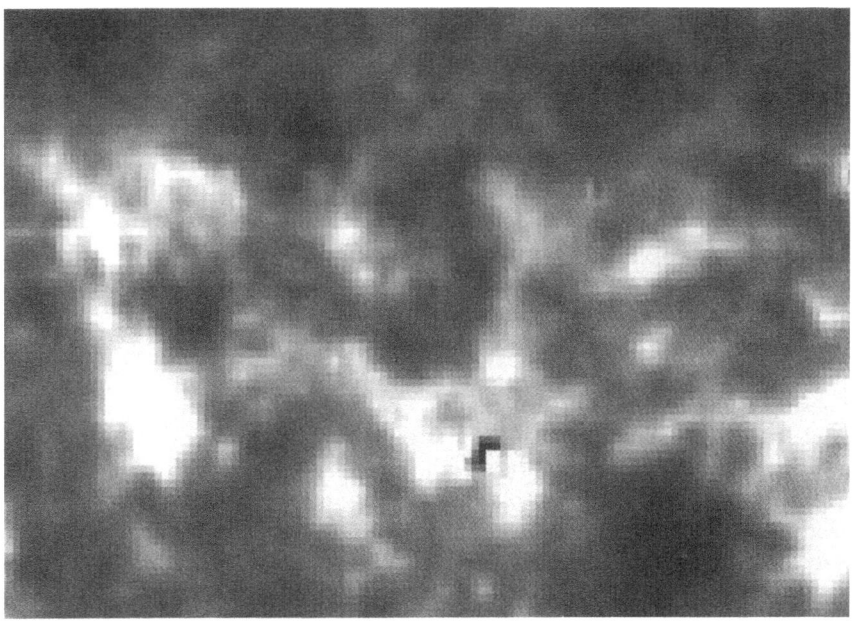

Fig. 8. HI emission in the direction of the Perseus arm, between $l = 103°$ and $l = 123°$ and at a velocity of $-30\,\mathrm{km\,s^{-1}}$ [103]

fractal quality, and there is some hope of understanding interstellar turbulence this way [89]. The method is not without ambiguity, however: a random distribution of unresolved clumps in a cloud may have the same D_H as a turbulent gas even when there is no turbulence [92].

Grayscale maps of the emission from molecular clouds illustrate the difficult task of quantifying their complex structure. The map of CO emission in Orion that was made by Bally et al. [98,99], for example, shows a lacy network of clumps, filaments, holes and tunnels, all surrounded by a high pressure HII region from an old epoch of star formation and all enclosing a new generation of stars that are currently forming.

Cloud spectra show evidence for unresolved clumpiness too. The spectra indicate a greater clumpiness or filling factor in the cloud cores [100,101] than on the periphery. The spectra also show some characteristics of magnetic waves [102], which are occasionally observed directly (cf. Sect. 2.4-2.5).

Figure 8 shows another example of clumpy filamentary structure from a recent study using the 140-foot telescope at NRAO in Greenbank, West Virginia [103]. The emission shown is in a $2\,\mathrm{km\,s^{-1}}$ velocity interval at $V_{\mathrm{LSR}} = -30\,\mathrm{km\,s^{-1}}$ in the direction of the Perseus arm. The longitude range is from 103 to 120 degrees and the latitude range is from -6 to $+8$ degrees. The pixels measure 10 by 20 arcmin in latitude and longitude, respectively (the beam resolution is 20 arcmin). The dark spot is self-absorption in front of the bright continuum source Cas A, a supernova remnant. The gas is highly textured with

169

clumps and filaments and various types of fluff. Similar structures are present along the entire velocity interval of the survey, from local gas at slightly positive velocities to distant gas in the outer spiral arm (beyond the Perseus arm) at $V_{LSR} = -80\,\mathrm{km\,s}^{-1}$ and further. Some of the brighter clumps can be associated with molecular clouds and HII regions, but most of the faint emission has no known connection with other observed interstellar or stellar features.

Visual inspection of Fig. 8 suggests that hierarchical structure in this region is not regular. There are no sharp boundaries, and no obvious large scale gradients. Considering the kinematic distance, the smallest structures that can be resolved are about 5 pc in diameter.

The hierarchical, or fractal, textured structure of interstellar clouds is probably caused by a combination of non-linear processes related to turbulence. These processes include fragmentation of larger clouds driven by self-gravity or thermal instabilities [104 − 106], sweeping and segmenting of gas by high pressure sources, especially stars with a power law distribution of luminosities, fracturing during cloud collisions [107], and transient compression by hydrodynamic [108] or hydromagnetic turbulence [109 − 110]. Presumably some distinction between these mechanisms will be revealed by the motions of the gas.

2.4 Gas Motions

Relative velocities between different parts of the interstellar medium tend to increase with separation. For the ISM as a whole this increase was discussed by Larson [111]; for the internal velocity dispersions inside clouds, it was discussed by the same author in a later paper [112]. Numerous other studies followed ([113] and references therein). The correlation takes the form of a power law, $V \propto D^\kappa$ with κ between $1/3$ and $1/2$ for velocity V and separation D. Kolmogorov's [114] model of turbulence has $\kappa = 1/3$ because the energy density cascades from large to small scales at an average rate proportional to $(\rho V^2)(V/D)$ for each scale D. When ρ is constant, V^3/D is constant too. For compressible turbulence ρ is not constant but may be larger on smaller scales [115], in which case $\kappa > 1/3$; this could explain the observed value of $\kappa \sim 1/2$. A value of $\kappa = 1/2$ also follows from the virial theorem at constant pressure with or without turbulence; i.e., $V^2 = 0.8\,G\rho R^2$ from the virial theorem and $\rho V^2 = $ constant from the pressure constraint; then $V^2 \propto R$ for cloud radius R [116 − 118]. A more general explanation for the correlations in terms of properties of turbulence is in Henriksen [119].

The relative velocities between clouds [120 − 122] are generally supersonic because internal cloud temperatures tend to be very low, such as 10 K to 100 K, and relative velocities of several kilometers per second correspond to energies with effective temperatures of several thousand degrees. The velocity-separation relation discussed above implies that the relative velocities between clouds are faster than their internal rms dispersions. If this internal dispersion is close to or slightly larger than the internal Alfvén speed, because much of the internal motion is magnetic for example, and if the internal dispersion is also comparable to the virial speed for a self-gravitating cloud, then the kinetic energy of random cloud motions exceeds the internal cloud binding energy. This implies that collisions should be somewhat shredding or shattering if they occur at the full rms speed [123]. Then the ISM as a whole would be very dissipative and its bulk properties, including the velocity dispersion, would be transient unless the gas is constantly stirred by energy from bright stars.

The ISM could be less dissipative if collisions between small near-neighbor clouds or clumps occurred at less than the full rms speed of the larger cloud or ISM region that encloses these small clouds or clumps. This situation might arise because of the velocity-separation correlation, which implies that smaller regions have smaller rms speeds. If the relative cloud speed decreases with decreasing separation, then most collisions between *near*-neighbors could be slow and not severely dissipative. In this case, the important relative speed for determining the shattering or coalescing properties of clouds is the relative speed at a separation of one collision mean free path. If this speed is comparable to or less than the internal rms speed of a typical cloud, then random collisions should lead to coalescence.

The ISM might be able to *regulate* itself via dissipation to stay close to this threshold [124]. If we imagine that as in turbulence, all motions satisfy the relation $v_{rel} \sim As^\kappa$ for relative velocity v_{rel} and separation s, and we consider a separation between clumps equal to the grazing collision mean free path, $\lambda = (4\pi n R^2)^{-1}$ for space density of clumps n and radii R, then the effective relative velocity is $v_{rel} = A(4\pi n R^2)^{-\kappa}$. The internal dispersion in a clump is much smaller than this, $v_{int} = AR^\kappa$. Thus

$$\frac{v_{rel}}{v_{int}} = \frac{1}{(4\pi n R^3)^\kappa} = \frac{1}{(3f)^\kappa} \qquad (2)$$

for clump volume filling factor f. Taking $\kappa = 1/2$ and $f \sim 0.1$, we get $v_{rel}/v_{int} \sim 1.8$, which is about the threshold between collisional coalescence and destruction for self-gravitating clumps [123]. This result implies that some collisions are destructive and some are sticking on the scale where $f \sim 0.1$. Note that if part of the pressure that confines the individual clouds is from collisions with other clouds, then $(v_{rel}/v_{int})^2 \sim \rho_c/\langle\rho\rangle = 1/f$ for cloud and average densities ρ_c and $\langle\rho\rangle$. Thus the index κ, which was shown above to equal $1/2$ for virialized motions at constant pressure, can also equal $1/2$ without gravity if collisions or momentum exchanging interactions of some type help to confine the clumps.

The velocity distribution function for interstellar clouds may be characterized as having a Gaussian core and an exponential tail, somewhat analogous to a hyperbolic secant function. This form was determined first using interstellar absorption lines from diffuse clouds on the lines of sight to nearby stars [125]. A similar form has been determined more recently from average spectra of individual clouds; Falgarone & Phillips [126] fit these average spectra to two Gaussians, and showed that the broad component is consistently three times wider than the narrow component. Early observations of broad spectral line wings that are not related to star formation were made by Blitz et al. [127, 128].

The Gaussian core motions of molecular line profiles, which are usually virialized [112, 129 − 131] (although see [37, 132]), are probably the result of turbulent [133, 134] and magnetic [110, 135 − 138] motions driven by stellar energy [139, 140], clump interactions [141], and gravity. Wave-like motions have been observed directly in some cases [142, 143], as well as torsional motions [144, 145]. Turbulence is apparently pervasive in many phases of the interstellar medium [146 − 148]. Even on the largest scales, turbulence and individual cloud motions are responsible for much of the thickness of the Galactic gas layer [149].

The origins of the exponential tail in the cloud velocity distribution function and the broad line wings in individual molecular clouds (i.e., far from embedded stellar sources) are not known. Plausible suggestions are that they result from two components at different temperatures or dispersions, mixed so completely that both are present on most lines of sight [127, 150]; magneto-hydrodynamic processes related to turbulent intermittency [126] or non-linear waves [110], and pervasive clump collisions with the core of the velocity profile coming from the shocked regions between colliding clumps and the wing of the velocity profile coming from the unshocked parts of the clumps [151]. All of these models may have applications somewhere.

The gas that emits in the linewings of molecular cloud spectra has apparently the same local (i.e., internal clump) density as the core-emitting gas [96], although the average density of the linewing gas (averaged over the dense clumps in a large region) may be less than the average for the line-core gas by a factor of ∼ 10 [110, 152]. The excess motion in the wings apparently does not heat the cloud much [153], which seems to suggest that magnetic processes such as waves or non-colliding interactions between clumps are involved.

2.5 Magnetism

The average magnetic field strength in the local Milky Way is about 1.6×10^{-6} Gauss, and the random component of the field has a strength of about 5×10^{-6} Gauss [154]. Similar and higher values are observed in other galaxies [155 − 157]. The average magnetic pressure, $B^2/8\pi$, for $B = 1.6 \times 10^{-6}$ Gauss corresponds to an equivalent $nT = 740\,\mathrm{cm}^{-3}$ K, which is so low that the average field is easily pushed around by stellar pressures and cloud motions. The random component of the magnetic pressure, corresponding to 5×10^{-6} Gauss, is $7200\,\mathrm{cm}^{-3}$ K, which is approximately $1/3$ of the total pressure in all forms in the Galactic midplane. Because of likely field line distortions in the random component, this total magnetic pressure should be more isotropic than the average. It is also large enough to resist in a significant way the transient pressure fluctuations from stars and distant supernovae, and to resist random cloud motions. If the magnetic field strength is determined by a dynamo in most disk galaxies, then the observed variation in field strength from galaxy to galaxy suggests a variation in the total interstellar pressure.

Recent observations of magnetic field orientations in other galaxies suggest that there are two types of configurations [155]. In axisymmetric spiral fields, the magnetic orientation is in the same sense, inward or outward, along each arm, while in bisymmetric spiral fields, the orientation is opposite for the two arms, i.e. the field points inward in one arm and outward in the other arm. Theory suggests that these two configurations are different modes of a galactic dynamo [158 − 166]. A third configuration is observed for some highly inclined galaxies in which the field lines point outward, perpendicular to the disk [156, 167]. Such structures may result from explosions in the midplane [168].

Detailed observations by Beck [169] of the magnetic field structure in the galaxy NGC 6946 indicate that the degree of polarization is largest in the inter-arm regions and smallest in the arms. His explanation is that star formation in the arms creates a chaotic field there, and this decreases the degree of polarization even though the absolute field strength may be larger. The first magnetic Zeeman measurement in another galaxy was recently reported by Kazes et al. [170].

Magnetic field strengths in Galactic clouds have been summarized recently by Heiles et al. [171, 172] and Güsten & Fiebig [173]. The implications of such fields for star formation have been reviewed by Lisano [174], McKee et al. [175], Mestel [176], and Mouschovias [177, 178].

Acknowledgement: I am grateful to Drs. Richard Rand, Françoise Combes, and John Scalo for helpful comments.

References

2.1 Efremov, Yu.N. 1989, *Stellar Complexes*, (London: Harwood)
2.2 Efremov, Yu., N., Sitnik, T.G. 1988, *Sov. Astron.* 14, 347
2.3 Stark, A.A. 1979, Ph.D. Dissertation, Princeton University
2.4 Dame, T.M., Elmegreen, B.G., Cohen, R.S., Thaddeus, P. 1986, *Astrophys. J.* 305, 892
2.5 Combes, F. 1991, *Ann. Rev. Astron. Astrophys.* 29, 195
2.6 Kutner, M.L., Verter, F., Rickard, L.J. 1990, *Astrophys. J.* 365, 195
2.7 Casoli, C.F., Clausset, F., Viallefond, F., Combes, F., Boulanger, F. 1990, *Astron. Astrophys.* 233, 357
2.8 Jahoda, K., Lockman, F.J., McCammon, D. 1990, *Astrophys. J.* 354, 184
2.9 Reynolds, R.J. 1989, *Astrophys. J.* 339, L29
2.10 Spitzer, L. Jr., Jenkins, E.B. 1975, *Ann. Rev. Astron. Astrophys.* 13, 133
2.11 Lada, C.J., Margulis, M., Sofue, Y., Nakai, N., Handa, T. 1988, *Astrophys. J.* 328, 143
2.12 Boulanger, F., Bystedt, J., Casoli, F., Combes, F. 1984, *Astron. Astrophys.* 140, L5
2.13 Nakano, M., Ichikawa, T., Tanaka, Y.D., Nakai, N., Sofue, Y. 1987, *Publ. Astron. Soc. Japan* 39, 57
2.14 Koper, E., Dame, T.M., lsrael, F.P., Thaddeus, P. 1991, *Astrophys. J.* 383, L11
2.15 Elmegreen, B.G., Elmegreen, D.M. 1987, *Astrophys. J.* 320, 182
2.16 Grabelsky, D.A., Cohen, R.S., May, J., Bronfman, L., Thaddeus, P. 1987, *Astrophys. J.* 315, 122

2.17   Israel, F.P., de Graauw, Th., van de Stadt, H., de Vries, C.P. 1986, *Astrophys. J.* **303**, 186
2.18   Thronson, H. A. 1988, in *Molecular Clouds in the Milky Way and External Galaxies*, eds. R.L. Dickman, R.L. Snell, J.S. Young, (Berlin: Springer), p. 413
2.19   Hunter, D. 1992, in *Star Formation in Stellar Systems*, eds. G. Tenorio-Tagle, M. Prieto, F. Sanchez, Cambridge, Cambridge Univ. Press, in press
2.20   Tacconi, L.J., Young, J.S. 1990, *Astrophys. J.* **352**, 595
2.21   Boulanger, F., Viallefond, F. 1991, submitted to *Astron. Astrophys.*
2.22   Blitz, L., Israel, F.P., Neugebauer, G., Gatley, I., Lee, T.J., Beattie, D.H. 1981, *Astrophys. J.* **249**, 76
2.23   Rots, A.H., Bosma, A., van der Hulst, J.M., Athanassoula, E., Crane, P.C. 1990, *Astron. J.* **100**, 387
2.24   Tilanus, R.P.J., Allen, R.J. 1989, *Astrophys. J.* **339**, L57
2.25   Rand, R.J., Kulkarni, S.R. 1990, *Astrophys. J.* **349**, L43
2.26   Tilanus, R.P.J., Allen, R.J. 1991, *Astron. Astrophys.* **244**, 6
2.27   Wilson, C.D., Scoville, N. 1991, *Astrophys. J.* **370**, 184
2.28   Allen, R.J., Atherton, P.D., Tilanus, R.P.J. 1986, *Nature* **319**, 296
2.29   Lord, S.D., Kenney, J.D.P. 1991, *Astrophys. J.* **381**, 130
2.30   Planesas, P., Scoville, N., Myers, S.T. 1991, *Astrophys. J.* **369**, 364
2.31   Kaufman, M., Bash, F.N., Hine, B., Rots, A.H., Elmegreen, D.M., Hodge, P.W. 1989, *Astrophys. J.* **345**, 674
2.32   Brouillet, N., Baudry, A., Combes, F., Kaufman, M., Bash, F. 1991, *Astron. Astrophys.* **242**, 35
2.33   Wiklind, T., Rydbeck, G., Hjalmarson, A., Bergman, P. 1990, *Astron. Astrophys.* **232**, L11
2.34   Sato, F., Fukui, Y. 1978, *Astron. J.* **83**, 1607
2.35   McGee, R.X., Milton, J.A. 1964, *Austr. J. Physics* **17**, 128
2.36   Burton, W.B. 1971, *Astron. Astrophys.* **10**, 76
2.37   Issa, M., MacLaren, I., Wolfendale, A.W. 1990, *Astrophys. J.* **352**, 132
2.38   Garcia-Burillo, S. Combes, F., Gerin, M. 1992, *Astron. Astrophys.* , submitted
2.39   Westerlund, B.E., Mathewson, D.S. 1966, *Monthly Notices Roy. Astron. Soc.* **131**, 371
2.40   Meaburn, J. 1980, *Monthly Notices Roy. Astron. Soc.* **192**, 365
2.41   Berkhuijsen, E.M. 1971, *Astron. Astrophys.* **14**, 359
2.42   Sivan, J.P. 1974, *Astron. Astrophys. Suppl. Ser.* **16**, 163
2.43   Brand, P.W.J.L., Zealey, W.J. 1975, *Astron. Astrophys.* **38**, 363
2.44   Heiles, C. 1979, *Astrophys. J.* **229**, 533
2.45   Heiles, C. 1984, *Astrophys. J. Suppl. Ser.* **55**, 585
2.46   Cash, W., Charles, P., Bowyer, S., Walter, F., Garmine, G., Riegler, G. 1980, *Astrophys. J.* **238**, L71
2.47   McCray, R., Snow, T.P., Jr. 1979, *Ann. Rev. Astron. Astrophys.* **17**, 213
2.48   Ikeuchi, S. 1988, *Fund. Cosmic Physics* **12**, 255
2.49   Tenorio-Tagle, G., Bodenheimer, P. 1989, *Ann. Rev. Astron. Astrophys.* **26**, 145
2.50   Brinks, E., Bajaja, E. 1986, *Astron. Astrophys.* **169**, 14
2.51   Deul, E.L., Hartog, R.H. 1990, *Astron. Astrophys.* **229**, 362
2.52   Brinks, E., Braun, R., Unger, S.W. 1989, in *Structure and Dynamics of the Interstellar Medium*, IAU Colloquium 120, eds. G. Tenorio-Tagle, M. Moles, J. Melnick, Berlin, Springer, p. 524
2.53   Bruhweiler, F.C., Gull, T.R., Kafatos, M., Sofia, S. 1980, *Astrophys. J.* **238**, L27
2.54   Tenorio-Tagle, G. 1981, *Astron. Astrophys.* **94**, 338
2.55   Kamphuis, J., Sancisi, R., van der Hulst, T. 1991, *Astron. Astrophys.* **244**, L29
2.56   Chaboyer, B., Vader, J.P. 1991, *Publ. Astron. Soc. Pacific* **103**, 35
2.57   Irwin, J.A., Seaquist, E.R. 1990, *Astrophys. J.* **353**, 469
2.58   Wang, Q., Helfand, D. 1991, *Astrophys. J.* **370**, 541
2.59   Wang, Q., Helfand, D. 1991, *Astrophys. J.* **379**, 327
2.60   Dopita, M.A., Mathewson, D.J., Ford, W.L. 1985, *Astrophys. J.* **297**, 599
2.61   Wang, D., Hamilton, T., Helfand, D., Wu, X. 1991, *Astrophys. J.* **374**, 475
2.62   Shostak, G.S., Skillman, E.D. 1989, *Astron. Astrophys.* **214**, 33
2.63   Hunter, D.A., Gallagher, J.S. III 1991, *Astrophys. J.* **362**, 480
2.64   Hindman, J.V. 1967, *Aust. J. Phys.* **20**, 147
2.65   Bruhweiler, F.C., Fitzurka, M.A., Gull, T.R. 1991, *Astrophys. J.* **370**, 551

2.66   Lindblad, P.O., Grobe, K., Sandqvist, Aa., Schober, J. 1973, *Astron. Astrophys.* **24**, 309
2.67   Blaauw, A. 1984, Irish *Astron. J.* **16**, 141
2.68   Walter, F.M., Boyd, W.T. 1991, *Astrophys. J.* **370**, 318
2.69   Olano, C.A. 1982, *Astron. Astrophys.* **112**, 195
2.70   Elmegreen, B.G. 1982, in *Submillimeter Wave Astronomy*, eds. J.E. Deckman, J.P. Phillips, (Cambridge: Cambridge Univ. Press), p. 3
2.71   Pöppel, W. 1988, *Bol. Asoc. Argent. Astron.* **34**, 61
2.72   Pöppel, W.G.L., Marronetti, P., Benaglia, P. 1991, poster paper at the 21st IAU General Assembly
2.73   Comerón, F., Torra, J. 1991, *Astron. Astrophys.* **241**, 57
2.74   Elmegreen, B.G. 1991, in *Evolution of Interstellar Matter and Dynamics of Galaxies*, eds. B.W. Burton, P.O. Lindblad, J. Palouš, (Cambridge: Cambridge Univ. Press), in press
2.75   Taylor, D.K., Dickman, R.L., Scoville, N.Z. 1987, *Astrophys. J.* **315**, 104
2.76   Olano, C.A., Pöppel, W.G.L. 1987, *Astron. Astrophys.* **179**, 202
2.77   Innes, D.E., Hartquist, T.W. 1984, *Monthly Notices Roy. Astron. Soc.* **209**, 7
2.78   Vladilo, G. 1991, *Astrophys. J.* **372**, 494
2.79   Franco, G.A.P. 1990, *Astron. Astrophys.* **227**, 499
2.80   Reynolds, R.J., Ogden, P.M. 1978, *Astrophys. J.* **224**, 94
2.81   Cowie, L.L., Songaila, A., York, D.C. 1979, *Astrophys. J.* **230**, 469
2.82   Franco, J., Tenorio-Tagle, G., Bodenheimer, P., Rozyczka, M., Mirabel, I.F. 1988, *Astrophys. J.* **333**, 826
2.83   Alfaro, E.J., Cabrera-Cano, J., Delgado, A.J. 1991, *Astrophys. J.* **378**, 106
2.84   Shapiro, P.R., Field, G.B. 1976, *Astrophys. J.* **205**, 762
2.85   Norman, C.A., Ikeuchi, S. 1989, *Astrophys. J.* **345**, 372
2.86   Scalo, J.M. 1985, in *Protostars and Planets II*, eds. D.C. Black, M.S. Matthews, (Tucson: Univ. of Arizona), p. 201
2.87   Scalo, J.M. 1990, in *Physical Processes in Fragmentation and Star Formation*, eds. R. Capuzzo-Dolcetta, C. Chiosi, A. Di Fazio (Dordrecht: Kluwer), p. 151
2.88   Stutzki, J., Genzel, R., Graf, U., Harris, A., Sternberg, A., Güsten, R. 1991, in *Fragmentation of Molecular Clouds and Star Formation*, eds. E. Falgarone, F. Boulanger, G. Duvert, Dordrecht, Kluwer, p. 235
2.89   Falgarone, E., Phillips, T.G. 1991, in *Fragmentation of Molecular Clouds and Star Formation*, eds. E. Falgarone, F. Boulanger, G. Duvert (Dordrecht: Reidel), p. 119
2.90   Dickey, J.M., Garwood, R.W. 1989, *Astrophys. J.* **341**, 201
2.91   Houlahan, P., Scalo, J.M. 1992, *Astrophys. J. Suppl. Ser.* , in press
2.92   Dickman, R.L., Horvath, M.A., Margulis, M. 1990, *Astrophys. J.* **365**, 586
2.93   Lovejoy, S. 1982, *Science* **216**, 185
2.94   Beech, M. 1987, *Astrophys. Space Sci.* **133**, 193
2.95   Bazell, D., Desert, F.X. 1988, *Astrophys. J.* **333**, 353
2.96   Falgarone, E., Phillips, T.G., Walker, C.K. 1991, *Astrophys. J.* **378**, 186
2.97   Mandelbrot, B.B. 1977, *Fractals* (San Francisco: Freeman)
2.98   Bally, J., Langer, W.D., Stark, A.A., Wilson, R.W. 1987, *Astrophys. J.* **312**, L45
2.99   Bally, J. 1989, *Structure and Dynamics of the Interstellar Medium*, IAU Colloquium 120, eds. G. Tenorio-Tagle, M. Moles, J. Melnick, Berlin, Springer, p. 309
2.100 Kwan, J., Sanders, D.B. 1986, *Astrophys. J.* **309**, 783
2.101 Tauber, J.A., Goldsmith, P.F., Dickman, R.L. 1991, *Astrophys. J.* **375**, 635
2.102 Stenholm, L.G. 1990, *Astron. Astrophys.* **232**, 495
2.103 Chromey, F.R., Elmegreen, B.G. 1991, in preparation
2.104 Kolesnik, I.G. 1991, *Astron. Astrophys.* **243**, 239
2.105 Monaghan, J.J., Lattanzio, J.C. 1991, *Astrophys. J.* **375**, 177
2.106 Elphick, C., Regev, O., Spiegel, E.A. 1991, *Monthly Notices Roy. Astron. Soc.* **250**, 617
2.107 Nozakura, T. 1990, *Monthly Notices Roy. Astron. Soc.* **243**, 543
2.108 Passot, T., Pouquet, A., Woodward, P. 1988, *Astron. Astrophys.* **197**, 228
2.109 Carlberg, R.G., Pudritz, R.E. 1990, *Monthly Notices Roy. Astron. Soc.* **247**, 353
2.110 Elmegreen, B.G. 1990, *Astrophys. J.* **361**, L77
2.111 Larson, R.B. 1979, *Monthly Notices Roy. Astron. Soc.* **186**, 479
2.112 Larson, R.B. 1981, *Monthly Notices Roy. Astron. Soc.* **194**, 809

2.113 Solomon, P.M., Rivolo, A.R., Barrett, J., Yahil, A. 1987, *Astrophys. J.* **319**, 730
2.114 Kolmogorov, A. 1941, *Compt. Rend. Acad. Sci. URSS* **30**, 301
2.115 Fleck, R.C., Jr. 1983, *Astrophys. J.* **272**, L45
2.116 Chieze, J.P. 1987, *Astron. Astrophys.* **171**, 225
2.117 Fleck, R.C., Jr. 1988, *Astrophys. J.* **328**, 299
2.118 Elmegreen, B.G. 1989, *Astrophys. J.* **338**, 178
2.119 Henriksen, R.N. 1991, *Astrophys. J.* **377**, 500
2.120 Clemens, D.P. 1985, *Astrophys. J.* **295**, 422
2.121 Alvarez, H., May, J., Bronfman, L. 1990, *Astrophys. J.* **348**, 495
2.122 Stark, A.A., Brand, J. 1989, *Astrophys. J.* **339**, 763
2.123 Vazquez, E.C., Scalo, J.M. 1989, *Astrophys. J.* **343**, 644
2.124 Lattanzio, J.C., Henriksen, R.N. 1988, *Monthly Notices Roy. Astron. Soc.* **232**, 565
2.125 Munch, G., Zirin, H. 1961, *Astrophys. J.* **133**, 11
2.126 Falgarone, E., Phillips, T.G. 1990, *Astrophys. J.* **359**, 344
2.127 Blitz, L., Stark, A.A. 1986, *Astrophys. J.* **300**, L89
2.128 Blitz, L., Magnani, L., Wandel, A. 1988, *Astrophys. J.* **331**, L127
2.129 Dickman, R.L. 1975, *Astrophys. J.* **202**, 50
2.130 Lee, Y., Snell, R.L., Dickman, R.L. 1990, *Astrophys. J.* **355**, 536
2.131 Carpenter, J.M., Snell, R.L., Schloerb, F.P. 1990, *Astrophys. J.* **362**, 147
2.132 Maloney, P. 1990, *Astrophys. J.* **348**, L9
2.133 Dickman, R.L. 1985, in *Protostars and Planets II*, eds. D.C. Black, M.S. Matthews, Tucson, Univ. of Arizona, p. 150
2.134 Scalo, J.M. 1987, in *Interstellar Processes*, eds. D.J. Hollenbach, H.A. Thronson, Jr., Dordrecht, Reidel, p. 349
2.135 Arons, J., Max, C.E. 1975, *Astrophys. J.* **196**, L77
2.136 Morfill, G.E., Stenholm, L.G. 1980, *Astron. Astrophys.* **90**, 134
2.137 Myers, P.C., Goodman, A.A. 1988, *Astrophys. J.* **329**, 392
2.138 Fedorenko, V.N. 1990, in *Galactic and Intergalactic Magnetic Fields*, eds. R. Beck, P.P. Kronberg, R. Wielebinski, (Dordrecht: Kluwer), p. 153
2.139 Lada, C.J. 1985, *Ann. Rev. Astron. Astrophys.* **23**, 267
2.140 Snell, R.L., Dickman, R.L., Huang, Y.L. 1990, *Astrophys. J.* **352**, 139
2.141 Falgarone, E., Puget, J.L. 1986, *Astron. Astrophys.* **162**, 235
2.142 Shuter, W.L.H., Dickman, R.L., Klatt, C. 1987, *Astrophys. J.* **322**, L103
2.143 Verschuur, G.L. 1991, *Astrophys. Space Sci.* **185**, 137
2.144 Uchida, Y., Fukui, Y., Monishima, Y., Mizuno, A., Iwata, T., Takaba, H. 1991, *Nature* **349**, 140
2.145 Uchida, Y., Mizuno, A., Nozawa, S., Fukui, Y. 1990, *Publ. Astron. Soc. Japan* **42**, 69
2.146 Spangler, S.R., Gwinn, C.R. 1990, *Astrophys. J.* **353**, L29
2.147 Lazarian, A. 1991, in *Dynamics of Disc Galaxies*, ed. B. Sundelius, Göteborg, Chalmers University, p. 293
2.148 Lazarian, A. 1991, in *Fragmentation of Molecular Clouds and Star Formation*, eds. E. Falgarone, F. Boulanger, G. Duvert, Dordrecht: Kluwer, p. 65
2.149 Lockman, P.J., Gehman, C.S. 1991, *Astrophys. J.* **382**, 182
2.150 Siluk, R.S., Silk, J. 1974, *Astrophys. J.* **192**, 51
2.151 Keto, E.R., Lattanzio, J.C. 1989, *Astrophys. J.* **346**, 184
2.152 Magnani, L., Carpenter, J.M., Blitz, L., Kassim, N.E., Nath, B.B. 1990, *Astrophys. J. Suppl. Ser.* **73**, 747
2.153 Boreiko, R.T., Betz, A.L. 1991, *Astrophys. J.* **369**, 382
2.154 Rand, R.J., Kulkarni, S.R. 1989, *Astrophys. J.* **343**, 760
2.155 Beck, R. 1990, *Geophys. Astrophys. Fluid Dyn.* **50**, 3
2.156 Beck, R. 1991, in *The Interstellar Disk-Halo Connection in Galaxies*, ed. H. Bloemen, (Dordrecht: Kluwer), p. 267
2.157 Buczilowski, U.R., Beck, R. 1991, *Astron. Astrophys.* **241**, 47
2.158 Chiba, M., Tosa, M. 1989, *Monthly Notices Roy. Astron. Soc.* **238**, 621
2.159 Chiba, M., Tosa, M. 1990, *Monthly Notices Roy. Astron. Soc.* **244**, 714
2.160 Chiba, M. 1991, *Monthly Notices Roy. Astron. Soc.* **250**, 769
2.161 Fujimoto, M., Sawa, T. 1990, in *Galactic and Intergalactic Magnetic Fields*, eds. R. Beck, P.P. Kronberg, R. Wielebinski, (Dordrecht: Kluwer), p. 90
2.162 Krauss, M. 1990, in *Galactic and Intergalactic Magnetic Fields*, eds. R. Beck, P.P. Kronberg, R. Wielebinski, (Dordrecht: Kluwer), p. 97

2.163 Ruzmaikin, A. 1990, in *Galactic and Intergalactic Magnetic Fields*, eds. R. Beck, P.P. Kronberg, R. Wielebinski, (Dordrecht: Kluwer), p. 83

2.164 Donner, K.J., Brandenberg, A. 1990, *Astron. Astrophys.* **240**, 289

2.165 Mestel, L., Subramanian, K. 1991, *Monthly Notices Roy. Astron. Soc.* **248**, 677

2.166 Hanasz, M., Lesch, H., Krause, M. 1991, *Astron. Astrophys.* **243**, 381

2.167 Hummel, E., Beck, R., Dahlem, M. 1991, *Astron. Astrophys.* **248**, 23

2.168 Sofue, Y. 1987, *Publ. Astron. Soc. Japan* **39**, 547

2.169 Beck, R. 1991, *Astron. Astrophys.* **251**, 15

2.170 Kazes, I., Troland, T.H., Crutcher, R.M. 1991, *Astron. Astrophys.* **245**, L17

2.171 Heiles, C. 1990, in *Galactic and Intergalactic Magnetic Fields*, eds. R. Beck, P.P. Kronberg, R. Wielebinski, Dordrecht, Kluwer, p. 35

2.172 Heiles, C., Goodman, A.A., McKee, C., Zweibel, E. 1992, in *Protostars and Planets III*, eds. E.H. Levy, M.S. Matthews, Tucson, University of Arizona, in press

2.173 Güsten, R., Fiebig, D. 1990, in *Galactic and Intergalactic Magnetic Fields*, eds. R. Beck, P.P. Kronberg, R. Wielebinski, Dordrecht, Kluwer, p. 305

2.174 Lisano, S. 1989, *Rev. Mex. Astron. Astrophys.* **18**, 11

2.175 McKee, C., Zweibel, E., Heiles, C., Goodman, A.A. 1992, in *Protostars and Planets III*, eds. E.H. Levy, M.S. Matthews, Tucson, University of Arizona, in press

2.176 Mestel, L. 1990, in *Galactic and Intergalactic Magnetic Fields*, eds. R. Beck, P.P. Kronberg, R. Wielebinski, Dordrecht, Kluwer, p. 259

2.177 Mouschovias, T.Ch. 1990, in *Galactic and Intergalactic Magnetic Fields*, eds. R. Beck, P.P. Kronberg, R. Wielebinski, Dordrecht, Kluwer, p. 269

2.178 Mouschovias, T.Ch. 1991, in *The Physics of Star Formation and Early Stellar Evolution*, eds. C.J. Lada, N.D. Kylafis, (Dordrecht: Kluwer), p. 61

3 Large Scale Heating and Cooling

3.1 Energy Sources

The random motion of interstellar gas is dominated by the kinetic energy of individual clouds rather than thermal motions. This implies that bulk heating of the interstellar medium involves the stirring of individual clouds or the formation of new clouds at high speeds. The first subsection in this chapter summarizes the various mechanisms that have been proposed for such cloud fluid heating. The second subsection summarizes the dissipation processes for the energy of bulk cloud motions. Reviews of this topic may be found in Wang & Cowie [1] and McCray & Snow [2].

The amount of energy required to stir the cloud population is approximately the total content of kinetic energy density, $U_{\rm ISM} = \frac{3}{2}\rho a^2$, which is about 10^{-12} erg cm^{-3} for average density ρ and one-dimensional velocity dispersion a, divided by the dissipation time, $T_{\rm diss}$, which is on the order of 10^7 years, as discussed in Sect. 3.2. We can use this ratio to assess the relative importance of various types of energy input. In convenient units, the power necessary to maintain a steady ISM is approximately 0.1 eV/Myr (where Myr $= 10^6$ years).

On a galactic scale, giant cloud complexes that interact gravitationally or collide with each other and exchange momentum will make the ISM viscous, leading to a flow of angular momentum outward and a drift of mass inward (cf. Chap. 7). The mass that drifts inward gains kinetic energy from the gravitational binding energy of the galaxy, i.e., from orbital motion. Thus shear viscosity heats the fluid at the expense of orbital energy [3]. The energy input to the cloud fluid is approximately $\rho v_{\rm orbit}^2/T_{\rm visc}$ for viscous time $T_{\rm visc}$. Because $v_{\rm orbit} \approx 50\,a$ and $T_{\rm visc} \approx 1000\,T_{\rm diss}$, this energy source can be important, i.e., the ratio $\rho v_{\rm orbit}^2/T_{\rm visc}$ can be comparable to $U_{\rm ISM}/T_{\rm diss}$. Detailed studies of this heating mechanism were made by Simakov [4], Fukunaga & Tosa [5] and Gammie et al. [6].

Orbital energy also goes into the magnetic field because shear stretches the field. Then the field pumps energy into gas motions by the Rayleigh-Taylor instability [7]. However, this energy source is not likely to be competitive with others because the energy density in the field is only comparable to $U_{\rm ISM}$ and the shear time, which is the inverse of the Oort A constant, is longer than $T_{\rm diss}$ by a factor of at least 3.

Transient spiral density waves also tap galactic energy because the waves grow and wrap up with stellar self-gravity and galactic shear forces, and they impart some of this energy to the gas by gravitational scattering. The energy input to the gas is approximately $\rho\,\Delta\Psi A$ for average excess potential $\Delta\Psi$ from the arms and for Oort constant A. If we measure this spiral potential in terms of the total streaming velocity of the stars in the arms, which may be several tens of km s^{-1}, the net effect of transient waves could be large; both $\Delta\Psi$ and the shear time A^{-1} are larger than a^2 and $T_{\rm diss}$, respectively, by factors of 5 to 10. A recent numerical study confirms the importance of this heating mechanism [8]. Note that the waves must be transient or growing to have a net energy input to the cloud fluid because the energy of random motions can increase only for a time-changing potential; a steady spiral wave with constant amplitude, shape, and speed will remove as much energy from the fluid during decompression as it pumps into the fluid during the compression (if there is no significant spiral torque-driven accretion). However, even a steady wave should heat the cloud population inside the arms and cool it in the interarms. Then the velocity dispersion will be larger in the arms unless cloud coagulation or collisional cooling decrease it, or unless stellar energy sources contribute relatively more to heating in the interarm regions because of the lower density. In fact, observations suggest that the velocity dispersions in galactic arms are lower than in the interarm regions [9 − 11]. This is apparently explained by coalescence (and the associated energy dissipation) because in N-body simulations, the velocity dispersion of the clouds in the arms is less than in the interarm regions by the inverse square root of the number of particles in each cloud conglomerate [12].

The interstellar fluid is also stirred by the occasional impact of high velocity clouds from the halo [13] (cf. Sect. 4.5). If a cloud falls from one scale height, then it will acquire a velocity approximately equal to the rms dispersion a and the rate of energy gain by the ambient medium will be approximately $\frac{1}{2}a^2(d\rho/dt)$ for accretion rate $d\rho/dt$. This is a very small energy input averaged over the whole galaxy because $d\rho/dt$ is approximately ρ divided by the age of the galaxy, which is a much larger time than T_{diss}. If the clouds hit the disk at a very high speed V_{infall}, then the energy input can be larger, but not much larger because by the time the infalling mass M_{infall} mixes with disk material and slows to the rms speed a, it will have lost most of its initial energy to radiative shocks [13]. The mass of mixed material is approximately $M_{\mathrm{infall}}V_{\mathrm{infall}}/a$ by momentum conservation, so the net energy gain in this case is the mixing ratio V_{infall}/a times $\frac{1}{2}a^2(d\rho/dt)$. This gain can be large where the cloud hits the disk [14], but it is not important for the whole galaxy unless V_{infall}/a is comparable to the total disk gas mass divided by the mass that is accreted in a disk dissipation time.

These four energy sources, viscosity, magnetic field stretching by shear, transient spiral arms, and infalling clouds, are all derived from the gravitational binding energy of the galaxy (although the energy from infall can be related to supernovae if the gas was first pushed into the halo from the disk). They are all uncertain because they depend on unknown details of cloud dynamics. For example, theories of the viscous heating mechanism generally assume infinite cloud lifetimes and no magnetic forces. When there is a magnetic field, the collision mean free path can be much smaller than the geometric mean free path, and then the heating rate will be smaller too. Processes involved with magnetic field stretching and spiral wave generation are also uncertain and may apply differently in different galaxies.

Another source of gas stirring is from gravitational or other interactions with field stars, which move through the interstellar medium and make small tidal wakes and weak wind-related bow shocks. This heat source has been studied by Just & Kegel et al. [15 – 17].

A relatively large amount of energy comes from stars in the form of supernovae, winds, and HII regions. Supernovae explode with some 10^{51} ergs of energy at a rate of 1 per ~ 30 years in the Milky Way Galaxy. Most of this energy is radiated away in shock fronts and not available to cloud stirring, but a fraction ϵ, which is on the order of 10^{-2}, goes into gas motions [18,19] (see Sect. 4.4.3). The importance of this energy source can be estimated by comparing the energy density of cloud motions, U_{ISM}, with the energy input rate from supernovae: U_{ISM} equals 10^{51} erg/$(300\,\mathrm{pc})^3$, and the mean time between supernovae in each cubic $300\,\mathrm{pc}$ is about 3×10^4 years. Thus cloud stirring by supernovae can balance the dissipation rate if $\epsilon > 3 \times 10^4$ years/T_{diss}, which is likely.

The efficiency for heating by supernovae can be even higher than 10^{-2}, perhaps as high as 10^{-1} in a giant bubble [20], if the supernova are clustered together in OB associations, as observed for Type II supernovae [21]. However, in that case there could be a problem with the distribution of this energy throughout the disk: OB associations are separated by several hundred parsecs in the solar neighborhood, and this is larger than the typical bubble radius. The associations may also lose energy by forming chimneys and venting the hot gas into the galactic halo [22].

Supernova heating between OB associations can be important if there is not much gas there, which is probably true if most of the disk gas is concentrated into the same cloud complexes that form OB stars. An average inter-association density of 0.1 cm^{-3}, for example, would be easily heated, as in the model by McKee & Ostriker [23], by Type I supernova and a low fraction of Type II supernova. Such heating at low density should be very important between the spiral arms of our Galaxy, especially considering that the distribution of known supernova remnants is not strongly correlated with spiral structure [24]. If this observation of supernovae is not just a selection effect, then the interarm regions should have a high filling factor of hot gas.

Stellar winds and HII regions contribute an amount of energy to local gas motions that may be comparable to or larger than the input from supernovae (Sect. 4.4.1-4.4.2). OB stars are probably more confined to the midplane than Type I supernovae, and many of them are very close to the giant molecular clouds out of which they formed, so they can push on these clouds directly. Moreover, in OB associations, the gradual build-up of momentum that is imparted to the gas from several million years of expansion of the HII region and wind around a single O star can be much larger than the sudden momentum put into the same gas from that star when it explodes [25, 26]. This is because the density in the association environment is generally high and the supernova radiates away a larger fraction of its energy than it would in the ambient medium. The distribution of this stellar energy may be a problem, however, as it is for Type II supernovae. There is a lot of volume between OB associations where the gas is probably not stirred much by the stars in the associations, unless the gas density is very low and shock fronts or Alfvén waves can propagate there from the star-formation centers (or high-velocity clouds propagate there directly, via the halo). Perhaps type I supernova and stray or runaway O-type stars dominate the energy input to these regions.

The pressures on a giant molecular cloud from the associated HII regions, winds, and supernovae all average about $10^6 \, \mathrm{cm}^{-3} \, \mathrm{K}$ times Boltzmann's constant. This is a factor of 100 above the ambient interstellar pressure. The average column density of a typical molecular cloud corresponds to about 10 magnitudes of extinction, or some 150 solar masses per square parsec. The ratio of these two quantities, the pressure divided by the mass column density, is the average cloud acceleration. It equals about $1 \, \mathrm{km \, s}^{-1} \, \mathrm{Myr}^{-1}$. Thus in the typical lifetime of an association, $10 \, \mathrm{Myr}$, the cloud can acquire a velocity of about $10 \, \mathrm{km \, s}^{-1}$. Because most of the mass in the interstellar medium is in the form of giant molecular clouds and their low-density halos, this persistent acceleration (and the rocket-like acceleration of lower density gas away from the molecular clouds) could account for much of the observed cloud velocity dispersion (see also [27]).

Another source of velocity dispersion is gaseous self-gravity. A study of giant HI complexes in the inner Galaxy [28] concluded that ~ 0.64 of the HI mass is in giant complexes, many of which appear virialized with a one-dimensional rms speed of $\sim 5 \, \mathrm{km \, s}^{-1}$. Thus a large amount of kinetic energy could be in virialized motions. Suppose that cloud formation causes the gas to contract by a factor of 4 in density, which is reasonable for these HI clouds. Then they release a factor $1 - 4^{-1/3} = 0.37$ of their potential energy, of which approximately half goes into kinetic energy, according to the virial theorem. Thus $(0.37/2) \times 0.64 = 0.12$ of the kinetic energy of the low-density neutral medium may come from gravitational binding energy. Note that continued gravitational contraction may sustain a high velocity dispersion in a cloud, so gravity acts like a heat source, but in fact the total energy is dissipated and this lost energy has to be replaced when the cloud disperses (as required for a steady-state ISM).

3.2 Energy Dissipation

Physical and magnetic interactions between interstellar clouds lead to energy losses in shock fronts and strong magnetic waves. Kinetic energy can also be transferred from one region to another by random cloud motions (analogous to conduction) and Alfvén wave radiation. This section calculates energy dissipation and transfer rates for these processes.

First consider the oscillation of a single cloud back and forth on a magnetic field line. Observations of pervasive clumpy structure and magnetism in the ISM suggest that many clouds and clumps could be oscillating in this way, tied to their immediate neighborhoods by magnetic field lines that become entangled in the field lines of other clouds and clumps. One can imagine clouds trapped like this, as if in a net, with random motions from gravitational and magnetic interactions that are sustained for a dissipation time.

If the oscillation of a predominantly neutral cloud is driven by its magnetic field, e.g., through magnetic stresses generated by the random motions of other clouds, then the ions in the cloud will follow the field (they actually drift transverse to both the field and the frictional force, according to the vector product of these two quantities, but for the purposes of calculating the viscous force on the neutrals we can usually consider the ions as locked to the field lines). Suppose that the ions move as

$$v_i = V \cos \omega t. \tag{3}$$

The neutrals follow because of ion-neutral collisions. Ignoring gravity and pressure on very small scales, the equation of motion for the neutrals is

$$\frac{\partial v_n}{\partial t} = -\omega_{in} (v_n - v_i) \tag{4}$$

where

$$\omega_{in} = n_i \langle \sigma_{in} a_{th} \rangle \frac{m_r}{m_n} \approx \frac{n_n^{1/2}}{1.6 \, \text{Myr}} \tag{5}$$

is the ion-neutral collision rate for an ionization fraction

$$\frac{n_i}{n_n} = 10^{-7} \left(\frac{n_n}{10^4} \right)^{-1/2} \tag{6}$$

and $m_r \approx m_n$ is the reduced mass for ion-neutral collisions with m_n the neutral molecular mass; σ_{in} is the ion-neutral collision cross section and a_{th} is the thermal speed. The solution to the neutral particle equation of motion at $\omega_{in} t \gg 1$ is

$$v_n(t) = \frac{\omega_{in} V}{\omega^2 + \omega_{in}^2} (\omega_{in} \cos \omega t + \omega \sin \omega t). \tag{7}$$

The energy dissipation rate equals the average frictional force integrated over the relative distance between the ions and neutrals as they drift for one period T. For total cloud mass M, this dissipation rate is:

$$\frac{dE}{dt} = \frac{1}{T} \int_0^T \text{Force} \cdot ds = \int_0^T M \frac{dv_n}{dt} \cdot (v_n - v_i) \frac{dt}{T} \approx -\frac{MV^2 \omega^2}{2\omega_{in}} \tag{8}$$

for $\omega \ll \omega_{in}$ and $t \gg \omega_{in}^{-1}$. This result implies that a fraction $2\pi \omega / \omega_{in}$ of the clump kinetic energy gets converted into heat during each full oscillation.

Note that if the oscillation is from virialized motions in a cloud and the frequency of oscillation is approximately the gravitational frequency, $\omega \approx (G\rho)^{1/2}$, then $2\pi \omega / \omega_{in} \approx 0.12$, independent of parameters. Thus virialized clump motions driven by magnetic stresses can last for only about 8 oscillations or 8 free fall times in the cloud before their energy is converted into heat and radiation. Of course, the cloud will probably contract because of this dissipation, and, according to the virial theorem, half of the gravitational binding energy that is released during the contraction will go into more clump motions. In this way a cloud can slowly dissipate its energy and contract, keeping its velocity dispersion at the virial theorem value. If the boundary pressure remains constant, then contraction in this sense means increasing central concentration.

Whether clouds really contract this slowly depends on how smoothly the clumps oscillate. The dissipation rate for a single clump is proportional to $V^2\omega^2$, which is the mean squared acceleration. If clump motions are jerky, with large accelerations followed by force-free drifts, then the dissipation rate is much higher for the same average clump speed.

To calculate the energy dissipation following a sudden impulse in the motion of a clump, suppose that the ions jerk from $v = 0$ to $v = V$ on a time scale t_j, so that

$$v_i = V\left(1 - e^{-t/t_j}\right). \tag{9}$$

The equation of motion for the neutral particles is now solved for this ion motion, and the total energy turned into heat from $t = 0$ to $t = \infty$ becomes

$$\Delta E = \begin{cases} \dfrac{1}{2}\dfrac{MV^2}{\omega_{in}t_j} & \text{if} \quad \omega_{in}t_j \gg 1, \\[2mm] \dfrac{1}{2}MV^2 & \text{if} \quad \omega_{in}t_j \ll 1. \end{cases} \tag{10}$$

These energy losses are larger than the losses during one period of a smooth oscillation by the ratios $1/2\pi\omega t_j$ and $\omega_{in}/2\pi\omega$, respectively. Note that if the acceleration is extremely rapid $(\omega_{in}t_j \ll 1)$, the energy lost in accelerating a clump up to speed V is equal to the final kinetic energy, so twice this energy has to be put into the neutrals by the ions. This factor of two will appear again for the energy loss in shock fronts (cf. Sect. 4.2.2).

Cloud collisions are impulsive, so they dissipate energy in shock fronts at a rate comparable to the case $\omega_{in}t_j \lesssim 1$. When two clouds collide, a shock front forms on each side of the compressed interface. This shock can radiate a large fraction of the relative energy. Instead of actually calculating the jump conditions for the shocks and integrating the energy lost over all possible cloud collision angles, we can determine the energy lost during the collision more simply by considering only the conservation of momentum.

Two clouds with speed V that collide at an angle θ and then stick together will emerge from the collision with a speed $V\cos(\theta/2)$. The average of the square of this emergent speed over all solid angles is $\frac{1}{2}V^2$, so the average energy lost per mass doubling collision per cloud is $\frac{1}{2}MV^2 - \frac{1}{2}M\left(\frac{1}{2}V^2\right) = \frac{1}{4}MV^2$.

We can do the same calculation in the rest frame of one of the clouds. Then the relative speed of the other cloud is $V_{rel} = 2^{1/2}V$ on average, and we have to use a reduced mass M_r for the other cloud, which equals $\frac{1}{2}M$ in this case. After the collision, both clouds are at rest (in this frame of one cloud). The result is a total loss of all the relative energy of the collision, or a loss of $\frac{1}{2}M_rV_{rel}^2 = \frac{1}{2}MV^2$, which is $\frac{1}{4}MV^2$ for each cloud, as derived above.

For an ensemble of sticky clouds with a mass-dependent space density $n(M)$ from lower to upper mass limits M_L to M_U, the energy loss rate per unit volume is

$$\Lambda = \frac{1}{2}\int_{M_L}^{M_U} n(M_1)\,dM_1 \int_{M_L}^{M_U} n(M_2)\,dM_2 \times \frac{1}{2}M_rV_{rel}^2 \times \pi\left(R_1 + R_2\right)^2 V_{rel}. \tag{11}$$

The factor $1/2$ in front of the integral is to avoid repetition of cloud pairs. We can use $V_{rel} = 2^{1/2} V$ if all clouds have the same speed, and

$$\langle V_{rel}^3 \rangle = 2^{3/2} \frac{8}{3} \left(\frac{2}{3\pi} \right)^{1/2} a_{3D}^3 = 3.47 \, a_{3D}^3 \tag{12}$$

for a Maxwell-Boltzmann distribution of speeds with a three-dimensional rms dispersion a_{3D}, assuming V_{rel} is independent of mass.

In this expression for Λ, we assumed that all of the mass in two colliding clouds comes to the same final velocity after the collision, even for grazing collisions. This is a reasonable approximation, to within a factor of 2, if the shocks that form between the clouds do not cool, or if the magnetic pressure behind the front is important. Then the post-shock gas spreads in the transverse direction and the shocks engulf almost all of both clouds, regardless of impact parameter. If the shocked regions cool and only the overlapping parts mix, then the energy lost per collision and the collision rate depend on the impact parameter, which must be integrated too. This uncertainty in the effective collision cross section for dissipation is sometimes resolved [23] by using $\frac{1}{2}$ of the grazing collision rate in Λ, i.e., by taking a cross section of $\frac{1}{2}\pi(R_1 + R_2)^2$ in the integral. This correction factor of $\frac{1}{2}$ is carried through here to the final expression, where it appears in the denominator of equation (13), below.

The cooling rate can be simplified if we assume a cloud mass distribution $n(M) \propto M^{-\alpha}$ and a cloud radius $R \propto M^\beta$, and then integrate over a mass range specified by the ratio $X = M_U/M_L$. Then

$$\Lambda = \frac{0.87 \, \rho^2 a_{3D}^3}{2\sigma} \, L(\alpha, \beta, X) \qquad \text{with} \qquad \begin{cases} \rho = \int M \, n(M) \, dM, \\ \sigma = \dfrac{\int \frac{M}{\pi R^2} \, n(M) \, dM}{\int n(M) \, dM}, \end{cases} \tag{13}$$

where ρ is the mean mass density and σ is the average mass column density of a cloud in the ensemble. For $X = 10^3$, $\alpha = 1.5$, and $\beta = 1/3$ (constant cloud density) or $\beta = 1/2$ (constant cloud column density), numerical integration gives $L = 1.0$ and $L = 2.9$, respectively. As α gets larger the cooling rate gets larger, with the coefficient L increasing to 10.4, 30, and 300 as α increases to 1.8, 2, and 3, respectively. This increase in L is a result of the increase in cloud number density and collision rate for large α.

When the magnetic field is important during the collision, some of the kinetic energy of the clouds can be converted into energy of the compressed field. This compressed energy can be returned to the clouds if they separate after the collision because the magnetic pressure can accelerate the clouds back up to nearly their original speeds. In this case the energy lost per collision is less than half of the initial energy (the factor of $1/2$ was discussed above), but not much less than $1/2$ unless the collision is nearly at the Alfvén speed inside the cloud and the shock is very weak. An estimate for the energy loss factor in a magnetic collision was made [29] by integrating over the collision angle between two clouds, and by integrating over the two angles for the relative field orientation in the cloud whose energy loss is to be determined. For each set of angles, the energy lost behind the oblique shock front that forms between the two clouds was calculated. The average energy lost per collision was then determined by integrating again over impact parameter. The result indicates that the energy loss factor decreases from 0.5 for the case of no magnetic field or for infinitely strong (i.e., fast) shocks with a magnetic field, to a value of 0.43 for a shock speed equal to 10 times the Alfvén speed inside the cloud, to 0.375 at a shock speed of 4 times the internal Alfvén speed, and so on. If this stored magnetic energy does not return to the kinetic energy of the cloud but is dissipated by magnetic oscillations and damping for two clouds that ultimately stick, then the effective value is 0.5 again, i.e., the energy lost per cloud per mass doubling collision is 0.5 times $\frac{1}{2}MV^2$, as discussed above.

The energy dissipation time is $\frac{1}{2}\rho a_{3D}^2/\Lambda$, or $1.1\sigma/(\rho a_{3D}L)$. For an ISM composed of diffuse clouds with an average B-V color excess of 0.1 mag., giving $\sigma = 1.65\times10^{-3}$ gm cm^{-2}, a mean density of 1 atom cm^{-3}, giving $\rho = 2.2\times10^{-24}$ gm cm^{-3}, and a three-dimensional velocity dispersion of $10\,\mathrm{km\,s^{-1}}$, the dissipation time with $L = 2$ is 14 million years. This is a much shorter time than the dissipation time from purely molecular cloud interactions (even though molecular clouds often dominate the mass), because molecular clouds have a higher σ by a factor of ~ 20 and a lower a_{3D} by a factor of ~ 2. Molecular cloud motions are probably damped by conversion of their energy into diffuse cloud motions, through gravitational and magnetic interactions, and then by the collisional dissipation of the diffuse cloud motions. The timescale for this dissipation of molecular cloud motion is limited by the first step, which takes approximately the Jeans time in the ambient medium for magnetic field strengths inside molecular clouds that are equal to the virial theorem values [30]. The result is several times 10^7 years in the solar neighborhood, which is only a factor of 2 or 3 larger than the diffuse cloud dissipation time. Magnetic damping of molecular cloud motions (via conversion into diffuse cloud motions) is much faster than their direct collisional damping.

This conversion of molecular cloud motions into surrounding diffuse cloud motions is presumably accomplished by magnetically dragging the diffuse clouds along with the molecular clouds and by creating a gravitational wake in the diffuse cloud fluid. Such magnetic transfer of energy is also important for diffuse clouds and other gas, so we discuss it briefly here. The point is that the energy of internal or bulk cloud motions can be transferred from one place to another by Alfvén waves without much of an actual loss to radiation. This happens when the magnetic field lines that thread the clumps in a cloud are forced to move because of the oscillation of these clumps or the motion of the whole cloud, and then this motion travels as a wave to the gas outside the cloud, causing this external gas to move also. The time scale for a significant amount of the cloud's internal or bulk momentum to be transferred outside is the time it takes an Alfvén wave to travel over a distance that includes an amount of mass equal to the cloud's mass. The rate at which energy is transferred is about twice the rate at which momentum is transferred because energy is proportional to the square of the momentum, i.e., if momentum decreases as

$e^{-\omega t}$ then energy decreases approximately as $e^{-2\omega t}$. Thus the energy transfer time scale is given by [29]

$$T_{\text{wave}} \approx \frac{M}{4\pi R^2 \rho_e v_{\text{Ae}}}, \tag{14}$$

where the external density is ρ_e, the external Alfvén speed is v_{Ae}, the cloud's mass is M and its radius is R. The denominator here contains one factor of 2 from converting momentum loss to energy loss and another factor of 2 to include the total surface area for the radiation of Alfvén waves, i.e., from both sides of the cloud. If the external field varies as r^{-3} and the external density is constant, then the effective value of v_{Ae} is about $B/(4\pi\rho_e)^{1/2}$ for cloud surface field B [29].

The energy transfer rate from Alfvén wave radiation is

$$\Lambda_{\text{wave}} = \frac{\rho a_{3D}^2}{2 T_{\text{wave}}}. \tag{15}$$

Note that wave energy can also enter the cloud from outside and that some of the emergent radiation might scatter back into the cloud after reflection off external clouds (for a discussion of wave reflection, see [31]). These inputs lower the loss rate for the total cloud energy. A proper treatment of Alfvén wave radiative losses should include such wave-cloud and wave-wave scatterings.

Alfvén waves can also dissipate their energy of motion by ion-neutral collisional viscosity, as shown at the beginning of this section. The loss rate there was found to depend on the mean squared acceleration of the gas in the wave, and to be larger when the motion is jerky rather than smooth. One of the most important energy losses connected with Alfvén waves occurs after they steepen by non-linear effects (Sect. 4.1); then the acceleration at the front surface of the wave is much larger than the average acceleration, and the total dissipation can be very large. An Alfvén wave steepens because the Alfvén speed is higher at the wave crest than at the trough. Steepening causes it to form a shock front at the leading part of the wave. A strong wave can dissipate its energy before it travels much more than several wavelengths. The rate at which an Alfvén wave of frequency ω and amplitude b, travelling parallel to a magnetic field of strength B, steepens into a shock front (one front per wave crest) is approximately $\omega(b/B)^2$. The rate at which a magnetic compression wave of frequency ω and amplitude b steepens if it travels perpendicular to the field is $\sim \omega b/B$. Thus the energy loss rate from non-linear magnetic waves is approximately [32]

$$\Lambda_{\text{steep}} \approx \omega \frac{b^2}{8\pi} \left(\frac{b}{B}\right)^i, \tag{16}$$

where $i = 1$ or 2 for perpendicular or parallel propagation, respectively. If cloud internal motions are from strong Alfvén waves, then this energy loss can dominate all others, but if the magnetic motions in a cloud are not smooth waves but the oscillation of discrete clumps, then the loss rate can be much less, as discussed at the beginning of this section.

The rotational energy of clouds can also be transferred to the external medium by magnetic forces. The time scale for this transfer is again approximately equal to one-half of the mass in the cloud divided by the rate at which Alfvén waves propagating away from the cloud enclose new ambient mass. A review of this process is in [33]. Observations of spinning filaments that are interpreted to be angular momentum sinks for collapsing cloud cores have been discussed by Uchida et al. [34, 35].

3.3 Magnetic Diffusion

Magnetic energy is lost from a cloud by diffusion, which occurs in the neutral ISM because the charged particles, such as ions, electrons, and charged grains, and the neutral atoms and molecules drift relative to each other. The magnetic field stays with the charged particles. This drift creates a friction that can heat a cloud by what is essentially $P\,dV$ compressional energy if the cloud is confined by an external pressure, or by the release of gravitational binding energy if the cloud is bound by self-gravity.

From Maxwell's equation of inductance, $\partial B/\partial t = -c\,\nabla \times E$ for speed of light c, and the electron equation of motion, $(m_e/e)\,\partial v_e/\partial t = E + v_e \times B/c \approx 0$ for negligible electron mass m_e, we can determine the rate of change of magnetic field strength

$$\frac{\partial B}{\partial t} = \nabla \times (v_e \times B)$$
$$= \nabla \times (v_n \times B) + \nabla \times (\Delta v \times B) \tag{17}$$

where $\Delta v = v_e - v_n \approx v_i - v_n$ is the drift velocity between electrons and neutrals, which is approximately the drift velocity between ions and neutrals.

The ion-neutral drift velocity comes from the ion equation of motion, which is

$$\rho_i \frac{D v_i}{Dt} = -\nabla P_i + \rho_i\, g + \frac{1}{4\pi}\,(\nabla \times B) \times B - n_i n_n m_r \alpha_{in} \Delta v \tag{18}$$

for ion pressure P_i and mass density ρ_i, neutral particle density n_n, ion density n_i, reduced ion-neutral mass m_r (which is approximately equal to the neutral mass for light neutral molecules, H_2, and heavy ions, e.g., HCO^+ or C^+), and for ion-neutral collision rate, α_{in} (in $cm^3\,s^{-1}$) from thermal plus drift motions, equal to $\sigma_{in} a_{th} \sim 2 \times 10^{-9}$ for subthermal relative speeds [36] and $\sim 10^{-15} \Delta v$ for superthermal speeds [37], in cgs units. The gravitational acceleration g is from the cloud.

These equations ignore the contribution to the total viscosity on neutrals that results from charged grains. This viscosity has been studied by Elmegreen et al. [38 − 40] and Umebayashi & Nakano et al. [41 − 43]. Small charged grains gyrate around the field like ions and then they follow the field lines and viscously couple the field to the neutrals. If there are very tiny grains in a cloud, and if the average neutral density is not too large, then this source of viscosity can dominate the ion collisions. However if there are no tiny grains, then the gyration of the remaining large grains stops at only moderate densities and this source of viscosity becomes small.

For a cloud in virial equilibrium with the magnetic pressure a significant fraction of the total pressure, P, and an ion density much less than the local neutral density, ρ_n ($\sim n_n m_r$), only the last two terms in the ion equation of motion are important and they give the drift velocity

$$\Delta v \approx \frac{(\nabla \times B) \times B}{4\pi \rho_n n_i \alpha_{in}}. \tag{19}$$

In a uniform cloud the local neutral density in this equation equals approximately the average density, $\rho_n \approx \rho$. In a clumpy cloud the local neutral density equals approximately the density in the clumps (see below).

The ion density can be determined from observations [44, 45] and the theory of molecular cosmic ray ionization [46, 47] with charge exchange, radiative and dissociative recombination, and recombination on grains [38, 41]. Then the local ion density is proportional to the square root of the neutral density, $n_i \sim 10^{-5} n_n^{1/2}$, for quantities in cgs units. This gives $n_i \alpha_{in} = n_n^{1/2}/1.6\,\text{Myr}$, as discussed in Sect. 3.2.

The term $\nabla \times (v_n \times B)$ in the equation for $\partial B/\partial t$ causes the field lines to move as if they were frozen into the neutral fluid, and the term $\nabla \times (\Delta v \times B)$ causes the diffusion of field lines away from the neutrals. Because the ions generally form and recombine on a very short time scale, the individual ions that at any instant comprise the ion fluid do not actually lag behind the neutrals very far before they become neutral themselves, nor do the ions really stick to the field lines as they gyrate; instead they drift in a direction perpendicular to both the magnetic field and the viscous force, $\rho n_i \alpha_{in} \Delta v$, in order to maintain the current that supports the cloud. Usually the current speed is much less than Δv.

The expression for the drift velocity Δv can be substituted into the induction equation to give the equation of evolution for the field strength. If we write the second term as $-B/\tau_{\text{diff}}$ for diffusion time τ_{diff}, then

$$\frac{\partial B}{\partial t} = \nabla \times (v_n \times B) - \frac{B}{\tau_{\text{diff}}} \tag{20}$$

where

$$\tau_{\text{diff}} \approx \frac{4\pi \rho_n n_i \alpha_{in}}{k^2 B^2}, \tag{21}$$

and the derivative ∇ has been replaced by a wavenumber k.

The diffusion time has a simple expression when the magnetic field strength in the cloud equals the virial equilibrium value. This gives $kB \sim 2\pi G^{1/2}\rho$ if $k \sim R^{-1}$ for cloud radius R. Then

$$\tau_{\text{diff}} \sim \frac{n_i \alpha_{in}}{\pi G \rho} \sim \frac{780}{n_n^{1/2}} \quad [\text{Myr}] \tag{22}$$

188

for mean molecular weight $\rho_n/n_n = 3.9 \times 10^{-24}$ g and $n_i \alpha_{in} = 2 \times 10^{-14} n_n^{1/2} s^{-1}$. Note that τ_{diff} scales with $n_n^{-1/2}$, as does the gravitational time scale, $\tau_{grav} \sim (G\rho)^{-1/2}$, so the ratio is the constant

$$\frac{\tau_{diff}}{\tau_{grav}} \sim 12. \tag{23}$$

This implies that magnetic diffusion is longer than gravitational collapse regardless of density (as long as the ion density scales as $n_n^{1/2}$). This relation is also discussed in a review by Shu et al. [48].

A recent study [49] of diffusion in clumpy clouds indicates that the diffusion time is longer than for a uniform cloud at the same average density and magnetic field strength, by the inverse of the square root of the volume filling factor of the clumps, f. Thus the above ratio becomes $\sim 12 f^{-1/2}$, where f can be less than 0.1.

Acknowledgement: I am grateful to Dr. Richard Rand for helpful comments.

References

3.1   Wang, Z., Cowie, L.L. 1988, *Astrophys. J.* **335**, 168
3.2   McCray, R., Snow, T.P., Jr. 1979, *Ann. Rev. Astron. Astrophys.* **17**, 213
3.3   Lynden-Bell, D., Pringle, J.E. 1974, *Monthly Notices Roy. Astron. Soc.* **168**, 603
3.4   Simakov, S.G. 1988, *Sov. Astron. Letters* **14**, 382
3.5   Fukunaga, M., Tosa, M. 1989, *Publ. Astron. Soc. Japan* **41**, 241
3.6   Gammie, C.F., Ostriker, J.P., Jog, C.J. 1991, *Astrophys. J.* **378**, 565
3.7   Parker, E.N. 1966, *Astrophys. J.* **145**, 811
3.8   Thomasson, M., Donner, K.J., Elmegreen, B.G. 1991, *Astron. Astrophys.* **250**, 316
3.9   Kutner, M.L., Verter, F., Rickard, L.J. 1990, *Astrophys. J.* **365**, 195
3.10 Kutner, M.L., Verter, F., Rickard, L.J. 1991, in *IAU Symp. 146, Dynamics of Galaxies and their Molecular Cloud Distributions*, eds. F. Combes, F. Casoli, Dordrecht, Kluwer, p. 29
3.11 Rydbeck, G., Bergman, P., Hjalmarson, A., Thomasson, M., Wiklind, T. 1991, in *Dynamics of Disc Galaxies*, ed. B. Sundelius, Göteborg, Chalmers University, p. 235
3.12 Thomasson, M., Donner, K.J., Elmegreen, B.G. 1991, in *Dynamics of Disc Galaxies*, ed. B. Sundelius, Göteborg, Chalmers University, p. 279
3.13 Tenorio-Tagle, G. 1981, *Astron. Astrophys.* **94**, 338
3.14 Mirabel, I.F., Morras, R. 1990, *Astrophys. J.* **356**, 130
3.15 Just, A., Kegel, W.H. 1990, *Astron. Astrophys.* **232**, 447
3.16 Jacoby, S., Just, A., Kegel, W.H. 1990, *Astron. Astrophys.* **237**, 461
3.17 Deiss, B.M., Just, A., Kegel, W.H. 1990, *Astron. Astrophys.* **240**, 123
3.18 Chevalier, R.A. 1974, *Astrophys. J.* **188**, 501
3.19 Trimble, V. 1989, *Structure and Dynamics of the Interstellar Medium*, IAU Colloquium 120, eds. G. Tenorio-Tagle, M. Moles, J. Melnick, Berlin, Springer, p. 530
3.20 MacLow, M.-M., McCray, R. 1988, *Astrophys. J.* **324**, 776
3.21 Huang, Y.-L., Thaddeus, P. 1986, *Astrophys. J.* **309**, 804
3.22 Norman, C.A., Ikeuchi, S. 1989, *Astrophys. J.* **345**, 372
3.23 McKee, C.F., Ostriker, J.P. 1977, *Astrophys. J.* **218**, 148
3.24 Li, Z., Wheeler, J.C., Bash, F.N., Jeffreys, W.H. 1991, *Astrophys. J.* **378**, 93
3.25 Lada, C.J., Elmegreen, B.G., Blitz, L. 1979, in *Protostars and Planets*, ed. T. Gehrels, (Tucson: University of Arizona), p. 341

3.26 Yorke, H.W., Tenorio-Tagle, G., Bodenheimer, P., Rozyczka, M. 1989, *Astron. Astrophys.* **216**, 207

3.27 Oort, J., Spitzer, L., Jr. 1955, *Astrophys. J.* **121**, 6

3.28 Elmegreen, B.G., Elmegreen, D.M. 1987, *Astrophys. J.* **320**, 182

3.29 Elmegreen, B.G. 1985, *Astrophys. J.* **299**, 196

3.30 Elmegreen, B.G. 1981, *Astrophys. J.* **243**, 512

3.31 Vasquez, B.J. 1990, *Astrophys. J.* **356**, 693

3.32 Zweibel, E.G., Josafatsson, K. 1983, *Astrophys. J.* **270**, 511

3.33 Mouschovias, T.Ch. 1991, in *The Physics of Star Formation and Early Stellar Evolution*, eds. C.J. Lada, N.D. Kylafis, (Dordrecht: Kluwer), p. 61

3.34 Uchida, Y., Fukui, Y., Monishima, Y., Mizuno, A., Iwata, T., Takaba, H. 1991, *Nature* **349**, 140

3.35 Uchida, Y., Mizuno, A., Nozawa, S., Fukui, Y. 1990, *Publ. Astron. Soc. Japan* **42**, 69

3.36 Spitzer, L., Jr. 1978, *Physical Processes in the Interstellar Medium*, Wiley Interscience

3.37 Chernoff, D.F. 1987, *Astrophys. J.* **312**, 143

3.38 Elmegreen, B.G. 1979, *Astrophys. J.* **232**, 729

3.39 Elmegreen, B.G. 1986, in *Light on Dark Matter*, ed. F.P. Israel, Dordrecht, Reidel, p. 265

3.40 Elmegreen, B.G., Fiebig, D. 1992, *Astron. Astrophys.* submitted

3.41 Umebayashi, T., Nakano, T. 1980, *Publ. Astron. Soc. Japan* **32**, 4

3.42 Umebayashi, T., Nakano, T. 1990, *Monthly Notices Roy. Astron. Soc.* **243**, 90

3.43 Nishi, R., Nakano, T., Umebayashi, T. 1991, *Astrophys. J.* **368**, 181

3.44 Wootten, A., Snell, R., Glassgold, A.E. 1979, *Astrophys. J.* **234**, 876

3.45 Guelin, M., Langer, W.D., Wilson, R.W. 1982, *Astron. Astrophys.* **107**, 107

3.46 Oppenheimer, M., Dalgarno, A. 1974, *Astrophys. J.* **192**, 29

3.47 de Jong, T., Dalgarno, A., Boland, W. 1980, *Astron. Astrophys.* **91**, 68

3.48 Shu, F.H., Adams, F.C., Lisano, S. 1987, *Ann. Rev. Astron. Astrophys.* **25**, 23

3.49 Elmegreen, B.G., Combes, F. 1992, *Astron. Astrophys.* in press.

4 Explosions

4.1 Nonlinear Waves and Shock Fronts

Shocks form in a travelling wave because the propagation speed is larger where the temperature is larger, and the temperature is larger in the wave crest because of the adiabatic invariance of P/ρ^γ. This follows from the equations of motion and the continuity equations, which are nonlinear. The equation of motion for a fluid is

$$\rho\frac{\partial v}{\partial t} + \rho v \cdot \nabla v = -\nabla P \qquad (24)$$

and the continuity equation is

$$\frac{\partial \rho}{\partial t} + v \cdot \nabla \rho = -\rho \nabla \cdot v. \qquad (25)$$

Another equation is needed to relate pressure to density; for an adiabatic fluid, this is

$$P = K\rho^\gamma \qquad (26)$$

for constant K and ratio of specific heats γ. Every term in these equations is non-linear, except $\partial \rho/\partial t$, because the terms are either products of two variables, ρ and v, or their derivatives, or because the exponent γ exceeds 1. As a result, strong perturbations change their shape as they travel. Weak perturbations, such as everyday sound waves in the air, can propagate for a long distance without changing shape because in each product of two terms, only one term varies (v) and the other is about constant (ρ).

Strong sound waves are not common in everyday experience. The Concord supersonic jet makes strong sound waves as it moves and pushes the air aside; these waves quickly steepen into shocks, but this generally occurs over the oceans where there are few people to hear the shocks. The NASA space shuttle makes a clear double sonic boom (one from the nose and one from the tail) when it passes overhead near the California coast on its re-entry towards Edwards Air Force base. High speed bullets can make shock waves too. But other examples are rare; it takes a lot of force to compress the air so much that the non-linearity of the wave equations becomes important.

Non-linear steepening effects in other types of waves are more common. Water waves, for example, steepen as they approach the shore because their height becomes comparable to the depth of the water and the gravitational potential energy becomes important. This potential energy makes the crests of the waves move faster than the troughs, and so leads to the steepening. Then the waves turn over at the top and crash on themselves. The non-linearity also causes the water and various flotsam to be pushed along with the waves toward the shore for a short distance. After many such non-linear waves pass, the flotsam is eventually washed ashore. These waves differ from those in the deep ocean where the flotsam just bobs around when a wave passes.

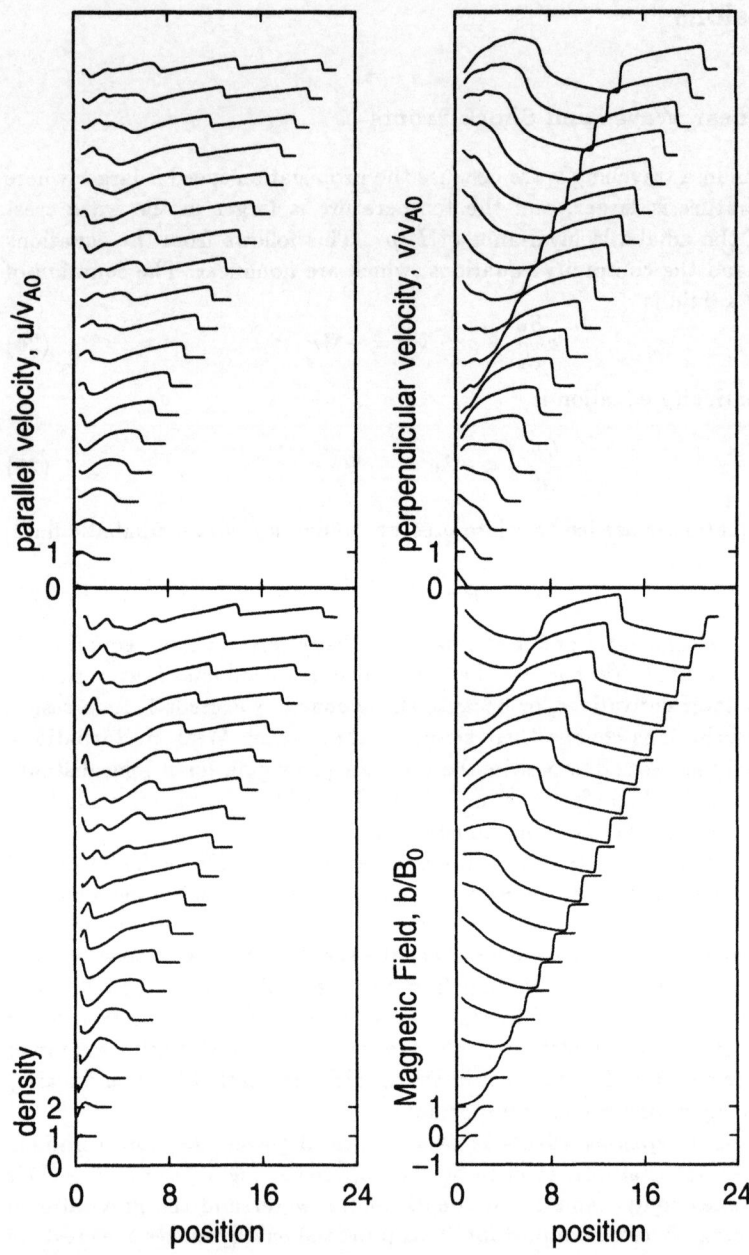

Fig. 9. A strong magnetic wave is shown moving to the right and steepening into a shock front. The curves are separated by a time interval equal to the position interval divided by the Alfvén velocity

Waves near the shore also interact in a very non-linear fashion, in the sense that they do not pass through each other but splash when they cross, sending water high into the air. The splash comes from the bulk motion of the water in each wave, which, as kinetic energy, gets converted into gravitational potential energy by the ram pressure of the moving fluids at the wave contact point. In very deep water, waves that are much higher than those near the shore just pass through each other and keep travelling without a change of form because the linear solutions to the wave equation apply there, and these solutions are additive.

Similar processes occur for sound and magnetosonic waves in interstellar space, where non-linearities are much more common. It takes only about a factor of two variation in the pressure to make a sound or magnetic wave that is so strong that it quickly steepens into a shock, and interstellar pressures vary by factors of 100 or more on relatively short time scales because of supernova explosions, passing bright stars, cloud collisions, and so on. Strong waves or fast-moving regions that interact in space can create density peaks and transverse motions in the splash regions between them, just as strong waves in water cause splashes. Moreover, non-linear wave motions of one type, such as transverse magnetic waves, can couple to non-linear wave motions of another type, such as longitudinal (i.e., compressional or sonic) waves.

An example of a transverse magnetic wave that steepens as it travels is shown in Fig. 9. This figure is a numerical solution to the non-linear magnetic fluid equations for plane polarized motion, made on a Lagrangian grid with 200 cells per wavelength. A transverse wave is excited by simple harmonic motion at position 0. The unperturbed quantities are denoted by subscript 0, and the Alfvén velocity is denoted by v_A. The leading edge of the wave starts with a smooth sinusoidal shape, but quickly steepens into a discontinuity, which is the shock front (see also [1 − 3]).

Physically a shock forms because the propagation speed is larger in some parts of the wave profile than in others. For a sound wave, this is because $P \propto \rho^\gamma$ and the sound speed is $a = (\gamma P/\rho)^{1/2}$ (not to be confused with the one-dimensional velocity dispersion in Sect. 3.1, $a = (P/\rho)^{1/2}$, which used the same notation). Thus a increases with ρ for $\gamma > 1$ and the wave moves faster in the crest than in the trough. For a magnetic wave the propagation speed is $v_A = B/(4\pi\rho)^{1/2}$. If the wave is compressional and moving across the field lines, and if diffusion is slow, then $B \propto \rho$ because of flux freezing; then $v_A \propto \rho^{1/2}$ and the propagation speed is higher in the wave crest, as for sound. If the wave is transverse and moving parallel to the field lines, then ρ is about constant and $v_A \propto B$, which is largest where the field is most distorted, at, for example, the leading and trailing edges of each sine wave in the field lines.

These variations in the propagation speed cause the wave shapes to change with time, but if the propagation speed were the only contributing factor, then the shape would change symmetrically, steepening at both edges. A travelling wave steepens only at the leading edge because the inertial term in the equation of motion, $\rho v \cdot \nabla v$, acts like a resistance to the wave's propagation, causing the compressed material to get thrown forward. If the wave were not moving but standing between two reflecting boundaries, then both edges could steepen symmetrically.

4.2 Discontinuous Fronts

4.2.1 Jump Conditions.
A wave will steepen until the leading edge becomes about as thin as the mean free path for momentum and energy exchanges between fluid particles. In general there will be several mean free paths involved, one for the conversion of translational motions from one direction into another, one for the conversion of translational motion into molecular rotational or vibrational motions, one for the collisional dissociation or ionization of molecules, and so on. When the fluid changes significantly over the scale of a mean free path at the leading edge of a wave, then this edge is a shock front. The equations of motion are still satisfied at a shock front, but the derivatives become very large and difficult to specify because they depend on the mean free paths. An easier way to describe the behavior of a fluid at a shock front is to integrate the equations through the front, so that only the endpoints of the integrals remain. These endpoints are the pre- and post-shock values of the variables, and the equations, instead of being continuous, become jump conditions.

In this section we derive the jump conditions for a magnetic oblique shock [4]. Other reviews of astrophysical shocks are in Ostriker & McKee [5], Shull & Draine [6], and in the book by Chakrabarti [7].

We begin with the equations for a continuous magnetic fluid. These are the equation of motion for total velocity vector w (we use w now because v will be the transverse component below),

$$\rho \frac{\partial w}{\partial t} + \rho w \cdot \nabla w = -\nabla P - \frac{1}{8\pi} \nabla B^2 + \frac{1}{4\pi} B \cdot \nabla B, \tag{27}$$

the continuity equation for density,

$$\frac{\partial \rho}{\partial t} + w \cdot \nabla \rho = -\rho \nabla \cdot w, \tag{28}$$

the continuity equation for magnetic flux,

$$\frac{\partial B}{\partial t} = \nabla \times (w \times B), \tag{29}$$

and the energy equation,

$$\frac{\partial U}{\partial t} + w \cdot \nabla U + (U + P)\nabla \cdot w = G, \tag{30}$$

where $U = P/(\gamma - 1)$ is the internal energy density for γ equal to the ratio of specific heats, i.e., $5/3$ if there are only translational degrees of freedom and less

194

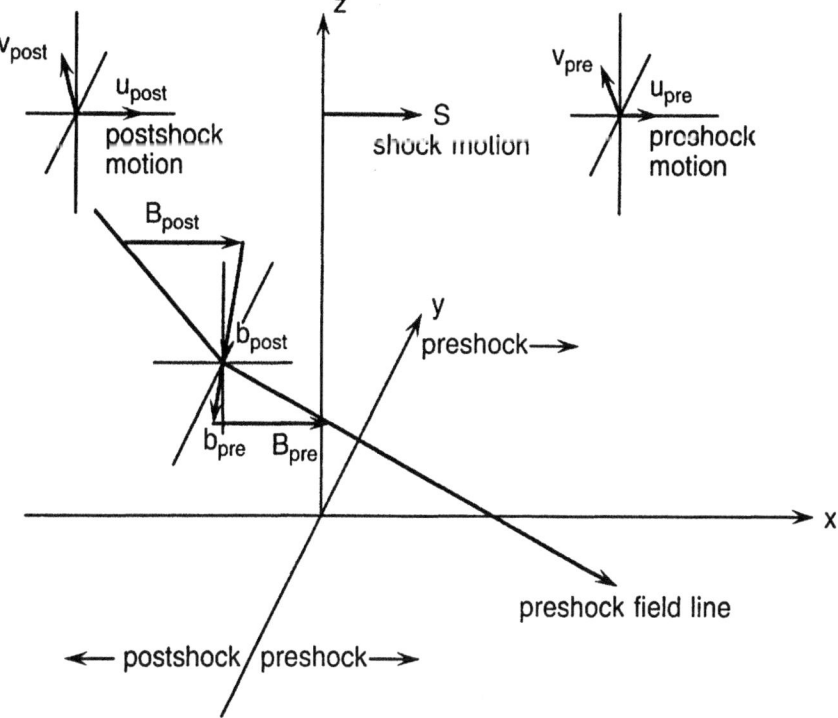

Fig. 10. Diagram of the structure of a magnetic shock, showing the notation used in the text.

than $5/3$ if there are other internal energies; G is the rate of change of internal energy density from radiation.

Now suppose that a high pressure disturbance moves in the x direction at a fixed speed S, and that the wave profile is constant in time (see Fig. 10). This would be the case for a steady shock. Then we can replace the time derivative $\partial/\partial t$ by a convective derivative

$$\frac{\partial}{\partial t} \rightarrow -S \frac{\partial}{\partial x}. \tag{31}$$

When this substitution is made, all of the equations reduce to equations of the form

$$\frac{\partial F}{\partial x} = 0, \tag{32}$$

which implies that the quantity F is constant. Thus F is the same ahead of the front as behind the front and the jump condition is

$$F_{\text{pre}} = F_{\text{post}}, \tag{33}$$

where *pre* and *post* stand for pre-shock and post-shock conditions.

For example, assume that all quantities vary only in the x direction ($\partial/\partial y = \partial/\partial z = 0$), which is the direction the shock is moving. The component of $\boldsymbol{w}$ in the x direction is u. Then the continuity equation is

$$\frac{\partial \rho}{\partial t} + \rho \frac{\partial u}{\partial x} + u \frac{\partial \rho}{\partial x} = 0. \tag{34}$$

After substituting S, this equation reduces to a jump condition as follows:

$$-S \frac{\partial \rho}{\partial x} + \rho \frac{\partial u}{\partial x} + u \frac{\partial \rho}{\partial x} = -\frac{\partial \rho S}{\partial x} + \frac{\partial \rho u}{\partial x} = \frac{\partial \rho(u - S)}{\partial x} = 0, \tag{35}$$

so the jump condition is

$$\{\rho(u - S)\}_{\text{pre}} = \{\rho(u - S)\}_{\text{post}}. \tag{36}$$

This is the condition for the conservation of mass in the front.

For the equation of motion in the x direction,

$$\rho \frac{\partial u}{\partial t} + \rho u \frac{\partial u}{\partial x} + \frac{\partial P}{\partial x} + \frac{1}{8\pi} \frac{\partial b^2}{\partial x} = 0, \tag{37}$$

we get, after substitution of S

$$-\rho S \frac{\partial u}{\partial x} + \rho u \frac{\partial u}{\partial x} + \frac{\partial}{\partial x} \left(P + \frac{b^2}{8\pi} \right) = (\rho u - \rho S) \frac{\partial u}{\partial x} + \frac{\partial}{\partial x} \left(P + \frac{b^2}{8\pi} \right)$$

$$= \frac{\partial}{\partial x} \left(\rho u(u - S) + P + \frac{b^2}{8\pi} \right) = 0, \tag{38}$$

so the jump condition is

$$\left\{ \rho u(u - S) + P + \frac{b^2}{8\pi} \right\}_{\text{pre}} = \left\{ \rho u(u - S) + P + \frac{b^2}{8\pi} \right\}_{\text{post}}. \tag{39}$$

Here we have used the first jump condition to eliminate $\partial(\rho u - \rho S)/\partial x$ and we denote the perpendicular component of the field by b (cf. Fig. 10). We have also used the parallel component of the field convection equation,

$$\frac{\partial B}{\partial x} = 0, \tag{40}$$

which gives a constant x-component of the field $\boldsymbol{B}$. The result is the jump condition for conservation of momentum parallel to the direction of propagation of the shock.

Similarly, the perpendicular component of the equation of motion, which is a vector equation,

$$\rho \frac{\partial v}{\partial t} + \rho u \frac{\partial v}{\partial x} - \frac{B}{4\pi} \frac{\partial b}{\partial x} = 0 \tag{41}$$

reduces as

$$-\rho S \frac{\partial v}{\partial x} + \rho u \frac{\partial v}{\partial x} - \frac{B}{4\pi} \frac{\partial b}{\partial x} = \frac{\partial \rho(u - S)v}{\partial x} - \frac{1}{4\pi} \frac{\partial Bb}{\partial x} = 0, \tag{42}$$

which becomes the jump condition for perpendicular momentum

$$\left\{\rho v\,(u-S) - \frac{Bb}{4\pi}\right\}_{\text{pre}} = \left\{\rho v\,(u-S) - \frac{Bb}{4\pi}\right\}_{\text{post}}. \tag{43}$$

Here we have again used the previous jump conditions.

The perpendicular component of the magnetic convection equation is

$$\frac{\partial b}{\partial t} + u\frac{\partial b}{\partial x} + b\frac{\partial u}{\partial x} - B\frac{\partial v}{\partial x} = 0 \tag{44}$$

which becomes

$$-S\frac{\partial b}{\partial x} + \frac{\partial bu}{\partial x} - \frac{\partial Bv}{\partial x} = -\frac{\partial bS}{\partial x} + \frac{\partial bu}{\partial x} - \frac{\partial Bv}{\partial x} = 0. \tag{45}$$

This reduces to the jump condition for magnetic flux conservation in the perpendicular direction

$$\left\{b(u-S) - Bv\right\}_{\text{pre}} = \left\{b(u-S) - Bv\right\}_{\text{post}}. \tag{46}$$

These last two jump conditions for perpendicular momentum and magnetic flux can be rewritten as

$$\frac{b_{\text{post}}}{b_{\text{pre}}} = \frac{\dfrac{\varPhi^2}{\rho_{\text{pre}}} - \dfrac{B^2}{4\pi}}{\dfrac{\varPhi^2}{\rho_{\text{post}}} - \dfrac{B^2}{4\pi}}, \qquad v_{\text{post}} - v_{\text{pre}} = -\frac{B}{4\pi\,\varPhi}\left(b_{\text{post}} - b_{\text{pre}}\right), \tag{47}$$

where $\varPhi = \{\rho(S-u)\}_{\text{pre}} = \{\rho(S-u)\}_{\text{post}} > 0$ is the mass flux through the front. When $\varPhi^2/\rho_{\text{post}} - B^2/4\pi > 0$, the first of these two equations implies that b_{post} exceeds b_{pre} (because $\rho_{\text{post}} > \rho_{\text{pre}}$) and that they are in the same plane regardless of v, and the second equation implies that v_{post} exceeds v_{pre} in the direction of $b_{\text{pre}} - b_{\text{post}}$, which generally puts v_{pre} and v_{post} in different planes.

Finally, the energy equation can be added to the inner product of w and the momentum equation, i.e., u times the x (parallel) component of the momentum equation, v_y times the y component and v_z times the z component, to give an equation for the change in the total energy

$$\begin{aligned}
&\left\{U_{\text{tot}}(u-S) + P_{\text{tot}}u - \frac{B}{4\pi}b\cdot v\right\}_{\text{pre}} = \\
&\left\{U_{\text{tot}}(u-S) + P_{\text{tot}}u - \frac{B}{4\pi}b\cdot v\right\}_{\text{post}} + \int G\,dx
\end{aligned} \tag{48}$$

where

$$\begin{aligned}
U_{\text{tot}} &= \tfrac{1}{2}\rho\left(u^2 + v^2\right) + \frac{P}{\gamma-1} + \frac{b^2}{8\pi}, \\
P_{\text{tot}} &= P + \frac{b^2}{8\pi},
\end{aligned} \tag{49}$$

are the total energy density and transverse pressure. The first term in the brackets of the energy condition is the convection of energy density into the front, the second term is the $P\,dV$ energy input, and the third term is the energy put into the kink in the field lines at the front. The integral of the energy radiation rate G is over the depth of the front, so the difference between the *pre* and *post* shock brackets is a function of position behind the front. When there is no radiation, the total energy of all types entering the front equals the total energy behind the front. With radiation, the temperature at various positions behind the front has to be determined by the history of heating and cooling for each parcel of gas, and this involves a numerical solution to the flow with a cooling function and the inclusion of chemical reactions and various types of heating.

The energy jump condition with $G = 0$ is useful for limiting cases, such as very young shocks that have not yet radiated much energy, or very old shocks in which the gas has returned to its original temperature. In this latter case, one can use $\gamma = 1$ as an approximation, and then the energy jump condition becomes

$$\{P(u - S)\}_{\text{pre}} = \{P(u - S)\}_{\text{post}}, \tag{50}$$

which reduces to $\{P/\rho\}_{\text{pre}} = \{P/\rho\}_{\text{post}}$, or $a_{\text{pre}} = a_{\text{post}}$ (as expected for $\gamma = 1$) because $\rho(u - S)$ is constant.

4.2.2 A Comment About Energy Conservation in Shocks. Sect. 3.2 showed that the energy lost in accelerating a gas suddenly up to speed V without changing its temperature or volume is equal to the final kinetic energy, so twice this final energy has to be applied. The same factor of 2 applies to shock fronts. First consider a shock with no energy loss from radiation.

Imagine that a piston is pushing a gas in a tube of length L with a pressure P_{post}; a shock advances through this tube with speed S and the gas behind the shock moves with the speed $u_{\text{post}} \equiv u$. The gas speed ahead of the shock is $u_{\text{pre}} = 0$. Ignore magnetic fields. Then the conservation of mass and momentum imply

$$\rho_{\text{pre}} S = \rho_{\text{post}}(S - u), \tag{51.1}$$

$$P_{\text{post}} = P_{\text{pre}} + \rho_{\text{post}} u(S - u). \tag{51.2}$$

The total energy per unit area inside the tube is the kinetic and thermal energy in the shocked part, of length L_{post}, plus the thermal energy in the unshocked part, of length L_{pre}. (Note that $L_{\text{post}}\rho_{\text{post}} = (L - L_{\text{pre}})\rho_{\text{pre}}$ by mass conservation). The energy per unit area in the whole tube is therefore

$$\frac{\text{Energy}}{\text{Area}} = \left(\tfrac{1}{2}\rho_{\text{post}}u^2 + \frac{P_{\text{post}}}{\gamma - 1}\right) L_{\text{post}} + \frac{P_{\text{pre}}}{\gamma - 1} L_{\text{pre}}, \tag{52}$$

so the rate of change of this energy is

$$\frac{d\text{Energy}/dt}{\text{Area}} = \left(\tfrac{1}{2}\rho_{\text{post}}u^2 + \frac{P_{\text{post}}}{\gamma - 1}\right)(S - u) - \frac{P_{\text{pre}}}{\gamma - 1} S, \tag{53}$$

because dL_{post}/dt is the rate of growth of the post-shock region, $S - u$, and S is the rate at which L_{pre} decreases (note that the whole column of gas gets shorter with time).

Now, the energy jump condition in Sect. 4.2.1 says that

$$\left(\tfrac{1}{2}\rho_{post}u^2 + \frac{P_{post}}{\gamma - 1}\right)(S - u) - P_{post}u = \frac{P_{pre}}{\gamma - 1}S, \tag{54}$$

from which it follows, using the above equation, that

$$\frac{d\mathrm{Energy}/dt}{\mathrm{Area}} = P_{post}u. \tag{55}$$

This is a sensible result, indicating that the energy of the gas increases at a rate given by the product of the applied pressure and the velocity of this application; i.e., the power input is the force times the speed at which this force is applied. Note that this speed of application is not the shock speed, S, because this is only the speed of the front and not of the mass in the front. The speed of the mass in the front is u (this would also be the speed of a piston that drives the shocked column of gas). In this example there is no radiative loss, so all of the applied energy is present in the gas.

Now consider a shock with energy loss, to make an analogy with the rapid acceleration of a cloud from Sect. 3.2. For an isothermal shock, the thermal energy density is $\tfrac{3}{2}\rho a^2$ (now $\gamma = 1$ so a is both the sound speed and the one-dimensional dispersion), so the energy increases at the rate

$$\frac{d\mathrm{Energy}/dt}{\mathrm{Area}} = \left(\tfrac{1}{2}\rho_{post}u^2 + \tfrac{3}{2}\rho_{post}a^2\right)(S - u) - \tfrac{3}{2}\rho_{pre}a^2 S, \tag{56}$$

but the terms involving a^2 cancel because of the jump condition for mass conservation, giving

$$\frac{d\mathrm{Energy}/dt}{\mathrm{Area}} = \left(\tfrac{1}{2}\rho_{post}u^2\right)(S - u) = \tfrac{1}{2}\rho_{pre}u^2 S. \tag{57}$$

The applied power is the same as before,

$$\frac{\mathrm{Power\ Input}}{\mathrm{Area}} = P_{post}u = \left(P_{pre} + \rho_{post}u(S - u)\right)u = P_{pre}u + \rho_{pre}u^2 S. \tag{58}$$

Now we see that for a strong shock, for which P_{pre} is relatively unimportant, the rate of change of energy in the gas, which is $\tfrac{1}{2}\rho_{pre}u^2 S$, is one-half of the applied power, which is $\rho_{pre}u^2 S$. The other half of the applied power is radiated away. This factor of $1/2$ is the same as in ΔE for jerky cloud motions (Sect. 3.2).

4.2.3 Jump Conditions for Limiting Cases. Solutions for the jump conditions, such as the compression factor, must generally be found numerically by iteration because the jump conditions are non-linear like the fluid equations. But some limiting cases have simple solutions. When there is no magnetic field and all motions are parallel to the shock motion ($v = 0$), the pressure and density jumps have the solutions

$$\frac{P_{post}}{P_{pre}} = \frac{2\gamma}{\gamma+1}\mathcal{M}^2 - \frac{\gamma-1}{\gamma+1}, \tag{59.1}$$

$$\frac{\rho_{pre}}{\rho_{post}} = \frac{2}{\gamma+1}\frac{1}{\mathcal{M}^2} + \frac{\gamma-1}{\gamma+1}, \tag{59.2}$$

where $\mathcal{M}$ is the Mach number, given by

$$\mathcal{M}^2 = \frac{\rho_{pre}u_{pre}^2}{\gamma P_{pre}} = \frac{u_{pre}^2}{a_{pre}^2} \tag{60}$$

in the shock frame ($S = 0$). For a non-radiative shock and for molecules with only translational degrees of freedom, we take $\gamma = \frac{5}{3}$ and get

$$\frac{P_{post}}{P_{pre}} = \frac{5}{4}\mathcal{M}^2 - \frac{1}{4}, \tag{61.1}$$

$$\frac{\rho_{pre}}{\rho_{post}} = \frac{3}{4}\mathcal{M}^{-2} + \frac{1}{4}. \tag{61.2}$$

Note that the first of these can be used to get the shock speed if the shock-driving pressure is known:

$$u_{shock} \equiv u_{pre} = \left(\frac{4}{3}\frac{P_{post}}{\rho_{pre}}\right)^{1/2}, \tag{62}$$

where we have assumed a strong shock and are in the shock frame. The second relation gives $\rho_{post} = 4\rho_{pre}$ for strong shocks, independent of $\mathcal{M}$.

The ratio of the post-shock pressure to the density can be used to get the post-shock temperature or sound speed:

$$a_{post}^2 = \frac{\gamma P_{post}}{\rho_{post}} = \frac{5}{16}u_{shock}^2. \tag{63}$$

The post-shock random motions that determine this sound speed are a combination of isotropic scattering of the incoming flow and a $P\,dV$ energy input from the factor of 4 compression.

For a radiative shock that has the same temperature in the post-shock and pre-shock regions, we take $\gamma = 1$ and get

$$\frac{P_{post}}{P_{pre}} = \frac{\rho_{post}}{\rho_{pre}} = \mathcal{M}^2. \tag{64}$$

Now we consider simple cases for a magnetic shock. If the pre-shock field is parallel to the direction of motion of the shock ($b_{pre} = 0$) and all motions are parallel to this direction ($v = 0$), then the field lines stay parallel to this direction behind the shock ($b_{post} = 0$).

If the pre-shock field is perpendicular to the direction of motion of the shock ($B = 0$), and all motions are parallel to the shock direction ($v = 0$), then the field compresses in direct proportion to the density for any γ:

$$\frac{b_{\text{post}}}{b_{\text{pre}}} = \frac{\rho_{\text{post}}}{\rho_{\text{pre}}}. \tag{65}$$

This implies that small, transverse irregularities in the pre-shock field (i.e., $b_{\text{pre}} \ll B_{\text{pre}}$) will amplify in the shock, possibly giving $b_{\text{post}} \sim B_{\text{post}} - B_{\text{pre}}$; thus, shock fronts pump energy into the transverse component of the field.

In the case of a transverse field, an isothermal shock front has much less compression when a magnetic field is present than when a magnetic field is not present because the field easily dominates the pressure behind the shock. Using the jump conditions for mass, parallel momentum, and perpendicular magnetic flux in the shock frame $(S = 0)$,

$$\{\rho u\}_{\text{pre}} = \{\rho u\}_{\text{post}}, \tag{66.1}$$

$$\left\{ \rho \left(\frac{a^2}{\gamma} + u^2 + \tfrac{1}{2} v_A^2 \right) \right\}_{\text{pre}} = \left\{ \rho \left(\frac{a^2}{\gamma} + u^2 + \tfrac{1}{2} v_A^2 \right) \right\}_{\text{post}}, \tag{66.2}$$

$$\left\{ \frac{b}{\rho} \right\}_{\text{pre}} = \left\{ \frac{b}{\rho} \right\}_{\text{post}}, \tag{66.3}$$

we can derive an equation for the density compression factor $X = \rho_{\text{post}}/\rho_{\text{pre}}$ in terms of the sound speed a, the shock speed $u_{\text{shock}} \equiv u_{\text{pre}}$ and the pre-shock Alfvén speed $v_{A,\text{pre}}$:

$$\frac{a^2}{\gamma}(X - 1) + u_{\text{shock}}^2 \left(\frac{1}{X} - 1 \right) + \tfrac{1}{2} v_{A,\text{pre}}^2 \left(X^2 - 1 \right) = 0, \tag{67}$$

which, for $u_{\text{shock}} > v_{A,\text{pre}} > a$ and moderately large X becomes

$$\frac{\rho_{\text{post}}}{\rho_{\text{pre}}} \approx 2^{1/2} \frac{u_{\text{shock}}}{v_{A,\text{pre}}}. \tag{68}$$

Thus the compression factor goes as the first power of the Alfvén Mach number, and not as the second power of the sonic Mach number, when the post-shock magnetic pressure dominates the post-shock thermal pressure.

4.2.4 Applications to Colliding Clouds. It is interesting to see an example where the presence of a magnetic field makes an important difference in the dynamics. We consider here the gravitational collapse of compressed gas between two colliding clouds, and show that the collapse time is less than the duration of the compressed phase when magnetic fields are absent, but that the collapse time is longer than the compressed phase when magnetic fields are present.

In a non-magnetic isothermal collision, the compression factor is $\rho_{\text{post}}/\rho_{\text{pre}} \sim \mathcal{M}^2$ for mach number $\mathcal{M}$ of the collision, so the gravitational collapse time in the shocked gas, $(G\rho_{\text{post}})^{-1/2}$, is

$$\tau_{\text{grav,post}} = \frac{\tau_{\text{grav,pre}}}{\mathcal{M}}, \tag{69}$$

where $\tau_{\text{grav,pre}}$ was the collapse time in the uncompressed gas, before the clouds collided. The duration of the dense shocked phase is the re-expansion time

$$\tau_{\text{dense}} \sim \frac{R}{a} \qquad (70)$$

for cloud radius R and sound speed a. Thus the ratio of the gravitational collapse time to the duration of the dense phase is

$$\frac{\tau_{\text{grav,post}}}{\tau_{\text{dense}}} = \frac{\tau_{\text{grav,pre}}}{R/c} \frac{1}{\mathcal{M}} \equiv \frac{\lambda_{\text{Jeans,pre}}}{R} \frac{1}{\mathcal{M}} \qquad (71)$$

where $\lambda_{\text{Jeans,pre}} \sim \tau_{\text{grav,pre}} a$ is the initial Jeans length in the cloud. Because of the $1/\mathcal{M}$ dependence in this equation, clouds that are initially smaller than their Jeans length and so stable against collapse can become unstable during the collision and begin to form stars. Such instabilities are found in numerical simulations of colliding non-magnetic clouds [8].

For a magnetic cloud the situation is very different. The density compression is less, $\rho_{\text{post}}/\rho_{\text{pre}} \sim \mathcal{M}$, so the gravitational collapse time is longer than for a non-magnetic cloud,

$$\tau_{\text{grav,post}} = \frac{\tau_{\text{grav,pre}}}{\mathcal{M}^{1/2}}, \qquad (72)$$

and the duration of the dense phase is shorter

$$\tau_{\text{dense}} \sim \frac{R}{v_{A,\text{post}}} \sim \frac{R}{v_{A,\text{pre}}} \frac{1}{\mathcal{M}^{1/2}}; \qquad (73)$$

here $v_{A,\text{post}}$ is the Alfvén speed in the shocked region and $v_{A,\text{pre}}$ is the Alfvén speed in the unshocked cloud; because $B \propto \rho$, $v_A \propto \rho^{1/2} \propto \mathcal{M}^{1/2}$. The ratio of the gravitational collapse time to the duration of the dense phase in the magnetic case is

$$\frac{\tau_{\text{grav,post}}}{\tau_{\text{dense}}} = \frac{\tau_{\text{grav,pre}}}{R/v_{A,\text{pre}}} \equiv \frac{\lambda_{\text{Jeans,pre}}}{R}, \qquad (74)$$

independent of $\mathcal{M}$. Now the pre-collision Jeans length is $\tau_{\text{grav,pre}} v_{A,\text{pre}}$. With a magnetic field, the dynamics of the collision does little to enhance the collapse of the cloud (unless the collision is parallel to the field). Nevertheless, an increase in the magnetic diffusion rate during the collision may enhance star formation anyway [9, 10].

4.2.5 Pressure Fronts in a Cloudy Gas. Shock fronts in the interstellar medium travel quickly through the low density intercloud gas and slowly through the clouds, with a speed roughly proportional to the inverse square root of the density, as discussed in Sect. 4.2.3. The clouds are accelerated up to the post-shock flow velocity, but they are also compressed by the high post-shock pressure and shredded by evaporation, surface instabilities, and turbulence as the post-shock flow moves around them.

Much effort has been directed toward understanding such cloud-shock interactions. Among the earliest work was a study by Sgro [11], who considered shock propagation in a cloudy medium and modelled the Cas A supernova remnant as an example. Chevalier & Theys [12] then found that radiative shocks are unstable when they overtake a cloud, in the sense that transverse flows in the distorted shock front lead to cloud growth. Ikeuchi & Spitzer [13] studied the scattering of a shock by a cloud.

Perhaps the biggest conceptual leap came when McKee & Cowie [14] suggested that supernova shocks would go around a cloud and propagate mostly through the low density intercloud medium. They also studied cloud evaporation in the hot remnant gas [15]. This led to the McKee & Ostriker [16] model of a three-phase medium after it was realized that such intercloud propagation produced remnants that are large enough to overlap [17]. Subsequent studies have been reviewed by Ostriker & McKee [5].

The response of a cloud to compression by a shock was first modelled numerically by Woodward [18], who suggested (following Dibai [19]) that star formation could be triggered in the cloud this way. Numerous other studies of such cloud implosions followed, the most recent of which are by Tenorio-Tagle & Rozyczka [20, 21], Falle & Giddings [22], Klein et al. [23], Bedogni & Woodward [24], and Kimura & Tosa [25 − 28].

Shocked cloud models were also applied to Herbig-Haro objects by Schwartz [29] and many others (see Sect. 4.5), and to molecular abundances in clouds by Elitzur & Watson [30], Draine & Katz [31], any many others.

There is another aspect of interstellar pressure fronts that is not addressed by these shock-cloud interaction models because it does not involve only the conventional type of shock. This other aspect was motivated by observations of spiral density wave shocks in galaxies (cf. Chap. 5), which, in spite of an intercloud medium that is often too hot to shock, still contain numerous individual clouds that have merged into ragged, dense fronts [32, 33], and by observations of high latitude clouds near the sun [34], which contain clumps with such large relative motions that the clouds appear to be unbound and short-lived. In these and many other cases, the pressure fronts comprise not only smooth shocks in atomic or molecular gas, but also moving clouds (or tiny clumps) that collide with and scatter off of other clouds (or clumps) in an irregular fashion. The individual clouds in these fronts can either be destroyed by mutual interactions [35], producing a more-or-less continuous ridge of dense cloudy debris, or they can remain in tact with an rms speed comparable to the front speed if soft magnetic forces rather than direct collisions mediate the transfer of momentum and kinetic energy behind the front [36].

We consider here this large-scale cloud collection process, and write the analog of the shock jump conditions for a situation in which a high pressure disturbance sweeps up individual clouds by forcing cloud-cloud collisions and magnetic field line entanglements (cf. Sect. 4.2.6). Such fronts are like shocks, but clouds replace molecules as carriers of momentum and energy. The jump conditions derived above are still useful, but they should be modified because the cloud mean free path is long enough to mix up the clouds behind the front. This differs from a molecular shock front where the collision mean free path is very short compared to the post-shock layer thickness. When mixing is important in a cloud collision front, cooling, magnetic diffusion, self-gravity, and other processes that operate in the bulk of the collected material become part of the jump conditions. For example, when cloud collisions remove energy or the magnetic field diffuses away, the whole front collapses with the loss of internal pressure. This adds compressional energy to the layer and changes the post-shock density even when the velocity is constant. There are likely to be other forces too when clouds collect in bulk along a moving pressure front, such as strong self-gravity and inertial forces that resist the internal collapse. We show here how the jump conditions can be modified to account for these changes. A more detailed discussion is in [36].

It is convenient here to go into the frame of the pre-shock gas, or rest frame, rather than the shock or general frame as in the previous sections, because the motion of the post-shock gas is directly observed by a Doppler shift from this rest frame. In general, the frame of the shock front is useful for shocks that are driven by predictable pressures (e.g., supernova remnants), because the shock speed depends on this pressure. The frame of the trailing edge of the layer is useful for shocks that are driven at predictable velocities (e.g., spiral density waves). Note that in the first case, the gas collapses toward the front of the layer as it cools and in the second case the gas collapses toward the trailing edge as it cools. The post-shock quantities actually differ in the two cases when this time-dependent collapse occurs, so the proper choice of a reference frame is important if the equations are to remain simple.

We consider here a high pressure disturbance that is exerted on the trailing surface of a compressed layer of clouds. This trailing surface is assumed to move through the ISM like a piston at a fixed velocity u_0, and to scoop up cloudy material into a layer of variable thickness L. This thickness increases as material is added, so the total velocity of the leading edge of the disturbance is $u_0 + dL/dt$. In the notation of Sect. 4.2.1, $S = u_0 + dL/dt$, $u_{\text{pre}} = 0$, $u_{\text{post}} = u_0$, $\rho_{\text{pre}} = \rho_0$, $\rho_{\text{post}} = \rho_1$, $b_{\text{pre}} = b_0$, and $b_{\text{post}} = b_1$. We also consider only motions parallel to the front's motion ($v = 0$) and a magnetic field parallel to the front's plane ($B = 0$). The jump condition for mass conservation is then

$$\rho_0 \left(u_0 + \frac{dL}{dt} \right) = \frac{d(\rho_1 L)}{dt}. \tag{75}$$

Note that the time derivative $d\rho_1/dt$ has been included in the post-shock part to allow for bulk cooling and the associated contraction. This would not be necessary for a molecular shock even in the post shock frame because the cooling length in a molecular shock is much smaller than the total thickness of the post-shock layer (except at very early times).

Similarly, the jump condition for magnetic flux is

$$b_0 \left(u_0 + \frac{dL}{dt} \right) = \frac{d(b_1 L)}{dt} + \frac{b_1 L}{\tau_{\text{d}}}. \tag{76}$$

Here we have also included $b_1 L/\tau_{\text{d}}$ to allow for magnetic diffusion at the rate $1/\tau_{\text{d}}$, and, because of mixing and bulk layer expansion and contraction, db_1/dt is included too.

The jump condition for parallel momentum is

$$P_0 + \frac{b_0^2}{8\pi} + \rho_0 u_0 \left(u_0 + \frac{dL}{dt} \right) = P_1 + \frac{b_1^2}{8\pi} - \frac{\pi}{2} G \rho_1^2 L^2 - \frac{(\rho_1 L)^2}{2\pi} \frac{d^2}{dt^2} \frac{1}{\rho_1}. \quad (77)$$

Most of these terms can be recognized from the momentum equation derived in Sect. 4.2.1, which was for a conventional shock, but now we have added self-gravity and inertial forces from contraction during cooling (the last two terms on the right hand side). This momentum equation can also be viewed as the equation of motion for internal adjustments of a pressure-bounded, self-gravitating layer.

The jump condition for energy is

$$U_0 \left(u_0 + \frac{dL}{dT} \right) + \psi \rho_0 \left(u_0 + \frac{dL}{dT} \right) + P_{\text{tot},0} u_0 = \frac{d(U_1 L)}{dt} - \frac{P_1 L}{\rho_1} \frac{d\rho_1}{dt} + \Lambda L, \quad (78)$$

where the total pre-shock energy density is

$$U_0 = \tfrac{1}{2} \rho_0 u_0^2 + \frac{P_0}{\gamma - 1} + \frac{b_0^2}{8\pi}, \quad (79)$$

the total post-shock energy density is

$$U_1 = \tfrac{3}{16} \rho_1^3 L^2 \left(\frac{d}{dt} \frac{1}{\rho_1} \right)^2 + \frac{P_1}{\gamma - 1} + \frac{b_1^2}{8\pi}, \quad (80)$$

and the total pre-shock pressure is

$$P_{\text{tot},0} = P_0 + \frac{b_0^2}{8\pi}. \quad (81)$$

The average post-shock cooling function, $\Lambda \approx \rho_1^2 a_1^3 / \sigma_c$, should be appropriate for clump collisions with clump column density σ_c, as described in Sect. 3.2. The gravitational potential of the layer is

$$\psi = 2\pi G R_{\text{width}} \rho_1 L, \quad (82)$$

where R_{width} is the width of the layer. The time derivatives of ρ_1 enter because the contraction of the layer during bulk cooling adds a $-P \, dV/dt$ energy term.

Solutions to these equations for diffuse and molecular cloud collision fronts were given in [36]. They appear to be relevant to spiral density wave shocks and other moving cloud fronts because of the pervasive clumpy structure of interstellar gas and the generally long mean free paths for clump or cloud collisions. In fact, many of the pressure fronts generally described as shocks could also be cloud collision fronts. For example, the high latitude molecular clouds observed by Magnani et al. [34] could be moving through the ambient medium, accumulating tiny clumps and other clouds as they go. Some could be driven by the pressure from the Sco-Cen association, and the accumulated pieces could be the clumps that are now observed inside the high latitude clouds. If this is the case, then these clouds should be bound by the ram pressure of the ambient clumps that hit them, and not necessarily by self-gravity, so they can have internal velocity dispersions from this pressure that exceed the virial values. In fact, the expected correlation between LSR cloud speed and internal dispersion was found in the data of Magnani et al. [34], as shown in Fig. 2 of [36]. Cloud collisions in fronts such as these are frequent enough to make the internal pressure somewhat uniform with time, and the observed clouds can be bound on average even if their velocity dispersions exceed the virial theorem values.

4.2.6 A Collision Cross Section for Magnetic Clouds. An important part of the concept of cloud collision fronts, as opposed to molecular shock fronts, is the assumption that the interstellar magnetic field connects all of the incident clouds together and that purely magnetic entanglements, in addition to direct cloud collisions, can convert the high pressure behind the front into a cloud acceleration inside the front. The mean free path for cloud collisions can be less than the geometric mean free path when magnetic fields are present [37]. This magnetic mean free path depends slightly on how kinked the field lines are at each clump, so we calculate it here for two cases.

Consider first a smooth deformation of the magnetic field lines that are inside and outside of a cloud resulting from a collision between that cloud, or between the field lines that emerge from the cloud, and other clouds or other field lines that are already in a moving compression front. The incident cloud moves at the speed v relative to the front and its unperturbed field strength is B_0. Suppose that this field line deformation gives a component of the field in the direction of motion equal to $b = b_1 \sin kx$ for wavenumber k and position x parallel to the front. Then the tension in this curved field line is $\mathbf{B} \cdot \nabla B / 4\pi = B_0 (db/dx)/4\pi = B_0 b_1 k \cos kx / 4\pi$. The spatial extension of this field line in the y direction, perpendicular to the front, is given by $b/B_0 = dy/dx$, so $y = -b_1 \cos kx/(kB_0)$ and the tension is therefore $-(B_0^2/4\pi)k^2 y$. If we multiply this by the volume of the flux tube that threads the cloud, $V = \pi R^2 \lambda$ for $\lambda = 2\pi/k$, the result equals the force on the gas in the flux tube, which is the mass M times the acceleration, $d^2 y/dt^2$. Thus we get the equation of motion for the flux tube,

$$M \frac{d^2 y}{dt^2} = -\frac{V B_0^2 k^2 y}{4\pi},$$
(83)

and the solution

$$y(t) = y_{max} \cos kx \sin \omega t \quad \text{for} \quad \omega^2 = \frac{V B_0^2 k^2}{4\pi M}.$$
(84)

This solution corresponds to an initial velocity $v = dy/dt$, from which we get a maximum extension,

$$y_{max} = \frac{v}{\omega} = \frac{v}{B_0 k} \left(\frac{4\pi M}{V} \right)^{1/2}.$$
(85)

In this case of a nearly uniform distortion of the field line, M/V is the average density ρ_0 on the flux tube and then $y_{max} k = v/v_A$, which is the same as for an Alfvén wave with speed $v_A = B_0/(4\pi\rho_0)^{1/2}$. Note that $k = 2\pi/\lambda$ for

λ equal to the mean separation between the cloud anchors at the ends of the flux tube; λ would be the geometric mean free path of field clouds if only these two anchor clouds were available for the magnetic interaction. Thus,

$$y_{\max} = \frac{\lambda}{2\pi} \frac{v}{v_A} \qquad (86)$$

for this limit of smooth field line distortions.

The result is slightly different when the field lines are straight everywhere between the clouds and tightly kinked inside the clouds, with all of the mass near the kink. Then $\boldsymbol{B} \cdot \nabla B/4\pi = B_0 \Delta b/(4\pi \Delta R)$ for field line distortion Δb inside a cloud of thickness $\Delta R = 2R$. This tension equals $-B_0\left(2y\, B_0/(\lambda/2)\right)/(4\pi\, 2R) = -B_0^2\, 2y/(4\pi\lambda R)$ if the anchored parts of the field lines are separated by the distance λ and the cloud with the distortion hits midway between these anchors. Thus the equation of motion is

$$M\frac{d^2 y}{dt^2} = -\frac{V B_0^2\, 2y}{4\pi\lambda R}, \qquad (87)$$

and the solution for the moving cloud position is

$$y(t) = y_{\max} \sin \omega t \quad \text{for} \quad \omega^2 = \frac{2V B_0^2}{4\pi\lambda R M} = \frac{2 B_0^2}{4\pi\lambda R \rho_c} \qquad (88)$$

for individual cloud density $M/V = \rho_c$. This corresponds to a maximum extension

$$y_{\max} = \frac{v}{\omega} = \left(\frac{\lambda}{2} R\right)^{1/2} \frac{v}{v_{Ac}}, \qquad (89)$$

where v_{Ac} is equal to the Alfvén speed inside the cloud. This case of kinked field lines is essentially the one considered in reference [37].

The distance $y_{\max}$ represents the length of the overshoot of one moving cloud relative to two other clouds whose field lines intersect the first cloud. In general, there will be other clouds too and the actual stopping distance will be smaller. One way to define a stopping distance is to make it the average length of the overshoot at which the number of clouds in the volume swept out by the distorted field lines equals 1 [37]. Generally, the moving cloud will not hit exactly between two field clouds, but at some impact parameter b relative to one of them. Then $y_{\max}$ is given approximately [37] by substituting b for $\lambda/2$ in the above equation, so that

$$y_{\max} \approx \frac{b}{\pi} \frac{v}{v_A} \quad \text{and} \quad y_{\max} \approx (bR)^{1/2} \frac{v}{v_{Ac}} \qquad (90)$$

in the smooth and kinked cases, respectively. The average volume swept out by the flux tube (see Figs. 4 and 7 in [37]) is the tube thickness, $2R$, multiplied by $\int y_{\max}\, db$, where the integral is from 0 to $b(y_{\max})$. This should be multiplied by the cloud number density, n_c, and set equal to 1 to solve for the average value of $y_{\max}$ in the random cloud distribution case. The result is

$$\frac{y_{max}}{\lambda} = \left(\frac{3}{4}\frac{v}{v_A}f\right)^{1/2} \tag{91}$$

in the case of a smooth deformation, and

$$\frac{y_{max}}{\lambda} = \frac{3}{4}\pi^{1/3}\left(\frac{v}{v_{Ac}}f\right)^{2/3} = \frac{3}{4}(\pi f)^{1/3}\left(\frac{v}{v_A}\right)^{2/3} \tag{92}$$

in the case of locally straight field lines. Here we have set $v_{Ac}/v_A \sim (\rho/\rho_c)^{1/2} \sim f^{1/2}$ for cloud filling factor f. This second result also follows from [37].

Both of these results suggest that the average stopping distance from purely magnetic interactions in a cloud collision front is comparable to the geometric collision mean free path in the front, λ. To simplify these expressions, we can take $v \approx v_A\left(\frac{1}{2}\rho/\rho_{pre}\right)^{1/2}$ for front speed v where the post-shock magnetic pressure is in balance with the incident shock ram pressure; we also define $C = \rho/\rho_{pre}$ to be the shock compression factor. Then $y_{max}/\lambda = 0.73\left(f^2 C\right)^{1/4}$ and $0.87\left(fC\right)^{1/3}$ for the smooth and kinked cases. For $f \sim 0.1$ and $C \sim 10$, these equal 0.41 and 0.87, respectively. The actual mean free path combines both physical and magnetic collisions, and is approximately $\lambda/(1 + \lambda/y_{max})$, which is 0.29 and 0.46 times the geometric mean free path λ for these parameters.

4.3 Continuous Magnetic Pressure Fronts

The jump conditions for a magnetic shock are useful only when the ions and neutrals are so tightly coupled by collisions that they have approximately the same velocities. Then the neutral fluid stays with the magnetic field and the magnetic forces contribute to its motion. When the fractional ionization is low, as in most molecular clouds, and the shock speed is high, as in regions of active star formation, the ions and neutrals may not collide fast enough to keep the same flow speed. Then the ions can move out ahead of the neutrals in a magnetic pressure pulse, and their streaming motion relative to the neutrals will heat, and perhaps ionize, the neutrals in a continuous fashion [38]. There could still be a shock front in the neutrals if the ion-neutral momentum transfer rate is high enough, but generally the two fluids will change their state more continuously than in the case of a simple shock, and then the jump conditions derived above are no longer relevant. Instead we have to use the full fluid equations for each species, preserving the time derivatives (instead of making the substitution $\partial/\partial t = -S\partial/\partial x$) and the spatial derivative (instead of taking $\partial/\partial x = \{\}_{pre} - \{\}_{post}$).

This section writes these fluid equations (from [39]). Solutions and applications may be found in papers by Draine et al. [40], Chernoff et al. [41, 42], and Flower et al. [43]. Molecular emission line profiles from continuous shocks were calculated by Brand et al. [44 – 47]; their stability was analyzed by Wardle [48 – 50], and applications to the Cygnus loop were made by Graham et al. [51, 52]. Reviews of this topic are in [53, 54].

The basic point about two-fluid or multiple-fluid flows, where the fluids may be the electron, ion, and neutral gases, for example, is that different equations are written for the derivatives of each fluid, and interactions between the fluids are included as source terms. For the magnetic, partially ionized fluid in many interstellar regions, the ion and electron velocities and densities (but not temperatures) can be taken equal (or else strong electric fields will make them equal).

The magnetic field enters only into the equations for the charged component. The neutral fluid reacts indirectly to the magnetic forces through collisions (i.e., viscosity) with the charged particles. Chemical reactions, recombinations and ionizations, etc., are also likely to be important in astronomical flows, so the elements in each fluid can interchange. Thus we need separate equations for the number density and mass density of each fluid type, as well as equations for the rate of change of momentum and energy for each fluid and for the conservation of magnetic flux (ignoring Ohmic diffusion).

The equation for the rate of change of the number density of some species α that flows at the velocity w_f of fluid f (i.e., ions or electrons flowing with the ion fluid velocity, or neutrals flowing with the neutral fluid velocity) is

$$\frac{\partial n_\alpha}{\partial t} + \nabla \cdot (n_\alpha w_f) = N_\alpha, \tag{93}$$

where N_α is the rate per unit volume at which species α is created by chemical or other reactions [39].

The equation for the combined mass density of all of the species α in the fluid f is

$$\frac{\partial \rho_f}{\partial t} + \nabla \cdot (\rho_f w_f) = S_f, \tag{94}$$

where S_f is the rate per unit volume at which mass is being added to the fluid f by chemical and other reactions; S_f is generally equal to the sum over all $N_\alpha m_\alpha$ for species α in fluid f, where m_α is the mass of the particle of type α.

The equation of the rate of change of momentum (which is the force equation) should be written separately for the neutrals and ions because the magnetic field contributes only to the latter. For the neutrals, it is

$$\frac{\partial \rho_n w_n}{\partial t} + w_n \cdot \nabla \rho_n w_n + \rho_n w_n \nabla \cdot w_n + \nabla P_n = F_n - \rho_n \nabla \Phi, \tag{95}$$

where Φ is the gravitational potential and F_n is the rate at which the momentum per unit volume is changed by collisions and chemical reactions. For ion-neutral collisions,

$$F_n = n_i n_n m_r \alpha_{in}(w_i - w_n), \tag{96}$$

where α_{in} is the ion-neutral collision rate, equal to 1.9×10^{-9} for subthermal relative speeds and approximately $10^{-15} |w_i - w_n|$ for superthermal relative speeds [42], in cgs units. There are also collisions between neutrals and charged grains [38, 55] and between neutrals and electrons, which are relatively unimportant because of the small electron mass.

The momentum equation for the ion plus electron fluid is similar, but contains also the magnetic forces,

$$\frac{\partial \rho_i \, w_i}{\partial t} + w_i \cdot \nabla \rho_i \, w_i + \rho_i \, w_i \nabla \cdot w_i + \nabla \left(P_i + P_e + \frac{B^2}{8\pi} \right) - \frac{1}{4\pi} B \cdot \nabla B \qquad (97)$$
$$= F_i + F_e - \rho_i \, \nabla \Phi.$$

The collision force F_i is equal and opposite F_n for ion-neutral collisions, but when grain collisions are also considered (and grains are treated as an additional fluid) some of these two forces must also include grain-gas viscosity.

The equation of conservation of magnetic flux is

$$\frac{\partial B}{\partial t} = \nabla \times (w_i \times B). \qquad (98)$$

Note that this equation includes the possibility of magnetic flux loss by ion-neutral slip (because the ion velocity and not the neutral velocity is used here — see Sect. 3.3), but it ignores diffusion by ion-electron collisions, which is Ohmic diffusion, because such collisions are infrequent.

The last equation is for thermal energy, which for all fluids has the general form

$$\frac{\partial}{\partial t} \left(\frac{P_f}{\gamma - 1} + n_f \theta_f \right) + w_f \cdot \nabla \left(\frac{P_f}{\gamma - 1} + n_f \theta_f \right) + \left(\frac{\gamma P_f}{\gamma - 1} + n_f \theta_f \right) \nabla \cdot w_f = G_f, \qquad (99)$$

where $P_f/(\gamma - 1)$ for $\gamma = \frac{5}{3}$ is the energy density from translational motions in three dimensional space, θ_f is the mean internal energy per particle due to rotation, vibration and electronic excitation, and G_f is the rate of change per unit volume of the thermal energy content of fluid f due to collisions, radiation, chemical reactions, etc. The first two terms are the convective derivative of the thermal energy density. The third term on the left contains, as one part of it, the thermal energy density $P_f/(\gamma - 1) + n_f \theta_f$, which increases with decreasing volume as a result of the density change with volume, and it contains, as a second part, the energy gained from adiabatic compression, which is P_f times $-\nabla \cdot w_f$. The sum of P_f and $P_f/(\gamma-1)$ is $\gamma P_f/(\gamma-1)$. The adiabatic compression does not change the internal energy of each particle, only the translational energy. Note that the magnetic field does not occur here explicitly; it is brought into the energy equation when the ion velocity vector is written in terms of its components, using the momentum equation.

For a steady ($\partial/\partial t = 0$), plane parallel ($\partial/\partial z = \partial/\partial y = 0$) flow in the x direction, as considered in Sect. 4.2.1, we write u for the velocity parallel to the flow ($= w_x$) and v for the velocity in the transverse direction, and we write B for the magnetic field strength in the parallel direction and b for the field strength in the perpendicular direction. Then the equations of momentum for the neutrals and ions are, respectively:

$$\frac{d}{dx}(\rho_n w_n u_n + P_n e_x) = F_n - \rho_n \nabla \Phi, \qquad (100.1)$$

$$\frac{d}{dx} \left(\rho_i \, w_i u_i + \left[P_i + P_e + \frac{b^2}{8\pi} \right] e_x \right) - \frac{B}{4\pi} \frac{db}{dx} = F_i + F_e - \rho_i \, \nabla \Phi, \qquad (100.2)$$

and the equations of energy for the neutrals and the ions in the parallel and perpendicular directions are

$$\frac{d}{dx}\left(u_n\left[\tfrac{1}{2}\rho_n u_n^2 + \tfrac{5}{2}P_n + n_n\theta_n\right]\right) = G_n + F_{nx}u_n - \tfrac{1}{2}S_n u_n^2 - \rho_n u_n \frac{d\Phi}{dx}, \qquad (101.1)$$

$$\frac{d}{dx}\left(\tfrac{1}{2}\rho_n v_n^2 u_n\right) = \mathbf{F}_n \cdot \mathbf{v}_n - \tfrac{1}{2}S_n v_n^2 - \rho_n \mathbf{v}_n \cdot \nabla\Phi, \qquad (101.2)$$

$$\frac{d}{dx}\left(u_i\left[\tfrac{1}{2}\rho_i u_i^2 + \tfrac{5}{2}(P_i + P_e) + n_i\theta_i + \frac{b^2}{4\pi}\right]\right) =$$

$$G_i + G_e + (F_{ix} + F_{ex})u_i - \tfrac{1}{2}S_i u_i^2 + \frac{B}{4\pi}\left(b_y\frac{dv_y}{dx} + b_z\frac{dv_z}{dx}\right) - \rho_i u_i\frac{d\Phi}{dx}, \qquad (101.3)$$

$$\frac{d}{dx}\left(\tfrac{1}{2}\rho_i v_i^2 u_i\right) =$$

$$(\mathbf{F}_i + \mathbf{F}_e) \cdot \mathbf{v}_i - \tfrac{1}{2}S_i v_i^2 + \frac{B}{4\pi}\left(v_y\frac{db_y}{dx} + v_z\frac{db_z}{dx}\right) - \rho_i \mathbf{v}_i \cdot \nabla\Phi. \qquad (101.4)$$

A discussion of the source terms, S, F, and G, is in reference [39]. Without the source terms, these equations reduce to the jump conditions in Sect. 4.2.1, if we write those jump conditions in the form $\{\}_{\text{pre}} - \{\}_{\text{post}} = d/dx$. Note that the energy condition in Sect. 4.2.1 is the same as the sum of the energy conditions for each component in this discussion.

4.4 Common Explosions

4.4.1 Expanding HII Regions. When a star is born in a cold cloud, the radiation from that star gradually increases in peak frequency as the photosphere heats up and may eventually become energetic enough to ionize hydrogen. Then an HII region forms. Even without ionization, the new star warms the surrounding gas, which appears for a time as an extended far-infrared source or infrared nebula. With this heating comes an increase in pressure, followed by an expansion of surrounding gas away from the star. The largest pressures and fastest expansions surround the hottest and most luminous stars, which are the most massive ones. But even low mass stars can create expansion zones, usually in the form of bipolar winds or jets that are driven by non-linear magnetic waves or other energy sources near the photosphere or disk-star interface (see reviews in [56,57]).

The expansion of ionized gas into a region of uniform density is relatively simple to calculate because the transition between the high and low pressure zones is sharp as a result of the short mean free path for Lyman limit photons in neutral hydrogen. The usual derivation assumes that the pressure in the HII region is given by

$$P_{\text{II}} = \rho_{\text{II}} a_{\text{II}}^2 \qquad (102)$$

where ρ and a are the density and one-dimensional velocity dispersion. As the HII region expands and the density decreases, the local recombination rate decreases and with it the residual neutral hydrogen density inside the nebula. This decreases the opacity in the HII region and so allows more total hydrogen mass to be ionized. If the total ionization rate is constant, S, then the density n_{II} and radius R satisfy the equation $S = 4\pi\alpha n_{II}^2 R^3/3$ for recombination rate α to all hydrogen electron orbitals but the ground state (because a ground state recombination emits another ionizing photon). This implies that the density in the expanding HII region scales with radius as

$$\rho_{II} = \rho_I \left(\frac{R_0}{R}\right)^{3/2}, \tag{103}$$

where ρ_I is the neutral atomic hydrogen mass density surrounding the HII region and R_0 is the initial radius at the time when all of the recombinations occurred at the density ρ_I. Then if

$$\frac{dR}{dt} = \left(\frac{P_{II}}{\rho_I}\right)^{1/2} \tag{104}$$

for an isothermal shock in a swept-up neutral shell, the radius increases with time as

$$R = R_0 \left(1 + \frac{7}{4}\frac{a_{II}t}{R_0}\right)^{4/7}. \tag{105}$$

This is the classical solution for the expansion of an HII region.

Actually the shock decelerates as it expands, and at late times there is a dense neutral shell behind the shock. This implies that there is a pressure gradient inside the shell (because of the deceleration), and that the pressure is larger at the leading edge where the shock is located than at the trailing edge where the HII region is located. So the pressure that drives the shock is not P_{II} but the larger pressure where the shock is.

To see this, we have to derive the motion by setting the total force equal to the rate of change of total momentum for velocity v:

$$4\pi R^2 P_{II} \approx \frac{d}{dt}\left(\frac{4}{3}\pi R^3 \rho_I v\right) = 4\pi R^2 \rho_I v^2 + \frac{4}{3}\pi R^3 \rho_I \frac{dv}{dt}. \tag{106}$$

Knowing the solution, we can assume a power law $v \propto R^{-3/4}$, in which case

$$\frac{dv}{dt} = \frac{1}{2}\frac{dv^2}{dR} = -\frac{3}{4}\frac{v^2}{R}; \tag{107}$$

then we get from the total force equation above,

$$v = \frac{dR}{dt} = \left(\frac{4}{3}\frac{P_{II}}{\rho_I}\right)^{1/2}. \tag{108}$$

The solution to this equation is

$$R = R_0 \left(1 + \frac{7\left(\frac{4}{3}\right)^{1/2} a_{II} t}{4R_0}\right)^{4/7}. \tag{109}$$

This equation implies that at late times an HII region should be slightly larger than the radius given by the classical expression because the momentum of the swept-up neutral gas pushes it ahead more than the HII region pressure alone.

In a cloudy medium, a hot star ionizes and accelerates the nearby clouds and the debris fills the neighborhood with a more uniform density [58, 59]. The homogenization radius is almost the same as the classical Strömgren radius, $R([\text{pc}]) \approx 1.05 \, R_0 (a_{\text{II}} t / R_0)^{4/7}$. After a main sequence lifetime [59], $t_{\text{ms}} \approx 4.4 \, S_{49}^{-0.25}$ Myr, for UV photoionization rate S_{49} in units of 10^{49} ionizations per second, the radius becomes

$$R \sim 60 \, n^{-2/7} \text{ pc}, \tag{110}$$

which is nearly independent of the ionizing luminosity for stars earlier than B0 [59].

The total energy in gas motions at this time is $\frac{1}{2} M V^2$, or

$$E \sim 1.5 \times 10^{49} \, S_{49}^{0.5} \, n^{-3/7} \text{ erg.} \tag{111}$$

Note that this has an inverse dependence on the ambient density (cf. Chap. 6).

Many HII regions first appear near the edges of molecular clouds in a blister configuration [60, 61] because the associated massive stars apparently form near these edges. This structure is easiest to observe in the solar neighborhood, where the spatial resolution is good (e.g., [62, 63]), but similar morphologies have been found in the Large [64] and Small [65] Magellanic Clouds too. In a blister, the flow of ionized gas away from the cloud is not spherically symmetric as in the classical Strömgren analysis, but open to one side and divergent. Such structure has been called a champagne flow by Tenorio-Tagle et al. [66 − 68], who modelled the dynamics in detail. The emission lines from such a flow were studied by Rubin et al. [69]. Other types of expansions into media with density gradients were considered by Franco et al. [70].

An interesting way [71] to determine if a poorly resolved source has a champagne flow is to measure the velocity difference between the narrow H recombination lines, which presumably originate on the cold neutral side of the ionization front where the velocity dispersion is low, and the broad H recombination lines, which presumably originate in the ionized gas where the dispersion is high. If, in addition, the recombination lines from C^+, which come mostly from the dense neutral side, are strong as a result of stimulated emission, then the neutral matter is closest on the line of sight and the ionized matter is furthest. In this configuration, the broad H lines should be redshifted relative to the narrow H lines if there is a champagne flow. Orello et al. [71] found 3 out of 5 observed sources to be champagne flows this way.

Models of expanding HII regions with embedded supernova [72] and stellar winds [59, 73] are also more realistic than the spherically symmetric case. No model is detailed enough yet to simulate all of the structure seen in the highest resolution images [74, 75].

4.4.2 Classical Wind Bubbles. If energy is injected continuously into a region which expands inside a shock front, the equations that determine the expansion are

$$\frac{dMv}{dt} = 4\pi R^2 P, \tag{112.1}$$

$$P = \tfrac{2}{3}U = \frac{2E}{4\pi R^3}, \tag{112.2}$$

$$\frac{dE}{dt} = L - 4\pi R^2 Pv, \tag{112.3}$$

for energy density U, energy E, and luminosity L. These equations have the similarity solution

$$R = 0.76 \left(\frac{Lt^3}{\rho_0}\right)^{1/5} \tag{113}$$

[76, 77]. For a weak wind, the radius of the cavity is less than the homogenized HII region radius, and this wind solution applies with ρ_0 equal to the homogenized density. For a strong wind, the radius is large and includes many clouds, which evaporate and cool it. Then the pressure from the wind decreases and the radius drops to equal the homogenized HII radius. The two radii then expand together [59]. Detailed calculations of the structure of a wind-driven bubble are in Weaver et al. [78] and Koo & McKee [79]. Bubble expansion into a nonuniform medium was discussed by Icke [80].

Observations of wind-driven bubbles around Wolf-Rayet stars are discussed by Chu et al. [81], Niemela et al. [82], and others. Observations of other types of wind bubbles are discussed by Schneps et al. [83], Zhang et al. [84], and others. A review of interstellar bubbles is in Dyson [85].

OB association can drive a large bubble into the galactic disk because of combined long-term pressures from many supernova and stellar winds. The equations that govern this expansion are often approximated by the pure wind solution given above, but with a total luminosity and time scale that are appropriate for the association rather than a single star [86, 87]. The assumption used for this approximation is that the energy from multiple supernova and other OB-star pressures is injected into the interstellar medium in a continuous fashion, rather than in a single burst. This continuous injection can make the efficiency for conversion into forced cloud motions much higher, perhaps by a factor of 10 [87], than in the burst case. The supershells and chimneys produced by such long-term expansions are discussed in Sect. 4.5.

4.4.3 Supernovae. If energy is injected into the interstellar medium instantaneously, the equations that govern the motion are:

$$v = \left(\frac{\gamma + 1}{2}\frac{P_{cav}}{\rho_0}\right)^{1/2}, \tag{114.1}$$

$$P_{cav} = K\frac{(\gamma - 1)E}{\tfrac{4}{3}\pi R^3}, \tag{114.2}$$

where $K = 1.53$ to correct for the pressure distribution inside the cavity and the fraction of the total SN energy that is thermal [88]. These equations have the solution [89]

$$R = \left(\frac{2.02\,E t^2}{\rho_0}\right)^{1/5}.$$

(115)

For a young supernova, the shock will be strong and adiabatic (if it is too young for radiative losses), and then the density jump at the shock front will be a factor of 4. This gives a thick shell. After a radiative cooling time in the shell, the gas there will cool and the shell density will increase to maintain pressure equilibrium. Then the shell will be thin. There is still hot gas in the interior of the remnant, however, because the density there is too low to cool the gas quickly, and this hot gas pushes on the shell as it slows down. Thus there is no true snowplow phase (momentum conserving phase) until the hot interior cools, but instead there is a pressure driven-snowplow (PDS) phase [90].

The equation of motion for the PDS phase is

$$\frac{d}{dt}\left(\tfrac{4}{3}\pi R^3 \rho_0 v\right) = 4\pi R^2 P$$

(116)

for pressure $P = (\gamma - 1)E/(\tfrac{4}{3}\pi R^3)$. Cioffi et al. [90] find that the thermal energy in the cavity, E, decreases because of cooling, and that the pressure can actually be written as

$$P_* \sim \frac{1.92}{R_*^5\, t_*^{4/9}}$$

(117)

where $*$ quantities are normalized to the values at the start of the PDS phase, i.e., $P_* = P/P_{\text{PDS}}$, and

$$R_{\text{PDS}} = 14.0\,\frac{E_{51}^{2/7}}{n_0^{3/7}\,\zeta^{1/7}} \qquad [\text{pc}],$$

(118.1)

$$v_{\text{PDS}} = 413\,E_{51}^{1/14}\,n_0^{1/7}\,\zeta^{3/14} \quad [\text{km s}^{-1}],$$

(118.2)

$$t_{\text{PDS}} = 1.33 \times 10^4\,\frac{E_{51}^{3/14}}{n_0^{4/7}\,\zeta^{5/14}} \quad [\text{y}],$$

(118.3)

for metallicity ζ ($= 1$ for solar). Cioffi et al. [90] find that a convenient solution to the shell motion is a power law, $R \propto t^\eta$, where $\eta = 0.4$ in the Sedov solution before the PDS stage and $\eta \sim 0.3$ during the PDS phase. For $t > t_*$, they derive

$$R_* = \left(\tfrac{4}{3}t_* - \tfrac{1}{3}\right)^{0.3},$$

(119.1)

$$v_* = \left(\tfrac{4}{3}t_* - \tfrac{1}{3}\right)^{-0.7}.$$

(119.2)

The SNR merges with the ambient ISM when $v = \beta a$ for ambient rms speed a and $\beta \sim 1$. We can solve for t_* at this time using the above solution $v_*(t_*)$, and then solve for R_*, which becomes the merging radius (in units of R_{PDS})

$$R_{*,\text{merge}} = 4.93 \left(\frac{E_{51}^{1/14} n_0^{1/7} \zeta^{3/14}}{\beta a_{10}} \right)^{3/7} \tag{120}$$

for a_{10} in units of $10 \,\text{km s}^{-1}$. At this time the kinetic energy of the shell is $\frac{1}{2} M v^2$ or

$$E_{\text{shell}} = \frac{1}{2} \frac{4}{3} \pi R_{*,\text{merge}}^3 \rho_0 (\beta a)^2$$
$$4.5 \times 10^{49} \, E_{51}^{0.95} \, n_0^{-0.10} \, \zeta^{-0.15} \, (\beta a_{10})^{0.71} \; [\text{erg}]. \tag{121}$$

Note again the negative power dependence on ambient density.

High mass supernovae presumably expand into the HII region - wind cavity of the O star and, closer to the star, into the old stellar wind debris (see reviews in [91 − 93] and see [94] for a discussion of the interaction of SN 1987A with the red giant envelope of the progenitor star). According to McKee [91], if the cavity from the previous bubble is larger than about 20 pc, the supernova remnant may not show much X-ray luminosity: inside the cavity the density is too low to radiate, and when the supernova shock finally hits the bubble wall, it will move too slowly to excite X-rays.

Massive-star supernovae may also have remnants with curved filaments (e.g., the Cygnus Loop) because of the initial spherical stratification of the environment from the bubble and red giant wind [95]. These filaments are apparently not pre-existing clouds because such clouds, when hit by the shock, generally take a more cometary structure, pointing toward the source of the explosion rather than transverse to it [21]. Calculations of the expansion of a supernova in a wind-blown bubble are in Chevalier & Liang [96], and numerical simulations are in Tenorio-Tagle et al. [97, 98]. Observations of this effect are summarized by Strom [99]. Detailed studies of regions of high mass star formation that contain a variety of pressure sources, including supernova, stellar winds, and ionization, are in Laval et al. [100] and Junkes et al. [101].

Low mass supernovae explode into cloudy media because there is not enough ionization from the pre-supernova star to homogenize it. The clouds that remain should evaporate and cool the remnant gas. Several things can happen to the clouds. They can get crushed, which means that shocks enter from all sides; this could trigger star formation. They can also get accelerated because the frontal shock is strongest and because of magnetic tension in the swept-back field lines. In that case a bow shock forms around the cloud, giving the appearance of a cometary globule [18, 21]. Also, various instabilities (e.g., Kelvin-Helmholtz, Rayleigh-Taylor) erode mass from the edges of the cloud [23]. A review of many of these processes is in Ostriker & McKee [5].

When the interior of the supernova cavity is still hot, conduction (or cooling) can transfer mass to (or from) the hot gas from (to) the clouds. Mass added to the hot gas can cool it and stall the remnant. Note that the electrons conduct heat and the ions conduct momentum (viscosity). If radiative losses are important at the surface of a cloud, then the gas condenses onto the cloud rather than evaporates from the cloud. Suppose that n_i is the intercloud hydrogen density, R is the cloud radius in parsecs, and T_{i7} is the intercloud temperature in units of 10^7 K, then the effects of electron collisions in the conductive interface of a cloud of radius R can be measured by the saturation parameter [91, 102]

$$\sigma = 1.67 \frac{\lambda_{ee}}{R} - 0.393 \frac{T_{i7}^2}{n_i R},$$ (122)

where λ_{ee} is the electron-electron collision mean free path. There are three important cases: when $\sigma > 1$ there is saturated conduction or free electron streaming; when $\sigma > 0.03$ there is cloud evaporation, and when $\sigma < 0.03$ the intercloud gas condenses onto the cloud.

As discussed in Sect. 3.1, supernovae play an important role in the maintenance of interstellar turbulence and random cloud motions. They also heat a large fraction of the interstellar volume to temperatures in excess of 10^5 K. The morphology of the gas that results from such intense heating and stirring is not known, but calculations by Salpeter [103], Cox & Smith [17], McKee et al. [16, 104], Cioffi & Shull [105], White & Long [106], and others suggest that random and clustered supernovae easily form pervasive hot regions that interact with the lower temperature and more massive neutral component at shock fronts and conductive interfaces. Supernovae explosions inside dense clouds were discussed recently by Draine & Woods [107]. Explosions with magnetic fields in the surrounding gas were considered by Insertis & Rees [108] and Ferrier et al. [109]. A review of the variation of supernova rate with galactic Hubble type is in van den Bergh [110].

4.5 Supershells and Chimneys

Many of the small scale explosions discussed in the previous section have an analog on larger scales where collective pressures from many massive stars in OB associations or star complexes energize the interstellar medium out to distances comparable to or larger than the galactic scale height (see reviews in [111, 112] and see Sect. 2.2 for observations). These pressures are not particularly high compared to the pressure in a single HII region or supernova, but they are persistent, lasting for more than several times 10^7 years in the case of the largest structures. The result of such pressurization can be the formation of a supershell [113 − 115] or chimney [116] that carries material into the halo [117 − 120]. Collisions between high velocity clouds and the galactic plane produce similar disturbances [121]. If the source of pressure is localized and a large shell forms, then giant molecular clouds can grow in the accumulated material along the perimeter of the expanding structure, and whole new OB associations can form in these clouds [121 − 125].

Computer simulations of the large scale expansion of interstellar gas perpendicular to the plane and into the halo have been made by Tomisaka et al. [126, 127], MacLow et al. [87, 128], Igumentshchev et al. [129], and Tenorio-Tagle et al. [130]. These authors suggest that a bubble formed by multiple supernova in an OB association can blow out into the halo if the density gradient perpendicular to the disk is large, as for a Gaussian distribution of gas instead of an exponential one [87, 130]. They also suggest that the amount of mixture of bubble gas with halo gas depends on the sound speed in the halo. If the halo is hot, then the hot bubble material can mix with the halo on time scales of 10^8 years, but if the halo is cold, then the bubble and halo are separated by a shocked shell that inhibits mixing [130]. The formation of clouds high up in the fountain flow was discussed by Houck & Bregman [131]. Reviews of these processes are in Rozyczka [132], and Cox [133], who points out that a thick disk of HI or ionized gas, as observed for our Galaxy [134, 135], could quench superbubble mixing with the halo.

Observations of the chimney-type structures, or "worms", that result from superbubble blowout were discussed by Heiles et al. [136 − 139], Li & Ikeuchi [140 − 141], and van den Bergh et al. [142 − 143]. Numerous observations of the edge-on galaxy NGC 891 show these structures [144 − 148], which are also in the galaxies NGC 4565 [147], NGC 4631 [149], NGC 2363 [150], and NGC 3628 [151]. A blowout model for a region in the LMC was recently proposed by Wang & Helfand [152].

Characteristics of the expansion of superbubbles parallel to the galactic plane were calculated by Tenorio-Tagle et al. [125], Palouš et al. [153], and Franco et al. [124]. Shear and Coriolis forces distort the growing shell, making an ellipse that eventually closes up after an epicyclic period. The accretion of ambient gas is greatest at the apices of the ellipse because of transverse flows inside the shock; this is where most of the matter should collect into new molecular clouds. Magnetic forces were included by Ferrier et al. [109], who showed that the gas in an expanding magnetic bubble tends to flow away from the magnetic axes and toward the magnetic equator, and that the transverse magnetic pressure behind the shock is greatest at the equator. These effects make the shell thicker at the equator than along the direction of the field.

The interaction between infalling clouds and the galactic plane has been modelled by Tenorio-Tagle et al. [121, 124, 154, 155], who predicted that supershells and rings of molecular clouds can be formed this way. Evidence for this process has been discussed by Alfaro et al. [156], who found, in the distribution of star clusters, wide and shallow depressions in the Galactic plane, including one near the Sun which they call the "Big Dent". These depressions correspond to the locations of giant star complexes, so the cloud collisions may have triggered star formation [156]. Observations of cloud-Galaxy collisions on a smaller scale were discussed by Odenwald [157], Mirabel et al. [158] and Meyerdierks [159, 160].

High velocity gas in the galaxies M101 [161], NGC 628 [162], and NGC 4258 [163] indicate that cloud-plane collisions could be common in galaxies, although some of this gas could be tidal debris from a collision with another galaxy [164].

Shells and holes in other galaxies sometimes appear larger than the local galactic scale height. The formation process for such large holes is difficult to imagine if the pressure comes from a centralized source, such as an OB association, and not from a large cloud collision, for example. As soon as a bubble gets as large as a scale height, the hot gas that drives it should vent into the halo and the pressure should drop [139, 165]. If the shell has a large velocity at that time, then it will continue to expand in a snowplow phase until it eventually stalls at a radius larger than a scale height. This would explain the observation of large holes. If the velocity at blowout is small, however, then the snowplow phase will not enlarge the radius much and the hole should stay small. The largest holes could be a problem for OB association-driven models unless multiple internal sites of star formation continuously enlarge it in spite of a continuous blowout to the halo, or unless there is another source of pressure that is not diminished by blowout. An example of such a pressure is radiation pressure from field stars on dust grains. In a hollow cavity several hundred parsecs in size, formed, for example, by the pressures from an OB association, there will be an additional outward pressure from the near-uv radiation emitted by enclosed field stars. This extra pressure can exceed the gas pressure and drive the cavity to much larger radii [166]. Because radiation pressure is always directed radially away from the stellar sources, it cannot vent to the halo when the cavity is larger than a scale height. Radiation pressure also has the interesting property that the acceleration of optically thin gas is independent of the gas density, so very large kinetic energies can be deposited into dense shells.

4.6 Bowshocks and Jets

One of the more interesting new developments on interstellar shocks is the recognition that jets [167] from young stars, and that moving compact HII regions around young O stars [168], commonly make bowshocks in the dense clouds where they reside. A bowshock is a shock that curves around a moving object or stream, and because of this curvature, contains a continuum of different obliquities and compression factors. Numerical models of jets by Norman et al. [169] and Tenorio-Tagle & Rozyczka [170] show the bowshock structure that is often identified [171,172] with Herbig-Haro objects [173] in regions of star formation. Models of bowshocks around moving compact HII regions are in van Buren et al. [174,175], and reviews of these compact HII regions are in Churchwell [176,177]. Observations of the dense molecular gas that is associated with bowshock IIII regions are in Churchwell et al. [178,179], who find densities and masses high enough to confine the HII region, and relative velocities between the ionized and molecular gas that are consistent with the bowshock interpretation.

Acknowledgement: I am grateful to Dr. Chris McKee for helpful comments.

References

4.1 Whang, Y.C. 1991, *Astrophys. J.* **377**, 250
4.2 Whang, Y.C. 1991, *Astrophys. J.* **377**, 255
4.3 Whang, Y.C. 1991, *Astrophys. J.* **381**, 559
4.4 Cohen, R.H., Kulsrud, R.M. 1974, *Phys. Fluids*, **17**, 2215
4.5 Ostriker, J.P., McKee, C.F. 1988, *Rev. Mod. Phys.* **60**, 1
4.6 Shull, J.M., Draine, B.T. 1987, in *Interstellar Processes*, eds. Hollenbach, D.J., Thronson, H.A., Jr., (Dordrecht: Reidel), p. 283
4.7 Chakrabarti, S.K. 1990, *Theory of Transonic Astrophysical Flows*, (Singapore, World Scientific)
4.8 Nelson, A.H. 1992, in *Evolution of Interstellar Matter and Dynamics of Galaxies*, eds. B.W. Burton, P.O. Lindblad, J. Palouš, (Cambridge: Cambridge Univ. Press), in press
4.9 Elmegreen, B.G. 1989b, in *Evolutionary Phenomena in Galaxies*, eds. J.E. Beckman, B.E.J. Pagel, (Cambridge: Cambridge University Press), p. 83
4.10 Elmegreen, B.G. 1991, in *Evolution of Interstellar Matter and Dynamics of Galaxies*, eds. B.W. Burton, P.O. Lindblad, J. Palouš, (Cambridge: Cambridge Univ. Press), in press
4.11 Sgro, A.G. 1975, *Astrophys. J.* **197**, 621
4.12 Chevalier, R.A., Theys, J.C. 1975, *Astrophys. J.* **195**, 53
4.13 Ikeuchi, S., Spitzer, L., Jr. 1984, *Astrophys. J.* **283**, 825
4.14 McKee, C.F., Cowie, L.L. 1975, *Astrophys. J.* **195**, 715
4.15 Cowie, L.L., McKee, C.F. 1977, *Astrophys. J.* **211**, 135
4.16 McKee, C.F., Ostriker, J.P. 1977, *Astrophys. J.* **218**, 148
4.17 Cox, D.P., Smith, B.H. 1974, *Astrophys. J.* **189**, L105
4.18 Woodward, P.R. 1976, *Astrophys. J.* **207**, 484
4.19 Dibai, E.A. 1958, *Sov. Astron.* **2**, 429
4.20 Tenorio-Tagle, G., Rozyczka, M. 1986, *Astron. Astrophys.* **155**, 120
4.21 Rozyczka, M., Tenorio-Tagle, G. 1987, *Astron. Astrophys.* **176**, 329
4.22 Falle, S.A.E.G., Giddings, J.R. 1989, in *Supernova Shells and their Birth Events*, ed. W. Kundt, Lect. Notes Phys. **316**, p. 63
4.23 Klein, R.I., McKee, C.F., Collela, P. 1990, in *The Evolution of the Interstellar Medium*, ed. L. Blitz, San Francisco, Astronomical Society of the Pacific, p. 117
4.24 Bedogni, R., Woodward, P.R. 1990, *Astron. Astrophys.* **231**, 481
4.25 Kimura, T., Tosa, M. 1987, *Astrophys. Space Sci.* **129**, 261
4.26 Kimura, T., Tosa, M. 1988, *Monthly Notices Roy. Astron. Soc.* **234**, 51
4.27 Kimura, T., Tosa, M. 1990, *Monthly Notices Roy. Astron. Soc.* **245**, 365
4.28 Kimura, T., Tosa, M. 1991, *Monthly Notices Roy. Astron. Soc.* **251**, 664
4.29 Schwartz, R.D. 1978, *Astrophys. J.* **223**, 884
4.30 Elitzur, M., Watson, W.D. 1978, *Astrophys. J.* **222**, L141
4.31 Draine, B.T., Katz, N. 1986, *Astrophys. J.* **306**, 655
4.32 Cowie, L.L. 1980, *Astrophys. J.* **236**, 862
4.33 Yuan, C., Wang, C.Y. 1982, *Astrophys. J.* **252**, 508
4.34 Magnani, L., Blitz, L., Mundy, L. 1985, *Astrophys. J.* **295**, 402
4.35 Keto, E.R., Lattanzio, J.C. 1990, *Astrophys. J.* **346**, 184
4.36 Elmegreen, B.G. 1988, *Astrophys. J.* **326**, 616
4.37 Clifford, P., Elmegreen, B.G. 1983, *Monthly Notices Roy. Astron. Soc.* **202**, 629
4.38 Draine, B.T. 1980, *Astrophys. J.* **241**, 1021
4.39 Draine, B.T. 1986, *Monthly Notices Roy. Astron. Soc.* **220**, 133
4.40 Draine, B.T., Roberge, W.G., Dalgarno, A. 1983, *Astrophys. J.* **264**, 485
4.41 Chernoff, D.F., Hollenbach, D.J., McKee, C.F. 1982, *Astrophys. J.* **259**, L97
4.42 Chernoff, D.F. 1987, *Astrophys. J.* **312**, 143
4.43 Flower, D.R., Pineau des Forets, G., Hartquist, T.W. 1985, *Monthly Notices Roy. Astron. Soc.* **216**, 775
4.44 Brand, P.W.J.L., Moorhouse, A., Burton, M.G., Geballe, T.R., Bird, M., Wade, R. 1988, *Astrophys. J.* **334**, L103
4.45 Brand, P.W.J.L., Toner, M.P., Geballe, T.R., Webster, A.S., Williams, P.M., Burton, M.G. 1989, *Monthly Notices Roy. Astron. Soc.* **236**, 929

4.46   Smith, M.D., Brand, P.W.J.L. 1990, *Monthly Notices Roy. Astron. Soc.* **242**, 495

4.47   Smith, M.D., Brand, P.W.J.L. 1990, *Monthly Notices Roy. Astron. Soc.* **243**, 498

4.48   Wardle, M. 1990, *Monthly Notices Roy. Astron. Soc.* **246**, 98

4.49   Wardle, M. 1991, *Monthly Notices Roy. Astron. Soc.* **250**, 523

4.50   Wardle, M. 1991, *Monthly Notices Roy. Astron. Soc.* **251**, 119

4.51   Graham, J.R., Wright, G.S., Geballe, T.R. 1991, *Astrophys. J.* **372**, L21

4.52   Graham, J.R., Wright, G.S., Hester, J.J., Longmore, A.J. 1991, *Astron. J.* **101**, 175

4.53   Draine, B.T. 1991, in *Fragmentation of Molecular Clouds and Star Formation*, eds. E. Falgarone, F. Boulanger, G. Duvert, Dordrecht: Kluwer, p. 185

4.54   Chernoff, D.F., McKee, C.F. 1990, in *Molecular Astrophysics*, eds. T.W. Hartquist, Cambridge, Cambridge Univ. Press, p. 360

4.55   Philipp, W., Hartquist, T., Havnes, O. 1990, *Monthly Notices Roy. Astron. Soc.* **243**, 685

4.56   Pudritz, R.E., Pelletier, G., Gomez de Castro, A.I. 1991, in *The Physics of Star Formation and Early Stellar Evolution*, eds. C.J. Lada, N.D. Kylafis, (Dordrecht: Kluwer), 539

4.57   Natta, A., Giovanardi, C. 1991, in *The Physics of Star Formation and Early Stellar Evolution*, eds. C.J. Lada, N.D. Kylafis, (Dordrecht: Kluwer), 595

4.58   Elmegreen, B.G. 1976, *Astrophys. J.* **205**, 405

4.59   McKee, C.F., van Buren, D., Lazareff, B. 1984, *Astrophys. J.* **278**, L115

4.60   Zuckerman, B. 1973, *Astrophys. J.* **183**, 863

4.61   Israel, F.P. 1978, *Astron. Astrophys.* **70**, 769

4.62   Gomez, J.F., Torrelles, J.M., Tapia, M., Roth, M., Verdes Montenegro, L., Rodriguez, L.F. 1990, *Astron. Astrophys.* **234**, 447

4.63   Wilson, T.L., Johnston, K.J., Mauersberger, R. 1991, *Astron. Astrophys.* **251**, 220

4.64   Ye, T., Turtle, A.J., Kennicutt, R.C., Jr. 1991, *Monthly Notices Roy. Astron. Soc.* **249**, 722

4.65   McCall, M.L., Hill, R., English, J. 1990, *Astron. J.* **100**, 193

4.66   Tenorio-Tagle, G. 1979, *Astron. Astrophys.* **71** 59

4.67   Bodenheimer, P., Tenorio-Tagle, G., Yorke, H.W. 1979, *Astrophys. J.* **233**, 85

4.68   Yorke, H.W., Bodenheimer, P., Tenorio-Tagle, G. 1982, *Astron. Astrophys.* **108**, 25

4.69   Rubin, R.H., Simpson, J.P., Haas, M.R., Erickson, E.F. 1991, *Astrophys. J.* **374**, 564

4.70   Franco, J., Tenorio-Tagle, G., Bodenheimer, P. 1990, *Astrophys. J.* **349**, 126

4.71   Orello, J.S., Phillips, J.A., Terzian, Y. 1991, *Astrophys. J.* **383**, 693

4.72   Yorke, H.W., Tenorio-Tagle, G., Bodenheimer, P., Rozyczka, M. 1989, *Astron. Astrophys.* **216**, 207

4.73   Dyson, J.E. 1977, *Astron. Astrophys.* **59**, 161

4.74   Yusef-Zadeh, F. 1990, *Astrophys. J.* **361**, L19

4.75   Hester, J.J., et al. 1991, *Astrophys. J.* **369**, L75

4.76   Avedisova, V.S. 1972, *Sov. Astron.* **15**, 708

4.77   Castor, J., McCray, R., Weaver, R. 1975, *Astrophys. J.* **200**, L107

4.78   Weaver, R., McCray, R., Castor, J., Shapiro, P., Moore, R. 1977, *Astrophys. J.* **288** 377

4.79   Koo, B.C., McKee, C.F. 1992, *Astrophys. J.* **388**, in press

4.80   Icke, V. 1988, *Astron. Astrophys.* **202**, 177

4.81   Chu, Y.H., Treffers, R.R., Kwitter, K.B. 1983, *Astrophys. J. Suppl. Ser.* **53**, 937

4.82   Niemela, V.S., Cappa de Nicolau, C.E. 1991, *Astron. J.* **101**, 572

4.83   Schneps, M.H., Haschick, A.D., Wright, E.L., Barrett, A.H. 1981, *Astrophys. J.* **243**, 184

4.84   Zhang, C.Y., Green, D.A. 1991, *Astron. J.* **101**, 1006

4.85   Dyson, J.E. 1989, in *Structure and Dynamics of the Interstellar Medium*, IAU Colloquium 120, eds. G. Tenorio-Tagle, M. Moles, J. Melnick, Berlin, Springer, p. 137

4.86   McCray, R., Kafatos, M. 1987, *Astrophys. J.* **317**, 190

4.87   MacLow, M.-M., McCray, R. 1988, *Astrophys. J.* **324**, 776

4.88   Chevalier, R.A. 1974, *Astrophys. J.* **188**, 501

4.89   Sedov, L.I. 1959, *Similarity and Dimensional Methods in Mechanics*, (New York: Academic Press)

4.90   Cioffi, D.F., McKee, C.F., Bertschinger, E. 1988, *Astrophys. J.* **334**, 252

4.91   McKee, C.F. 1988, in *Supernova Remnants and the Interstellar Medium*, eds. R.S. Roger, T.L. Landecker, (Cambridge: Cambridge University Press), p. 205

4.92 Shull, P. Jr., Dyson, J., Kahn, F. 1988, in *Supernova Remnants and the Interstellar Medium*, eds. R.S. Roger, T.L. Landecker, (Cambridge: Cambridge University Press), p. 231

4.93 Chevalier, R.A. 1990, in *Supernovae*, ed. A.G. Petschek, New York, Springer, p. 91

4.94 Chevalier, R.A., Emmering, R.T. 1989, *Astrophys. J.* **342**, L75

4.95 Charles, P.A., Kahn, S.M., McKee, C.F. 1985, *Astrophys. J.* 295, 456

4.96 Chevalier, R.A., Liang, E.P. 1989, *Astrophys. J.* **344**, 332

4.97 Tenorio-Tagle, G., Bodenheimer, P., Franco, J., Rozyczka, M. 1991, *Monthly Notices Roy. Astron. Soc.* **244**, 563

4.98 Tenorio-Tagle, G., Rozyczka, M., Franco, J., Bodenheimer, P. 1991, *Monthly Notices Roy. Astron. Soc.* **251**, 318

4.99 Strom, R.G. 1990, in *Neutron Stars and their Birth Events*, ed. W. Kundt, NATO ASI Ser., Ser. C, Math. Phys. Sci. **300**, p. 263

4.100 Laval, A., Rosado, M., Bodesteix, J., Georgelin, Y.P., Le Coarer, E., Marcelin, M., Viale, A. 1992, *Astron. Astrophys.* **253**, 213

4.101 Junkes, N., Fürst, E., Reich, W. 1992, *Astron. Astrophys.* , in press

4.102 McKee, C.F., Begelman, M.C. 1990, *Astrophys. J.* **358**, 392

4.103 Salpeter, E.E. 1976, *Astrophys. J.* **206**, 673

4.104 Cowie, L.L., McKee, C.F., Ostriker, J.P. 1981, *Astrophys. J.* **247**, 908

4.105 Cioffi, D.F., Shull, J.M. 1991, *Astrophys. J.* **367**, 96

4.106 White, R.L., Long, K. 1991, *Astrophys. J.* **373**, 543

4.107 Draine, B.T., Woods, D.T. 1991, *Astrophys. J.* **383**, 621

4.108 Insertis, F.M., Rees, M.J. 1991, *Monthly Notices Roy. Astron. Soc.* **252**, 82

4.109 Ferrier, K.M., MacLow, M.-M., Zweibel, E.G. 1991, *Astrophys. J.* **383**, 602

4.110 van den Bergh, S. 1990, *Publ. Astron. Soc. Pacific* **102**, 657

4.111 McCray, R., Snow, T.P., Jr. 1979, *Ann. Rev. Astron. Astrophys.* **17**, 213

4.112 Tenorio-Tagle, G., Bodenheimer, P. 1989, *Ann. Rev. Astron. Astrophys.* **26**, 145

4.113 Heiles, C. 1979, *Astrophys. J.* **229**, 533

4.114 Bruhweiler, F.C., Gull, T.R., Kafatos, M., Sofia, S. 1980, *Astrophys. J.* **238**, L27

4.115 Kolesnik, I.G., Silich, S.A. 1990, *Astrophys. J.* **30**, 178

4.116 Norman, C.A., Ikeuchi, S. 1989, *Astrophys. J.* **345**, 372

4.117 Shapiro, P.R., Field, G.B. 1976, *Astrophys. J.* **205**, 762

4.118 Bregman, J.N. 1980, *Astrophys. J.* **237**, 280

4.119 Kahn, F. 1989, in *Structure and Dynamics of the Interstellar Medium*, IAU Colloquium 120, eds. G. Tenorio-Tagle, M. Moles, J. Melnick, Berlin, Springer, p. 474

4.120 Shapiro, P.R., Benjamin, R.A. 1991, *Publ. Astron. Soc. Pacific* **103**, 923

4.121 Tenorio-Tagle, G. 1981, *Astron. Astrophys.* **94**, 338

4.122 Ögelman, H.B., Maran, S.P. 1976, *Astrophys. J.* **209**, 124

4.123 McCray, R., Kafatos, M. 1987, *Astrophys. J.* **317**, 190

4.124 Franco, J., Tenorio-Tagle. G., Bodenheimer, P., Rozyczka, M., Mirabel, I.F. 1988, *Astrophys. J.* **333**, 826

4.125 Tenorio-Tagle, G., Palouš, J. 1987, *Astron. Astrophys.* **186**, 287

4.126 Tomisaka, K. 1990, *Astrophys. J.* **361**, L5

4.127 Tomisaka, K., Ikeuchi, S. 1986, *Publ. Astron. Soc. Japan* **38**, 697

4.128 MacLow, M.-M., McCray, R., Norman, R.L. 1989, *Astrophys. J.* **337**, 141

4.129 Igumentshchev, I.V., Shustov, B.M., Tutukov, A.V. 1990, *Astron. Astrophys.* **234**, 369

4.130 Tenorio-Tagle, G., Rozyczka, M, Bodenheimer, P. 1990, *Astron. Astrophys.* **237**, 207

4.131 Houck, J.C., Bregman, J.N. 1990, *Astrophys. J.* **352**, 506

4.132 Rozyczka, M. 1989, in *Structure and Dynamics of the Interstellar Medium*, IAU Colloquium 120, eds. G. Tenorio-Tagle, M. Moles, J. Melnick, Berlin, Springer, p. 463

4.133 Cox, D.P. 1989, in *Structure and Dynamics of the Interstellar Medium*, IAU Colloquium 120, eds. G. Tenorio-Tagle, M. Moles, J. Melnick, Berlin, Springer, p. 500

4.134 Jahoda, K., Lockman, F.J., McCammon, D. 1990, *Astrophys. J.* **354**, 184

4.135 Reynolds, R.J. 1989, *Astrophys. J.* **339**, L29

4.136 Heiles, C. 1984, *Astrophys. J. Suppl. Ser.* **55**, 585

4.137 Heiles, C. 1989, in *Structure and Dynamics of the Interstellar Medium*, IAU Colloquium 120, eds. G. Tenorio-Tagle, M. Moles, J. Melnick, Berlin, Springer, p. 484

4.138 Koo, B.C., Heiles, C., Read, W.T. 1990, in *The Interstellar Disk Halo Connection in Galaxies*, ed. J. Bloemen, (Dordrecht: Reidel), p. 114

4.139 Heiles, C. 1991, *Astrophys. J.* **354**, 483

4.140 Li, F., Ikeuchi, S. 1989, *Publ. Astron. Soc. Japan* 41, 221

4.141 Li, F., Ikeuchi, S. 1990, *Astrophys. J. Suppl. Ser.* 73, 401

4.142 van den Bergh, S. 1990, *Astron. Astrophys.* 231, L27

4.143 van den Bergh, S., McClure, R.D. 1990, *Astrophys. J.* 359, 277

4.144 Keppel, J.W., Dettmar, R.J., Gallagher, J.S., III., Roberts, M.S. 1991, *Astrophys. J.* 374, 507

4.145 Rand, R.J., Kulkarni, S.R., Hester, J.J. 1990, *Astrophys. J.* 352, L1

4.146 Dettmar, R.J. 1990, *Astron. Astrophys.* 232, L15

4.147 Sukumar, S., Allen, R.J. 1991, *Astrophys. J.* 382, 100

4.148 Hummel, E., Dahlem, M., van der Hulst, J.M., Sukumar, S. 1991, *Astron. Astrophys.* 246, 10

4.149 Hummel, E., Dettmar, R.J. 1990, *Astron. Astrophys.* 236, 33

4.150 Roy, J.R., Boulesteix, J., Joncas, G., Grundseth, B. 1991, *Astrophys. J.* 367, 141

4.151 Reuter, H.P., Krause, M., Wielebinski, R., Lesch, H. 1991, *Astron. Astrophys.* 248, 12

4.152 Wang, Q., Helfand, D.J. 1991, *Astrophys. J.* 379, 327

4.153 Palouš, J., Franco, J., Tenorio-Tagle, G. 1990, *Astron. Astrophys.* 227, 175

4.154 Tenorio-Tagle, G. 1980, *Astron. Astrophys.* 86 61

4.155 Tenorio-Tagle, G., Franco, J., Bodenheimer, P., Rozyczka, M. 1987, *Astron. Astrophys.* 179, 219

4.156 Alfaro, E.J., Cabrera-Cano, J., Delgado, A.J. 1991, *Astrophys. J.* 378, 106

4.157 Odenwald, S. 1989, *Astrophys. J.* 325, 320

4.158 Mirabel, I.F., Morras, R. 1990, *Astrophys. J.* 356, 130

4.159 Meyerdierks, H. 1991, *Astron. Astrophys.* 251, 269

4.160 Meyerdierks, H. 1992, *Astron. Astrophys.* 253, 515

4.161 van der Hulst, J.M., Sancisi, R. 1988, *Astron. J.* 95, 1354

4.162 Kamphuis, J., Briggs, F. 1992, *Astron. Astrophys.* 253, 335

4.163 Rubin, V.C., Graham, J.A. 1990, *Astrophys. J.* 362, L5

4.164 Combes, F. 1991, *Astron. Astrophys.* 243, 109

4.165 Tenorio-Tagle, G. 1988, in *Star Bursts and Galaxy Evolution*, eds. T.X. Thuan, T. Montmerle, J. Tran Thanh Van, (Gif sur Yvette: Editions Frontieres), p. 37

4.166 Elmegreen, B.G., Chaing, W.H. 1982, *Astrophys. J.* 253, 666

4.167 Mundt, R. 1988, in *Formation and Evolution of Low Mass Stars*, eds. A.K. Dupree, M.T.V.T. Lago (Dordrecht: Kluwer), p. 257

4.168 Wood, D.O., Churchwell, E. 1991, *Astrophys. J.* 372, L99

4.169 Norman, M.L., Winkler, K.H., Smarr, L., Smith, M.D. 1982, *Astron. Astrophys.* 113, 285

4.170 Tenorio-Tagle, G., Rozyczka, M. 1991, *Astron. Astrophys.* 245, 616

4.171 Poetzel, R., Mundt, R., Ray, T.P. 1989, *Astron. Astrophys.* 224, L13

4.172 Zealey, W.J., Mundt, R., Ray, T.P., Sandell, G., Geballe, T., Taylor, K.N.R., Williams, P.M., Zinnecker, H. 1989, *Proc. Astron. Soc. Aust* 8, 62

4.173 Schwartz, R.D. 1983, *Ann. Rev. Astron. Astrophys.* 21, 209

4.174 van Buren, D., MacLow, M.-M., Wood, D.O.S., Churchwell, E. 1990, *Astrophys. J.* 353, 550

4.175 MacLow, M.-M., van Buren, D., Wood, D.O.S., Churchwell, E. 1991, *Astrophys. J.* 369, 395

4.176 Churchwell, E. 1990, *Astron. Astrophys. Rev.* 2, 79

4.177 Churchwell, E. 1991, in *The Physics of Star Formation and Early Stellar Evolution*, eds. C.J. Lada and N.D. Kylafis, (Dordrecht: Kluwer), 221

4.178 Cesaroni, R., Walmsley, C.M., Kömpe, C., Churchwell, E. 1991, *Astron. Astrophys.* 252, 278

4.179 Churchwell, E., Walmsley, C.M., Wood, D.O.S. 1992, *Astron. Astrophys.* 253, 541

5 Gas Flow in Spiral Density Waves

5.1 An Introduction to Stellar Density Waves

Interstellar gas dynamics is driven not only by localized pressures such as supernova explosions and cloud collisions, but also by large-scale gravitational forces from waves in the stellar population. These stellar waves can be excited by a variety of mechanisms. Some are transient, following local instabilities in the stars or gas, or following tidal interactions with other galaxies, and some are more steady, resulting from wave reflections and self-amplification in the disk.

The gas plays an important role in the formation of stellar spirals. It adds a gravitational force to that from the stars [1 − 4], which increases the spiral amplification and allows the spirals to penetrate resonances at which individual stellar orbits randomly scatter or purely stellar waves turn evanescent, and it dissipates some of the excess energy from the wave, which holds down the amplitude of the spiral so that self-gravity does not destroy it [5 − 7]. The gas also reacts to the stellar spirals by forming a shock, which sometimes appears as a dust lane, and by forming or collecting young stars downstream from the shock, which highlights the spiral arms.

In this section, we review some of the theories of stellar spiral formation, and in the next section we discuss the gas reaction to these spirals. This is the order in which these topics were originally studied when the gas was treated as a passive tracer of stellar wave activity. Now the gaseous and stellar waves are usually viewed as inseparable parts of a single phenomenon, with the gas playing an active role in the generation and maintenance of the spiral. Reviews of the first decade of density wave theory are in references [8] and [9]. More recent reviews of various aspects of spiral waves are in references [10 − 15].

Galactic disks have strong *two* arm spirals because the difference between the angular frequency of rotation, Ω, and *one-half* the epicyclic frequency, $\kappa/2$, is approximately constant, to within a factor of two or so, over a large part of the disk, as pointed out by Lindblad [16]. This is a result of the mass distribution in the combined disk and halo. It implies that in a reference frame rotating around the galaxy at a rate $\Omega - \kappa/2$, the orbits in most of the disk are nearly closed loops centered on the nucleus of the galaxy. Lindblad postulated that orbit crowding in these loops made two symmetric spiral arms, but he did not consider the gravitational force from these arms and how this force would deflect the orbits and change the loop structure. Nevertheless, the basic reason why most galaxies have two prominent arms and not one, for example, is still believed to be related to this near constancy of $\Omega - \kappa/2$. In contrast, a Keplerian disk has a constant $\Omega - \kappa$, and one-arm spirals are preferred [17, 18].

Lin & Shu [1, 19] modified the basic idea of Lindblad and included the gravity of the arms. This led to a dispersion relation for *density waves* in the stellar system. The derivation of this dispersion relation is straightforward for linear perturbations in a gas. The dispersion relation for stars is more complex

224

because of a difference in the treatment of pressure, but the basic result is about the same. Here we follow the derivation [8] of the dispersion relation for linear density waves in a gas. We begin with the equations of motion, continuity, and gravitational potential for a fluid in inertial coordinates:

$$\rho\frac{D\boldsymbol{w}}{Dt} = \rho\left(\frac{\partial\boldsymbol{w}}{\partial t} + \boldsymbol{w}\cdot\nabla\boldsymbol{w}\right) = -\nabla P - \rho\nabla\Psi_{\mathrm{T}}, \qquad (123.1)$$

$$\frac{\partial\rho}{\partial t} + \nabla\cdot\rho\boldsymbol{w} = 0, \qquad (123.2)$$

$$\nabla^2\Psi = 4\pi G\rho. \qquad (123.3)$$

Considering the derivative of the unit vectors with respect to θ, the equation of motion has the radial and angular components

$$\rho\left(\frac{\partial w_r}{\partial t} + w_r\frac{\partial w_r}{\partial r} + w_\theta\frac{\partial w_r}{r\partial\theta} - \frac{w_\theta^2}{r}\right) = -\frac{\partial P}{\partial r} - \rho\frac{\partial\Psi_{\mathrm{T}}}{\partial r}, \qquad (124.1)$$

$$\rho\left(\frac{\partial w_\theta}{\partial t} + w_r\frac{\partial w_\theta}{\partial r} + w_\theta\frac{\partial w_\theta}{r\partial\theta} + \frac{w_\theta w_r}{r}\right) = -\frac{1}{r}\frac{\partial P}{\partial\theta} - \rho\frac{1}{r}\frac{\partial\Psi_{\mathrm{T}}}{\partial\theta}. \qquad (124.2)$$

The total potential Ψ_{T} is given by the sum of the average galactic potential and the spiral, so that

$$\nabla\Psi_{\mathrm{T}} = \Omega(r)^2 r\boldsymbol{e}_r + \nabla\Psi; \qquad (125)$$

$\Omega(r)$ is the rotation rate.

For in-plane motions, we can integrate through the plane and replace the space density ρ by the column density σ. We can also approximate the fluid as adiabatic with sound speed $a = (\partial P/\partial\rho)^{1/2}$, so ∇P becomes $a^2\nabla\sigma$. Then small perturbations in column density σ_1 around the equilibrium value σ_0, and small radial and azimuthal velocity perturbations v_r and v_θ around the zero order velocity, which is azimuthal, $w_\theta = \Omega r$, can be found for forced oscillations by a two-arm logarithmic spiral potential

$$\Psi = \psi\cos(2\Omega_{\mathrm{p}}t - \Phi) \quad\text{where}\quad \Phi = \frac{2\ln\left(\dfrac{r}{r_0}\right)}{\tan i} + 2\theta \qquad (126)$$

for pattern speed Ω_{p}. The wavenumber for this perturbation is the radial derivative of the phase Φ, which is

$$k = \frac{2}{r\tan i} \qquad (127)$$

for spiral pitch angle $i > 0$ for a trailing spiral. (This is a different convention for i than in [8], where $2\Omega_{\mathrm{p}}t + \Phi$ is written for the phase of the spiral, and $\Phi = 2\ln(r/r_0)/\tan i - 2\theta$. Then $i < 0$ for a trailing spiral.)

To solve these equations, we first write them in terms of linear combinations of the perturbation variables:

$$\sigma_0 \left(\frac{\partial v_r}{\partial t} + \Omega \frac{\partial v_r}{\partial \theta} - 2\Omega v_\theta \right) = -a^2 \frac{\partial \sigma_1}{\partial r} - \sigma_0 \frac{\partial \Psi}{\partial r}, \qquad (128.1)$$

$$\sigma_0 \left(\frac{\partial v_\theta}{\partial t} + \Omega \frac{\partial v_\theta}{\partial \theta} - 2B v_r \right) = -a^2 \frac{1}{r} \frac{\partial \sigma_1}{\partial \theta} - \sigma_0 \frac{1}{r} \frac{\partial \Psi}{\partial \theta}, \qquad (128.2)$$

$$\frac{\partial \sigma_1}{\partial t} + \Omega \frac{\partial \sigma_1}{\partial \theta} + \frac{1}{r} \frac{\partial r \sigma_0 v_r}{\partial r} + \frac{1}{r} \frac{\partial \sigma_0 v_\theta}{\partial \theta} = 0. \qquad (128.3)$$

Here B is the Oort function, equal to $-\Omega - \frac{1}{2} r \, d\Omega/dr$ for $\Omega > 0$.

Now the cosine function for Ψ can be substituted into the equations of motion, and these equations, along with the continuity equation, can be reduced to a single equation for each variable. This gives the solutions

$$\sigma_1 = \frac{-k^2 \psi \sigma_0 \cos(2\Omega_p t - \Phi)}{\kappa^2 - 4(\Omega_p - \Omega)^2 + k^2 a^2}, \qquad (129.1)$$

$$v_r = \frac{2k\psi(\Omega - \Omega_p) \cos(2\Omega_p t - \Phi)}{\kappa^2 - 4(\Omega_p - \Omega)^2 + k^2 a^2}, \qquad (129.2)$$

$$v_\theta = \frac{k\psi\kappa^2 \sin(2\Omega_p t - \Phi)}{2\Omega(\kappa^2 - 4(\Omega_p - \Omega)^2 + k^2 a^2)}. \qquad (129.3)$$

where $\kappa = (-4B\Omega)^{1/2}$ is the epicyclic frequency. Here we have assumed that $2/r \ll \partial/\partial r$ so azimuthal derivatives have been ignored (see below).

Note that in the center of an arm, the potential has a local minimum so $2\Omega_p t - \Phi = \pi$ with $\psi > 0$. Then $\sigma_1 > 0$ so the density is a maximum, $v_r < 0$ inside corotation ($\Omega > \Omega_p$) so the radial streaming motion is inward, and $v_\theta = 0$. Between the arms, $2\Omega_p t - \Phi = 0$ so $\sigma_1 < 0$ and the density is a minimum, $v_r > 0$ inside corotation so the radial motion is outward, and $v_\theta = 0$ again. At the leading and trailing edges of the arm, $2\Omega_p t - \Phi = \pi/2$ and $3\pi/2$ respectively, so $\sigma_1 = 0$, $v_r = 0$ and $v_\theta > 0$ and < 0, respectively. Thus the tangential motion slows down as it approaches an arm from the concave side inside corotation, and it speeds up on the other side as it leaves the arm. This variation in azimuthal speed is the result of the conservation of angular momentum during radial motions that are induced by the spiral potential, which exerts a force mostly in the radial direction (in the tight winding approximation).

Evidently the density response to a spiral potential is in phase with the potential, in the sense that the density maximum occurs at the potential minimum. Thus, the density can both produce the spiral potential and respond to it. The inverse problem is to find the potential Ψ that results from a spiral density perturbation. This comes from the solution to Poisson's equation for Ψ, and gives

$$\Psi = \frac{-2\pi G \sigma_1}{|k|} \qquad (130)$$

when the spiral arms are tightly wound.

Now we can substitute ψ into the solution for σ_1 and this gives the so-called Lin-Shu-Kalnajs dispersion relation for 2-arm density waves in the gas:

$$4(\Omega_p - \Omega)^2 = -2\pi G \sigma_0 |k| + k^2 a^2 + \kappa^2. \qquad (131)$$

This equation gives the local spiral arm pitch angle, k, for any pattern speed, Ω_p, as a function radius, through the radial dependencies of Ω, κ, σ_0, and a.

The stellar system does not have a pressure like the gas, but the stars still have a random component of the motion with a nearly Gaussian distribution function, and the equation for the wave dispersion relation can be derived in a manner analogous to the above derivation [20], with about the same numerical coefficients. This stellar dispersion relation for a 2-arm spiral is [20, 21]

$$4\left(\Omega_p - \Omega\right)^2 = -2\pi G\sigma_0 |k| \mathcal{F}_\nu(x) + \kappa^2 \tag{132}$$

where

$$\mathcal{F}_\nu(x) = \frac{1-\nu^2}{x}\left\{1 - \frac{\nu\pi}{\sin\nu\pi}\frac{1}{2\pi}\int_{-\pi}^{\pi} e^{-x(1+\cos s)}\cos\nu s\, ds\right\} \tag{133}$$

is a reduction factor and where $x = (ka/\kappa)^2$ and $\nu = 2(\Omega_p - \Omega)/\kappa$. Reduction factors are plotted in [20]. The density response for stars has a similar change from that for a gas,

$$\sigma_1 = \frac{-k^2\psi\sigma_0\cos(2\Omega_p t - \Phi)}{\kappa^2 - 4(\Omega_p - \Omega)^2}\mathcal{F}_\nu(x). \tag{134}$$

It is easy to verify that if ψ from the solution to Poisson's equation is substituted into this equation for σ_1, then the stellar dispersion relation results. A comparison to the gaseous density response suggests that [8]

$$\mathcal{F}_\nu(x) \approx \frac{\kappa^2 - 4(\Omega_p - \Omega)^2}{\kappa^2 - 4(\Omega_p - \Omega)^2 + k^2 a^2}, \tag{135}$$

which is also evident from Table 1A in [20].

These equations are modified in an important way when the azimuthal forces from the spiral are included from the $\partial\Psi/(r\partial\theta)$ terms; these terms appear as $2/r$ and were ignored in comparison to $\partial/\partial r$ in the above derivation. When they are included, the dispersion relation for gas becomes [21]

$$4\left(\Omega_p - \Omega\right)^2 = -2\pi G\sigma_0 k_{\text{tot}}(1+\chi) + k_{\text{tot}}^2 a^2(1+\chi) + \kappa^2, \tag{136}$$

where $k_{\text{tot}} = (k^2 + 4/r^2)^{1/2}$, $\chi = (\mathcal{J}k_c/k_{\text{tot}})^2$, $k_c = \kappa^2/(2\pi G\sigma_0)$, and

$$\mathcal{J} = \frac{4}{rk_c\kappa}(2A\Omega)^{1/2} \tag{137}$$

for Oort constant $A = -\frac{1}{2}r\, d\Omega/dr$.

The dispersion relation for density waves implies that the orbits of stars are deflected by the gravity of the spiral arms, but the average orbits still appear as closed loops in a rotating system with angular frequency Ω_p, which is the spiral pattern speed. This pattern speed is not generally equal to $\Omega - \kappa/2$ as in the Lindblad theory, but larger than this. The difference between Ω_p and $\Omega - \kappa/2$ for the reference frame of the closed loops is made up by the gravitational and effective pressure forces from the spirals. This force must go to zero, and the spirals must vanish, when $\Omega_p = \Omega - \kappa/2$. This determines the absolute limits to the radii of the arms, and probably the actual limits in the outer parts of some galaxies (the inner limits to some arm systems may not be at a resonance but at a wave reflection point instead).

Although Lin & Shu [19] were trying to explain persistent spiral arms in galaxies, or quasi-stationary spiral structure, as envisioned by Lindblad [16], their success in this first effort was limited to the introduction of spiral waves, which are more persistent than material arms but which still change their shape with time. This evolution of shape was pointed out by Toomre [22] who showed that the Lin-Shu-Kalnajs dispersion relation implies a wave group velocity, $\partial\Omega_p/\partial k$, which causes trailing spiral waves inside the corotation radius to wrap up with time. This proposal, along with the interaction models by Toomre & Toomre [23], led to various alternate theories in which spiral structure is transient. Such structure is often present in N-body computer simulations [24].

The irregular or asymmetric arms seen in these computer simulations are presumably the result of local instabilities amplified by self-gravity in the disk [11, 25]. Symmetric, two-arm spirals in these transient models are always driven by symmetric perturbations such as bars and companion galaxies. Recall that even though a companion galaxy is on one side of the disk, the tidal force from this companion tends to stretch the disk along the line between them, and compress the disk perpendicular to this line. This is enough of a symmetric jolt to perturb the orbits in a regular fashion, after which gravity in the disk amplifies the perturbation to make symmetric spirals. Idealized models in [11] illustrate this process well. More realistic models of the short term response to a tidal interaction, with detailed comparisons to observations, are in [26].

An alternative to the transient or driven-spiral theories was proposed by Lin and collaborators [21, 27 − 31]. These authors realized that the group motion of the spiral waves was not necessarily the end of the spiral pattern. They proposed a model of standing waves, in which the incoming (i.e., wrapping-up) spirals would refract near the central regions because of the bulge and get converted into longer-wavelength outward-moving waves. The outward moving waves would then reflect at corotation in the middle part of the galaxy and return inward, with some amplification, to reinforce the original spiral. In this way, the galaxy could generate a standing wave pattern, or wave mode, and this would solve the spiral persistence problem. The pattern speed would be determined by the condition that the second incoming wave exactly reinforce the first incoming wave, which is a quantum condition for the number of wavelengths between the refraction point in the center and the reflection-amplification point at corotation in the outer part. Such a wave would also continue to propagate outward outside of corotation, until it reaches the outer Lindblad resonance where $\Omega_p = \Omega + \kappa/2$, as discussed above.

The main point of this modal theory is that an incoming wave has to turn back near the center and get amplified again at corotation. But the exact source of this turnaround, and the exact nature of the amplifier, are not understood well. The wave could reflect in the center instead of refract (a reflected wave comes off as a leading wave with about the same angle as the incident trailing wave, while a refracted wave comes off at a different angle than the incident wave, and can still be trailing), in which case the amplification at corotation has to turn a leading wave into a trailing wave, as would be the case for the swing-amplifier discussed by Toomre [11] and Goldreich & Tremaine [32]. Such an amplifier applied to a wave mode is called a type II WASER [33 – 34]. If the wave refracts into another trailing wave near the center, then the original (type I) WASER amplifier may be more relevant for corotation [30], although this is weaker than the swing amplifier and more sensitive to parameters (Q). Observations of M81, for example, suggest that the reflected waves are leading and have the same pitch angle as the trailing waves [15], in which case the swing amplifier might apply.

One of the problems with the modal theory proposed in references [21, 27 – 30], as pointed out by Toomre [9], is that, for the original WASER amplifier, a galaxy has to be too close to the threshold of ring-forming instabilities near the corotation resonance to be reasonable, and that real galaxies are much further from this threshold than Lin and collaborators proposed. With too much stability near corotation, waves cannot propagate there (there is a forbidden zone) and the WASER would be weak. There would also not be a strong wave on the other side, out to the outer Lindblad resonance. There are essentially three solutions to this problem. One has the wave tunnel through the forbidden zone and get a small amount of WASER amplification even if Q is slightly larger than 1 [9, 22]. Another suggests that the long range force of gravity sends the wave through the forbidden zone [35] without an exponential decrease in amplitude, in a manner similar to that for a spiral wake [36], and then it can amplify in this corotation zone by the swing mechanism. A third suggests that galaxy disks really are close to the threshold of ring instabilities, primarily because of the cold gas [2].

An important consideration for all of these cases, and for strong wave amplification in general, is the value of the stability parameter Q near the corotation resonance. This parameter strongly influences the stability of the interstellar medium and cloud formation too, so we discuss it in some detail here and in Chap. 6.

In 1960, Safronov [37] studied the stability of disks and found that a dimensionless parameter, written today as $a\kappa/(\pi G\sigma)$, determines when the self-gravity of a ring-like perturbation can overcome the Coriolis and centrifugal forces and lead to instability for sufficiently short wavelengths. Here a is the radial component of the velocity dispersion, κ is the epicyclic frequency, and σ is the mass column density. In 1964, Toomre [38] derived a similar dimensionless parameter for a stellar system, for which the factor π is replaced by 3.36. This parameter is called Q. In most galaxies, Q for the stars lies between approximately 1.5 and 2 for the visible part of the disk [39 – 42]. For the gas, Q is very close to its threshold value of 1 [43]. When $Q < 1$, the corresponding gas or star component is unstable to collapse in the radial direction, i.e., to form rings, so the effective value of Q, including everything such as magnetic forces, is not likely to be much less than 1. For a combined gas+star system, the effective Q value is less than both the pure gaseous and pure stellar values [2], and can be approximately 1 (or less than 1) even when each component is separately stable. Such effective Q values were applied to spiral waves by Bertin & Romeo [4]. They demonstrated that the effective value of Q, and, consequently, the amplification of spiral waves in a disk, is strongly influenced by the gas, even though the total gas column density in the disk is less than the total stellar column density.

The closeness of Q to the threshold value of 1 would be a remarkable coincidence if it were not continuously adjusted to this value all over the disk by various processes that operate in the gas, such as star formation, heating, cooling, and radial migration driven by spiral torques (see Chap. 7). As pointed out by Goldreich & Lynden-Bell [25], a gas disk with too low a Q value should be so unstable that the resulting excess of star formation will heat the gas and increase Q. Conversely, if Q is too high, star formation will slow down and allow the gas to cool, thereby decreasing Q. Many of the details of how Q regulates itself in this scenario are not known, but it seems likely that the regulation is controlled by the formation of giant cloud complexes in large scale gravitational instabilities, or by the radial migration of gas through spiral-driven angular momentum transfer; both of these processes are sensitive to Q.

Whatever the thermostat that regulates the gaseous velocity dispersion and Q, the observations by Kennicutt [43] allow for a combined gas+star Q value that is much lower near the halfway point of a galaxy disk than the pure stellar value, and this suggests that waves of all types should have a strong amplifier at corotation. Then self-amplified wave modes should exist whenever a disk has a means for reflecting or refracting the incoming waves near the center. Such reflection or refraction could result from a dense bulge of stars near the center, because a bulge creates large values of κ and c and small values of σ in the inner disk, all of which increase Q [44]. The reflector could also result from a lack of gas in the inner disk, as observed in our Galaxy, or perhaps from a bar.

The modal theory suggests that any galaxy with enough gas and a massive enough disk (compared to the halo) to keep Q low via some thermostat, and with a wave reflector in the central regions, can act like a resonant cavity, amplifying any tiny perturbation, internal or external, to create a global standing wave pattern or wave mode. Computer simulations of standing wave patterns generated entirely from self-amplified noise in an N-body disk are in [45] and [46]. Galaxies could create these modes entirely on their own, as in the simulations, amplifying internal perturbations such as giant clouds or local instabilities [27] and their wakes [36], or the perturbation for the mode could come from weak tidal arms in the outer part of the disk, generated by a companion galaxy. Galaxies without the ability to reflect incoming waves (and without bars) would have only transient spirals. Moreover, any galaxy, if hit hard enough by a companion, can fill its entire disk with a transient spiral that is not a mode. This latter situation is likely to apply to some of the Arp [47] interacting galaxies.

Whether standing wave patterns actually exist in galaxies is still a matter of some debate [48], although it is generally accepted that most arms are stellar waves of some sort. What is more important for these lectures is the influence that stellar waves or wave modes have on the gas, and this is the topic of the next section.

5.2 Gas Flow in Stellar Density Waves

5.2.1 General Flow Properties.
A fluid that flows at high speed in a region where the gravitational potential energy changes by an amount comparable to the fluid's kinetic energy can partition itself into two regimes separated by a sharp transition. In one regime the gas streams quickly and has a low density, and in the other regime the gas streams slowly and has a high density. This separation is commonly observed in a bathtub where the water that flows from the faucet and then streams quickly in a thin layer along the nearby tub floor suddenly plows into the deeper pool of collected water in the high part of the basin. There is a jump in height from the fast stream to the slow stream, at which point the kinetic energy in the fast stream is converted into potential energy in the pool.

Similar jumps in water speed and depth occur in a rocky river bed. If there is a depression in the bed, the water will flow down at high speed into the potential valley, making a thin stream in the falling region, and then it will climb up the other side of the depression and meet the more stagnant river water on the other side, forming a sharp transition where the two streams meet. If you watch the flow in a rocky river you will see that the jump from the thin fast stream to the thick slow stream occurs on the downstream (far) side of potential minima (holes) and on the upstream (near) side of potential maxima (rocks).

The same type of transition occurs for the continuous component of gas in a spiral density wave, except that the Coriolis force in the rotating galaxy causes the material to slow down in azimuthal speed as it approaches and enters a potential minimum (the arm) and speed up in azimuth as it leaves the potential minimum (cf. Sect. 5.1). The transition in a galaxy still occurs when the thin fast-moving gas slows down and meets denser more stagnant gas, but because of this Coriolis effect in a galaxy, this transition occurs on the upstream side of the potential minimum — just opposite to the case of a river. This corresponds to the inside edge of a spiral arm in the inner regions of the galaxy. A better analogy between density waves and hydrodynamic flows is with a de Laval nozzle, because the gas streamlines constrict and then open up as they flow through a spiral arm [49].

Galaxies also have self-gravity in the fluid, unlike the water in a river, and this self-gravity can sometimes dominate the imposed spiral potential and cause the density transition to be smoother than in the non-self-gravitating case, and to occur nearly anyplace in the arm, instead of just at the inside edge [3].

The equations generally used to determine the gas flow in an imposed spiral density wave are, from Roberts [50], the equation of motion,

$$\rho \left(\frac{\partial w}{\partial t} + w \cdot \nabla w + 2\Omega_{\mathrm{p}} \times w - \Omega_{\mathrm{p}}^2 r \right) = \rho g - \nabla P, \qquad (138)$$

for density ρ and total velocity w in a coordinate system rotating at the pattern speed Ω_p, and for gravitational acceleration g and pressure P; the mass continuity equation,

$$\frac{\partial \rho}{\partial t} + \nabla \cdot \rho w = 0, \tag{139}$$

the equation of state,

$$P = \rho a^2 \tag{140}$$

and the equation for gravity,

$$g = -\Omega(r)^2 r - \nabla \Psi, \tag{141}$$

which includes separate contributions from the overall galactic gravity and from the two-arm spiral potential (in the pattern frame)

$$\Psi = \psi \cos \Phi \qquad \text{where} \qquad \Phi = \frac{2 \ln \left(\dfrac{r}{r_0} \right)}{\tan i} + 2\theta \tag{142}$$

for spiral pitch angle i and galactic angular coordinate θ increasing in the direction of rotation. Gas self-gravity can be included if there is a contribution to g that satisfies Poisson's equation,

$$\nabla \cdot g_{\text{self}} = -4\pi G \rho. \tag{143}$$

These equations can be solved for small perturbations around the mean flow relative to the spiral potential by writing

$$\begin{aligned} \rho &= \rho_0 + \rho_{\text{spiral}} \\ w &= (\Omega - \Omega_p) r \, e_\theta + w_{\text{spiral}}. \end{aligned} \tag{144}$$

It is also convenient to go into spiral coordinates, where η increases in the direction locally perpendicular to the spiral, and ξ increases in the parallel direction,

$$\begin{aligned} \eta &= \ln \left(\frac{r}{r_0} \right) \cos i + \theta \sin i, \\ \xi &= -\ln \left(\frac{r}{r_0} \right) \sin i + \theta \cos i. \end{aligned} \tag{145}$$

The resulting linear equations must be solved numerically if ψ is large. Roberts [50] assumed that the solution is periodic in θ for a two-arm spiral. Then for strong ψ there must be a shock in the flow. In this case, the solution is found by assuming a shock position and various post-shock quantities, integrating to the next shock, $\pi + \theta$ away, and checking to see if the variables satisfy the shock jump conditions compared to the initial assumption. The solution is then iterated until the jump conditions are satisfied. Figure 11 shows a typical solution. The shock is on the inner edge of the spiral potential minimum and would presumably correspond to a visible dust lane on the inner side of the arm. In a strong bar, similar shocks form on the leading edges [51 – 53].

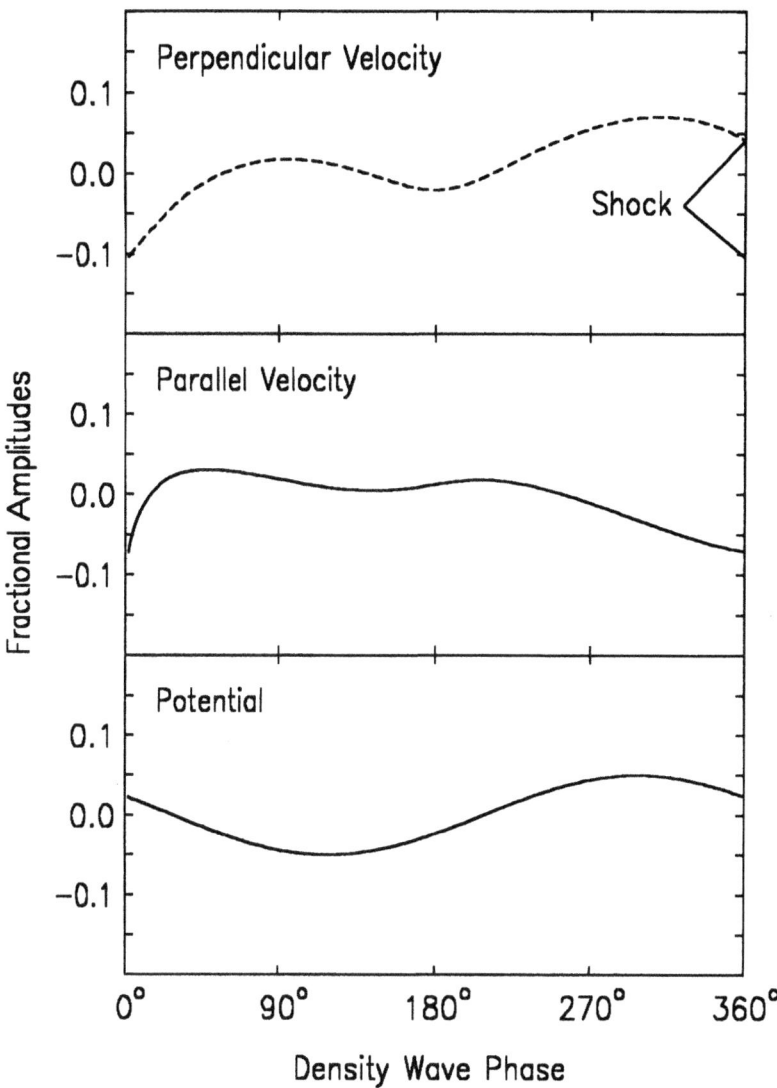

Fig. 11. The solution for the gas flow in a spiral potential with a relative amplitude of 5%

How important is the shock front to the solution? For a hydrodynamic gas it is very important — there are no periodic solutions without it when ψ is large. This may seem strange at first because the stars have periodic orbits in the spiral but they have no shock. The periodicity for the stars results from the closeness of $\Omega - \Omega_p$ to $\kappa/2$, and from the ability of spiral gravity to deflect the orbits so that they close in the frame rotating at Ω_p, even though $\Omega - \Omega_p - \kappa/2$ differs slightly from zero. This was discussed in the preceding section. It follows that if the gas were composed of ballistic particles like the

233

stars, with no effective pressure to offset the force balance that gives such closed orbits, then the gas would have periodic solutions too, without a shock. This was demonstrated by Roberts & Steward [54], who showed N-body cloud flow simulations with and without cloud collisions. The results were nearly the same in the two cases: the cloudy gas responded to a density wave with a peak compression near the centers of the arms, just as the stars did. A shock forms in the gas when the individual clouds in the fluid collide often enough to communicate pressure variations during the relatively short time scale for the flow in the wave.

This difference between ballistic clouds and pressurized gas has led to the suggestion [55] that giant molecular clouds, which are somewhat ballistic because of their high internal densities, should respond to a spiral density wave differently than diffuse clouds, which have a higher collision frequency and are better connected to each other hydrodynamically and magnetically than molecular clouds. The molecular clouds were proposed to have their density ridge in the center of an arm, as the stars do, and the diffuse clouds were proposed to have their peak on the inner edge, where Roberts [50] found the hydrodynamic shock. This implies that only the diffuse and well-connected cloud population contributes to the dust lanes (which may or may not be molecular depending on the amount of self-shielding). It also implies that even if the star-forming molecular clouds live forever, most of them will appear downstream from the shock front in the direction of the gas flow through the arm. This relative location will give the impression that there is a time sequence for molecular cloud or star formation triggering, i.e., from galactic shock to molecular clouds, when in fact this spatial separation is not a time sequence but a consequence of the differential flow for the two components. Now, there probably really is a time sequence of cloud formation in some regions of spiral arms, as discussed in Sect. 6.5, but the appearance of star formation in dense cloud complexes downstream from a dustlane should be much more general than the occurrence of triggering.

What is the effect of the interstellar magnetic field? A magnetic field is important on length scales over which it is curved or compressed because the force it exerts depends on the gradients and curls of the field strength vector. The magnetic field also tends to follow the gas, so it contributes to the pressure and makes the gas better connected than if it were composed only of ballistic clouds (cf. Sect. 4.2.6). As a result, dense molecular clouds should be connected to diffuse clouds by magnetic tension and compression forces, and both components of the interstellar gas should flow together on average. Calculations [56] suggest that this "average" applies to length scales comparable to the Jeans length in the ambient medium, or actually to the inverse of the Jeans critical wavenumber, which is smaller by 2π and equal to $\sim 500\,\mathrm{pc}$ locally. This follows for a magnetic field strength comparable to the virial value in a molecular cloud, which gives a stopping time from gas drag comparable to the Jeans collapse time in the surrounding medium.

The interstellar magnetic field can therefore keep ballistic molecular clouds from straying too far from the diffuse gas and other molecular clouds that are connected to them on the same field lines. This has the effect of preserving the integrity of the fluid for large-scale and slow changes, making it act like a continuous gas for the average flow around a density wave. But it also allows a separation between the dense and diffuse components whenever there is a rapid or sharp transition in the pressure. The pressure jump still acts most strongly on the diffuse component, but the reaction of this diffuse component to the pressure takes some time (the Jeans time) to be transmitted to the dense molecular clouds. Thus the molecular clouds can overshoot a shock front and act somewhat ballistically as they enter an arm, giving a separation between the diffuse cloud dust lane and the molecular cloud-star formation front mentioned above, but then these molecular clouds can rejoin their old neighbor diffuse clouds after a while and emerge from the arm as a single fluid. Such a separation can also pump random motions into both of the cloud fluids at the expense of spiral wave energy, which is ultimately derived from galactic orbital energy.

The exact flow pattern for a multi-component magnetic fluid in a density wave has not yet been calculated, so comparisons between theory and observations are not yet possible, but one might expect a generally similar flow for the diffuse and dense gas clouds in our Galaxy, giving the same general trends on the longitude-velocity diagram for these two components, and also expect a separation between diffuse and dense gas at a shock front. This separation might correspond to a difference in the HI and CO distributions or velocities at the longitudes where the line of sight is tangent to a shock (if there is a shock in our Galaxy). This would presumably be the longitude at the inside edge of an arm, towards small longitudes in the first quadrant (although for self-gravitating gas flows the shock can occur closer to the midpoint of the arm). What should be observed is a jump in the HI velocity because of the shock, but no comparable jump for dense CO clouds in a small longitude range around this HI jump.

An exception to this shock difference might arise if the diffuse cloud shock is so strong and the resulting dust lane so thick that incident molecular clouds are prevented from penetrating it by gas or magnetic field drag alone. This would happen if the mass column density through the shock is comparable to the column density in a molecular cloud, which in our galaxy corresponds to about 10 magnitudes of extinction. Presumably strong density waves in gas rich galaxies would have such trapping [57].

High resolution observations of dustlanes, HI, and CO in other galaxies have provided some evidence for a separation between diffuse and dense cloud components in the density wave flow. The diffuse shock is not always atomic, or at least it is not always evident as an emission ridge in the HI maps, but when it is not, the front still shows up sometimes as a dustlane or kink in the velocity flow. Then the relative position between the shock and the CO emission ridge can be determined. When this comparison was made for M81 [58,59] and M83 [60,61], it was found that the CO ridges are displaced downstream from the dustlanes by about 300 pc. High resolution observations of the structures of these dustlanes might indicate the presence of small clumps that are not virialized, whereas the CO ridges might contain small clouds that are virialized.

In other galaxies, such as M51 [62] and NGC 1068 [63], the CO emission ridge is in the same place as the shock, which is evident from the abrupt shift in the CO velocity. The difference between these two cases may be partly the result of a difference in the density wave strength, as discussed above, and partly the result of a difference in the total interstellar pressure, which contributes to the threshold column density and cloud mass for the conversion of atomic gas into molecular gas [64]. The ballistic flow of molecular clouds is also evident from the way they crowd together and make giant complexes at the ends of the bar in M83 [65].

5.2.2 A Decrease in Shear in Spiral Arms.

An interesting and important property of spiral density waves is that the shear in the stellar or gaseous fluid is lower in the arms than between the arms and also lower than the average shear if there were no arms. This has important consequences for cloud and star formation. The decrease in shear in spiral arms can be derived from simple considerations of angular momentum conservation or from the linear density wave solutions σ_1 and v_θ given in Sect. 5.1 [66].

Consider first the conservation of angular momentum in an annulus of thickness $2D \ll r$ around a galaxy centered on the radius r, when the annulus gets thinner because of some radially directed compressive force. Suppose the inner radius changes from $r - D$ to $r - D + d$ and the outer radius changes from $r + D$ to $r + D - d$. Assume first a flat rotation curve with velocity V. Then the velocity at the outer part of the annulus increases from V to $V(r + D)/(r + D - d) \approx V(1 + d/r)$ by angular momentum conservation, and the velocity at the inner part of the annulus changes from V to $V(1 - d/r)$ for the same reason. The velocity gradient in the annulus after the compression is

$$\frac{dV}{dr} = \frac{V\left(1 + \frac{d}{r}\right) - V\left(1 - \frac{d}{r}\right)}{2D - 2d} = \frac{V}{r} \frac{\frac{d}{D}}{1 - \frac{d}{D}} \tag{146}$$

and the Oort A constant in the annulus is

$$A_{\text{ann}} = \frac{1}{2}\left(\frac{V}{r} - \frac{dV}{dr}\right) \approx A_0 \left(1 - \frac{\frac{d}{D}}{1 - \frac{d}{D}}\right) \tag{147}$$

for A_0 equal to the average Oort constant at that radius, $\frac{1}{2}V/r$.

Now, at the same time, the column density in the annulus increases as

$$\sigma_{\text{ann}} = \sigma_0 \frac{2D}{2D - 2d} = \sigma_0 \frac{1}{1 - \frac{d}{D}} \tag{148}$$

so

$$\frac{\frac{d}{D}}{1 - \frac{d}{D}} = \frac{\sigma_{\text{ann}}}{\sigma_0} - 1, \tag{149}$$

in which case

$$\frac{A_{\text{arm}}}{A_0} = 2 - \frac{\sigma_{\text{ann}}}{\sigma_0}. \tag{150}$$

236

It is illustrative to write this result in terms of an arm/interarm contrast, A, for a sinusoidal density distribution, which has

$$\sigma_{\text{interarm}} = \sigma_0 - (\sigma_{\text{ann}} - \sigma_0) = 2\sigma_0 - \sigma_{\text{ann}}. \tag{151}$$

This gives

$$A = \frac{\sigma_{\text{ann}}}{2\sigma_0 - \sigma_{\text{ann}}} \quad \text{so that} \quad \frac{\sigma_{\text{ann}}}{\sigma_0} = \frac{2A}{1 + A}, \tag{152}$$

and the ratio of the perturbed to the unperturbed Oort constants is

$$\frac{A_{\text{ann}}}{A_0} = \frac{2}{1 + A}. \tag{153}$$

This result illustrates how the shear decreases with a radial compression, because $A = -\frac{1}{2} r \, d\Omega/dr$ is a measure of the rate of shear.

For a more general rotation curve, of the form $V \propto r^\alpha$, this ratio equals

$$\frac{A_{\text{ann}}}{A_0} = \frac{2}{1 + A} \left(\frac{1 - \alpha A}{1 - \alpha} \right), \tag{154}$$

where $A_0 = \frac{1}{2}(1 - \alpha)V/r$. This implies that if $\alpha > 0$ for a typically rising rotation curve, then the relative decrease in shear is even more dramatic than in the case of a flat rotation curve, because $1 - \alpha A$ can be negative for finite A.

The shear rate A_{arm} can be negative also for a strong (non-sinusoidal) perturbation, because when σ_{arm} locally exceeds $2\sigma_0$, the A value, given above as $A_0(2 - \sigma_{\text{arm}}/\sigma_0)$, becomes less than zero.

The same result is obtained from the linear theory of spiral density waves. If we write, from Sect. 5.1, $\sigma_1 = -S_1 \cos(2\Omega_p t - \Phi)$ and $v_\theta = V_1 \sin(2\Omega_p t - \Phi)$, then $V_1 = S_1 \kappa^2/(2\Omega k \sigma_0)$ and in the center of a spiral arm, $A_{\text{arm}} = \frac{1}{2}(\Omega - \alpha\Omega - V_1 k)$ for rotation curve $V \propto r^\alpha$. After substituting $\kappa^2 = 2\Omega^2(1 + \alpha)$, this gives

$$A_{\text{arm}} = A_0 \left\{ 1 - \left(\frac{S_1}{\sigma_0} \right) \left(\frac{1 + \alpha}{1 - \alpha} \right) \right\}. \tag{155}$$

But $S_1/\sigma_0 = (A - 1)/(A + 1)$, so

$$\frac{A_{\text{arm}}}{A_0} = 1 - \left(\frac{A - 1}{A + 1} \right) \left(\frac{1 + \alpha}{1 - \alpha} \right) = \frac{2}{1 + A} \left(\frac{1 - \alpha A}{1 - \alpha} \right). \tag{156}$$

These results imply that strong arms have low (or negative) shear and that low-density clouds forming in them are not easily broken apart, at least not until they enter the interarm region. It also implies that galactic rotation curves should show rising velocities inside spiral arms and that for strong arms with $\alpha \approx 0$ on average, the extrapolation of these rising parts to low galactic radii should give a line that goes through the origin of the velocity-radius curve, that is, the extrapolated velocity should be nearly zero at zero radius. This is true for many cases, proving that there is little shear in the arm (e.g., NGC 2998 in

[67]- see also Fig. 1 in [68]). A local decrease in the rate of shear has also been observed by Comerón & Torra [69] for the spiral arm in the solar neighborhood.

A decrease in shear corresponds to an increase in the epicyclic frequency. Balbus & Cowie [70] derived an instability condition for gas in spiral arms, considering compression and rarefaction in the shock, and found that the effective epicyclic frequency for perturbations is $\kappa(\sigma/\sigma_0)^{1/2}$, which is larger in the arms than the average value. This increase has important implications for the spiral arm value of the instability parameter Q (cf. Sect. 6.3.1).

5.2.3 A Decrease in Tidal Forces in Spiral Arms.

Galactic tidal forces are also relatively low in spiral arms. This too has important consequences for gas dynamics and cloud formation. Suppose that the potential from an arm in comoving coordinates is again

$$\Psi = \psi \cos \Phi \qquad \text{where} \qquad \Phi = \frac{2 \ln \left(\dfrac{r}{r_0} \right)}{\tan i} + 2\theta \qquad (157)$$

where the convention is $i > 0$ for a trailing spiral, as in Sect. 5.1, and θ is increasing in the direction of rotation.

The total accelerations in the radial and azimuthal directions are

$$g_r = -\Omega^2 r - \frac{\partial \Psi}{\partial r} = -\Omega^2 r - \psi' \cos \Phi + \frac{2\psi}{r \tan i} \sin \Phi, \qquad (158.1)$$

$$g_\theta = -\frac{1}{r} \frac{\partial \Psi}{\partial \theta} = \frac{2\psi}{r} \sin \Phi, \qquad (158.2)$$

where we include the background galactic acceleration in the $\Omega^2 r$ term, and define $\psi' = \partial \psi / \partial r$.

If we transform to spiral coordinates, η in the locally perpendicular direction and ξ parallel to the spiral,

$$\eta = \ln \left(\frac{r}{r_0} \right) \cos i + \theta \sin i, \qquad (159.1)$$

$$\xi = -\ln \left(\frac{r}{r_0} \right) \sin i + \theta \cos i, \qquad (159.2)$$

then the accelerations become

$$g_\perp = g_r \cos i + g_\theta \sin i, \qquad (160.1)$$

$$g_\parallel = -g_r \sin i + g_\theta \cos i. \qquad (160.2)$$

Now we find the tidal acceleration gradients in the η and ξ directions considering that the cloud corotates with the galaxy (which introduces the Ω^2 term):

238

$$T_\perp = \Omega^2 + \frac{\partial g_\perp}{\partial \eta} = \Omega^2 + \cos i \frac{\partial g_\perp}{\partial r} + \frac{\sin i}{r} \frac{\partial g_\perp}{\partial \theta}$$

$$= \Omega^2 + \cos^2 i \frac{\partial g_r}{\partial r} + \frac{\sin^2 i}{r} \frac{\partial g_\theta}{\partial \theta} + \cos i \sin i \left(\frac{\partial g_\theta}{\partial r} + \frac{1}{r} \frac{\partial g_r}{\partial \theta} \right), \qquad (161.1)$$

$$T_\parallel = \Omega^2 + \frac{\partial g_\parallel}{\partial \xi} = \Omega^2 - \sin i \frac{\partial g_\parallel}{\partial r} + \frac{\cos i}{r} \frac{\partial g_\parallel}{\partial \theta}$$

$$= \Omega^2 + \sin^2 i \frac{\partial g_r}{\partial r} + \frac{\cos^2 i}{r} \frac{\partial g_\theta}{\partial \theta} - \cos i \sin i \left(\frac{\partial g_\theta}{\partial r} + \frac{1}{r} \frac{\partial g_r}{\partial \theta} \right). \qquad (161.2)$$

To evaluate these quantities, consider part of a galaxy with a rotation curve $V \propto r^\alpha$ and a small spiral pitch angle so that $\sin i \ll \cos i$. Also consider relatively large spiral arm gradients, so that $1/r \ll \partial/\partial r$. Then, in the center of a spiral arm where $\Phi = \pi$,

$$T_{\perp,\text{arm}} = 2\Omega^2(1 - \alpha) + \psi'' - \frac{4\psi}{r^2 \tan^2 i}, \qquad (162.1)$$

$$T_{\parallel,\text{arm}} = \Omega^2 + \sin^2 i \big((1 - 2\alpha)\Omega^2 + \psi''\big), \qquad (162.2)$$

where $\psi'' = \partial^2\psi/\partial r^2$ and we have assumed for the second equation that $r \sin i\, \partial/\partial r \gg 1$. Similarly, between the arms ($\Phi = 0$)

$$T_{\perp,\text{interarm}} = 2\Omega^2(1 - \alpha) - \psi'' + \frac{4\psi}{r^2 \tan^2 i}, \qquad (163.1)$$

$$T_{\parallel,\text{interarm}} = \Omega^2 + \sin^2 i \big((1 - 2\alpha)\Omega^2 - \psi''\big). \qquad (163.2)$$

Now we have to evaluate $\psi'' - 4\psi/(r^2 \tan^2 i)$. The relation between spiral-perturbed potential and perturbed density is given by Poisson's law:

$$\nabla^2 \Psi = 4\pi G \rho_{\text{spiral}}. \qquad (164)$$

This gives

$$-\psi'' - \frac{\psi'}{r} + \frac{4\psi}{r^2 \tan^2 i} = 4\pi G \rho_0 \frac{A - 1}{A + 1} \qquad (165)$$

in the arm center, for ρ_0 equal to the average total density at r and A equal to the peak arm to interarm density contrast for a sinusoidal perturbation.

This expression may be evaluated for $\psi(r)$ given $\rho_0(r)$ and $A(r)$, but at the center of a modal oscillation along the arm, where $\psi' = 0$ (a modal oscillation is a region of high spiral arm strength compared to neighboring regions along the same arm, presumably resulting from the interference between inward-trailing and outward-leading spirals — see examples in [71, 72]), we can substitute the left-hand side directly into the expression for $T_\perp$ to get

$$T_{\perp,\text{arm}} = 2\Omega^2(1 - \alpha) - 4\pi G \rho_0 \frac{A - 1}{A + 1}, \qquad (166.1)$$

$$T_{\perp,\text{interarm}} = 2\Omega^2(1 - \alpha) + 4\pi G \rho_0 \frac{A - 1}{A + 1}. \qquad (166.2)$$

This result implies $T_{\perp,\text{arm}} \ll T_{\perp,\text{interarm}}$. We also note that $\psi'' < 0$ at this modal amplitude peak, in which case $T_{\parallel,\text{arm}} \ll T_{\parallel,\text{interarm}}$ from above. The first of these two tidal force inequalities is also true if ψ is constant with radius, but the second inequality becomes an equality in this case.

This derivation demonstrates that the tidal force on a cloud from the background galaxy is less in an arm than in the interarm region. What does this say about the critical cloud density for tidal stability? Tidal stability requires $GM/R^2 > TR$ for cloud radius R, or

$$\rho_{\text{cloud}} > \frac{3T}{4\pi G}. \tag{167}$$

Thus we find that tidally bound clouds must satisfy

$$\rho_{\text{cloud,arm}} > \frac{3\Omega^2(1-\alpha)}{2\pi G} - 3\rho_0 \frac{A-1}{A+1}, \tag{168.1}$$

$$\rho_{\text{cloud,interarm}} > \frac{3\Omega^2(1-\alpha)}{2\pi G} + 3\rho_0 \frac{A-1}{A+1}. \tag{168.2}$$

Note that the first term here is the usual tidal density limit. It can differ from the axially symmetric value if the cloud does not corotate with the galaxy as assumed here, but has some other rotation rate dependent on the local spiral arm flow. The second term shows the effect of a spiral density wave.

A typical value in the disk of our Galaxy for the first term in this inequality corresponds to an atomic hydrogen density of around $3\,\text{cm}^{-3}$, while the second term is about three times larger. These are such low densities that both the average and spiral-arm tidal forces are too low to affect superclouds much, because the average density in a supercloud is about $10\,\text{cm}^{-3}$. The critical densities are not too low to affect the cloud's structure in the interarm region, however, because there the critical density may exceed the actual cloud density. This result suggests that superclouds should become tidally unbound when they enter the interarm region (considering that they started gravitationally bound in the arms). The results also suggest that tidal forces should not affect dense molecular clouds in the same way. Giant molecular clouds and their pieces should survive the interarm transit if their star formation is not too intense.

It seems possible that superclouds can continuously accrete interarm gas at the upstream side of a spiral arm, with some gravitational focusing of this gas onto the superclouds. Then they will process this gas internally into giant molecular clouds (GMC's) because of the enhanced rate of energy dissipation and large self-gravity there. In some galaxies or regions of galaxies, these GMC's will form stars while they are still inside the superclouds, giving the nested, hierarchical structure discussed in Chap. 1. But in other regions the GMC's will emerge from the downstream side of the superclouds before star formation begins. In either case, the GMC's will emerge from the superclouds without tidal disruption, although tidal forces will separate their centers. The lower density gas around each GMC, which occupies most of the volume of the supercloud, will be dispersed downstream by tidal forces.

This scenario predicts a supercloud structure that stays in a spiral arm for a much longer time than the arm flow-through time because of accretion on one side and dispersal at the same rate on the other side. Presumably the whole collection of these superclouds in an arm will appear as a lumpy dust lane. It also predicts that when the dust lane flow-through time is longer than the collapse time the GMC's with star formation will be inside the superclouds, and when the dust lane flow-through time is shorter than the collapse time, the GMC's with star formation will be downstream from the superclouds, giving the appearance of a triggered ridge of star formation following a density wave shock. Both types of structures seem to be observed in various galaxies.

Another implication of the tidal force variation discussed above is that the tidal force inside an arm will be slightly less at the peak of a modal oscillation ($\psi'' < 0$) than between the peaks. Thus supercloud formation may prefer the peaks, in which case star formation should roughly delineate the underlying modal interference pattern (e.g., see M100 in [71]).

5.2.4 A Comparison of Rates for Gas Processes in Spiral Density Waves. Several processes happen simultaneously as the gas flows through a spiral arm. For considerations involving gas collapse into giant cloud complexes, we should compare the rates for these various processes. We consider here the rates of energy dissipation in the gas, gravitational instability, shear, and arm traversal. A more detailed discussion of gravitational instabilities in spiral arms is in Sect. 6.5.

The dissipation rate in the ambient medium is, from cloud collisions,

$$\omega_{\text{diss}} \sim n_{\text{c}} \pi R_{\text{c}}^2 a \tag{169}$$

for cloud number density n_{c}, velocity dispersion a, and spherical cloud radius R_{c} (Sect. 3.2). The gravitational instability rate is

$$\omega_{\text{grav}} \sim (4\pi G\rho)^{1/2}. \tag{170}$$

The ratio of these two rates is

$$\frac{\omega_{\text{diss}}}{\omega_{\text{grav}}} \approx \left(\frac{P}{G\sigma_{\text{c}}^2} \right)^{1/2} \tag{171}$$

for cloud mass column density $\sigma_{\text{c}} = M/\pi R^2$ and average pressure $P = \rho a^2$. This ratio depends on the mean properties of the clouds in the fluid; structurally diffuse clouds have high values of this ratio and self-gravitating clouds have low values (see review in [73]). The average value in the local interstellar medium is about 1, with molecular clouds having lower values and diffuse clouds higher values. Then the dissipation time from cloud collisions, which is dominated by diffuse clouds, is approximately equal to the gravitational free-fall time in the ambient medium — both are around 2×10^7 years. The ratio of dissipation time to gravitational collapse time varies from the arm to the interarm region as self-gravitating clouds are converted into diffuse clouds, and vice versa.

The shear rate in the ambient medium is

$$\omega_{\text{shear}} = A = -\tfrac{1}{2}r\frac{d\Omega}{dr} \sim \tfrac{1}{2}\Omega \qquad (172)$$

for Oort constant A. The dissipation of random kinetic energy in a density perturbation occurs before shear destroys the perturbation if $\omega_{\text{shear}} < \omega_{\text{diss}}$. In the solar neighborhood, $A \sim 0.015\,\text{km\,s}^{-1}\,\text{pc}^{-1}$ and $n_c\pi R_c^2 a \sim 0.05\,\text{km\,s}^{-1}\,\text{pc}^{-1}$, so

$$\frac{\omega_{\text{diss}}}{\omega_{\text{shear}}} \sim 3. \qquad (173)$$

This result implies that in the ambient medium, the dissipation of turbulent energy occurs faster than shear, so cloud formation by condensation and collapse of the ambient gas is possible. If the ratio were less than 1, then shear would destroy each density perturbation before internal dissipation could cool it, and the spontaneous collapse of gas into dense clouds would be rare.

In a spiral density wave, ω_{diss} and ω_{grav} increase, while ω_{shear} decreases. This change makes the ratio of the formation rate (from dissipation and gravity) to the shear rate larger in an arm than in a region without spiral arms, and so the dissipation of energy leading to gas collapse and cloud formation is much more important in the arm. To demonstrate this, suppose that the average arm/interarm density contrast is A for a sinusoidal wave, in which case the shear rate in the arm is

$$\omega_{\text{shear}} = \frac{2\omega_{\text{shear,average}}}{1+A} = \frac{\Omega}{1+A}, \qquad (174)$$

as discussed in Sect. 5.2.2. The density of the clouds in the fluid scales with $2A/(1+A)$, so if the average cloud radius and velocity dispersion do not change when the clouds enter an arm, then

$$\omega_{\text{diss,arm}} = \frac{2A\omega_{\text{diss,average}}}{1+A}. \qquad (175)$$

Now it follows that

$$\left(\frac{\omega_{\text{diss}}}{\omega_{\text{shear}}}\right)_{\text{arm}} = A\left(\frac{\omega_{\text{diss}}}{\omega_{\text{shear}}}\right)_{\text{average}} \sim 10, \qquad (176)$$

where we take $A \sim 3$. Thus dissipation becomes much more important than shear in a spiral arm.

Similarly,

$$\left(\frac{\omega_{\text{grav}}}{\omega_{\text{shear}}}\right)_{\text{arm}} = \left(\frac{A(1+A)}{2}\right)^{1/2}\left(\frac{\omega_{\text{grav}}}{\omega_{\text{shear}}}\right)_{\text{average}} \sim 7, \qquad (177)$$

so the gravitational rate also increases relative to the shear rate in an arm.

What limits the dissipation and collapse of gas in a spiral density wave if it is not shear? Why doesn't all of the gas collapse and make star formation catastrophic? The limiting factor to star formation is probably the short flow-through time in the arm. The arm flow-through rate is taken to be 2π divided by the arm-to-arm flow time, which is $\pi/(\Omega - \Omega_{\text{p}})$ for a two arm spiral, and divided by the fraction of the time spent in an arm, which is $\mathcal{A}/(1 + \mathcal{A})$ from the continuity equation:

$$\omega_{\text{flow}} = \frac{2(1 + \mathcal{A})(\Omega - \Omega_{\text{p}})}{\mathcal{A}}. \tag{178}$$

This compares to the shear rate as

$$\frac{\omega_{\text{shear}}}{\omega_{\text{flow}}} = \frac{\Omega \mathcal{A}}{2(\Omega - \Omega_{\text{p}})(1 + \mathcal{A})^2} \sim \frac{\mathcal{A}}{(1 + \mathcal{A})^2}, \tag{179}$$

where we have set $\Omega - \Omega_{\text{p}} \sim \frac{1}{2}\Omega$ for a typical position inside corotation. If an arm is strong, then the total amount of shear inside the arm is small; i.e., $\omega_{\text{shear}}/\omega_{\text{flow}} \ll 1$. This result implies that what limits the dissipation of energy and collapse into stars in a spiral density wave arm is the short flow-through time, not the shear.

For non-linear waves, which are more appropriate for the gas, the same results follow from the use of

$$\omega_{\text{shear}} \approx \omega_{\text{shear,average}}\left(2 - \frac{\sigma_{\text{arm}}}{\sigma_0}\right) \quad , \quad \omega_{\text{diss}} \approx \omega_{\text{diss,average}}\left(\frac{\sigma_{\text{arm}}}{\sigma_0}\right),$$
$$\omega_{\text{grav}} \approx \omega_{\text{grav,average}}\left(\frac{\sigma_{\text{arm}}}{\sigma_0}\right)^{1/2} \quad , \quad \omega_{\text{flow}} \approx \omega_{\text{flow,average}}\left(\frac{\sigma_0}{\sigma_{\text{arm}}}\right), \tag{180}$$

for the local rates in the arm. Here we use the local density σ_{arm}, which can be much larger than the limit of twice the average density σ_0 given by the sinusoidal approximation (i.e., when $\mathcal{A}$ is infinite). Note that the gravitational instability rate scales with the square root of the column density if the scale height of the gas layer does not change, and with the first power of the column density if the scale height is largely determined by the gas gravity in the direction perpendicular to the disk.

Now we can address the important question of density-wave triggered star formation: how much *extra* collapse of gas into stars can a spiral arm trigger compared to what would have occurred in the same gas without the arm? To estimate this, we compare the dissipation rate (energy dissipation is the rate-limiting process in gravitational collapse), or equivalently, the cloud-cloud collision rate ($\omega_{\text{diss}} \approx \omega_{\text{collision}}$) to the flow-through rate, which determines how much time the gas has available for dissipation and collapse:

$$\frac{\omega_{\text{diss}}}{\omega_{\text{flow}}} \sim \left(\frac{\omega_{\text{diss}}}{\omega_{\text{shear}}}\right)_{\text{average}} \left(\frac{\mathcal{A}}{1 + \mathcal{A}}\right)^2. \tag{181}$$

Here we have assumed again that $\Omega \sim 2(\Omega - \Omega_{\text{p}})$ and we have written the dissipation-to-flow ratio in terms of the average dissipation-to-shear ratio.

This result suggests that low amplitude spiral density waves, which have $\mathcal{A} \sim 1$, have a dissipation rate comparable to the flow-through rate and ~ 3 times faster than the shear rate. Shear limits the collapse in this case because there are no tidal force variations in such weak arms. The collapse also has very little correlation with the arms because the flow-through rate is so large.

The result is presumably a pattern of star formation that is very broad and centered on, but not entirely confined to, the arms.

High amplitude waves ($\mathcal{A} \approx 3$) have a dissipation or cloud collision rate ~ 10 times the shear rate in the arms and ~ 2 times the flow-through rate, while the flow rate is ~ 5 times the shear rate in the arms, so the flow (and the corresponding variation in tidal forces) limits the total amount of collapse to star formation (instead of shear limiting the collapse), and this collapse is highly confined to the arms (because $\omega_{diss} > \omega_{flow}$). Thus strong waves tightly confine star formation to the arms.

In both cases the dissipation and gravity rates are only slightly larger than the limiting rates for the *corresponding* resistance to collapse, i.e., for a weak wave the dissipation rate is only ~ 3 times the shear rate, and shear is the limiting factor for collapse, and for a strong wave, the dissipation rate is only ~ 2 times the flow-through rate, and the flow is the limiting factor. This result suggests that the interstellar medium is only marginally unstable to form stars whether or not there is a density wave, and that the presence of a wave does very little to affect the overall star formation rate. The wave organizes star formation into the spiral pattern by putting most of the gas in the arms and by increasing $\omega_{diss}/\omega_{flow}$ there, and it changes the limiting factor for collapse from shear in the case of no waves or weak waves to the flow-through time in the case of strong waves. Strong arms produce narrow and bright star formation ridges at the positions of the density wave shocks because $\omega_{diss} > \omega_{flow}$, giving the appearance of a substantial triggering effect even if there is not much of an efficiency change (i.e., a factor of ~ 2 is plausible), and weak arms produce only ragged star formation structures because $\omega_{diss} < \omega_{flow}$, giving the impression that very little star formation has been triggered whether or not this is the case. The ratio of the dissipation or gravitational collapse time to the flow-through time controls the confinement of the gas and the amount of star formation in the arms.

This apparent variation in triggering with arm strength is somewhat deceptive, and has led to the idea that density waves trigger substantial amounts of star formation [50, 74]. In fact density waves may trigger very little excess star formation in normal galaxies [75 − 80], although this point is very uncertain. For example, for M51, Lees & Lo [80] claim that the data are consistent with very little or no perceptible triggering, while Knapen et al. [81] find a variation in the ratio of Hα to CO from the arms to the interarm regions, which implies that there is triggering. Studies of other galaxies by this latter group, in Cepa et al. [82, 83], find more evidence for arm-interarm variations in the Hα/CO or Hα/HI ratios, again leading to the conclusion that there is triggering. Cluster age progressions in our Galaxy [84] and color gradients in spiral arms of other galaxies [85] have also been attributed to density wave triggering. Furthermore, the giant HII regions in the spiral arms of our Galaxy, W49A and W51A, have much larger efficiencies than the smaller Orion molecular cloud, which is more of an interarm feature [86].

Yet the evidence against triggering seems persuasive too. In NGC 6946, for example, one arm shows evidence for an excess star formation efficiency, while

244

the other arm does not [87], and in M33 the efficiency in molecular clouds is the same regardless of position relative to the arms [88]. Also in many other galaxies, even those with early types and very little gas, or with a variety of spiral structures, the average star formation efficiency is almost always the same in the main disk [78, 79, 89 – 92]. This result apparently holds for some hot spot galaxies too [93], although higher nuclear efficiencies might be more common [92].

Evidently global studies of efficiency variations are measuring a different property of star formation than direct comparisons between the arm and interarm regions. The global studies are relatively easy to do with present day techniques, but they may not tell us much about density wave triggering and they may be inaccurate. Conversely, the arm-interarm comparisons may tell us the most about triggering, but they still suffer from a lack of angular resolution and sensitivity. Moreover, the star formation *efficiency* in cloud complexes may not be the thing to measure when looking for triggering effects. Efficiency may be more the result of cloud destruction by the stars it forms than of a triggered initiation of cloud or star formation by large scale processes.

The organizational effect of the wave is important for another feature of galaxies: the presence of the largest cloud complexes in spiral arms. The excess density and relatively long flow-through time in strong spiral arms probably promotes gravitational instabilities and cloud coalescence leading to prominent $10^7 \, M_\odot$ clouds and star complexes (cf. Sect. 6.5). But this again could be more of a restructuring of the gas than a direct triggering of excess star formation (Sect. 7.4). The basic point is that such instabilities are not catastrophic in normal galaxies: they do not lead to a tremendous excess of star formation following cloud collapse because the energy in these complexes usually cannot dissipate very rapidly compared to the arm flow-through time. The clouds disperse too quickly because of the large tidal forces that arrive with the interarm region, and then only the dense molecular clouds that have had time to form produce new stars, while most of the lower density gas gets dispersed. The scenario could be different for starburst galaxies because of the higher gas density and velocity dispersion there (cf. Sect. 7.5).

Acknowledgement: Helpful comments by Drs. Chi Yuan, Richard Rand, William Roberts, and Françoise Combes are appreciated.

References

5.1 Lin, C.C., Shu, F.H. 1966, *Proc. Nat. Acad. Sci.*, **55**, 229
5.2 Jog, C.J., Solomon, P.M. 1984, *Astrophys. J.* **276**, 127
5.3 Lubow, S.H., Balbus, S.A., Cowie, L.L. 1986, *Astrophys. J.* **309**, 496
5.4 Bertin, G., Romeo, A.B. 1988, *Astron. Astrophys.* **195**, 105
5.5 Kalnajs, A.J. 1972, *Astrophys. Letters* **11**, 41
5.6 Shu, F.H. 1985, in *IAU Symp. 106, The Milky Way Galaxy*, eds. H. van Woerden, R.J. Allen, W.B. Burton, Dordrecht, Reidel, p. 530

5.7   Lubow, S. 1984, *Astrophys. J.* **307**, L39
5.8   Rohlfs, K. 1977, *Lectures on Density Wave Theory*, Berlin, Springer-Verlag
5.9   Toomre, A. 1977, *Ann. Rev. Astron. Astrophys.* **15**, 437
5.10 Bertin, G. 1980, *Phys. Rep.* **61**, 1
5.11 Toomre, A. 1981, in *The Structure and Evolution of Normal Galaxies*, eds. S.M. Fall, D. Lynden-Bell, Cambridge, Cambridge University, p. 111
5.12 Athanassoula, E. 1984, *Phys. Rep.* **114**, 319
5.13 Sellwood, J.A. 1987, in *Proc. 10th European Regional Meeting of the of the IAU, Evolution of Galaxies*, ed. J. Palouš, Prague, Czechoslovakian Academy of Sciences, p. 249
5.14 Bertin, G. 1991, in *IAU Symp. 146, Dynamics of Galaxies and their Molecular Cloud Distributions*, eds. F. Combes, F. Casoli, Dordrecht, Kluwer, p. 93
5.15 Elmegreen, B.G. 1991, in *IAU Symp. 146, Dynamics of Galaxies and their Molecular Cloud Distributions*, eds. F. Combes, F. Casoli, Dordrecht, Kluwer, p. 113
5.16 Lindblad, B. 1963, *Stockholm Obs. Ann.* **22**, No. 5
5.17 Kato, S. 1983, *Publ. Astron. Soc. Japan* **35**, 249
5.18 Adams, F.C., Ruden, S.P., Shu, F.H. 1989, *Astrophys. J.* **347**, 959
5.19 Lin, C.C., Shu, F.H. 1964, *Astrophys. J.* **140**, 646
5.20 Lin, C.C., Yuan, C., Shu, F.H. 1969, *Astrophys. J.* **155**, 721
5.21 Lau, Y.Y., Bertin, G. 1978, *Astrophys. J.* **226**, 508
5.22 Toomre, A. 1969, *Astrophys. J.* **158**, 899
5.23 Toomre, A., Toomre, J. 1972, *Astrophys. J.* **178**, 623
5.24 Miller, R.H., Prendergast, K.H., Quirk, W.J. 1970, *Astrophys. J.* **161**, 903
5.25 Goldreich, P., Lynden-Bell, D. 1965, *Monthly Notices Roy. Astron. Soc.* **130**, 97
5.26 Elmegreen, D.M., Sundin, M., Elmegreen, B., Sundelius, B. 1991, *Astron. Astrophys.* **244**, 52
5.27 Lin, C.C. 1970, in *IAU Symp. 38, The Spiral Structure of Our Galaxy*, eds. W. Becker, G. Contopoulos, Dordrecht, Reidel, p. 377
5.28 Lin, C.C. 1971, in *Highlights of Astronomy* **2**, 88
5.29 Lin, C.C., Lau, Y.Y. 1975, *SIAM Journal of Applied Mathematics* **129**, 352
5.30 Mark, J.W.K. 1976, *Astrophys. J.* **205**, 363
5.31 Bertin, G., Lin, C.C., Lowe, S.A., Thurstans, R.P. 1989, *Astrophys. J.* **338**, 104
5.32 Goldreich, P., Tremaine, S. 1978, *Astrophys. J.* **222**, 850
5.33 Lin, C.C., Lau, Y.Y. 1979, *Studies in Applied Math.* **60**, 97
5.34 Lin, C.C., Thurstans, R.P. 1984, in *Proceedings of a Course on Plasma Astrophysics*, Varenna, Italy, ESA SP-207), p. 121
5.35 Pellat, R., Tagger, M., Sygnet, J.F. 1990, *Astron. Astrophys.* **231**, 347
5.36 Julian, W.H., Toomre, A. 1966, *Astrophys. J.* **146**, 810
5.37 Safronov, V.S. 1960, *Annales d'Astrophysique* **23**, 979
5.38 Toomre, A. 1964, *Astrophys. J.* **139**, 1217
5.39 van der Kruit, P.C., Freeman, K.C. 1986, *Astrophys. J.* **303**, 556
5.40 Bottema, R., van der Kruit, P.C., Freeman, K.C. 1987, *Astron. Astrophys.* **178**, 77
5.41 Bottema, R. 1988, *Astron. Astrophys.* **197**, 105
5.42 Bottema, R. 1989, *Astron. Astrophys.* **221**, 236
5.43 Kennicutt, R.C. 1989, *Astrophys. J.* **344**, 685
5.44 Lau, Y.Y., Lin, C.C., Mark, J.W.K. 1976, *Proceedings of the National Academy of Sciences* **73**, 1379
5.45 Thomasson, M., Elmegreen, B.G., Donner, K.J., Sundelius, B. 1990, *Astrophys. J.* **356**, L9
5.46 Elmegreen, B.G., Thomasson, M. 1992, *Astron. Astrophys.* in press
5.47 Arp, H. 1966, *Astrophys. J. Suppl. Ser.* **14**, 1
5.48 Fridman, A.M. 1989, *Sov. Astron.* **33**, 684
5.49 Soukup, J.E., Yuan, C. 1981, *Astrophys. J.* **246**, 376
5.50 Roberts, W.W. 1969, *Astrophys. J.* **158**, 123
5.51 Sanders, R.H., Huntley, J.M. 1976, *Astrophys. J.* **209**, 53
5.52 Roberts, W.W., Huntley, J.M., van Albada, G.D. 1979, *Astrophys. J.* **233**, 67
5.53 Sanders, R.H., Tubbs, A.D. 1980, *Astrophys. J.* **235**, 803
5.54 Roberts, W.W., Steward, G.R. 1987, *Astrophys. J.* **314**, 10
5.55 Elmegreen, B.G. 1985, in *IAU Symp. 115, Star-Forming Regions*, eds. M. Peimbert, J. Jugaku, Dordrecht, Reidel, p. 457

5.56 Elmegreen, B.G. 1981, *Astrophys. J.* **243**, 512

5.57 Elmegreen, B.G. 1988, *Astrophys. J.* **326**, 616

5.58 Kaufman, M., Bash, F.N., Hine, B., Rots, A.H., Elmegreen, D.M., Hodge, P.W. 1989, *Astrophys. J.* **345**, 674

5.59 Brouillet, N., Baudry, A., Combes, F., Kaufman, M., Bash, F. 1991, *Astron. Astrophys.* **242**, 35

5.60 Lord, S.D., Kenney, J.D.P. 1991, *Astrophys. J.* **381**, 130

5.61 Wiklind, T., Rydbeck, G., Hjalmarson, A., Bergman, P. 1990, *Astron. Astrophys.* **232**, L11

5.62 Rand, R.J., Kulkarni, S.R. 1990, *Astrophys. J.* **349**, L43

5.63 Planesas, P., Scoville, N., Myers, S.T. 1991, *Astrophys. J.* **369**, 364

5.64 Elmegreen, B.G. 1989, *Astrophys. J.* **338**, 178

5.65 Kenney, J.D.P., Lord, S.D. 1991, *Astrophys. J.* **381**, 118

5.66 Elmegreen, B.G. 1987, *Astrophys. J.* **312**, 626

5.67 Rubin, V.C., Ford, W.K., Jr., Thonnard, N. 1980, *Astrophys. J.* **238**, 471

5.68 Elmegreen, B.G. 1987, in *Physical Processes in Interstellar Clouds*, eds. G.E. Morfill, M. Scholer, Dordrecht, Reidel, p. 1

5.69 Comerón, F., Torra, J. 1991, *Astron. Astrophys.* **241**, 57

5.70 Balbus, S.A., Cowie, L.L. 1985, *Astrophys. J.* **297**, 61

5.71 Elmegreen, B.G., Elmegreen, D.M., Seiden, P.E. 1989, *Astrophys. J.* **343**, 602

5.72 Elmegreen, B.G., Elmegreen, D.M., Montenegro, L. 1992, *Astrophys. J. Suppl. Ser.* **78**, in press

5.73 Elmegreen, B.G. 1992, in *Protostars and Planets III*, eds. E.H. Levy, M.S. Matthews, Tucson, University of Arizona, in press

5.74 Roberts, W.W., Roberts, M., Shu, F.H. 1975, *Astrophys. J.* **196**, 381

5.75 Elmegreen, B.G., Elmegreen, D.M. 1986, *Astrophys. J.* **311**, 554

5.76 McCall, M.L., Schmidt, F.H. 1986, *Astrophys. J.* **311**, 548

5.77 Stark, A.A., Elmegreen, B.G., Chance, T. 1987, *Astrophys. J.* **322**, 64

5.78 Thronson, H.A., Tacconi, L., Kenney, J., Greenhouse, M.A., Margulis, M., Tacconi-Garman, L., Young, J. 1989, *Astrophys. J.* **344**, 747

5.79 Wiklind, T., Henkel, C. 1989, *Astron. Astrophys.* **225**, 1

5.80 Lees, J.F., Lo, K.Y. 1991, *Astrophys. J.* submitted

5.81 Knapen, J.H., Beckman, J.E., Cepa, J., van der Hulst, T., Rand, R.J. 1992, *Astrophys. J.* in press

5.82 Cepa, J., Beckman, J.E. 1989, *Astrophys. Space Sci.* **156**, 289

5.83 Cepa, J., Beckman, J.E. 1990, *Astrophys. J.* **349**, 497

5.84 Sitnik, T.G. 1991, *Sov. Astron. Letters* **17**, 61

5.85 Hodge, P., Jaderlund, E., Meakes, M. 1990, *Publ. Astron. Soc. Pacific* **102**, 1263

5.86 Sievers, A.W., Mezger, P.G., Gordon, M.A., Kreysa, E., Haslam, C.G.T., Lemke, R. 1991, *Astron. Astrophys.* **251**, 231

5.87 Tacconi, L.J., Young, J.S. 1990, *Astrophys. J.* **352**, 595

5.88 Wilson, C.D., Scoville, N. 1991, *Astron. J.* **101**, 1293

5.89 Wiklind, T., Henkel, C. 1990, *Astron. Astrophys.* **227**, 394

5.90 Bajaja, E., Krause, M., Dettmar, R.J., Wielebinski, R. 1991, *Astron. Astrophys.* **241**, 411

5.91 Devereux, N.A., Young, J.S. 1991, *Astrophys. J.* **371**, 515

5.92 Sage, L., Solomon, P.M. 1991, *Astrophys. J.* **379**, 392

5.93 Jackson, J.M., Eckart, A. Cameron, M., Wild, W., Ho, P.T.P., Pogge, R.W., Harris, A.I. 1991, *Astrophys. J.* **375**, 101

6 Gas Instabilities

The interstellar gas is far from stable. Various forces, particularly from stars, magnetic fields, and gravity, continuously try to disrupt and compress the gas. Here we consider thermal, gravitational, and Parker [1] instabilities in various environments, building from the simple to the more complicated applications. Kelvin-Helmholtz instabilities [2] will not be discussed. Note that the linearized nature of these calculations, and the homogeneous initial conditions that are assumed, are not always consistent with the chaotic and clumpy structure of the real ISM. A more thorough review is in [3].

6.1 Thermal Instability

We start with the most basic of equations for gas dynamics, the equations of motion, continuity, and energy:

$$\rho \frac{Dv}{Dt} = -\nabla P, \tag{182.1}$$

$$\frac{D\rho}{Dt} = -\rho \nabla \cdot v, \tag{182.2}$$

$$\frac{DP}{Dt} = \frac{\gamma P}{\rho} \frac{D\rho}{Dt} + (\gamma - 1)(\Gamma - \Lambda), \tag{182.3}$$

where Γ and Λ are heating and cooling rates in $\mathrm{erg\,cm^{-3}\,s^{-1}}$ and $D/Dt = \partial/\partial t + v \cdot \nabla$. We also assume a perfect gas and write

$$P = \rho a^2. \tag{183}$$

Now consider an equilibrium $v = 0$, $\rho = \mathrm{constant}$, and $P = \mathrm{constant}$, and perturb it using variables $\hat{\rho}$, etc. Then substitute $\rho + \hat{\rho}$ for ρ and so on, and separate the first order quantities ($\hat{\rho}$) from the zero order (ρ) quantities and second order ($\hat{\rho}v$) quantities to get equations that are entirely first order:

$$\rho \frac{\partial v}{\partial t} = -\nabla \hat{P}, \tag{184.1}$$

$$\frac{\partial \hat{\rho}}{\partial t} = -\rho \nabla \cdot v, \tag{184.2}$$

$$\frac{\partial \hat{P}}{\partial t} = \gamma a^2 \frac{\partial \hat{\rho}}{\partial t} + (\gamma - 1)(\hat{\Gamma} - \hat{\Lambda}), \tag{184.3}$$

$$\hat{P} = \hat{\rho} a^2 + 2\rho a \hat{a}. \tag{184.4}$$

These equations have constant coefficients and therefore exponential solutions in space and time. We choose general solutions of the form $e^{\omega t + i k \cdot x}$ for wavenumber k and growth rate ω. Then the spatial derivatives can be replaced by ik and the time derivatives by ω. We also change the notation slightly, writing $\hat{\rho} e^{\omega t + i k \cdot x}$ for the more general $\hat{\rho}$ used above, so that $\hat{\rho}$ is now just the wave amplitude.

After this substitution and some reduction, the first two equations alone give

$$\omega^2 = -k^2 \frac{\hat{P}}{\hat{\rho}},$$ (185)

which is the dispersion relation. The ratio $\hat{P}/\hat{\rho}$ comes from the energy equation. Evidently negative $\hat{P}/\hat{\rho}$ gives instability and the perturbed quantities grow as

$$\exp\left[k\left(-\frac{\hat{P}}{\hat{\rho}}\right)^{1/2}t\right].$$ (186)

To determine $\hat{P}/\hat{\rho}$, we need to know Γ and Λ. Suppose that in equilibrium $\Gamma = \Lambda = \Lambda_0$ and that near equilibrium

$$\Gamma \propto \rho^r a^s \quad \text{and} \quad \Lambda \propto \rho^l a^m.$$ (187)

Then

$$\hat{\Gamma} - \hat{\Lambda} = \Lambda_0 \left[(r-l)\left(\frac{\hat{\rho}}{\rho}\right) + (s-m)\left(\frac{\hat{a}}{a}\right)\right].$$ (188)

Now substitute $\hat{a}/a = \frac{1}{2}(\hat{P}/P - \hat{\rho}/\rho)$ from above and solve for $\hat{P}/\hat{\rho}$ to get:

$$\frac{\hat{P}}{\hat{\rho}} \equiv \gamma_{\text{eff}} a^2 = \gamma a^2 \left[\frac{\omega - \omega_c(2l + s - m - 2r)}{\omega + \gamma \omega_c(m - s)}\right],$$ (189)

where

$$\omega_c = \frac{(\gamma - 1)\Lambda_0}{2\gamma P}$$ (190)

is a cooling rate.

Evidently, $\hat{P}/\hat{\rho} < 0$, or $\gamma_{\text{eff}} < 0$, for some values of positive ω when either $(2l + s - m - 2r) > 0$ or $(m - s) < 0$. The most important case is the first, which occurs when the pressure is nearly uniform because of fluid motions; this gives instability when

$$2l - m > 2r - s.$$ (191)

These conditions for instability are the same as those derived in a different way by Field [4], the first, for constant pressure perturbations ($\hat{P} = 0$), and the second for constant density perturbations ($\hat{\rho} = 0$). For example, Field [4] writes

$$\mathcal{L} = \frac{\Lambda}{\rho} - \frac{\Gamma}{\rho}$$ (192)

and gives an instability condition

$$\left(\frac{\partial \mathcal{L}}{\partial T}\right)_P < 0.$$ (193)

The result is the same as ours because

$$\mathcal{L} = \Lambda_0 \left(\rho^{l-1} T^{m/2} - \rho^{r-1} T^{s/2}\right)$$ (194)

from our definitions of Γ and Λ (normalized to the equilibrium ρ and T), so

$$\left(\frac{\partial \mathcal{L}}{\partial T}\right)_P = \frac{\partial}{\partial T}\Lambda_0 \left(K^{l-1} P^{l-1} T^{m/2+1-l} - K^{r-1} P^{r-1} T^{s/2+1-r}\right) \quad (195)$$

where we have defined K so that $\rho = KP/T$. This last equation can be simplified to

$$\left(\frac{\partial \mathcal{L}}{\partial T}\right)_P = \left[\left(\frac{m}{2}+1-l\right) - \left(\frac{s}{2}+1-r\right)\right]\frac{\Lambda}{T}, \quad (196)$$

which is less than 0 (instability) if

$$\frac{m}{2} - l - \frac{s}{2} + r < 0 \quad \text{or} \quad 2l - m > 2r - s. \quad (197)$$

This is the same as the result obtained directly from the equations of motion, continuity, and energy.

In the late 1960's and early 1970's, the thermal instability [4,5] was the basis for the two phase model of the ISM [6]. Today, thermal instabilities are discussed in the context of shocked gas [7] and shells [8], cooling flows in galaxies [9,10], and the general ISM (see review in Begelman [11]). They are probably active in the warm (10^3 K) HI phase and in the moderately hot (10^5 K) phase. Recent cloud formation and coalescence models based on this instability were developed by Parravano et al. [12 − 16], Lioure & Chieze [17], Kolesnik [18], Elphick et al. [19], and others. Magnetic effects were considered by Vandakurov [20].

Thermal-like instabilities may also operate in a *cloudy* fluid [21 − 23], where a in the above equations is the cloud dispersion and not the thermal sound speed. Recall that for cloud collisional cooling (Chap. 3), $\Lambda \propto \rho^2 a^3/\sigma_c$, so $l = 2$ and $m = 3$. Thus instability at fixed pressure requires $r < (1+s)/2$.

Recall also that for HII regions and windy bubbles (Chap. 4), the energy input into an expanding shell is

$$E_{\text{HII}} \sim 1.5 \times 10^{49} S_{49}^{0.5} n^{-3/7} \quad [\text{erg}] \quad (198)$$

so that

$$\Gamma_{\text{HII}} = E_{\text{HII}} R_{\text{OB}} \propto n^{-3/7}. \quad (199)$$

Similarly, for a supernova remnant, the energy input is

$$E_{\text{SNR}} = 4.5 \times 10^{49} E_{51}^{0.95} n^{-0.10} \zeta^{-0.15} (\beta a_{10})^{0.71} \quad [\text{erg}] \quad (200)$$

so that

$$\Gamma_{\text{SNR}} = E_{\text{SNR}} R_{\text{SNR}} \propto n^{-0.10} a^{0.71} \quad (201)$$

for OB star formation rates and supernova rates R. In both cases, $r \ll \frac{1}{2}(1+s)$ if the HII region or supernova rates are independent of gas density, and then gas stirred by HII regions and SNR should be thermally unstable in a macroscopic sense.

Heating in the general ISM should also be quite patchy, while cooling occurs wherever the clouds are and can be more uniform. This patchiness to the heating gives an even stronger thermal instability in the regions between the patches (e.g., between the OB associations) where cooling from cloud collisions dominates stirring from massive stars. Balbus [24] showed that the instability condition changes for a region out of equilibrium, when $\Gamma = EA$ at the initial time (see also [25]). For the present case of cloudy dissipation, instability then requires

$$r < \frac{1 + sE}{2E}. \tag{202}$$

When cooling initially dominates heating, $E < 1$, and instability seems inevitable.

6.2 Gravitational Instability

The gravitational instability is similar to the thermal instability in the sense that the result is a condensation of ambient gas into a cloud complex. With self-gravity, the equation of motion is

$$\rho \frac{D\boldsymbol{v}}{Dt} = -\nabla P + \boldsymbol{g}\rho \tag{203}$$

where the acceleration $\boldsymbol{g}$ satisfies

$$\nabla \cdot \boldsymbol{g} = -4\pi G\rho. \tag{204}$$

Using the same perturbation analysis as before, the dispersion relations for different geometries become

$$\omega^2 = -k^2 \frac{\hat{P}}{\hat{\rho}} + 4\pi G\rho, \tag{205.1}$$

$$\omega^2 = -k^2 \frac{\hat{P}}{\hat{\rho}} + 2\pi G\sigma k, \tag{205.2}$$

$$\omega^2 = -k^2 \frac{\hat{P}}{\hat{\rho}} + 4\pi G\rho\big(1 - kRK_1(kR)\big) \approx -k^2 \frac{\hat{P}}{\hat{\rho}} + 2Gk^2\mu \ln\left(\frac{2}{kR}\right), \tag{205.3}$$

in 3, 2, or 1 dimensions, respectively, where ρ, σ, and μ are the density, column density and mass per unit length, respectively, and the last expression is for $kR \ll 2$.

The gravitational accelerations used for these equations were found by direct integrations over the mass distributions. For a plane, we integrated over force lines for which $g = 2G\mu/r$ and $d\mu = \hat{\rho}dz$.

$$g_x(x_0) = \int_{-\infty}^{\infty} dx \int_{-H}^{H} dz \frac{2G\hat{\rho}\, e^{ikx}\,(x - x_0)}{(x - x_0)^2 + z^2}. \tag{206}$$

For the center line of a cylinder in the one dimension problem, we used

$$g_x(x_0) = \int_{-\infty}^{\infty} dx \int_{0}^{R} dr \frac{G\hat{\rho}\, e^{ikx} 2\pi r\,(x - x_0)}{\big((x - x_0)^2 + r^2\big)^{3/2}}. \tag{207}$$

251

Note that the thermal instability enters here too because it is always part of the equation of state. If the gas is thermally unstable ($\hat{P}/\hat{\rho} < 0$) then all gravitational instabilities also contain the thermal instability. This thermal instability part of the total instability has the effect of removing the minimum mass (the "Jeans mass") from the problem. Small length scales are unstable because cloud collisional cooling removes so much energy that the small amount of self-gravity on these scales can still be enough to drive a collapse. The minimum mass is set by erosion (or evaporation) and not pressure for a cloudy fluid, and the minimum length scale is the cloud collision mean free path. Thus gravitational-thermal instabilities can form low mass clouds even in regions with high equilibrium velocity dispersions, which is very different from the pure Jeans instability. Gravitational instabilities also tend to produce elongated or flattened structures during the collapse, because the eccentricity of a collapsing object increases with time [26].

An important new development in the theory of condensation instabilities is the inclusion of the expected variation of velocity dispersion a with perturbation size k. This inclusion is an attempt to simulate turbulence and the possible coupling between different length scales in a turbulent cascade [27]. If the power spectrum $P(k)$ of velocity fluctuations is sufficiently steep [28, 29], then there will be enough power on large scales (small k) to give a cloud stability, but not enough power on small scales to give the small clumps stability. Such a steep spectrum allows for the possibility of gravitational collapse into a star on a small scale while the whole cloud surrounding the clump is stable.

6.3 Instabilities in Galactic Disks

6.3.1 Ring Instabilities. There are several important instabilities on the scale of a galaxy. We first consider the radial collapse of gas to form rings in a differentially rotating disk (a galaxy) and then the azimuthal direction is discussed. This latter analysis includes the combined self-gravitational, thermal, and Parker instabilities [30].

First consider the problem of ring instabilities without magnetic fields. The equation of motion in a frame of reference rotating at the angular rate Ω_0 is

$$\rho \left(\frac{\partial w}{\partial t} + w \cdot \nabla w + 2\Omega_0 \times w - \Omega_0^2 r \right) = \rho g_T - \nabla P \qquad (208)$$

where the total gravitational acceleration is

$$g_T = g - \Omega(r)^2 r e_r \qquad (209)$$

for g in the gas, and the total velocity, including shear, is

$$w = v - 2A(r - r_0)e_\theta. \qquad (210)$$

for Oort constant $A = -\frac{1}{2}r \, d\Omega/dr > 0$, angular rate $\Omega > 0$, and radius r close to the fiducial radius r_0.

With purely radial perturbations $\hat{\rho}e^{\omega t + ikr}$, and so on, the components of the perturbed equation of motion become

$$\rho(\omega v_\theta - 2B v_r) = 0, \tag{211.1}$$

$$\rho(\omega v_r - 2\Omega v_\theta) = \rho\hat{g} - ik\hat{P}, \tag{211.2}$$

where $B = A - \Omega$. The gravitational acceleration from this perturbation in the radial direction is

$$\hat{g} = \int_{-\infty}^{\infty} dr \int_{-H}^{H} dz \, \frac{2G\hat{\rho}\, e^{ikr}\,(r - r_0)}{(r - r_0)^2 + z^2} = 2\pi i G\hat{\sigma} \tag{212}$$

for surface density perturbation $\hat{\sigma} e^{ikr}$. The continuity equation gives

$$\omega\hat{\rho} = -ikv_r\rho \quad \text{or} \quad \omega\hat{\sigma} = -ikv_r\sigma. \tag{213}$$

The energy equation has the same reduction as before,

$$\hat{P} = \gamma_{\text{eff}}\hat{\rho}a^2. \tag{214}$$

Thus the dispersion relation becomes

$$\omega^2 = -k^2 a^2 \gamma_{\text{eff}} + 2\pi G\sigma k - \kappa^2 \tag{215}$$

where $\kappa^2 = -4\Omega B$.

In the conventional problem, which considers only an isothermal or adiabatic gas ($\gamma_{\text{eff}} = \gamma = $ constant), this dispersion relation has a peak growth rate at

$$k = \frac{\pi G\sigma}{\gamma a^2} \tag{216}$$

and it is

$$\omega_{\text{peak}} = \frac{\pi G\sigma}{a\gamma^{1/2}}\left(1 - Q^2\right)^{1/2} \tag{217}$$

where

$$Q = \frac{a\kappa\gamma^{1/2}}{\pi G\sigma}. \tag{218}$$

Thus the instability condition is $Q < 1$. This Q parameter is sometimes called the Toomre Q parameter, because Toomre [31] first did the problem for a stellar disk (replace π by 3.36). Safronov [32] first did the gas problem shown here.

A magnetic field changes the radial collapse problem only a small amount. With a magnetic field, the equation of motion is

$$\rho\left(\frac{\partial w}{\partial t} + w \cdot \nabla w + 2\Omega \times w - \Omega^2 r\right) = \rho g_T - \nabla P - \frac{1}{8\pi}\nabla B^2 + \frac{1}{4\pi}B \cdot \nabla B, \tag{219}$$

where the magnetic field satisfies

$$\frac{\partial B}{\partial t} = \nabla \times (w \times B). \tag{220}$$

For radial perturbations and an azimuthal magnetic field, we have $B \cdot \nabla\hat{B} = 0$ and $\nabla(B + \hat{B})^2/8\pi = ikB\hat{B}/4\pi$, which gives the dispersion relation

$$\omega^2 = -k^2(c^2\gamma_{\text{eff}} + v_A^2) + 2\pi G\sigma k - \kappa^2 \tag{221}$$

for Alfvén speed $v_A = B/(4\pi\rho)^{1/2}$. This dispersion relation differs only slightly from the field-free case, and has the effect of increasing the velocity dispersion and the Q threshold, making the gas more stable in the radial direction.

The criterion for instability in this case is now

$$Q = \frac{(\gamma_{\text{eff}}a^2 + v_A^2)^{1/2}\kappa}{\pi G\sigma} < 1, \tag{222}$$

If γ_{eff} is small, as might be expected from the discussion in Sect. 6.1, then most of the stability against ring formation comes from the azimuthal magnetic field and not from the gas pressure.

6.3.2 The Parker Instability. The two components of the above equation of motion that are in the directions parallel to the field and perpendicular to the galactic plane give the Parker [1] instability. If we add to the pressure P a cosmic ray pressure P_{CR} that satisfies $B \cdot \nabla P_{\text{CR}} = 0$ because the cosmic rays are free to stream along the field, and if the gravitational acceleration perpendicular to the plane is constant, giving exponential distributions for the density, pressure, and field strength in the unperturbed state, then the growth rate ω of this instability is given by the solution to the equation

$$\omega^4 + \omega^2\left([v_A^2 + \gamma a^2]\left[k_p^2 + k^2 + \frac{1}{4H^2}\right]\right) + \gamma v_A^2 a^2 k^2\left(k_p^2 + k^2 - \frac{ra^2}{H^2 v_A^2\gamma}\right) = 0 \tag{223}$$

where $r = (1+\alpha+\beta)(1+\alpha+\beta-\gamma)-\alpha\gamma/2$ for $\alpha = 0.5v_A^2/c^2$ and $\beta = P_{\text{CR}}/P$, and where H is the scale height. The wavenumbers are k and k_p in the directions parallel to the field and perpendicular to the plane, respectively. The field lines are assumed to lie in the plane. This equation contains a stable oscillatory solution and an unstable condensation mode, which grows at about the rate a/H for $\alpha \approx \beta \approx 1$.

Modifications to the Parker instability resulting from the inclusion of other terms in the equations, such as rotation [33, 34], self-gravity [30, 35], variable gravity perpendicular to the plane [36], or higher orders [36 − 39], cannot stabilize the disk to this perpendicular motion. The result is inevitably a buckling of the field lines above the plane and a condensation of gas into the magnetic valleys. Whether this motion makes large clouds or small scale turbulence depends on how well coupled the different field lines are [40]. When gravity is included, they are well enough coupled to make large cloud complexes in the combined instability.

6.3.3 The Gravitational Instability with Magnetic Fields and Shear. An azimuthal magnetic field also helps drive condensation instabilities purely in the plane when there is rotation and shear. If we consider as above that the total velocity w can be divided into shear plus a perturbation v around shear,

$$w = v + 2A(r - r_0)e_y, \tag{224}$$

254

for $e_y = -e_\theta$, then we can write the equations in shearing coordinates with perturbations of the form $\hat{\rho}(t, z)\, e^{ik(y - 2Azt)}$ [41]. In this "shearing sheet" coordinate system the convention has been to take $\Omega < 0$, $A = +\frac{1}{2} r\, d\Omega/dr > 0$, and $B = A + \Omega < 0$. This procedure gives time-dependent coefficients in the perturbation equations and there are no purely exponential solutions, so the equations must be solved numerically in time. Recent solutions that combine the thermal, gravitational, and Parker instabilities are in [30].

An approximate method that avoids numerical integrations is to consider only the time when the shearing perturbation points in the radial direction, which is when $t = 0$. This is approximately the time of peak growth, and it corresponds to purely azimuthal motions, i.e., at $t = 0$ the perturbation is of the form e^{iky} for y in the azimuthal $(-e_\theta)$ direction. Then the equations reduce to a single equation for instantaneous growth rate ω, and this equation is the dispersion relation. If all of the linear terms in the above equations are considered, then this dispersion relation contains several instabilities that generally operate together [30, 42]. One of these instabilities is related to that of Balbus & Hawley [43], but for an azimuthal magnetic field (see also [44]). The other is related to the gravitational instability. To simplify this latter result, we consider the disk to be in vertical equilibrium and look only for the compressive motions in the plane. Then we get a result that corresponds to the gravitational-thermal instability discussed above, but now with rotation and azimuthal motions included [30]:

$$\omega^2 = -k^2 \gamma_{\text{eff}} a^2 + 2\pi G \sigma k - \frac{\omega^2 \kappa^2}{\omega^2 + k^2 v_A^2}. \tag{225}$$

This result should be compared with the dispersion relation for ring-like gravitational instabilities, derived above. There, the Alfvén speed was added to the velocity dispersion so the magnetic field resisted the unstable motion, as an additional pressure. Now the Alfvén speed does not add to the velocity dispersion because the unstable motion is parallel to the field, and the field exerts no pressure along its average direction. The magnetic field does affect the Coriolis force, however, so v_A appears with κ^2. This is because tension in the field resists the Coriolis force as a perturbation begins to grow and twist. The magnetic field completely removes the Q condition for instabilities in the azimuthal direction. All perturbations sufficiently large are unstable because at the threshold of instability, where $\omega = 0$, the κ^2 term vanishes. One expects shear viscosity to have a similar effect [45].

Figure 12 shows solutions to this equation for $r = 0$ and 2, $s = 0$, and for Q between 0.5 and 5. The effect of increasing Q is to lower the growth rate, but the peak rate only approaches 0, it does not become negative as Q goes to infinity (unlike the radial instability). The thermal instability is also evident in this figure, because when $r = 0$ there is no minimum wavelength for instability, i.e., no critical Jeans length. Thus the behavior of the solution is very different for the $r = 0$ and $r = 2$ cases.

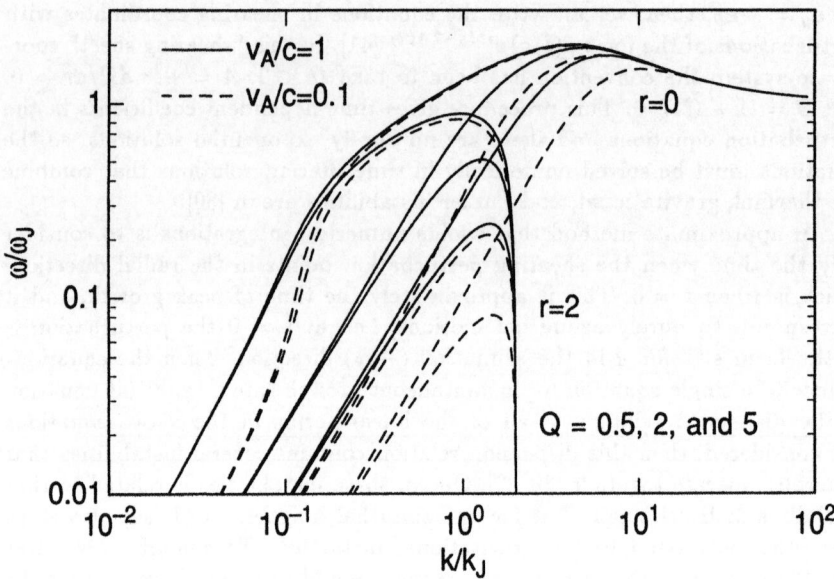

Fig. 12. Solutions to the dispersion relation $\omega(k)$ for azimuthal condensation instabilities parallel to the plane. These curves are for a heating function Γ with $r = 0$ or 2 and $s = 0$, an instability parameter Q between 0.5 and 5, and a relative Alfvén speed v_A/c equal to 0.1 or 1

6.4 Two Fluid Gravitational Instabilities

Jog & Solomon [46] considered a combined instability for gaseous and stellar fluids. The calculation involves separate equations of motion and continuity for the two fluids, and a combined Poisson equation for the perturbed potential. They obtained the dispersion relation for the growth rate ω as a function of the wavenumber k:

$$\omega^2 = -A + (A^2 - B)^{1/2}, \qquad (226)$$

where

$$A = \kappa^2 + k^2 \frac{a_s^2 + a_g^2}{2} - \pi G k \left(\sigma_s + \sigma_g \right) \qquad (227)$$

and

$$B = \left(\kappa^2 + k^2 a_s^2 - 2\pi G k \sigma_s \right) \left(\kappa^2 + k^2 c_g^2 - 2\pi G k \sigma_g \right) - \left(2\pi G k \sigma_s \right) \left(2\pi G k \sigma_g \right); \quad (228)$$

the subscripts s and g denote the stellar and gaseous fluids.

An expansion of $A^2 - B$ leads to a cancellation of all terms involving κ inside the radical (not considered in [46]). Then we get

$$\omega^2 = -\kappa^2 - \tfrac{1}{2} k^2 a_T^2 + \pi G k \sigma_T \left(1 + (1 + C)^{1/2} \right) \qquad (229)$$

where

$$C = \frac{1}{4}\left(\frac{ka_D^2}{\pi G\sigma_T}\right)^2 - \frac{\sigma_D}{\sigma_T}\left(\frac{ka_D^2}{\pi G\sigma_T}\right),$$

$$a_T^2 = a_s^2 + a_g^2 \quad , \quad a_D^2 = a_s^2 - a_g^2,$$

$$\sigma_T = \sigma_s + \sigma_g \quad , \quad \sigma_D = \sigma_s - \sigma_g. \tag{230}$$

The maximum growth rate can be determined by setting $d\omega^2/dk = 0$ and solving for k, and then using this k in $\omega(k)$. This equation for k is third order, but the real solution giving the maximum ω^2 can be found analytically with the usual expansions for a cubic equation. If we write this k in dimensionless form, $K = kc_T^2/(2\pi G\sigma_T)$, and also normalize ω to an average value $\omega_A = \pi G\sigma_T/(a_T^2/2)^{1/2}$, then the dispersion relation becomes

$$\left(\frac{\omega}{\omega_A}\right)^2 = -Q_A^2 - K^2 + K\left(1 + (1+C)^{1/2}\right), \tag{231}$$

where an average Q_A has been defined as $Q_A = \kappa(a_T^2/2)^{1/2}/(\pi G\sigma_T)$. Note that this dispersion relation is similar to the one-fluid relation, and that we can write an effective Q value using the epicyclic frequency. This gives

$$\left(\frac{\omega}{\omega_A}\right)^2 = \left\{K\left(1 + (1+C)^{1/2}\right) - K^2\right\}\left\{1 - Q_{eff}^2\right\} \tag{232}$$

where

$$Q_{eff}^2 = \frac{Q_A^2}{K\left(1 + (1+C)^{1/2}\right) - K^2}. \tag{233}$$

When the value of K at the maximum of ω^2 is used in this equation, then Q_{eff} has the property that it is less than 1 where there are unstable solutions, $\omega^2 > 0$, and greater than 1 when all solutions are stable, $\omega^2 < 0$, as in the one-fluid problem. Also, when $a_s = a_g = a$ and the two fluids act like a single fluid, then $a_D = 0$ and $C = 0$, giving $K = 1$ and $Q_{eff} = \kappa a/(\pi G\sigma_T)$, as expected. When $\sigma_g = 0$, so there is a pure stellar fluid, we get $K = a_T^2/(2a_s^2)$ and the usual $k = \pi G\sigma_s/a_s^2$, and we get $(1+C)^{1/2} = \pm(1 - \frac{1}{2}a_D^2/a_s^2)$; then there are two solutions, the usual gravitational instability for pure stars, which has $\omega = (\pi G\sigma_s/a_s)(1 - Q_s^2)^{1/2}$ for $Q_s = \kappa a_s/\pi G\sigma_s$, and an epicycle-sound wave solution, $\omega^2 = -\kappa^2 - k^2 a_g^2$, for the negligible mass of gas.

6.5 Instabilities in Galactic Spiral Shocks

Spiral density wave shocks compress the gas, and this leads to rapid cloud collisions, energy dissipation, and the possibility of gravitational collapse to giant cloud complexes. The spiral arm environment also has little shear and small galactic tidal forces, so these clouds are relatively free of disruption until they either flow into the interarm region, where the tidal forces are larger (Chap. 5), or they produce a large number of massive stars. Dense molecular clouds in which the new stars form are presumably made by continued dissipation and collapse in the cores of the larger, more loosely bound objects which form first.

The self-gravitational instability of gas in spiral arms has been discussed for two cases. One is for a collapse parallel to the arm, which produces separate cloud complexes inside the long and thin ridge of compressed gas [47], and the other is for collapse perpendicular to the arm, which produces a bulk compression of the whole ridge [48]. This second case is analogous to the radial or ring instability in a disk without a spiral, and has an analogous Q condition which, in effect, replaces the epicyclic frequency by $\kappa(\sigma/\sigma_0)^{1/2}$ and the mass column density by the spiral arm value, σ. The net change to Q is therefore a decrease from the average ring value by the ratio $(\sigma_0/\sigma)^{1/2}$ [48], which could possibly lead to enhanced cloud and star formation by a variety of processes related to low Q. If the velocity dispersion in the arm decreases too, then Q could be even lower, but if the magnetic field dominates the pressure in the arm, and the Alfvén speed scales with $\sigma^{1/2}$, then the product of v_A and κ in the numerator of Q will scale with σ in the same way as the denominator and Q will not change much from arm to interarm.

Here we consider in more detail the first of these two cases, in which a spiral shock front builds up a dense ridge of gas in time and then collapses into separate clouds. We want to determine if the total time for the build-up and collapse is less than the flow-through time in the arm. If it is, then the instability should be important.

The gas mass per unit length μ is assumed to increase with time as

$$\mu = \rho_0 v_s 2Ht \qquad (234)$$

for pre-shock density ρ_0, shock speed v_s, and scale height H in the galaxy. We assume that this ridge has a cylindrical shape with a radius R given by the equation

$$\mu = \pi R^2 \rho, \qquad (235)$$

and we assume that the density ρ in the ridge satisfies the magnetic shock jump condition (Sect. 4.2.3)

$$\frac{\rho}{\rho_0} = 2^{1/2}\frac{v_s}{v_A} \qquad (236)$$

for Alfvén speed in the pre-shock flow v_A. Then we have a self-gravitating ridge of gas, e.g., a dust lane, that gets more unstable with time as μ increases and eventually collapses when the unstable growth rate equals $1/t$.

We first write the equations of motion and continuity for perturbations μ_1 and v in the dustlane:

$$\mu\frac{\partial v}{\partial t} = -a^2\nabla\mu_1 + \mu g_1 - \mu\frac{v}{t}, \qquad (237)$$

$$\frac{\partial \mu_1}{\partial t} = -\mu\nabla \cdot v. \qquad (238)$$

These equations assume that the thermal and ram pressure boundary conditions behind and at the shock front, respectively, cancel [49], which is sometimes appropriate when there is no net acceleration or deceleration of the gas; thus the ridge acts as if it has a zero-pressure boundary condition. There is also an accretion term, $-\mu v/t$, which comes from the rate of change of the total transverse momentum per unit length in the perturbation, $\partial \mu v/\partial t = \mu \partial v/\partial t + \mu v/t$. We then assume perturbations of the form $\mu_1 = \hat{\mu} \cos(kx)$ and $v = \hat{v} \sin(kx)$ for distance x along the dustlane. We also substitute ω for the time derivative and ignore variations in μ during the growth of the perturbation, which is a very crude approximation (a numerical solution to the time-dependent quantities would be better). This gives a growth rate

$$\omega = -\frac{1}{2t} + \left(\frac{1}{4t^2} + 2G\mu k^2 \ln\left(\frac{2}{kR}\right) - a^2 k^2 \right)^{1/2}, \tag{239}$$

which has the peak value

$$\omega_{\text{peak}} = -\frac{1}{2t} + \left(\frac{1}{4t^2} + G\mu k_{\text{peak}}^2 \right)^{1/2} \tag{240}$$

at a wavenumber that satisfies

$$\frac{k_{\text{peak}} R}{2} = \exp\left[-0.5\left(1 + \frac{a^2}{G\mu} \right) \right]. \tag{241}$$

Now we set the maximum growth rate equal to $1/t$ and derive an equation that determines when the collapse will occur:

$$\left(\frac{G\mu}{a^2} \right)^2 \exp\left(-\frac{a^2}{G\mu} \right) = \frac{e}{\pi^2 2^{1/2}} (1 + \alpha + \beta) \frac{\rho_0}{\rho_{0T}} \frac{v_A v_s a_0^2}{a^4}. \tag{242}$$

Here we have assumed a scale height

$$H^2 = \frac{a_0^2(1 + \alpha + \beta)}{2\pi G\rho_{0T}} \quad \text{for} \quad \alpha = \frac{B^2}{8\pi P} \quad \text{and} \quad \beta = \frac{P_{\text{CR}}}{P}, \tag{243}$$

where a_0 is the velocity dispersion in the preshock gas. For typical parameters, $(1 + \alpha + \beta)\rho_0/\rho_{0T} \approx 1$, $v_A/a \approx 1$, $v_s/a \approx 3$, and $a/a_0 \approx 1$, this condition for instability implies that

$$\frac{G\mu}{a^2} \approx 1.17 \tag{244}$$

at the time of collapse, which corresponds to

$$t \approx \frac{G\mu}{a^2} \frac{a^2}{v_s a_0} \left(\frac{\pi \rho_{0T}}{2G\rho_0^2(1 + \alpha + \beta)} \right)^{1/2} \approx \frac{0.5}{(G\rho_0)^{1/2}} \approx 40 \, \text{Myr},$$

$$R \approx \left(\frac{G\mu}{a^2} \right)^{1/2} \left(\frac{1}{\pi 2^{1/2}} \frac{v_A}{v_s} \right)^{1/2} \frac{a}{(G\rho_0)^{1/2}} \approx \frac{0.30 \, a}{(G\rho_0)^{1/2}} \approx 170 \, \text{pc},$$

$$\lambda = \frac{2\pi}{k} = \pi R \exp\left[\tfrac{1}{2}\left(1 + \frac{a^2}{G\mu} \right) \right] \approx 1.4 \, \text{kpc}, \tag{245}$$

$$M = \mu\lambda \approx 1.9 \times 10^7 \, M_\odot, \quad \text{and}$$

$$A_{\text{dustlane}} = 1.5 \, \text{mag},$$

for $\rho_0 = 2.2 \times 10^{-24} \, \text{g cm}^{-3}$ and $a = 7 \, \text{km s}^{-1}$. Here A is the visual extinction in the ridge. Use of $\gamma_{\text{eff}} = 0.3$ would lower the effective a by $\sim 0.3^{1/2} \sim 0.5$ and therefore lower the length and mass scales for the smallest clouds that form; so would a value of $a/a_0 < 1$ [50]. In either case, the type of object that collapses directly from a spiral shock is not an individual molecular cloud but a much larger cloud with a lower density and higher mass. The characteristic masses of these larger clouds, and their predicted separation along spiral arms, are comparable to the observed masses and separations of the large clouds associated with star forming regions in spiral galaxies. The derived time scale for their formation is also small enough that they can appear in the arms before the gas flows out (cf. Chap. 5).

6.6 Collapse of Ellipsoids

Another way to study instability is to use the virial equation, which is the same as

$$\int_V r \cdot (\text{EoM}) \, dV \tag{246}$$

where EoM stands for equation of motion. This is useful when volume-averaged quantities are available.

An important problem is the collapse of a uniform pressure-bounded isothermal sphere, which has the equilibrium virial equation

$$4\pi R^3 P = \frac{3MkT}{\mu} - \frac{3GM^2}{5R}. \tag{247}$$

This equation may also be rewritten

$$P = \rho a^2 - \rho \frac{GM}{5R}. \tag{248}$$

For small P, ρ is small and R is large so $P \approx \rho a^2$. As P increases, R decreases and the gravity term becomes more important. For sufficiently large P, the gravity term begins to dominate (because of the $1/R$), and then further shrinkage leads to lower P. When this happens, we get P decreasing for increasing ρ, or $\hat{P}/\hat{\rho} < 0$ which is unstable, as discussed in Sect. 6.1. Thus there is a limiting stable mass, or a limiting boundary pressure for stability given by

$$P_{\text{max}} = \frac{1.4 \, a^8}{G^3 M^2} \tag{249}$$

where the 1.4 (instead of 3.15) comes from a numerical solution to the equation of hydrostatic equilibrium [51, 52].

Recently there has been an interest in the dynamic collapse problem (rather than static as in this derivation), in which converging flows driven by implosion (cloud collisions, ionized bright rims, etc.) trigger collapse [53, 54]. This process is apparently important when star formation is triggered by high pressures that completely surround a small cloud.

260

In the magnetic problem, there is another collapse condition in addition to a critical boundary pressure. The magnetic virial equation for a uniform cloud with a magnetic field is

$$4\pi R^3 \left(P + \alpha \frac{B^2}{8\pi} \right) = \frac{3MkT}{\mu} - \frac{3GM^2}{5R} \tag{250}$$

where α depends on the field distribution and boundary condition, and equals $2/3$ for the case where $B \propto R^{-3}$ outside the cloud [55]. A more general discussion of the virial theorem for magnetic clouds is in Zweibel [56].

Now we assume flux freezing:

$$B \propto \frac{1}{R^2} \tag{251}$$

and find that the magnetic and gravity terms both scale as $1/R$. This means that if the magnetic energy is sufficient to prevent collapse at some initial time, then it will always prevent collapse regardless of P, so there is no critical pressure.

Instability therefore requires two conditions [57]: (1) that the field be sufficiently weak:

$$\frac{B_0^3}{G^{3/2} \rho_0^2 M} < 280 \tag{252}$$

for initial field B_0 and initial density ρ_0 before contraction along field lines, and (2) that the external pressure be sufficiently large:

$$P > \frac{1.89\, a^8}{G^3 M^2 \left[1 - \left(\dfrac{M_c}{M} \right)^{2/3} \right]^3} \tag{253}$$

where M_c is the critical mass that satisfies the first condition,

$$M_c = \frac{B_0^3}{280\, G^{3/2} \rho_0^2}. \tag{254}$$

If $M \gg M_c$, then the magnetic field is relatively unimportant and the collapse condition becomes essentially $P > P_{\max}$, as in the non-magnetic problem. A discussion of these collapse conditions and their relation to magnetic diffusion is in [58].

The stability criterion for a rotating, magnetic cloud was given by Tomisaka et al. [59]. If we invert the second condition above (on P), giving $M_{\mathrm{mag}}(P, a, M_c)$, then the critical mass with rotation is found by numerical solution of the equilibrium states to be

$$M_{\mathrm{cr}}^2 = M_{\mathrm{mag}}^2 + \left(\frac{4.8\, aj}{G} \right)^2 \tag{255}$$

for specific angular momentum j. This implies that cloud collapse can be initiated by a reduction in magnetic flux (diffusion), a reduction in angular momentum, or an increase in boundary pressure. In the first case, M_c decreases, so M_{mag} decreases and M_{cr} decreases, at which point M can be greater than M_{cr} and collapse will follow. In the second case, M_{cr} decreases directly. In the third case, M_{mag} decreases and so M_{cr} decreases.

Acknowledgement: I am grateful to Drs. George Field, James H. Hunter, Jr., and John Scalo for helpful comments.

References

6.1 Parker, E.N. 1966, *Astrophys. J.* **145**, 811
6.2 Hunter, J.H., Jr., Whitaker, R.W. 1989, *Astrophys. J. Suppl. Ser.* **71**, 777
6.3 Elmegreen, B.G. 1990, in *The Evolution of the Interstellar Medium*, ed. L. Blitz, San Francisco, Astronomical Soc. of the Pacific, p. 247
6.4 Field, G.B. 1965, *Astrophys. J.* **142**, 531
6.5 Hunter, J.H., Jr. 1966, *Monthly Notices Roy. Astron. Soc.* **133**, 239
6.6 Field, G.B., Goldsmith, D.W., Habing, H.J. 1969, *Astrophys. J.* **155**, L149
6.7 Raymond, J.C., Wallerstein, G., Balick, B. 1991, *Astrophys. J.* **383**, 226
6.8 Heiles, C. 1989, *Astrophys. J.* **336**, 808
6.9 Nulsen, P.E.J. 1988, in *Cooling Flows in Galaxies*, ed. A.C. Fabian, (Dordrecht: Kluwer), 175
6.10 David, L.P., Bregman, J.N. 1988, in *Cooling Flows in Galaxies*, ed. A.C. Fabian, (Dordrecht: Kluwer), 199
6.11 Begelman, M.C. 1990, in *The Interstellar Medium in Galaxies*, eds. H.A. Thronson, Jr., J.M. Shull (Dordrecht: Kluwer), p. 287
6.12 Parravano, A. 1987, *Astron. Astrophys.* **172**, 280
6.13 Parravano, A. 1988, *Astron. Astrophys.* **205**, 71
6.14 Parravano, A. 1989, *Astrophys. J.* **347**, 812
6.15 Parravano, A., Rosenzweig, P., Teran, M. 1990, *Astrophys. J.* **356**, 100
6.16 Parravano, A., Mantilla, Ch.J. 1991, *Astron. Astrophys.* **250**, 70
6.17 Lioure, A., Chieze, J.P. 1990, *Astron. Astrophys.* **235**, 379
6.18 Kolesnik, I.G. 1991, *Astron. Astrophys.* **243**, 239
6.19 Elphick, C., Regev, O., Spiegel, E.A. 1991, *Monthly Notices Roy. Astron. Soc.* **250**, 617
6.20 Vandakurov, Yu. N. 1991 *Sov. Astron. Letters* **17**, 38
6.21 Struck-Marcell, C., Scalo, J.M. 1984, *Astrophys. J.* **277**, 132
6.22 Tomisaka, K. 1987, *Publ. Astron. Soc. Japan* **39**, 109
6.23 Elmegreen, B.G. 1989, *Astrophys. J.* **344**, 306
6.24 Balbus, S.A. 1986, *Astrophys. J.* **303**, L79
6.25 Hunter, J.H., Jr. 1970, *Astrophys. J.* **161**, 451
6.26 Smith, M.D. 1989, *Monthly Notices Roy. Astron. Soc.* **238**, 835
6.27 Leorat, J., Passot, T., Pouquet, A. 1990, *Monthly Notices Roy. Astron. Soc.* **243**, 293
6.28 Bonazzola, S., Falgarone, E., Hayvaerts, J., Perault, M., Puget, J.L. 1987, *Astron. Astrophys.* **172**, 293
6.29 Pudritz, R.E. 1990, *Astrophys. J.* **350**, 195
6.30 Elmegreen, B.G. 1991, *Astrophys. J.* **378**, 139
6.31 Toomre, A. 1964, *Astrophys. J.* **139**, 1217
6.32 Safronov, V.S. 1960, *Annales d'Astrophysique* **23**, 979
6.33 Shu, F.H. 1974, *Astron. Astrophys.* **33**, 55
6.34 Zweibel, E.G., Kulsrud, R.M. 1975, *Astrophys. J.* **201**, 63
6.35 Elmegreen, B.G. 1982, *Astrophys. J.* **253**, 634
6.36 Matsumoto, R., Horiuchi, T., Hanawa, T., Shibata, K. 1990, *Astrophys. J.* **356**, 259
6.37 Shibata, K., Tajima, T., Matsumoto, R., Horiuchi, T., Hanawa, T., Rosner, R., Uchida, Y. 1989, *Astrophys. J.* **338**, 471
6.38 Shibata, K., Tajima, T., Matsumoto, R. 1990, *Phys. Fluids B, Plasma Physics* **2**, 1989

6.39 Shibata, K., Matsumoto, R. 1991, *Nature* **353**, 633
6.40 Asséo, E., Cesarsky, C.J., Lachièze-Ray, M., Pellat, R. 1978, *Astrophys. J.* **225**, L21
6.41 Goldreich, P., Lynden-Bell, D. 1965, *Monthly Notices Roy. Astron. Soc.* **130**, 97
6.42 Tagger, M., Foglizzo, T, Elmegreen, B.G. 1992, in preparation
6.43 Balbus, S.A., Hawley, J.F. 1991, *Astrophys. J.* **376**, 214
6.44 Vishniac, E.T., Diamond, P. 1992, preprint
6.45 Hunter, J.H., Jr., Horak, T. 1983, *Astrophys. J.* **265**, 402
6.46 Jog, C.J., Solomon, P.M. 1984, *Astrophys. J.* **276**, 127
6.47 Elmegreen, B.G. 1979, *Astrophys. J.* **231**, 372
6.48 Balbus, S.A., Cowie, L.L. 1985, *Astrophys. J.* **297**, 61
6.49 Vishniac, E.T. 1983, *Astrophys. J.* **274**, 152
6.50 Rydbeck, G., Bergman, P., Hjalmarson, A., Thomasson, M., Wiklind, T. 1991, in *Dynamics of Disc Galaxies*, ed. B. Sundelius, Göteborg, Chalmers University, p. 235
6.51 Bonner, W.B. 1956, *Monthly Notices Roy. Astron. Soc.* **116**, 351
6.52 Ebert, R. 1955, *Z. Astrophys.* **37**, 222
6.53 Hunter, J.H., Jr. 1979, *Astrophys. J.* **233**, 946
6.54 Klein, R.I., Whitaker, R.W., Sandford, II., M.T. 1985, in *Protostars and Planets II*, eds. D.C. Black, M.S. Matthews, University of Arizona Press, 340
6.55 Spitzer, L. Jr. 1978, *Physics of Interstellar Matter*, Wiley, New York
6.56 Zweibel, E.G. 1990, *Astrophys. J.* **348**, 186
6.57 Mouschovias, T.C. and Spitzer, L. Jr. 1976, *Astrophys. J.* **210**, 326
6.58 Elmegreen, B.G. 1991, in *Evolution of Interstellar Matter and Dynamics of Galaxies*, eds. B.W. Burton, P.O. Lindblad, J. Palouš, (Cambridge: Cambridge Univ. Press), in press
6.59 Tomisaka, K., Ikeuchi, S., Nakamura, T. 1989, *Astrophys. J.* **341**, 220

7 Global Star Formation

The most difficult problems in interstellar gas dynamics span a wide range of scales and require a large number of concepts for their solutions. Global problems such as how the gas got its structure, how stars form, and how galactic disks evolve, are qualitatively different from more localized problems, such as the evolution of a supernova remnant or shock front. The large scale problems are likely to involve several feedback mechanisms that control the state of the gas on long time scales, and they could also include a mixture of forces that lead to a large number of coupled, non-linear equations.

In this Chapter, we review the following large scale problems: (1) The characteristic dependence of the star formation rate on the ~ 1.3 to 2 power of the average gas density $[1-3]$; (2) the column density threshold for star formation, $\sigma_{\text{thresh}} \sim 2$ to $8\,M_{\odot}\,\text{pc}^{-2}$ $[1,4,5]$, which may actually be a Q threshold, $Q < 1$, and the correlation between the maximum size of an HII region and $\sigma/\sigma_{\text{thresh}}$ $[7]$; (3) the exponential distribution of starlight intensity and the distribution of metallicity in a galactic disk; (4) the dependence of the rotation curve on the total galaxy luminosity, which is the integrated history of the star formation rate $[7]$; (5) the modulation of the star formation process by spiral density waves, and the near uniformity of the mean star formation efficiency in galactic disks, regardless of the presence or lack of spiral density waves and Hubble type as long as $\sigma > \sigma_{\text{thresh}}$, and (6) the occurrence of bursts in the global star formation rate, especially following galaxy interactions.

An important question is whether these large-scale properties of the interstellar medium can be explained in terms of the smaller-scale properties of star formation and gas dynamics, such as the observation of star formation and giant molecular clouds inside superclouds and the observation of shells and chimneys in the gas, or whether new physical processes are involved on the larger scales, including feedback mechanisms, which largely wash out the details of the small scale processes.

7.1 The Star Formation Rate and Surface Density Threshold

The observed dependence of the average star formation rate on the density or surface density of gas has been the subject of much theoretical speculation. There is a rough power law dependence such that the star formation rate per unit gas mass, which is the efficiency of star formation, increases with density. When only HI observations were used in the past, this efficiency was thought to increase in direct proportion to the density, but now with the inclusion of molecular gas, the power is much less, apparently around 0.3 $[1-3]$. Of course, this density dependence is an empirical relation, not a physical law, and it will not necessarily apply to regions where it has not been measured, such as star burst galaxies. Thus it is not really a predictive relation, although for similar galaxies one expects it to be similarly satisfied. In this respect it is somewhat like Bode's law of planetary positions. One hopes that it will eventually be explained by physical laws.

The observed increase in the star formation efficiency with density is not surprising. The star formation rate may be written as a product of some gas processing rate ω and the density ρ, so this result implies that ω increases slightly with ρ. Most processes that turn ambient gas into dense cloud complexes and stars have this property. Some exceptions to this rule are interesting, though. The Parker instability, for example, operates on a time scale that is independent of density and equal to the free fall time in the galactic disk. This would seem to rule out this instability as a mechanism for initiating star formation, but in fact the ratio of the gas density to the star density in the galactic midplane is nearly independent of radius in our Galaxy, and then the free fall rate, which depends primarily on the total gas+star density, ends up scaling with the gas density after all.

An essential point of this density scaling is that many of the processes that are involved with star formation are likely to be self-regulating. Then the dimensionless parameters that enter into the problem are about the same everywhere, such as the ratios of the various energy densities and time scales in the midplane of a galaxy. An example of such self-regulation is when star formation controls the velocity dispersion of the gas via supernova explosions and other pressurization processes, and yet the star formation rate is controlled by this velocity dispersion through the constraint that $Q = a\kappa/(\pi G\sigma) < 1$ for active star formation. This mechanism implies that when the dispersion is high, Q is high and the cloud and star formation rates decrease; then a decreases until Q is close to 1 again, at which point the star formation rate increases. This is a mechanism of feedback control that would tend to keep the gravitational collapse time, $(G\rho)^{1/2}$, close to both the energy dissipation time (which depends on the dispersion) and the shear time, which depends on the rotation curve (cf. Sect. 5.2.4).

Another feedback mechanism involving Q has to do with spiral arms, cloud heating, and mass accretion to the nucleus. When Q is low, spiral arms form more easily and with larger amplitudes than when Q is high. Because these arms have a time-changing potential as they grow or change shape, they can heat the cloud population [8] and increase Q, which limits the level of spiral arm generation. The arms also provide torques on the gas, and this drives mass accretion from the disk into the central regions. The accretion modifies Q by changing σ, the mass column density. Thus, there is a double feedback mechanism, one with Q and a from spiral arm heating and another with Q and σ from accretion. The result could again pin Q to a value near 1, the threshold of strong spiral instabilities, but not necessarily. If the spiral arms are too strong, then the rapid accretion from strong spiral arm torques could increase σ and decrease Q too much and make the arms grow ever stronger. This suggests that spiral torques could promote global accretion instabilities, as found by Saio & Yoshii [9].

Many other feedback mechanisms can be imagined. For example, there could be a magnetic dynamo in the galaxy that is driven by random torsional motions involved with interstellar turbulence (see refs. [2.154 − 2.162]). Such a dynamo could pin $B^2/(8\pi\rho a^2)$ close to 1 as magnetic stress limits the turbulence that generates the field.

Obviously, if feedback processes like these commonly operate, then many of the small-scale mechanisms that are involved with star formation lose their specific character and only one or two basic time scales will be left for the processing rate ω. This common time scale is probably a combination of the time scales that were discussed in Sect. 5.2.4, namely the gravitational time, ω_{grav}, the energy dissipation time, ω_{diss}, the shear time, ω_{shear}, and the arm flow-through time ω_{flow}. When there is feedback that keeps these time scales comparable, then all of the ω should together scale with a low power of density. A detailed calculation of $\omega\rho$ that combines gravity, shear, dissipation, and two magnetic instabilities in fact comes very close to the observed product of $\rho^{1.3}$ to $\rho^{1.5}$ for the star formation law [10].

The threshold column density for star formation [1, 4, 5] indicates that at a low enough σ the various feedback mechanisms that regulate the star formation rate begin to break down. Then the HII regions and star formation sites get much smaller [7, 12]. There are many possible reasons for this. The usual explanation (see also [12]) is that self-gravity becomes relatively unimportant compared to shear when the value of the stability parameter Q is greater than 1, which corresponds to $\sigma < \sigma_{\text{thresh}}$. This is a sensible result, but it does not explain why the velocity dispersion in the gas does not decrease to keep Q low when the star formation rate drops at the edge of a galaxy. That is, it does not explain how the value of Q can be pinned close to 1 in most of the galaxy and then increase to the stable regime above 1 in the outer parts. One way to get Q to increase at very low σ is to have a minimum velocity dispersion for the gas that is independent of the star formation rate. Possible origins for this minimum could be a pervasive heating of HI or HII from extragalactic photons, or cloud stirring by magnetic or shear instabilities (Chap. 6). Then a decrease in σ in the outer regions would result in a decrease in the star formation rate, but this decrease would not lower a much below the minimum value.

Another explanation for the threshold column density is that the gas disk begins to flare out at this point because the underlying stellar disk has a relatively low mass. Then the volume density drops precipitously. Such a drop would lower the gravity and dissipation rates but not change the shear rate. This offsets the balance of feedback regulators, making it difficult for clouds to form. Alternatively, the edge of the star formation disk could be a true edge to the galactic matter, resulting from an even larger edge before the gas collapsed into the galaxy [13].

7.2 Exponential Disks in Galaxies

Galactic disks tend to have a light distribution that is approximately exponential, $I(r) \propto e^{-r/r_0}$ for scale length r_0 equal to about $1/4$ of the Holmberg radius [14]. The origin of this distribution is not well understood, although there are several models.

In a model suggested by van der Kruit [13], the exponential distribution originated at the time of formation of the disk, when a nearly uniform-density and uniformly-rotating gas sphere collapsed in a fixed potential with a flat rotation curve until it reached an equilibrium in centrifugal force balance. This model also gives the outer disk cutoff at the right number of exponential scale lengths. A question then arises as to the origin of the HI disk beyond the optical edge. Van der Kruit suggested that this HI could have been accreted from another galaxy because outer HI disks are often unaligned relative to the main stellar disks. If this is the case, then one has to explain why the outer HI disk surface density follows the surface density of unseen matter as determined from the rotation curve [15, 16]. This latter observation suggests that the outer HI has always belonged to the galaxy.

Another model [17] is that the exponential disk results as an approximation to a true distribution in which the total mass surface density in the disk has the same radial distribution as the mass that determines the rotation curve $v(r)$, but that star formation, which determines the light distribution, depends on the difference between this column density distribution and a threshold column density. The total gas column density distribution then satisfies $\sigma(r) \approx v(r)^2/(Gr)$ for gravitational constant G. If only the part of the total disk mass that lies above a threshold column density σ_{thresh} is allowed to form stars, then a star formation and starlight distribution of the form $I(r) \propto v(r)^2/(Gr) - \sigma_{\text{thresh}}$ results. This distribution is very close to an exponential, especially for a flat or slowly rising rotation curve, and the scale length for this exponential has the correct proportion to the outer disk cutoff size.

The exponential disk could also be forced on a galaxy by a rearrangement of orbits in the presence of an inner bar potential [18, 19]. This is a dynamical model for disk structure that is not related to galaxy formation or star formation, but it relies on a bar or a former bar that has since been destroyed [20].

Another model is that the exponential disk arises as a result of a long history of viscous or spiral-torque driven mass accretion [9, 21 – 27]. The radial distribution of heavy elements has been shown to follow from this model too. The primary assumption for this model is that the star formation time scale is proportional to the viscous time everywhere in the disk. This is reasonable if there are feedback mechanisms involving star formation and cloud-collisional dissipation, or star formation and spiral arm formation, because both cloud collisions and spiral arms drive mass inflow. This model of viscous disk evolution introduces several concepts that have not yet been discussed in this lecture series, so the basic equations are reviewed here. We follow the analysis of Saio

& Yoshii [9], who consider the simultaneous evolution of the stellar and gaseous disks as the rotation curve changes with accretion.

In a disk undergoing a conversion of gas into stars, the equations of continuity for the gas and star populations are

$$\frac{\partial \sigma}{\partial t} + \frac{1}{r}\frac{\partial (rv\sigma)}{\partial r} = -\frac{\sigma}{t_*}, \tag{256.1}$$

$$\frac{\partial \sigma_*}{\partial t} + \frac{1}{r}\frac{\partial (rv_*\sigma_*)}{\partial r} = \frac{\sigma}{t_*}, \tag{256.2}$$

where t_* is the star formation time scale, v is the radial drift speed, and σ and σ_* are the gas and stellar surface densities.

The equations of angular momentum conservation (see derivation below) are

$$\sigma \frac{\partial \Omega}{\partial t} + \frac{\sigma v}{r^2}\frac{\partial (r^2\Omega)}{\partial r} = \frac{1}{r^3}\frac{\partial}{\partial r}\left(r^3\sigma v \frac{\partial \Omega}{\partial r}\right), \tag{257.1}$$

$$\frac{\partial \Omega_*}{\partial t} + \frac{v_*}{r^2}\frac{\partial (r^2\Omega_*)}{\partial r} = 0, \tag{257.2}$$

where the angular rotation rates are given by centrifugal force balance:

$$r\Omega^2 = \frac{L}{\sigma}\frac{\partial P}{\partial r} + \frac{\partial \Psi}{\partial r}, \tag{258.1}$$

$$r\Omega_*^2 = \frac{\partial \Psi}{\partial r}, \tag{258.2}$$

for disk thickness L. The gravitational potential is from Poisson's equation:

$$\nabla^2\Psi = 4\pi G\big((\sigma + \sigma_*)\delta(z) + \rho_{\text{halo}}\big), \tag{259}$$

where the use of δ assumes that all of the matter is in a thin disk except for the halo. Also, the turbulent pressure, viscosity, and star formation time scale are given by

$$LP = \nu\sigma\Omega \quad , \quad \nu = \nu_0\sigma^2\Omega^{-3} \quad , \quad t_* = \frac{\beta r^2}{\nu}. \tag{260}$$

The viscosity ν is assumed to be equal to the square of the maximum unstable length in a galaxy disk, k_c^{-2}, from Sect. 5.1, multiplied by the rotation rate Ω, and the pressure is the density σ/L times the squared velocity $(\Omega/k_c)^2$.

The angular momentum equation for the gas contains the viscosity ν. This equation can be derived as follows. The viscous stress inside a gas with velocity distribution $V(x)$ is $\rho\nu\, dV/dx$. In a rotating system, the angular velocity gradient should be used instead of the linear gradient, so the viscous stress is $\rho\nu r\, d\Omega/dr$. When this is multiplied by the area $2\pi rL$ of a cylinder in the disk with radius r and thickness L, and by the lever arm r, we get the viscous torque

$$T = 2\pi r^3 \nu\sigma \frac{\partial \Omega}{\partial r} \tag{261}$$

that is exerted by the inner part of the disk on the outer part at the radius r. The difference between the torques on two faces of a thin annulus between radii r and $r + \Delta r$ is the rate of change of the angular momentum of the gas in the annulus. The mass in the annulus, which moves slowly inward at the speed v, is conserved except for star formation. This mass is $\Delta M = 2\pi r \sigma \Delta r$, and aside from star formation, it satisfies the continuity equation $D \Delta M / Dt = 0$ where $D/Dt = \partial/\partial t + v \cdot \nabla$. The angular momentum in the annulus is $\Delta M r^2 \Omega$. Thus the equation of angular momentum conservation is

$$\frac{D \Delta M \, r^2 \Omega}{Dt} = \Delta M \frac{D r^2 \Omega}{Dt} = \Delta r \frac{\partial T}{\partial r}, \qquad (262)$$

from which the above evolution equation follows.

These equations were solved numerically as a function of time for a wide range of initial conditions by Saio & Yoshii [9]. They found that the disk evolves toward a nearly exponential distribution and that the rotation curve becomes approximately flat. This result illustrates the interplay between various physical processes in the interstellar medium as the large scale structure takes shape. The detailed connection between viscosity and star formation, as assumed for this model, is not well understood.

Viscous evolution and accretion of the ISM has also been shown to give a molecular ring [28−30], as observed in our Galaxy. Actually this gas distribution is not really a ring but part of the overall exponential distribution of the disk with a hole in the center, like the rim of a volcano perhaps. The origin of the ring might be more related to the formation of the hole inside this exponential distribution (by a bar perhaps) than any compression or radial motion of the disk gas.

7.3 A Universal Galaxy Rotation Curve

A remarkable property of galaxies is that the blue luminosity completely specifies the rotation curve between about one exponential scale length and the optical radius [7]. The luminosity gives the slope, the absolute velocity scale, and the absolute length scale. This is remarkable because it suggests that the past integrated history of star formation and gas abundance in the disk, which determine the luminosity of the galaxy, depend in part on the density and distribution of dark matter in the halo, which contributes to the rotation curve. One suspects that Q is involved because Q contains the epicyclic frequency κ and therefore the overall rotation curve, but the detailed connection is not understood.

7.4 Star Formation and Spiral Density Waves

The possible connection between star formation and spiral density waves was discussed in Chap. 5 and Sect. 6.5. There we showed how spiral density waves modify the gas flow, reducing shear and tidal forces and increasing the density.

We suggested that gravitational instabilities and collisional agglomeration of smaller clouds formed $10^7 \, M_\odot$ cloud complexes, and that most giant molecular clouds and star formation in spiral galaxies are in the cores of these complexes. We then suggested that the complexes are destroyed, in bulk, by large tidal forces in the interarm region, and to a lesser extent by star formation, but that many of the individual giant molecular clouds that form inside of them cannot be destroyed this way but may survive the interarm transit and continue to form stars. Spiral density waves trigger the formation of $10^7 \, M_\odot$ cloud complexes and giant HII regions, and they may also increase the efficiency of star formation, i.e., the rate of conversion of ambient interstellar gas into stars, on a macroscopic scale.

The key to understanding star formation, and how it gets modified by large scale flows such as density waves, *appears to be hidden in the details of energy dissipation*, which is a very complicated process (Chap. 3). Star formation can proceed only as the turbulent and magnetic energy in the interstellar medium is removed from the gas. Clouds do not collapse into stars at the free fall rate but at the energy dissipation rate. Sometimes both of these rates can be increased enormously by compression, on either small or large scales, and then star formation can be triggered, but even in the compressed regions, energy dissipation is often what limits star formation, not gravitational collapse.

The role of energy dissipation is somewhat less important for the formation of $10^7 \, M_\odot$ cloud complexes from the ambient medium than the ultimate formation of giant molecular clouds or stars because the process of supercloud formation is largely a geometric rearrangement from a flattened or elongated shape in the galactic midplane or dustlane to a spheroid (i.e., λ in Sect. 6.5 is much larger than the scale height H). Such rearrangement can proceed without energy dissipation and be driven only by self-gravity. Thus the calculation in Sect. 6.5, which considers only gravity, can give a reasonable time scale and cloud mass. However, the process of star formation inside the supercloud, or of giant molecular cloud formation inside of it, depends strongly on the dissipation of energy, which must occur before the associated core densities are reached. Star formation ultimately waits for energy dissipation, but supercloud formation may not.

This distinction between the dissipation rate ω_{diss} and the gravity rate ω_{grav} implies that superclouds can form in some situations without an associated increase in the star formation rate per unit gas mass. This might occur if the gravitational instability rate is large compared to the flow-through rate in a spiral arm, but the energy dissipation rate is small compared to this flow rate (cf. Sect. 5.2.4). Then superclouds can form but they cannot dissipate their energy to make stars. This situation may occur when

$$\frac{\omega_{\mathrm{diss}}}{\omega_{\mathrm{grav}}} \approx \left(\frac{P}{G\sigma_{\mathrm{c}}^2} \right)^{1/2} \ll 1 \tag{263}$$

or when all of the small clouds that make up the interstellar fluid are dense and individually self-gravitating (i.e., no diffuse clouds). Then the collision

270

time is much larger than the gravitational time and gravitational instabilities associated with a rearrangement of gas into a more spherical geometry can occur before the turbulent energy is dissipated. In Sect. 5.2.4, we considered only that $\omega_{\text{diss}} \sim \omega_{\text{grav}}$ in most regions, and we discussed spiral density wave triggers in the context of relative variations between ω_{flow}, ω_{diss}, and ω_{shear} because many of the large-scale properties of star formation can be explained in these terms. Other situations may arise in peculiar environments or in the outer regions of normal galaxies (cf. Sect. 7.5).

In the limit where the dissipation rate is much lower than the gravity rate, individual random cloud collisions play a more important role in building up large cloud complexes and in triggering star formation than do dissipative processes. For example, when the inner part of our Galaxy was first observed to be mostly molecular, the ISM there was assumed to be comprised almost entirely of self-gravitating molecular clouds, in which case the above inequality would be satisfied. This led to models of cloud formation involving random collisional agglomeration [31, 32] instead of dissipative instabilities, and to models of star formation involving direct impacts between clouds [33] instead of core collapse inside of newly-formed isolated clouds. At the present time, the origin of the molecular emission in the inner Galaxy is not well understood; some of it could be in the form of molecular diffuse clouds [34], which would increase ω_{diss} and possibly favor the dissipation rather than the random collision scenario. Nevertheless, the subcomponent of the ISM that is strongly self-gravitating could still have $\omega_{\text{diss}}/\omega_{\text{grav}} \ll 1$, and to the extent that this subcomponent behaves differently than the diffuse cloud component (cf. Sect. 5.2.1), many of the binary cloud processes discussed in these theories could still apply.

The star formation process seems to be independent of the large scale properties of a region, aside from the ratio $\sigma/\sigma_{\text{thresh}}$ and the value of the mean density [1]. The initial mass function does not appear to vary much from place to place either, except perhaps for an increase in the lower mass cutoff with increasing temperature, and the star formation efficiency seems to depend only on density as approximately $\rho^{0.3}$ once $\sigma > \sigma_{\text{thresh}}$ [1]. There is no detailed explanation for this universal density dependence and threshold aside from the notion that it may result from a combination of several simultaneous instabilities in the gas, which combine forces from self-gravity, magnetic fields, rotation, and pressure, with various heating and cooling processes [10], as discussed above. The σ threshold in this interpretation comes from the Q threshold for the onset of collapse in the radial direction (see also [12]).

Coincidentally, the time scale for propagating star formation, i.e., giant shell formation followed by collapse to new molecular clouds, is about the same as the time scale for spontaneous gravitational instabilities in the ambient medium [35]. There is also a threshold σ for strong triggering by this mechanism, which is about the same as the Q threshold for ambient instabilities [35]. This is because star formation always involves a struggle between self-gravity (and dissipation) and shear, and there is primarily one dimensionless parameter that governs the instability in the presence of these effects.

In summary, star formation is the end result of a long sequence of processes involving interstellar gas dynamics. These processes, which include energy dissipation, cloud collisions, self-gravitational collapse, spiral wave compression, and so on, often act together, with each depending on the other. Thus the slowest process tends to limit the overall star formation rate. This implies that *a variety of circumstances can lead to the same average star formation rate although different limiting factors actually determine the rate in different situations.* An example of such parallel limitation was given in Sect. 5.2.4.

7.5 Starburst Galaxies

The interstellar medium in a galaxy reacts to a collision with another galaxy by falling toward the nuclear region along strong spiral arms and by getting sprayed outward into intergalactic space along tidal arms. The strong inner arms can also form a bar [36, 37], and the bar can drive accretion, as observed by Ishizuki et al. [38, 39]. Yet in the midst of these rapid changes, the basic mechanism of star formation can be about the same in starburst galaxies as in normal galaxies, i.e., gravitational instabilities and cloud collisions can form the primary clouds and then condensation inside of these clouds can produce new stars. However, vastly different conditions in the interstellar media of starburst galaxies compared to normal galaxies (see reviews in [40 − 42]) can give these primary clouds very different properties, and this can affect how stars form in a qualitative and quantitative way.

For example, a high velocity dispersion in the ambient medium of an interacting galaxy (see observations in [43]) will cause the primary clouds that form by gravitational instabilities or gravitationally enhanced collisions to have a high internal velocity dispersion, and this can result in a situation where internal star formation does not disrupt the cloud because it is too tightly bound. Star formation tends to agitate local giant molecular clouds with a characteristic velocity of around $10 \, \mathrm{km \, s^{-1}}$, which is the expansion speed of an HII region. If the escape speed in a cloud is much larger than this, then star formation would presumably not disrupt it, or at least not very easily.

The velocity dispersion in a cloud that forms by a gravitational instability, Δv, is proportional to the velocity dispersion in the ambient medium out of which it forms, a. For instabilities in a disk, the wavenumber of the fastest growing perturbation is

$$k = \frac{\pi G \sigma}{a^2} \tag{264}$$

for mass column density σ, so the wavelength is $2\pi/k$ and the characteristic mass is

$$M \approx \sigma \left(\frac{\lambda}{2} \right)^2 = \frac{a^4}{G^2 \sigma}. \tag{265}$$

If we divide this mass by the radius $\lambda/2$, we get the virial velocity dispersion in the cloud

$$\Delta v = \left(\frac{GM}{5(\lambda/2)} \right)^{1/2} = \frac{a}{5^{1/2}} \approx 0.4\,a. \tag{266}$$

Now it follows that if gravitational instabilities initiate star formation by the formation of giant cloud complexes, and if an interaction agitates the interstellar medium in a galaxy and thereby increases its velocity dispersion, then the clouds that form stars in interacting galaxies may have large internal velocity dispersions and not be easily destroyed by the stars they form. This implies that star formation should occur at a much higher efficiency in each cloud in a starburst galaxy, and, because of the resulting larger density of massive stars in regions of star formation, the gas temperatures in star-forming clouds should be higher too. This could shift the initial mass function to a higher average mass because of the shift in the thermal Jeans mass. Such high efficiencies and shifts in the mass function are thought to be present in starburst galaxies [44].

Other differences in the properties of the interstellar medium in interacting galaxies can also lead to qualitative changes in the process of star formation. Vazquez & Scalo [45] suggest that a high velocity dispersion leads to the collisional destruction of individual clouds that make up the interstellar fluid, and that such destruction halts the star formation process until the gas cools down. After cooling, star formation begins very suddenly and proceeds at a high rate because of the high density. This model explains how the gas can accrete to the nuclear region without turning into stars before it gets there. Then, when it is there, it can form stars at a high rate. Other cloud collision models for starburst galaxies were discussed by Olson & Kwan [46,47], and a general model of galaxy interactions with star formation was in Mihos et al. [48]. Numerical simulations by Noguchi [49] suggest that a single interaction can give several starbursts in rapid succession. All of these models have reasonable gas dynamics on the large scale, but they are forced to assume detailed properties about star formation and cloud interactions that are not well understood or observed. Thus the models are uncertain at the present time. To some extent, we have to understand local star formation before we can make detailed applications to starburst galaxies, even if some of the conditions in starburst galaxies are very different from those near us.

References

7.1   Kennicutt, R.C. 1989, *Astrophys. J.* **344**, 685
7.2   Buat, V., Deharveng, J.M., Donas, J. 1989, *Astron. Astrophys.* **223**, 42
7.3   Tenjes, P., Haud, U. 1991, *Astron. Astrophys.* **251**, 11
7.4   Guiderdoni, B. 1987, *Astron. Astrophys.* **172**, 27
7.5   Phillipps, S., Edmunds, M.G., Davies, J.I. 1990, *Monthly Notices Roy. Astron. Soc.* **244**, 168
7.6   Caldwell, N., Kennicutt, R., Phillips, A.C., Schommer, R.A., *Astrophys. J.* **370**, 526
7.7   Persic, M., Salucci, P. 1991, *Astrophys. J.* **368**, 60
7.8   Thomasson, M., Donner, K.J., Elmegreen, B.G. 1991, *Astron. Astrophys.* **250**, 316
7.9   Saio, H., Yoshii, Y. 1990, *Astrophys. J.* **363**, 40

7.10   Elmegreen, B.G. 1991, *Astrophys. J.* **378**, 139

7.11   Vader, J.P., Vigroux, L. 1991, *Astron. Astrophys.* **246**, 32

7.12   Quirk, W.J. 1972, *Astrophys. J.* **176**, L9

7.13   van der Kruit, P.C. 1987, *Astron. Astrophys.* **173**, 59

7.14   Freeman, K.C. 1970, *Astrophys. J.* **160**, 811

7.15   Bosma, A. 1981, *Astron. J.* **86**, 1791

7.16   Carignan, C., Puche, D. 1990, *Astron. J.* **100**, 394

7.17   Seiden, P.E., Schulman, L.S., Elmegreen, B.G. 1984, *Astrophys. J.* **282**, 95

7.18   Hohl, F. 1971, *Astrophys. J.* **168**, 343

7.19   Pfenniger, D., Friedli, D. 1991, *Astron. Astrophys.* **252**, 75

7.20   Friedli, D., Pfenniger, D. 1991, in *IAU Symp. 146, Dynamics of Galaxies and Molecular Cloud Distributions*, eds. F. Combes, F. Casoli, Kluwer Academic Publishers, Dordrecht, p. 362

7.21   Silk, J., Norman, C. 1981, *Astrophys. J.* **247**, 59

7.22   Lin, D.N.C., Pringle, J.E. 1987, *Astrophys. J.* **320**, L87

7.23   Simakov, S.G. 1990, *Sov. Astron. Lett.* **16**, No. 4

7.24   Clarke, C.J. 1989, *Monthly Notices Roy. Astron. Soc.* **238**, 283

7.25   Yoshii, Y., Sommer-Larson, J. 1989, *Monthly Notices Roy. Astron. Soc.* **236**, 779

7.26   Sommer-Larson, J., Yoshii, Y. 1989, *Monthly Notices Roy. Astron. Soc.* **238**, 133

7.27   Sommer-Larson, J., Yoshii, Y. 1990, *Monthly Notices Roy. Astron. Soc.* **243**, 468

7.28   Icke, V. 1979, *Astron. Astrophys.* **78**, 21

7.29   Lesch, H., Biermann, P.L., Crusius, A., Reuter, H.P., Dahlem, M., Barteldrees, A., Wielebinski, R. 1990, *Monthly Notices Roy. Astron. Soc.* **242**, 194

7.30   Däther, M., Biermann, P.L. 1990, *Astron. Astrophys.* **235**, 55

7.31   Scoville, N.Z., Hersh, K. 1979, *Astrophys. J.* **229**, 578

7.32   Kwan, J. 1979, *Astrophys. J.* **229**, 567

7.33   Scoville, N., Sanders, D.B., Clemens, D.P. 1986, *Astrophys. J.* **310**, L77

7.34   Polk, K.S., Knapp, J.G., Stark, A.A., Wilson, R.W. 1988, *Astrophys. J.* **332**, 432

7.35   Elmegreen, B.G. 1992, in *Star Formation in Stellar Systems*, eds. G. Tenorio-Tagle, M. Prieto, F. Sanchez, Cambridge, Cambridge Univ. Press, in press

7.36   Noguchi, M. 1987, *Monthly Notices Roy. Astron. Soc.* **228**, 635

7.37   Gerin, M., Combes, F.& Athanassoula, E. 1990, *Astron. Astrophys.* **230**, 37

7.38   Ishizuki, S., Kawabe, R., Ishiguro, M., Okumura, S.K., Morita, K.I., Chikada, Y., Kasuga, T., Doi, M. 1990, *Nature* **344**, 224

7.39   Ishizuki, S., Kawabe, R., Ishiguro, M., Okumura, S.K., Morita, K.I., Chikada, Y., Kasuga, T., Doi, M. 1990, *Astrophys. J.* **355**, 436

7.40   Norman, C.A. 1990, in *Windows on Galaxies*, eds. G. Fabbiano, J.S. Gallagher, A. Renzini, (Dordrecht: Kluwer), p. 311

7.41   Pagel, B.E.J. 1991, in *Evolution in Astrophysics*, ed. E.J. Rolfe, (Paris: European Space Agency), p. 159

7.42   Telesco, C.M. 1988, *Ann. Rev. Astron. Astrophys.* **26**, 343

7.43   Casoli, F., Dupraz, C. Combes, F., Kazes, I. 1991, *Astron. Astrophys.* **251**, 1

7.44   Scalo, J.M. 1990, in *Windows on Galaxies*, eds. G. Fabbiano, J.S. Gallagher, A. Renzini, (Dordrecht: Kluwer), p. 125

7.45   Vazquez, E.C., Scalo, J.M. 1989, *Astrophys. J.* **343**, 644

7.46   Olson, K.M., Kwan, J. 1990, *Astrophys. J.* **349**, 480

7.47   Olson, K.M., Kwan, J. 1990, *Astrophys. J.* **361**, 426

7.48   Mihos, J.C., Richstone, D.O., Bothun, G.D. 1991, *Astrophys. J.* **377**, 72

7.49   Noguchi, M. 1991, *Monthly Notices Roy. Astron. Soc.* **251**, 360

Physics and Chemistry of Molecular Clouds

Reinhard Genzel

Max-Planck-Institut für extraterrestrische Physik, Garching, FRG

Introduction

We know now for more than a decade that about half of the interstellar matter in our Galaxy is in molecular form, primarily as H_2 molecules. The molecules are concentrated in clouds which reach masses of more than $10^6 \, M_\odot$. These Giant Molecular Clouds (GMC's) are fascinating objects. They are the most massive, physical entities in the Galaxy. They are the locations where new stars are born. They also play an important role in fueling the starburst and active nuclei of luminous external galaxies.

This review is an attempt to summarize what we know about the physics and chemistry of molecular clouds, about their structure, dynamics, thermal balance, and chemical composition, about stellar formation in them and about interstellar masers at their cores. The emphasis is on the principal ideas and concepts so that the material is especially suited for graduate students in Physics and Astrophysics.

The first chapter contains a "hands on" description of the tools of infrared and radio astronomy, including brief reviews of atomic and molecular spectroscopy, radiative transport, and interpretation of line and continuum emission in terms of physical parameters. Chapter 2 is devoted to the thermal balance of molecular clouds, with discussions of photon dominated regions, shocks, heating by cosmic rays and X-rays, and ambipolar heating. Again the direct confrontation of theoretical models with observational results is emphasized. Chapter 3 deals with chemistry and the composition of gas and dust. The structure and dynamics of molecular clouds is the topic of Chap. 4, along with a summary of observations of the best studied case, the Orion Molecular Cloud. Chapter 5 contains a discussion of circumstellar clouds and masers. Finally, while most of this review is devoted to the molecular interstellar medium in our own Galaxy, an outlook to recent investigations in external galaxies is included in the last chapter (Chap. 6).

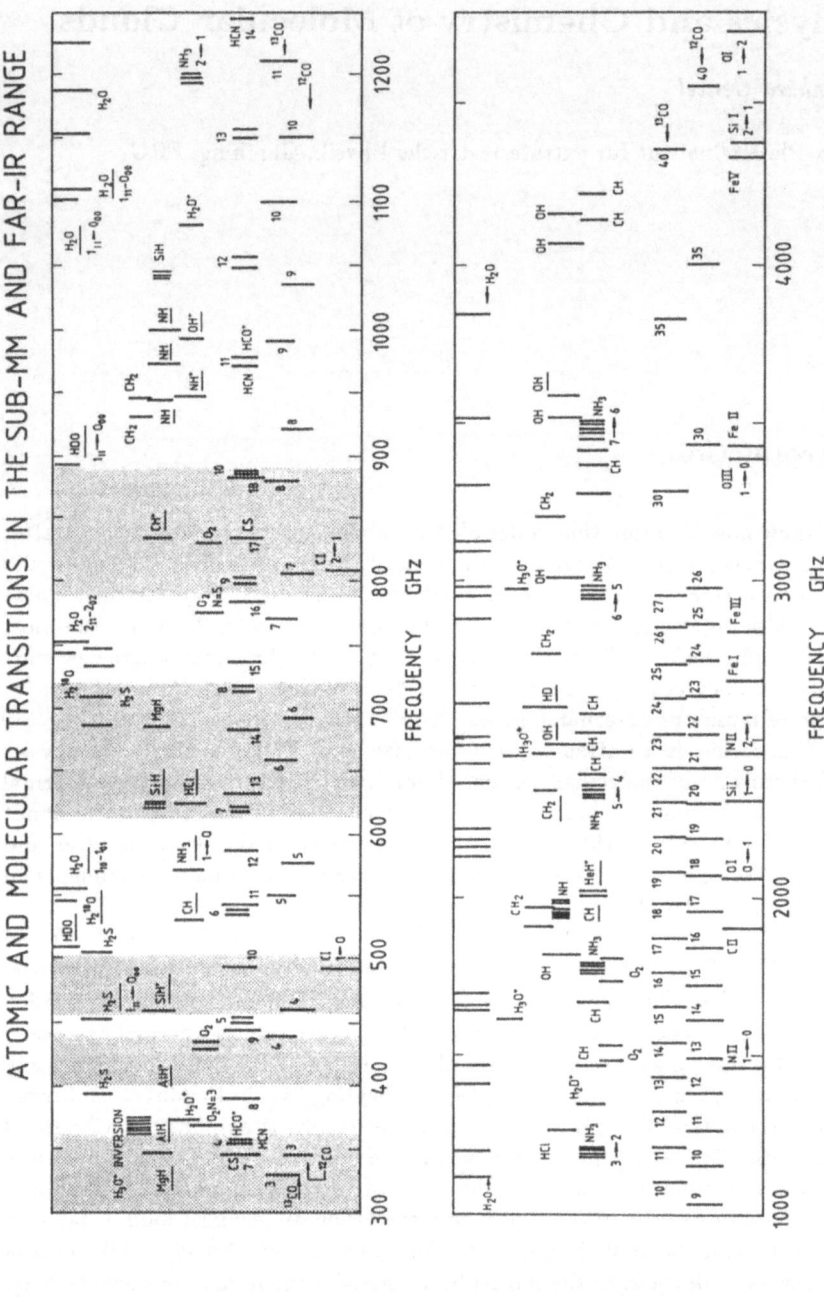

Fig. 1. Atomic and molecular transitions in the 60 μm to 1 mm range. The shaded regions represent the submillimetric atmospheric windows

1 Tools of Infrared and Radio Astronomy

We begin with a review of the tools of infrared and radio astronomy. Observations at infrared submillimeter and radio wavelengths ($1\,\mu$m to 1 m) are best suited for detailed studies of the dense interstellar medium (ISM), and of molecular clouds. Extinction by dust usually prevents meaningful measurements in the visible, ultraviolet, or soft X-ray region which are powerful tools for investigating the diffuse interstellar medium. Continuum and spectral lines, emission as well as absorption are all important and need to be discussed. Thermal Bremsstrahlung in $\approx 10^4$ K electron-ion plasma and synchrotron radiation as a result of gyration of fast electrons around magnetic field lines are the dominant continuum processes in the radio range ($\lambda \geq 1$ cm). At infrared and submillimeter wavelengths (a few μm to 1 mm) thermal emission from dust particles usually dominates. Spectroscopy of atomic and molecular transitions is particularly important for detailed studies of the physical conditions, composition and dynamics of dense molecular clouds. As an example, Fig. 1 shows spectral lines of astronomical interest that are contained in the $60\,\mu$m to 1 mm spectral band. Although the displayed lines represent only a small selection of all available transitions, it is immediately apparent how rich this wavelength range is for probing a wide range of physical and chemical parameters. In the following, I will give a brief and necessarily incomplete review of the basics of atomic and molecular spectroscopy in this wavelength range that are relevant to the astronomical application. This is followed by a discussion of radiative transport and excitation of line and continuum emission. For more detailed discussions, the reader is referred to the standard textbooks of Osterbrock (1989), Herzberg (1939, 1945), Townes & Schawlow (1955), Verschuur & Kellermann (1989), and the reviews by Winnewisser et al. (1974), Scoville (1984), Watson (1984, 1985), and Wynn-Williams (1984).

1.1 Summary of Infrared/Radio Spectroscopy

1.1.a Recombination Lines
Hydrogenic ions (H^+, He^+, He^{++}, C^+, etc.) with effective nuclear charge Z_{eff} have recombination transitions between main quantum numbers k' and k at wavelengths $\lambda_{k'k}$

$$\lambda_{k'k}^{-1} = R_A Z_{\text{eff}}^2 \left(\frac{1}{k^2} - \frac{1}{k'^2} \right). \tag{1}$$

$R_A = 1.097 \times 10^5 \left(1 - m_e/m_A \right) \text{cm}^{-1}$, m_e is the electron and m_A the ionic mass. For $k >$ a few and $k' \approx k$ these transitions are in the infrared and radio range. For $k' = k + \Delta k$ and $k', k \gg 1$,

$$\lambda_{k,\Delta k}^{-1} \approx 2 R_A Z_{\text{eff}}^2 \frac{\Delta k}{k^3}. \tag{2}$$

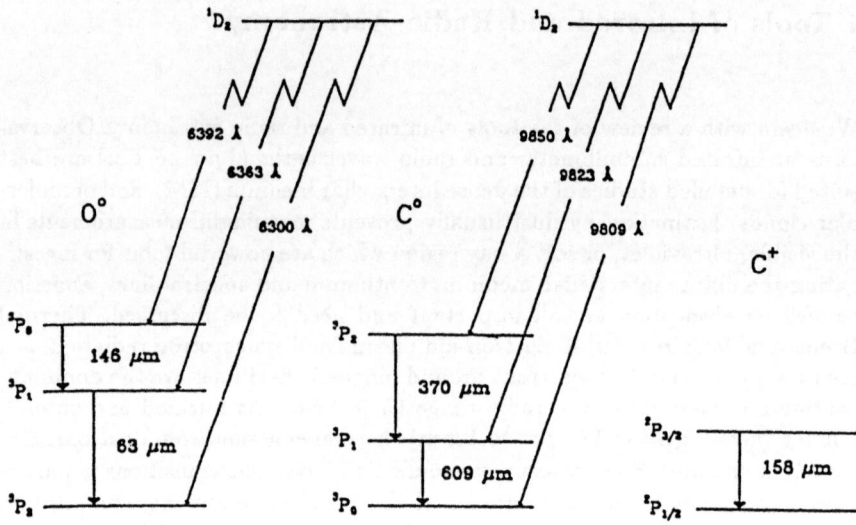

Fig. 2. Lowest energy levels and fine structure transitions of O^0, C^0 and C^+

$\Delta k = 1$ (α) transitions are stronger than $\Delta k = 2, 3, \ldots$ (β, γ, etc.) transitions.

1.1.b Fine Structure Lines

Atoms and ions with non-zero electronic, orbital angular momentum (L) and spin (S) in their ground state configuration can have low lying fine structure levels. As the magnetic spin-orbit splitting for light atoms is of the order of $\alpha_{fs}^2 \approx (1/137)^2$ times the electronic energy ($10-100$ eV), or about a few 10^{-3} to 10^{-1} eV, fine structure transitions lie in the $10-300\,\mu$m region. The multiplicity of fine structure levels depends on the total angular momentum ($J = L + S$) and thus the electronic configuration in the outermost shell.

For instance, atoms and ions with 1p or 5p electrons in their valence shell (C^+, N^{++}, Ne^+, Ar^+, Si^+, O^{+++}, S^{+++}) have fine structure doublets and a single fine structure transition in the ground state. Atoms and ions with 2p and 4p electrons (C^0, O^0, Si^0, S^0, N^+, O^{++}, S^{++}, Ne^{++}) have triplets and two fine structure transitions. Atoms and ions with 3p electrons or filled shells (Ne^0, Ar^0, He^0, O^+, S^+) have no ground state fine structure transitions. As a specific example, Fig. 2 shows the level diagrams of O^0, C^0, and C^+. Neutral atomic carbon has two fine structure transitions in the ground state, at $609\,\mu$m ($J = 1 \to 0$) and at $371\,\mu$m ($J = 2 \to 1$). Neutral oxygen also has two fine structure transitions (63 and $145\,\mu$m) while singly ionized carbon has one ($158\,\mu$m). Transitions between fine structure levels are magnetic dipole transitions ($\Delta J = \pm 1$, $\Delta L = 0$, $\Delta S = 0$) with transition probabilities typically α_{fs}^2 or 4 orders of magnitude, weaker than those of (electric dipole) permitted transitions. Infrared fine structure lines are, therefore, usually optically thin ($\tau \approx 10^{-2}$ to a few) and fine structure levels are easily excited by collisions. Table 1 is a list of important astronomical fine structure transitions, along with their wavelengths, Einstein A coefficients and critical densities (see Sect. 1.2.e).

278

As the moving mass in atomic fine structure is very close to the electronic mass $\left(\mu = m_A m_e/(m_A + m_e) \approx m_e \left(1 - m_e/m_A\right)\right)$ different isotopes cannot be spectroscopically resolved. In some cases, however, the isotopic nucleus has a nuclear spin and an observable splitting occurs because of hyperfine quadrupole splitting. This is the case, for instance, for the [^{13}CII] and [^{12}CII] fine structure transitions at 158 μm (Cooksy et al. 1986).

1.1.c Hyper-Fine Structure Lines

Magnetic interactions between nuclear and electronic spin can further split atomic (and molecular) energy levels. This hyper-fine splitting is about $m_e/m_H \approx 1/2000$ times smaller than fine structure. The most famous case is the hyper-fine structure in the 1s ground state of atomic hydrogen, leading to two states of total angular momentum $F = 1$ and $F = 0$ separated by 5.9×10^{-6} eV, corresponding to a frequency of 1.421 GHz, or a wavelength of 21 cm. Spin flip transitions between $F = 1 \rightarrow 0$ are extremely rare $\left(A_{21\,cm} = 2.87 \times 10^{-15}\,\mathrm{s}^{-1} = (10^7\,\mathrm{y})^{-1}\right)$, but the large column densities of atomic hydrogen make the 21 cm transition nevertheless easily detectable in interstellar space.

1.1.d Molecular Transitions

The fact that the light electrons move much faster than the heavy nuclei means that the motions of nuclei and electrons in molecules are more or less completely decoupled. The electrons see the Coulomb attraction of the basically stationary nuclei while the nuclei move in the time averaged electrostatic potential of the electrons. In the approximation of complete decoupling of the nuclear and electronic motions (Born-Oppenheimer approximation), the total energy of the molecular system can be written as

$$E_{tot} = E_{el}(R_e) + E_{vib} + E_{rot}, \tag{3}$$

where $E_{el}(R_e) \approx -5$ to -10 eV is the electronic binding energy at the equilibrium separation $R_e \approx 1$Å of the nuclei.

Diatomic Molecules. For the simplified case of a diatomic molecule $E_{vib}^{(v)} \approx \hbar\omega_{vib}(v + \frac{1}{2})$ is the vibrational energy with vibrational quantum numbers $v = 0, 1, 2, \ldots$. Most molecules have $\hbar\omega_{vib} \approx 0.1$ eV and the vibrational bands $\Delta v - \pm 1$ lie 3 to 10 μm. The rotational energy levels of a simple rigid diatomic or linear polyatomic rotor can be written as $E_{rot}^{(J)} = J(J + 1)\hbar^2/\Theta = B_e J(J + 1)$. $2\mu R_e^2$ is the moment of inertia; μ is the reduced mass of the system ($\mu = m_1 m_2/(m_1 + m_2)$ for diatomic molecules). $J = 0, 1, 2, \ldots$ is the rotational quantum number and $B_e = \hbar^2/\Theta$ is the rotational constant. If the molecule has a permanent (or rotationally induced) electric dipole moment, that is, a separation of the centers of charge and mass, strong electric dipole transitions are allowed with the selection rule $J_{upper} - J_{lower} = \Delta J = \pm 1$. Typical heavy top molecules such as CO, HCN, or CS have $\mu \approx 7m_H$ and $B_e \approx 3 \times 10^{-4}$ eV, resulting in the first rotational transition ($J = 1 \rightarrow 0$) at a wavelength of a few mm. Figure 3 shows the example rotational and ro-vibrational level diagram of CO. Diatomic molecules containing one hydrogen atom (hydrides), such as OH,

Table 1. Infrared fine structure lines [d]

Species	Excitational Potential [eV]	Ionization Potential [eV]	Transitions	λ [μm]	A [s^{-1}]	n_{crit} [cm^{-3}] [a]
C	-	11.26	^{3}P, $J = 1 \to 0$	609.1354	7.9(-8)	4.7(2)
						1. (4) for H$_2$ [a]
			$2 \to 1$	370.415	2.7(-7)	1.2(3)
CII	11.26	24.38	^{2}P, $J = \frac{3}{2} \to \frac{1}{2}$	157.7409	2.4(-6)	2.8(3)
						5. (3) for H$_2$
						50 for electrons
O	-	13.62	^{3}P, $J = 1 \to 2$	63.18372	8.95(-5)	4.7(5) $T_{300}^{-1/2}$ [b]
						7. (5) $T_{300}^{-1/2}$ for H$_2$
			$0 \to 1$	145.52547	1.7(-5)	9.5(4) $T_{300}^{-1/2}$
						$\geq$1. (5) $T_{300}^{-1/2}$ for H$_2$ [a]
Si	-	8.15	^{3}P, $J = 1 \to 0$	129.68173	8.25(-6)	2.4(4)
			$2 \to 1$	68.473	4.2(-5)	8.4(4)
SiII	8.15	16.35	^{2}P, $J = \frac{3}{2} \to \frac{1}{2}$	34.814	2.1(-4)	3.4(5)
S	-	10.36	^{3}P, $J = 1 \to 2$	25.245	1.4(-3)	1.9(6)
			$0 \to 1$	56.309	3.0(-4)	7.2(5) but H$_2$ [a]
Fe	-	7.87	^{5}D, $J = 3 \to 4$	24.0424	2.5(-3)	3.1(6)
			$3 \to 2$	34.7135	1.6(-3)	3. (6)
FeII	7.87	16.18	^{6}D, $J = \frac{7}{2} \to \frac{9}{2}$	25.9882	2.1(-3)	2.2(6)
			$\frac{5}{2} \to \frac{7}{2}$	35.491	1.6(-3)	3.3(6)

CH, HCl, have $\mu \approx m_H$ and thus have their first rotational transition in the submillimeter/far-infrared range. Ro-vibrational transitions of diatomic and linear polyatomic molecules in absorption have the selection rules $\Delta v = +1$ and $\Delta J = +1$ (R-branch), or $\Delta J = -1$ (P-branch). These transitions lie at 4.6 μm for the CO molecule. There is no $\Delta J = 0$ (Q-branch) for diatomic molecules with a zero electronic angular momentum (see Fig. 3). The important molecular hydrogen molecule, H$_2$, is a special case as it is a homonuclear molecule with a zero electric dipole moment. Rotational and ro-vibrational transitions are thus weak electric quadrupole transitions with the selection rules $\Delta v = \pm 1$, $\Delta J = 0$ (Q-branch), $\Delta J = \pm 2$ (S-branch) and $\Delta J = +2$ (O-branch) (see Fig. 4). These occur at a wavelength of about 2 μm. The lowest pure rotational transition of H$_2$ between $J = 2$ and $J = 0$, $S(0)$ is at 28 μm.

There are higher order effects that make the spectra more complicated. Anharmonicity in the potential results in overtone vibrational bands ($\Delta v = \pm 2$ etc., Fig. 3). Centrifugal distortion moves higher rotational states closer together. Coriolis forces couple the rotational and vibrational motion. Including these higher order effects the ro-vibrational energy levels can be written as

$$\frac{E_{Jv}}{h} = \nu_e \left(v + \tfrac{1}{2}\right) - x_e \nu_e \left(v + \tfrac{1}{2}\right)^2 + B_e\, J(J+1)$$
$$- D_e\, J^2 (J+1)^2 - \alpha_e \left(v + \tfrac{1}{2}\right) J(J+1). \tag{4}$$

Table 1. *(continued)*

Species	Excitational Potential [eV]	Ionization Potential [eV]	Transitions		$\lambda\,[\mu\mathrm{m}]$	$A\,[\mathrm{s}^{-1}]$	$n_{\mathrm{crit}}\,[\mathrm{cm}^{-3}]$ [c]
OIII	35.12	54.93	$^3\mathrm{P},\ J=$	$1\to0$	88.356	2.6(-5)	5.1(2)
				$2\to1$	51.815	9.8(-5)	3.6(3)
OIV	54.93	77.40	$^2\mathrm{P},\ J=\frac{3}{2}\to\frac{1}{2}$		25.87	5.2(-4)	1.0(4)
NII	14.53	29.60	$^3\mathrm{P},\ J=$	$1\to0$	203.5±0.8	2.1(-6)	48
				$2\to1$	121.89806	7.5(-6)	3.1(2)
NIII	29.60	47.45	$^2\mathrm{P},\ J=\frac{3}{2}\to\frac{1}{2}$		57.317	4.8(-5)	3. (3)
NeII	21.56	40.96	$^2\mathrm{P},\ J=\frac{1}{2}\to\frac{3}{2}$		12.81355	8.6(-3)	5.4(5)
NeIII	40.96	63.45	$^3\mathrm{P},\ J=$	$1\to2$	15.55	5.99(-3)	2.9(5)
				$0\to1$	36.02	1.2(-3)	4.2(4)
NeV	97.11	126.21	$^3\mathrm{P},\ J=$	$1\to0$	24.28	1.3(-3)	5.4(4)
				$2\to1$	14.32	4.6(-3)	3.8(5)
NeVI	126.21	157.93	$^2\mathrm{P},\ J=\frac{3}{2}\to\frac{1}{2}$		7.642	2.0(-2)	
ArII	15.76	27.63	$^2\mathrm{P},\ J=\frac{1}{2}\to\frac{3}{2}$		6.985274	5.3(-3)	1.9(5)
ArIII	27.63	40.74	$^3\mathrm{P},\ J=$	$1\to2$	8.99103	3.1(-2)	3.1(5)
				$0\to1$	21.84	5.2(-3)	3.5(4)
ArV	59.81	75.02	$^3\mathrm{P},\ J=$	$1\to0$	13.09	7.99(-3)	
				$2\to1$	7.903	2.7(-2)	
ArVI	75.02	91.01	$^2\mathrm{P},\ J=\frac{3}{2}\to\frac{1}{2}$		4.53	9.7(-2)	
SIII	23.33	34.83	$^3\mathrm{P},\ J=$	$1\to0$	33.482	4.7(-4)	2.0(3)
				$2\to1$	18.713	2.1(-3)	1.7(4)
SIV	34.83	47.30	$^2\mathrm{P},\ J=\frac{3}{2}\to\frac{1}{2}$		10.5105	7.7(-3)	5.6(4)
MgV	109.24	141.26	$^3\mathrm{P},\ J=$	$2\to1$	5.609	1.3(-1)	
				$1\to0$	13.54	2.2(-2)	
SiVI	166.77	205.05	$^2\mathrm{P},\ J=\frac{3}{2}\to\frac{1}{2}$		1.959	2.37	
SiVII	205.05	246.52	$^3\mathrm{P},\ J=$	$2\to1$	2.474	1.47	
				$1\to0$	6.515	1.9(-1)	
FeIII	16.18	30.65	$^5\mathrm{D},\ J=$	$2\to3$	33.04		
				$3\to4$	22.93		

a) If not otherwise specified critical density for collisions with atomic hydrogen: $n_{\mathrm{crit}} = A_{\mathrm{ul}}/\gamma_{\mathrm{ul}}$; collisional rates with H_2 may be different, see Monteiro, T. & Flower, D. 1987, *Monthly Notices Roy. Astron. Soc.* **228**, 101, for CI and OI excitation.

b) Kinetic temperature in units of 300 K.

c) Critical density for collisions with electrons/protons.

d) For additional fine structure lines, see Schmid-Burgk (1982, *Landolt-Börnstein*, **VI**, 2c, p. 115).

x_{e} is the anharmonicity parameter. Typically the centrifugal distortion parameter is $D_{\mathrm{e}} \approx 3 \times 10^{-6} \to 3 \times 10^{-5}\,B_{\mathrm{e}}$ and $\alpha_{\mathrm{e}} \approx 10^{-5}\,\nu_{\mathrm{e}}$ is a measure of the effect of the vibration on the rotor. It is also clear from the above discussion that vibrational frequency depends on isotopic mass as $\mu^{-1/2}$ $(\nu_v(^{12}\mathrm{CO})/\nu_v(^{13}\mathrm{CO}) \approx 1.02)$, rotational frequencies as μ^{-1} $(B_{\mathrm{e}}(^{12}\mathrm{CO})/B_{\mathrm{e}}(^{13}\mathrm{CO}) = 1.046)$. As a re-

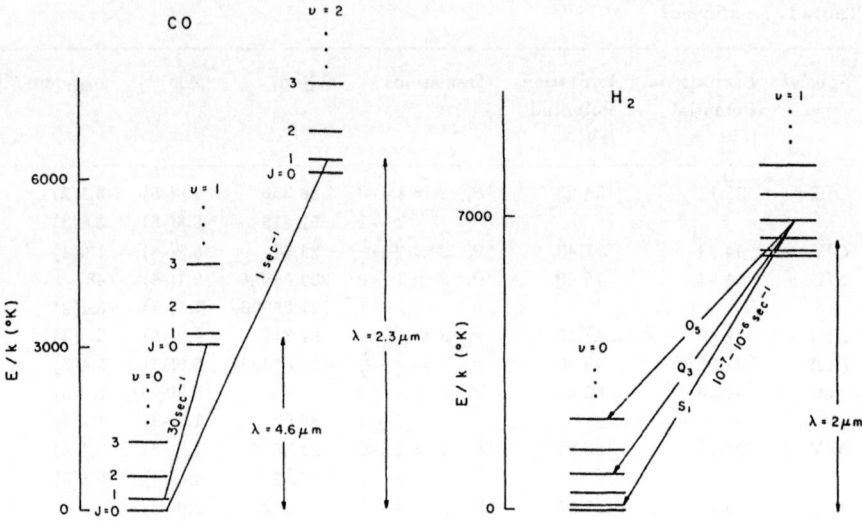

Fig. 3. Ro-vibrational levels of the CO molecule (adapted from Scoville 1984)

Fig. 4. Ro-vibrational levels of the H_2 molecule (adapted from Scoville 1984)

sult, the rotational transitions of isotopic species are easily resolved while the ro-vibrational bands of isotopic species overlap.

Polyatomic Molecules. In the case of non-linear, polyatomic molecules there are 3 different axes the molecule can rotate about. If a molecule has a three-fold (120° rotation symmetry) or greater, symmetry axis, the moments of inertia about two of the three principal axes must be identical and the molecule is called a symmetric top.

Examples are NH_3 ($B_1 = B_2 = 1.2 \times 10^{-3}$ eV, $B_3 = 8.2 \times 10^{-4}$ eV), CH_3CN or CH_3CCH. Rotational energy levels of symmetric tops have two quantum numbers, the total angular momentum J and its projection onto the B_3-axis, K ($K \leq J$). Figure 5 shows the case of the NH_3 molecule. Because of the symmetry of the molecule, radiative transitions are only allowed between J-states of a given K-stack: $\Delta J = \pm 1$, $\Delta K = 0$. Levels with $J = K$ are thus metastable with respect to radiative transitions. In the case of NH_3, each rotational level ($K \neq 0$) splits up further by about 10^{-4} eV, depending on whether the nuclear wavefunction is symmetric or antisymmetric with respect to the plane of the three hydrogens. This inversion symmetry leads to many transitions at ≈ 1.2 cm wavelength. Each of the inversion levels is further split by quadrupole/magnetic dipole, hyper-fine structure, resulting in several satellite transitions of each inversion line (Fig. 5). The relative line ratios of main and satellite lines are a measure of optical depth.

In the case of asymmetric tops, the moments of inertia about all three axes differ. None of the projections of the total rotational angular momenta onto these axes is then a good quantum number. Rotational levels are labeled by $J_{K^- K^+}$ where K^- and K^+ are approximate quantum numbers. This assumes

282

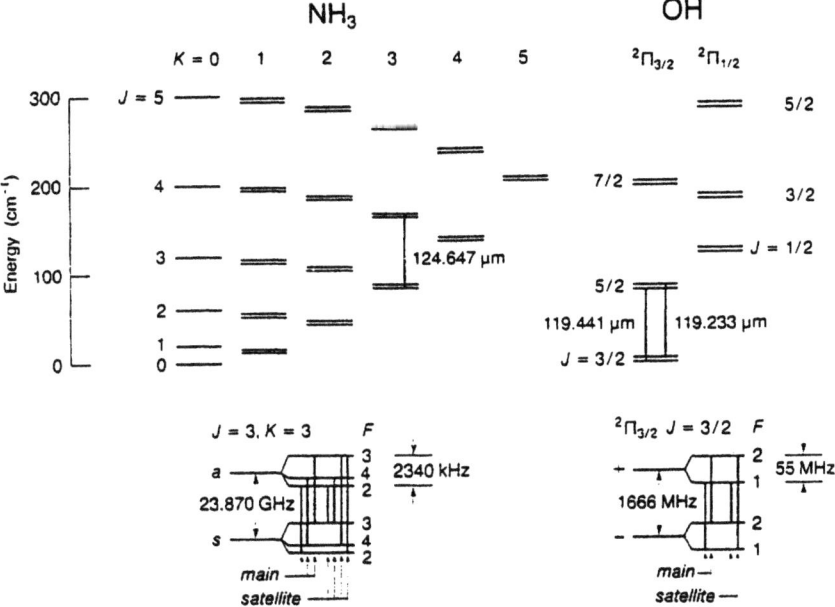

Fig. 5. Rotational levels of NH₃ (left) and OH (right) (adapted from Watson 1982)

that the two larger rotational constants (prolate case), or the two smaller rotational constants (oblate case) are identical, respectively. For asymmetric top molecules, such as H_2O, SO_2, or H_2S, transitions between different K-ladders are allowed.

Effect of Spin Statistics. For molecules with several identical atoms the effect of spin statistics on degeneracy and transition rules has to be taken into account. Consider for simplicity the case of the H_2 molecule that has two identical nuclei (protons) of spin $1/2$ each. Because of the Born-Oppenheimer approximation the total wave functions of the molecule can be written as a product of electronic and nuclear wave functions. The nuclear wave function must be antisymmetric with respect to exchange of the two nuclei,

$$
\begin{aligned}
P_{12}\tilde{\kappa}_n &= P_{12}\left[Y_{JM}(\theta,\phi)\,u_v(R)\,\tilde{\chi}(1,2)\right] \\
&= Y_{JM}(\theta,\phi-180°)\,u_v(R)\,\tilde{\chi}(2,1) = -Y_{JM}(\theta,\phi)\,u_v(R)\,\tilde{\chi}(1,2)
\end{aligned}
\tag{5}
$$

Y_{JM} is the rotational, u_v the vibrational, and $\tilde{\chi}$ the spin wavefunction. Here we have used the fact that an exchange of nuclei 1 and 2 corresponds to a 180° rotation in the plane connecting the two nuclei combined with an exchange of the spin coordinates. From the symmetry of the spherical harmonics one finds $Y_{JM}(\theta,\phi-180°) = (-1)^J Y_{JM}(\theta,\phi)$. The requirement of overall antisymmetry thus means that rotational states with *even* total angular momentum ($J = 0, 2, 4, \ldots$) must have an antisymmetric spin wavefunction, while *odd-J* states ($J = 1, 3, 5, \ldots$) must have a symmetric spin wavefunction. With two spin $1/2$

nuclei an antisymmetric spin wave function has total spin $S = 0$ (spins are antiparallel) and degeneracy $2S + 1 = 1$ (para-H_2). The symmetric spin states have $S = 1$ (parallel spins) and $2S + 1 = 3$ (ortho-H_2). Radiative transitions between para- and ortho-H_2 states (spin-flip transitions) are forbidden in lowest order. Similarly one can show for molecules with three identical protons (NH_3, PH_3, CH_3) that rotational states with $K = 3n$ ($n = 1, 2, 3, \ldots$) are ortho-states with statistical weight 2 while states with $K \neq 3n$ are para-states with statistical weight 1. $K = 0$ states also have statistical weight 1.

Molecules with Electronic Angular Momentum. The character of molecular spectra gets more complicated if there is a non-zero, orbital, or spin, *electronic* angular momentum. Consider the case of a linear molecule. Because of the symmetry of the system with respect to rotation about the molecular axis, the electronic orbital angular momentum projected onto that axis, Λ, is a constant of motion. States with $\Lambda = 0$ are called Σ states, $\Lambda = \pm 1$ are called Π states and so forth. If there is, in addition, electronic spin (Σ) the total electronic angular momentum is $\Omega = \Lambda + \Sigma$. If the coupling of orbital and spin angular momentum to the molecular axis is stronger than the spin-orbit coupling (Hund's case A), Ω is the sum of the projections Λ, Σ onto the molecular axis. The total angular momentum of the molecule J then is the vector sum of Ω (parallel to molecular axis) and the rotational angular momentum N (perpendicular to molecular axis). As an example, Fig. 5 shows the lowest rotational energy levels of the OH molecule ($\Lambda = 1$, $\Sigma = 1/2$) labeled as $^{2\Sigma+1}\Lambda_{\Omega}, J$ for the pure Hund's A coupling scheme.

There are two rotational ladders, one with $\Omega = 3/2$ and one with $\Omega = 1/2$. Coriolis interaction between rotational and electronic angular momenta removes the degeneracy of the states $\Lambda = \pm \Lambda$ and leads to a Λ-doubling of each of the rotational states. Finally the $I = 1/2$ spin of the proton results in hyperfine structure of each of the Λ-states. In reality OH is not pure Hund's case A, however. As a result, not only transition within each Ω-ladder are allowed, but in addition, cross-ladder transitions ($^2\Pi_{1/2} \longleftrightarrow {}^2\Pi_{3/2}$). If the coupling with the molecular axis is weak (Hund's case B), rotational (N) and electronic orbital angular momentum couple to a total orbital angular momentum (K) which then couples to the electronic spin (Σ) to the total angular momentum (J). States are now labeled by Σ, K, and J. For instance, CH can be best described as a Hund's case B molecule. Molecules with $\Lambda \neq 0$ may have a Q-branch ($\Delta J = 0$) in their ro-vibrational spectra.

Vibrational Transitions of Polyatomic Molecules. The vibrational motion of a molecule with N atoms can be described as the superposition of $3N - 6$ ($3N - 5$ for linear systems) normal vibrational modes. Their wavelengths depend on force constant k and vibrating mass m as $\lambda \sim (m/k)^{1/2}$. As the force constants of stretching modes are greater than bending modes, the former are at shorter wavelengths than the latter. Triple bonds are stronger than double bonds, and double bonds are stronger than single bonds, again leading to an obvious sequence of transitions from shorter to longer wavelengths. As a result of these rather general considerations, specific atomic combinations correspond

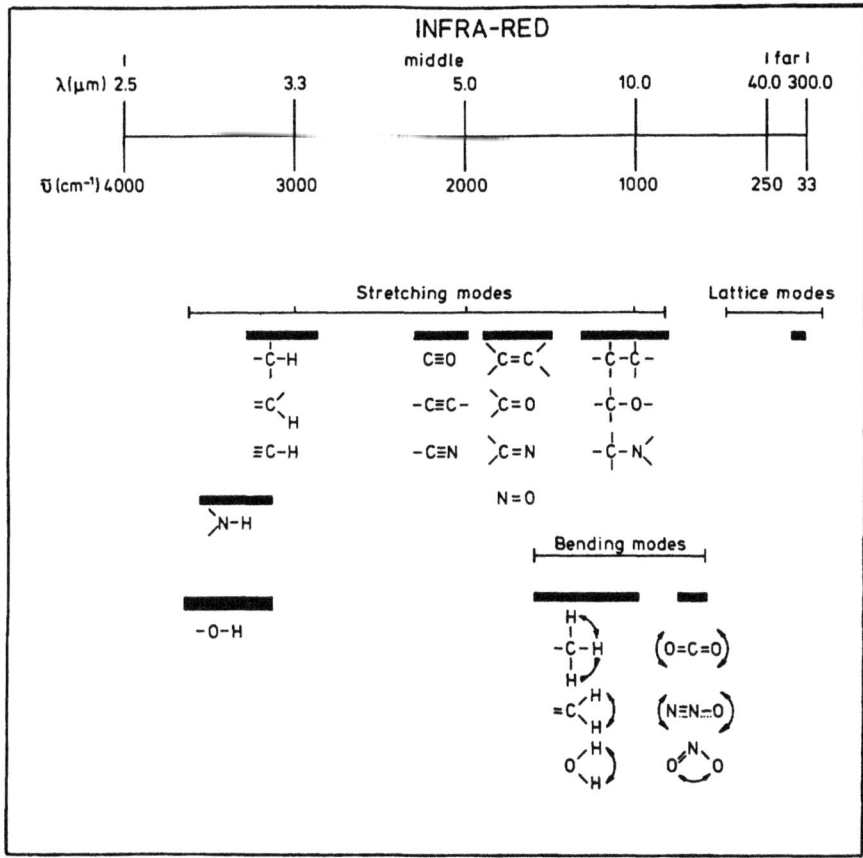

Fig. 6. Wavelengths of vibrational transitions of common diatomic groups (adapted from Allamandola 1984)

to specific vibrational wavelengths, dependent on the environment (this is particularly true for light-heavy combinations at the end of a chain).

For instance, O–H stretch vibrations are found in the 2.7 to 3 μm region, N–H at $\approx$ 3 μm, and C–H at 3 to 3.3 μm. C$\equiv$O, C$\equiv$C, C$\equiv$N stretch vibrations are at $\approx$ 4.7 μm, while C=C, C=O, C=N, and N=O stretch vibrations are at $\approx$ 5.5 μm, and C–C, C–O, and C–N stretch vibrations are between 7 and 11 μm. C–H, O–H, and N–H bending modes are located in the 8 to 12 μm range and N=O and C=O bending modes are at $\approx$ 15 μm. The locations of the various infrared bands are summarized in Fig. 6. For a vibrational mode to be infrared active, there must be a non-zero change in electric dipole moment induced during the vibrational motion. This is usually the case only for asymmetric stretching and bending modes. For example, of the 7 vibrational modes of the acetylene molecule (H–C$\equiv$C–H, 2 are double degenerate), only the antisymmetric C–H stretch at 3.04 μm and the cis-bend (both C's up while both H's are down) at 14 μm are infrared active. In the case of the planar triatomic molecule H_2O, the ν_2 bending mode at $6.3\,\mu m$ and the asymmetric ν_3-stretch at $2.7\,\mu m$ are much stronger than the more symmetric ν_1-stretch. If a molecule possesses degenerate normal modes in a plane, a linear combination of these degenerate modes corresponds to a rotational motion about an axis perpendicular to that plane. The so created vibrational angular momentum l leads to an interaction between rotational and vibrational motion and l-doubling of rotational states $(J_{tot} = J \pm l)$. Linear molecules with $l \neq 0$ have a Q-branch $(\Delta J = 0)$, in addition to the usual R-and P-branches.

Band-Heads. The rotational transition frequencies $(J \to J \pm 1)$ in the vibrational band $v' \to v > v'$ of a diatomic molecule can be written as

$$\nu = \nu_0 + \frac{1}{h}\left[(B_{v'} + B_v)\kappa + (B_v - B_{v'})\kappa^2\right] \tag{6}$$

where $\kappa = J+1\,(1, 2, \ldots)$ for the R-branch $(J \to J+1)$ and $\kappa = -J\,(-1, -2, \ldots)$ for the P-branch $(J \to J-1)$. This equation describes a parabola (the Fortrat-parabola). If $B_v \leq B_{v'}$ there will be a bunching of high-J transitions in the R-branch and $B_v \geq B_{v'}$ in the P-branch. This natural pile-up of high excitation ro-vibrational transition is called a *band-head* and can be used to sensitively detect highly excited gas.

1.1.e Transition Rates
The transition rate (s^{-1}) for spontaneous radiative transitions between an upper (u) and a lower (l) level is described by the Einstein A coefficient,

$$A_{ul} = \frac{64\pi^4 \mu_{ul}^2}{3h\lambda_{ul}^3} = 0.3\,\lambda_{100}^{-3}\,\mu_d^2 \quad [s^{-1}]. \tag{7}$$

μ_{ul} is a transition moment (matrix element) that has the dimensions of an electric dipole moment. μ_d is in units of Debye $(1\,\text{Debye} = 10^{-18}\,\text{esu cm})$. It represents the dipole moment of an electron at a distance of about $0.2\,\text{Å}$ from a proton. λ_{100} is the wavelength of the transition in units of $100\,\mu m$. The Einstein coefficient can also be written in terms of the *oscillator strength* f_{ul} of a transition,

$$A_{ul} = -6.7 \times 10^3 f_{ul}\,\lambda_{100}^{-2} \quad [s^{-1}]. \tag{8}$$

For a system with statistical weights g_u, g_l, in the upper and lower levels one has $f_{ul} = -(g_l/g_u)f_{lu}$. As a rough guide, molecular electronic transitions or rotational transitions of polar molecules have the largest moments, $\mu_{ul} \approx 1 \to 3\,\text{Debye}$. Important exceptions are non-polar molecules, such as CO $(\mu_{rot} \approx 0.112\,\text{Debye})$, or HD $(\mu_{rot} \approx 5.8 \times 10^{-4}\,\text{Debye})$. Recombination lines have $A_{k+1,k} \approx 6 \times 10^9/k^5 = 1.6 \times 10^4\,\lambda_{100}^{5/3}\,s^{-1}$ or $f \approx 1$. $\Delta v = 1$ vibrational transitions have $\mu_{vib} \approx 0.1\,\text{Debye}$, while the weaker overtone $(\Delta v = 2)$ transitions have $\mu_{vib}^{overtone} \approx 6 \times 10^{-3}\,\text{Debye}$. That is of the same magnitude as the transition moments of most fine structure lines $(\mu_{fs} \approx 3$ to $8 \times 10^{-3}\,\text{Debye})$.

Finally the quadrupole transitions of H_2 are the weakest with $\mu_{Quad}^{H_2} \approx 3 \times 10^{-6}\,\text{Debye}$. For rotational transitions, the exact value of μ_{ul} can be directly calculated from the permanent electric dipole moment, μ_{el}, that is listed in

Table 2. Selected list of molecular lines

MOLECULE	LOWEST ROTATIONAL LINE [μm]	ROTATIONAL DIPOLE MOMENT [Debye]	VIBRATIONAL TRANSITIONS [μm]
H_2	28.1	0	2.3
HD	112.1	5.8×10^{-4}	2.6
OH	119	1.7	2.7
NH	302		3.0
CH	149,562	1.46	3.5
HCl	479	1.11	3.3
CO	2600	0.112	4.8
CS	6110	2	7.8
HCN	3440	3	3.0, 4.9, 14
HCO^+	3360	3.3	
NH_3	524	1.5	2.9, 6.1, 10.5
H_2O	538,269	1.85	2.7, 6.3
HDO	590	1.84	3.0, 3.7, 7.1
H_2D^+	806		
H_3O^+	760→830	≈ 1	
O_2	5350	0	
SiO	6910	3.1	8.1

Table 2 for a number of astronomically interesting molecules. For diatomic and linear poly-atomic molecules one finds

$$\mu^2_{J+1,J} = \frac{J+1}{2J+3}\, \mu^2_{el}. \tag{9}$$

For symmetric tops one has

$$\mu_{(J,K)\to(J-1,K)} = \frac{J^2 - K^2}{J(2J+1)}\, \mu^2_{el}. \tag{10}$$

1.2 Radiative Transport and Excitation

In the following section, we will discuss the basic physics of radiative transport, excitation and interpretation of infrared/radio line, and continuum radiation.

1.2.a Simple Radiative Transport

Consider radiative transport of a line at frequency ν (wavelength λ) in a plane-parallel cloud. The change of specific line intensity I_ν [erg s^{-1} cm^{-2} Hz^{-1} sr^{-1}] in an infinitesimal element dz along the line of sight through the cloud is given by

$$\frac{dI_\nu}{dz} = \frac{h\nu}{4\pi}\phi(\nu)\, n_u\, A_{ul} - \frac{h\nu}{4\pi}\phi(\nu)\, I_u\, (n_u B_{ul} - n_l B_{lu})$$

$$= \epsilon_\nu - \kappa_\nu I_\nu. \tag{11}$$

n_u, n_l are the volume densities [cm^{-3}] of molecules in the upper and lower states of the transition, with statistical weights g_u and g_l. The line shape function

287

$\phi(\nu)$ gives the probability per frequency interval that a photon is emitted at ν $\left(\int\phi(\nu)\,d\nu = 1\right)$. It is easy to see that the first term in Eq. (11) represents the volume emissivity ϵ_ν, that is, the energy emitted per cm^3, sec, frequency interval and solid angle element through spontaneous emission. The second term takes into account the combined effects of absorption and stimulated emission. $B_{ul}I_\nu$ and $B_{lu}I_\nu$ $[s^{-1}]$ give the corresponding rates per sec per molecule for stimulated radiative processes $(A_{ul} = 2h\nu^3 c^{-2} B_{ul}$ and $B_{ul}/g_l = B_{lu}/g_u)$. The stimulated processes are characterized by the *absorption coefficient* κ_ν $[cm^{-1}]$, or *optical depth* τ_ν, given by

$$\tau_\nu(z) = -\int_0^z \kappa_\nu\,dz = \int_0^z \frac{\phi(\nu)\,A_{ul}c^2}{8\pi\nu^2}\left(\frac{g_u\,n_l}{g_l\,n_u} - 1\right)n_u\,dz. \qquad (12)$$

Another important quantity is the *source function* Σ_ν, defined as

$$\Sigma_\nu = \frac{\epsilon_\nu}{\kappa_\nu} = \frac{2h\nu^3}{c^2}\left[\frac{g_u n_l}{g_l n_u} - 1\right]^{-1} = \frac{2h\nu^3}{c^2}\left[\exp\left(\frac{h\nu}{kT_{ex}}\right) - 1\right]^{-1}. \qquad (13)$$

The so defined *excitation temperature* T_{ex} is an equivalent Planck temperature that describes the population of states l and u through a thermal (Boltzmann) population. With Eqs. (12) and (13) and an assumed spatially *constant* population (constant Σ_ν, T_{ex}) Eq. (11) can be easily integrated to give the familiar equation

$$I_\nu(\text{observed}) = I_\nu(\text{background})\exp(-\tau_\nu) + \Sigma_\nu\left(1 - \exp(-\tau_\nu)\right). \qquad (14)$$

τ_ν is now the optical depth at ν through the cloud. A standard application is the case when the background source emits independent of frequency (continuum intensity I_B) and the emission of the cloud is pure line radiation. In that case the line intensity in excess of the continuum ΔI_{line} is

$$\Delta I_{line} = I_\nu(\text{observed}) - I_B = \left(\Sigma_\nu(T_{ex}) - I_B\right)\left(1 - \exp(-\tau_\nu)\right). \qquad (15)$$

In real astronomical measurements one often deals with sources that do not fill the telescope beam. In that case Eq. (15) is modified by beam area filling factors Φ_{AL}, Φ_{AB} for line and background continuum respectively,

$$\langle\Delta I_{line}\rangle_{beam} = \left(\Phi_{AL}\Sigma_\nu(T_{ex}) - \Phi_{AB}I_B\right)\langle 1 - \exp(-\tau_\nu)\rangle. \qquad (16)$$

Brackets $\langle\rangle$ indicate averages over the beam. The line appears in emission, as long as $\Phi_{AL}\Sigma_\nu > \Phi_{AB}I_B$. Otherwise it is in absorption. For optically thin ($\tau_\nu \ll 1$) emission, $1 - \exp(-\tau_\nu) \to \tau_\nu$; for optically thick emission $1 - \exp(-\tau_\nu) \to 1$.

In the radio region one usually has $h\nu \ll kT$ so that the *Rayleigh-Jeans approximation* $\left(\exp(h\nu/kT) \approx 1 + h\nu/kT\right)$ is applicable. In that case it is convenient to express intensities in terms of *Rayleigh-Jeans radiation temperatures* $\tilde{T}_R$ [K],

$$I_\nu = \frac{2h\nu^3}{c^2} \left[\exp\left(\frac{h\nu}{kT_R}\right) - 1 \right]^{-1} = \frac{2k\nu^2}{c^2} \tilde{T}_R,$$

$$I_B = \frac{2h\nu^3}{c^2} \left[\exp\left(\frac{h\nu}{kT_B}\right) - 1 \right]^{-1} = \frac{2k\nu^2}{c^2} \tilde{T}_B, \qquad (17)$$

$$\text{and} \qquad \Sigma_\nu = \frac{2k\nu^2}{c^2} \tilde{T}_{ex}.$$

For $h\nu/k < T_R$, T_{ex} and T_B one has $\tilde{T} \approx T - \frac{1}{2}(h\nu/k)$. T_R and T_B are called (Planck)-*brightness* temperatures. Equation (16) turns into

$$\langle \Delta T_{line} \rangle = \left(\Phi_{AL} \tilde{T}_{ex} - \Phi_{AB} \tilde{T}_B \right) \langle 1 - \exp(-\tau_\nu) \rangle. \qquad (18)$$

Equation (12) becomes

$$\tau_\nu = \frac{hc^2}{8\pi\nu^2 k} A_{ul} \left(\frac{\nu}{\Delta\nu_0} \right) \left(\frac{N_u}{T_{ex}} \right) \qquad \text{for } h\nu \ll kT_{ex}. \qquad (19)$$

$N_u = \int_0^z n_u \, dz$ is the *column density* [cm^{-2}] of particles in the upper state through the cloud, $\Delta\nu_0$ is the equivalent width of the line $\phi(\nu) = 1/\Delta\nu_0$. Often one can approximate the beam shape as a circular Gaussian of full width at half maximum (FWHM) θ_B. If the (line or continuum) source can also be described by a two-dimensional Gaussian distribution of FWHM θ_{sx} and θ_{sy} (in the x and y coordinates), the filling factors in Eqs. (16) or (18) can be written as

$$\Phi_A = \frac{\theta_{sx}}{\sqrt{\theta_{sx}^2 + \theta_B^2}} \frac{\theta_{sy}}{\sqrt{\theta_{sy}^2 + \theta_B^2}}. \qquad (20)$$

Another case that occurs often is a disk shaped source of diameter θ_D. In that case one finds

$$\Phi_A = 1 - \exp \left[-(\ln 2) \left(\frac{\theta_D}{\theta_B} \right)^2 \right]. \qquad (21)$$

Radiative Transport with Intermixed Line and Continuum. An interesting special case occurs if line and continuum emission come from the same volume. Equation (11) is then replaced by the more general equation

$$\frac{dI_\nu}{dz} = -\kappa_{tot} I_\nu + \kappa_1 \Sigma_1 + \kappa_c \Sigma_c. \qquad (22)$$

Σ_1 and Σ_c are the source functions of line and continuum emission, κ_1 and κ_c are the absorption coefficients of line and continuum at ν and $\kappa_{tot} = \kappa_1 + \kappa_c$. For $\Sigma_c \approx \Sigma_1$ and no background continuum one can easily integrate Eq. (22) to obtain

$$\Delta I_{line} = \Sigma_1 (1 - \exp(-\tau_1)) \exp(-\tau_c). \qquad (23)$$

In comparison to Eq. (15), the line to continuum contrast is thus reduced by a factor $\exp(-\tau_c)$. Important applications are radio recombination lines in an optically thick HII region, or the measurement of far-infrared rotational emission or absorption of HD and H_2 in the presence of dust in the same cloud. For instance, the ground-state rotational transition of H_2 $(S(0)$ at $28\,\mu\text{m})$ has an Einstein A coefficient of only $2.9 \times 10^{-11}\,\text{s}^{-1}$. For normal dust to gas abundances $(10^{-2}$ in mass) and $T_{\text{gas}} = T_{\text{dust}}$ one then finds $\kappa(S(0))/\kappa_{\text{dust}}(28\,\mu\text{m}) \approx 1/200$. The largest possible contrast between line and neighboring continuum is 0.5% occurring in the case of optically thin dust emission. Similarly the ground rotational transition of HD at $112.1\,\mu\text{m}$ $(A = 2.54 \times 10^{-8}\,\text{s}^{-1})$ has $\kappa_{\text{HD}}/\kappa_{\text{dust}} \approx 0.1$ for $[\text{D}]/[\text{H}] \approx 10^{-5}$.

1.2.b Optical Depth

The discussion of the Einstein coefficient above has already shown that the optical depth depends on $\mu_{\text{ul}}^2 \propto (A_{\text{ul}}/\nu_{\text{ul}}^3)$ and is generally greater for electronic transitions or molecular rotational transitions than for molecular vibrational and atomic fine structure transitions. In addition the populations of the upper and lower levels and the abundance of the species are important for the evaluation of the optical depth. We can write Eq. (12) as

$$\tau = \frac{c^2}{8\pi} \left(\frac{A}{\nu^3}\right) \left(\frac{\nu}{\Delta\nu_0}\right) \left[\exp\left(\frac{h\nu}{kT_{\text{ex}}}\right) - 1\right] N_{\text{u}}. \tag{24}$$

In thermal equilibrium at temperature T the column density N_{u} can be expressed as

$$N_{\text{u}} = g_{\text{u}} \exp\left(\frac{-E_{\text{u}}}{kT}\right) \frac{N}{Q(T)}. \tag{25}$$

$Q(T) = \sum_i g_i \exp(-E_i/kT)$ is the partition function. E_i is the energy of level i and N the total number of atoms (molecules) per cm^2. For rotational levels of a diatomic of linear polyatomic molecule $Q(T_{\text{rot}}) = kT_{\text{rot}}/hB$ for a rotational constant B in units of Hz. For instance the CO molecule $(hB_{\text{CO}}/k = 2.8\,\text{K})$ has $Q = 18$ at $T = 50\,\text{K}$ and $Q = 360$ at $1000\,\text{K}$. T_{rot} is then called the *rotational temperature*. For rotational levels of a polyatomic, symmetric top molecule $(B_1 = B_2 \neq B_3)$ we have $Q(T_{\text{rot}}) = 1.8\,(kT_{\text{rot}}/hB_1)^{3/2}\,(B_1/B_3)^{1/2}$. For vibrational energy levels at a *vibrational temperature* T_{vib} one has $Q(T_{\text{vib}}) = (1 - \exp(-h\nu_{\text{vib}}/kT_{\text{vib}}))^{-1}$. Usually the vibrational partition function is close to 1. For instance, at $T_{\text{vib}} = 1000\,\text{K}$ the vibrational partition function of the CO molecule $(h\nu_{\text{vib}}/k = 3125\,\text{K})$ is 1.05.

In astronomical applications high-J or high-v states are usually subthermally populated. Realistic partition functions thus increase more slowly with temperature than in thermal equilibrium. $Q(T) = \text{const.}$ is often a good approximation. For most astronomical applications the line width $\Delta\nu_0$ is determined by Doppler broadening with a velocity width $\Delta v = c\,\Delta\nu_0/\nu$. For infrared transitions one usually has $h\nu > kT_{\text{ex}}$ and $N_{\text{u}} \ll N_1 \approx N_{\text{tot}}$ so that

$$\tau_{\text{IR}} \propto \mu_{\text{ul}}^2 \frac{N_{\text{tot}}}{\Delta v\, Q(T)}, \tag{26}$$

while for radio transitions $(h\nu/kT_{\text{ex}} \ll 1)$

$$\tau_{\text{rad}} \propto \mu_{\text{ul}}^2 \frac{h\nu_{\text{rad}}}{kT_{\text{ex}}} \frac{N_{\text{tot}}}{\Delta v \, Q(T)}. \qquad (27)$$

The ratio of infrared to radio optical depths of a given molecule (atom) then is simply

$$\frac{\tau_{\text{IR}}}{\tau_{\text{rad}}} \approx \left(\frac{\mu_{\text{IR}}}{\mu_{\text{rad}}}\right)^2 \frac{kT_{\text{ex}}}{h\nu_{\text{rad}}}. \qquad (28)$$

For typical interstellar clouds $kT_{\text{ex}}/h\nu_{\text{rad}}$ ranges between 10 and 100. Rotational transitions of molecules in the infrared/submillimeter range have thus one to two orders of magnitude greater optical depths than millimeter and centimeter transitions. For abundant molecules, such as CO, HCN, NH_3, H_2O, etc., far-infrared and submillimeter lines in dense molecular clouds can be very optically thick ($\tau \approx 10^2 \to 10^3$). For ro-vibrational transitions in the mid-IR $(\mu_{\text{IR}}/\mu_{\text{rad}})^2 \approx 10^{-2}$ so that they have similar optical depths as radio transitions.

1.2.c Rotational Emission of Linear Molecules

Diatomic or linear polyatomic molecules produce emission lines that are equidistant in frequency space ($\nu_{J\to J-1} = 2BJ$). For the abundant heavy top molecules in interstellar space (CO, CS, CN, HCN) these rotational transitions lie in the mm- or submm-band.

In the Rayleigh-Jeans regime, the optical depth at ν_J is proportional to $(\exp(h\nu_J/kT_{\text{ex}}) - 1) N_J \sim \nu_J g_J \sim J^2$. The integrated intensity and brightness temperature of spectral lines at ν_J in the optically thin and optically thick limits then are

$$\left. \begin{aligned} I_J &= \frac{2k\nu_J^2}{c^2} T_{\text{ex}} \tau(\nu_J) \, \Delta v \propto J^5 \\ T_J &= T_{\text{ex}} \tau(\nu_J) \propto J^2 \end{aligned} \right\}, \qquad \tau(\nu_J) \ll 1,$$

and $\qquad\qquad\qquad\qquad\qquad\qquad\qquad \left(\text{for } \dfrac{h\nu_J}{kT_{\text{ex}}} < 1\right). \quad (29)$

$$\left. \begin{aligned} I_J &= \frac{2k\nu_J^2}{c^2} T_{\text{ex}} \, \Delta v \propto J^3 \\ T_J &= T_{\text{ex}} \end{aligned} \right\}, \qquad \tau(\nu_J) \gg 1,$$

The maximum emission in a rotational ladder populated at temperature T_{rot} occurs approximately at $E_J = hBJ(J+1) \approx kT_{\text{rot}}$, or

$$J_{\text{max}} \approx \sqrt{\frac{T_{\text{rot}}}{hB/k}}. \qquad (30)$$

The abundant CO molecule ($hB/k = 2.76$ K), for instance, has its strongest line emission in the 1.3 mm $J = 2 \to 1$ transition for a cold cloud of $T = 10$ K. A warm cloud ($T = 50$ to 100 K), on the other hand, emits strongest in the mid-J ($J = 4$ to 7) transitions at submillimeter wavelengths. Finally, for CO

Table 3. Spectroscopic data for ro-vibrational transitions in some molecules (from Evans et al. 1991)

Molecule	Band	ν_0 [cm^{-1}]	μ_v [Debye]	A [s^{-1}]	S_v^i $\left[\dfrac{10^{-17}\,\mathrm{cm}^{-1}}{\mathrm{cm}^{-2}}\right]$	S $\left[\dfrac{10^{-17}\,\mathrm{cm}^{-1}}{\mathrm{cm}^{-2}}\right]$
C_2H_2	ν_5	729	0.22	3.9	2.91	1.46
C_2H_2	$\nu_4 + \nu_5$	1328	0.075	1.4	0.31	0.31
HCN	ν_2	712	0.13	1.3	1.0	0.51
OCS	ν_1	2062	0.41	154	14.4	14.4
NH_3	ν_2	968	0.24	5.2	2.35	2.35
HNC	ν_3	2024	0.13	15	1.5	1.5
CO		2143	0.108	12	1.0	1.0
^{13}CO		2096	0.107	11	1.0	1.0

line emission to be strongest at far-IR wavelengths ($\lambda \approx 50$ to $200\,\mu$m, $J = 20$ to 40), temperatures of about 1000 K are required.

1.2.d Ro-Vibrational Absorption Spectroscopy

Absorption spectroscopy of ro-vibrational molecular line against the continua of bright near-infrared and mid-infrared sources is a powerful tool for investigating circumstellar gas. The present discussion follows Evans et al. (1991). The *equivalent width* of an absorption line is defined as

$$\Delta\nu_{\mathrm{eq}} = \int \left(1 - \frac{I_\nu}{I_c}\right) d\nu \qquad [\mathrm{cm}^{-1}], \tag{31}$$

where I_ν is the (line + continuum) intensity at ν and I_c the (off line) continuum intensity. The equivalent width of a molecular ro-vibrational transition from the ground vibrational state ($v = 0 \rightarrow v = 1$, $J \rightarrow J \pm \Delta J$) is conveniently expressed as

$$\Delta\nu_{\mathrm{eq}} = \left(\frac{\nu_{J,v}}{\nu_{0,v}}\right) \frac{L(J, \Delta J)}{g_J} S_v N_J. \tag{32}$$

Here $\nu_{J,v}$ and $\nu_{0,v}$ are the frequencies of the transition under study and at the band center, respectively. $g_J = 2J + 1$ is the statistical weight and N_J the column density in the lower state. $S_v = \frac{1}{2}\sum_J \left(\int \kappa_\nu(J, v)\, d\nu / N_{\mathrm{mol}}\right)$ is the so-called *band-strength* ([cm^{-1}]/[cm^{-2}]) of the entire ro-vibrational band.

Band strengths or strengths for individual ro-vibrational lines are given in the appropriate molecular spectroscopy references but it is important to correct for deviations from room-temperature values that are usually in the laboratory spectroscopy literature (see Evans et al. 1991 for details). Table 3 lists the band strengths of some prominent ro-vibrational bands. $N_{\mathrm{mol}} = \sum_J N_J$ is the total molecular column density. Finally $L(J, \Delta J)$ is the Hönl-London factor of the transition. For diatomic molecules and stretching modes of linear polyatomic molecules $L(J, \Delta J) = J + \Delta J$. For bending modes of linear molecules $L(J, \Delta J) = J + 2$ for $\Delta J = +1$, $J - 1$ for $\Delta J = -1$ and $2J + 1$ for $\Delta J = 0$ (Q-branch). The above expression applies only to optically thin gas.

For optically thick gas with Gaussian line profile, line center optical depth τ_0 and intrinsic FWHM line width $\Delta\nu^0_{\text{FWHM}}$ a curve of growth relates equivalent width and $\Delta\nu^0_{\text{FWHM}}$

$$\Delta\nu_{\text{eq}} = 1.064\,\Delta\nu^0_{\text{FWHM}}\ln(1 + \tau_0) \tag{33}$$

for $\tau_0 \leq 10^2$.

1.2.e Excitation of Line Radiation

Collisional Excitation. Consider a simple system consisting of only two levels, spaced by $\Delta E_{\text{ul}} = h\nu$. Assume further that transitions l→u and u→l are triggered by collisions with a collision partner of volume density n. The collisions are described by rates per sec per molecule or atom C_{ul} and C_{lu},

$$C_{\text{ul}} \equiv C_{\text{u}\to\text{l}} = n\gamma_{\text{ul}} = n\,\langle\sigma_{\text{ul}}v\rangle$$

and $\tag{34}$

$$C_{\text{lu}} \equiv C_{\text{l}\to\text{u}} = C_{\text{ul}}\frac{g_{\text{u}}}{g_{\text{l}}}\exp\left(\frac{-h\nu}{kT_{\text{kin}}}\right).$$

σ_{ul} is the *cross section* [cm^2] for a collision u→l at velocity v, γ_{ul} [cm^3 s^{-1}] is the overall *collisional rate coefficient* u→l. It can be expressed as an average over all possible collision energies weighted by the distribution function that is assumed to be a Maxwell-Boltzmann distribution at kinetic temperature T_{kin},

$$\gamma_{\text{ul}} = \langle\sigma_{\text{ul}}v\rangle = \frac{4}{\sqrt{\pi}}\left(\frac{\mu}{2kT_{\text{kin}}}\right)^{\frac{3}{2}}\int_0^\infty dv\,\sigma_{\text{ul}}(v)\,v^3\exp\left(\frac{-\mu v^2}{2kT_{\text{kin}}}\right), \tag{35}$$

where μ is again the reduced mass of molecule and collision partner. Typical values of γ_{ul} for neutral-neutral collisions range between 10^{-11} and 10^{-10} cm^3 s^{-1}, while neutral-charged particle collisions have collisional rate coefficients of about 10^{-9} cm^3 s^{-1}.

The rate coefficients γ are usually obtained by detailed quantum mechanical calculations (see Flower 1987 for a review). Table 4 gives a list of references of cross sections for atom-atom, molecule-molecule, and atom-molecule collisions for a number of astronomically important species. Rate coefficients are typically uncertain by a factor of 2 for atoms and sometimes as much as an order of magnitude for collisions of complex atoms with H_2. Calculations often refer to collisions of molecules with helium atoms. Scaling to molecule-H_2 collisions is then done by the $\mu^{1/2}$ dependence apparent from Eq. (35) (resulting in a multiplication of the He-cross sections by 1.37). This neglects that the potentials of H_2 and He are quite different, however (quadrupole terms!).

In the specific case of collisions of electrons with ions it is customary to express cross sections and collisional rate coefficients in terms of a collision strength $\Omega_{\text{lu}}\ (= \Omega_{\text{ul}})$,

$$\gamma_{\text{ul}} = 8.63 \times 10^{-6}\frac{\Omega_{\text{lu}}}{g_{\text{u}}T_{\text{kin}}^{1/2}}\qquad[\text{cm}^3\ \text{s}^{-1}], \tag{36}$$

Table 4. Reference list for cross sections and rate coefficients for astrophysically important atoms/molecules

SPECIES	COLLISION PARTNER	REFERENCE
		A compendium of a fairly large number of references for <u>ion-electron</u> cross sections: Mendoza, C. in *Planetary Nebulae*, ed. D.R. Flower, 143 (Reidel, Dordrecht, 1983)
C^+	e	Haye & Nussbaumer, *Astr. Ap.* **134**, 193 (1984)
	H	Launay & Roueff, *J. Phys. B.* **10**, 879 (1977)
	H_2	Flower & Launay, *J. Phys. B.* **10**, 3673 (1977)
	H_2	Flower, *J. Phys. B.* **21**, L451 (1988)
C^0	e	Johnson et al., *J. Phys. B.* **20**, 2553 (1987)
	p	Roueff & Le Bourlot, *Astr. Ap.* **236**, 515 (1990)
	H	Launay & Roueff, *Astr. Ap.* **56**, 289 (1977)
	He	Schröder et al., *J. Phys. B.* **24**, 2487 (1991)
	H_2	Stämmler & Flower, *J. Phys. B.* **24**, 2343 (1991)
O^0	e	Le Dourneuf & Nesbet, *J. Phys. B.* **9**, L241 (1976)
	H	Launay & Roueff, *Astr. Ap.* **56**, 289 (1977)
	H	Federman & Shipsey, *Ap. J.* **269**, 791 (1983)
	He	Monteiro & Flower, *MNRAS* **228**, 101 (1987)
	H_2	Flower, *MNRAS* **242**, 1p (1990)
	H_2	Jaquet et al., *J. Phys. B.* (1991)
CO	He	Green & Chapman, *Ap. J. Suppl.* **37**, 169 (1978)
	He	McKee et al., *Ap. J.* **259**, 647 (1982)
	H_2	Flower & Launay, *MNRAS* **214**, 271 (1985)
	H_2	Schinke et al., *Ap. J.* **299**, 839 (1985)
CS OCS, HC_3N	He	Green & Chapman, *Ap. J. Suppl.* **37**, 169 (1978)
HCN	He	Green & Thaddeus, *Ap. J.* **191**, 653 (1974)
	H_2	Stutzki & Winnewisser, *Astr. Ap.* **144**, 1 (1985)
	H_2	Monteiro & Stutzki, *MNRAS* **221**, 33p (1986)
	H_2	Stutzki et al., *Ap. J.* **330**, L125 (1988)
OH	H_2	Dewangan et al., *J. Phys. B.* **19**, L747 (1986)
	H_2	Dewangan et al., *MNRAS* **226**, 505 (1987)
	H_2	Schinke & Andresen, *J. Chem. Phys.* **81**, 5644 (1984)
	H_2	Offer & Flower, *J. Phys. B.* **23**, L391 (1990)
H_2O	He	Green, *Ap. J. Suppl.* **42**, 103 (1980)
	He	Palma et al., *Ap. J. Suppl.* **68**, 287 (1988)
HDO	He	Green, *Ap. J. Suppl.* **70**, 813 (1989)
H_2CO	He	Garrison et al., *J. Chem. Phys.* **65**, 2193 (1976)
	He	Green et al., *Ap. J. Suppl.* **37**, 321 (1978)
HCO^+	He	Monteiro, *MNRAS* **210**, 1 (1984)
	H_2	Monteiro, *MNRAS* **214**, 419 (1985)

Table 4. *(continued)*

SPECIES	COLLISION PARTNER	REFERENCE
NH_3	He	Green, *NASA Techn. Memo* **83869** (1981)
	He	Billing & Diercksen, *Chem. Phys. Ltr.* **121**, 94 (1985)
	He	Davis, *J. Chem. Phys.* **95**, 411 (1985)
	H_2	Stutzki & Winnewisser, *Astr. Ap.* **144**, 1 (1985)
	H_2	Danby et al., *J. Phys. B.* **19**, 2891 (1986)
	H_2	Danby et al., *J. Phys. B.* **20**, 1039 (1987)
	H_2	Offer & Flower, *J. Phys. B.* **22**, L439 (1989)
	H_2	Offer & Flower, *J. Chem. Soc. Faraday Trans.* **86**, 1659 (1990)
CH_3CN	He	Green, *Ap. J.* **309**, 331 (1986)

$O(\Omega_{1u}) \approx 1$ and lists of collision strengths for specific ion-electron combinations are given in Osterbrock (1989) and Spitzer (1978).

Given rate coefficients γ_{ul}, Einstein coefficient A_{ul}, kinetic temperature T_{kin} and density n, it is then straightforward to calculate the populations n_u and n_l $(n_u + n_l = n_{tot})$ from detailed balance,

$$n_l(\gamma_{lu}n) = n_u(\gamma_{ul}n + A_{ul}),$$

$$\text{or,} \qquad \frac{n_u}{n_{tot}} = \frac{\dfrac{g_u}{g_l}\exp\left(\dfrac{-h\nu}{kT_{kin}}\right)}{1 + \dfrac{g_u}{g_l}\exp\left(\dfrac{-h\nu}{kT_{kin}}\right) + \dfrac{n_{crit}}{n}}. \qquad (37)$$

$n_{crit} = A_{ul}/\gamma_{ul}$ is called the *critical density*. The critical density is a key parameter for a given transition as it indicates at what density collisions can keep up with spontaneous radiative transitions. Equation (37) shows that for very high density $(n \gg n_{crit})$ the populations of the levels u, l are thermalized at kinetic temperature T_{kin}: $(n_u/n_l)_{thermal} = (g_u/g_l)\exp(-h\nu/kT_{kin})$. Below the critical density n_{crit}, however, spontaneous radiation rates u→l are much faster than collisions so that each collision l→u leads to photon emission; the population is subthermal: $n_u/n_l = (n/n_{crit})(n_u/n_l)_{thermal}$. With Eq. (37) the line flux for a simple two level system in the optically thin limit can then be easily expressed as

$$F_{line} = \frac{h\nu}{4\pi}A_{ul}\Omega\int n_u\,dz$$

$$n < n_{crit}: \qquad = \frac{h\nu}{4\pi}\gamma_{ul}\Omega\left(\frac{g_u}{g_l}\right)\exp\left(\frac{-h\nu}{kT_{kin}}\right)\int n_{tot}\,n\,dz \propto n^2, \qquad (38)$$

$$n > n_{crit}: \qquad = \frac{h\nu}{4\pi}A_{ul}\Omega\left(\frac{g_u}{g_l}\right)\exp\left(\frac{-h\nu}{kT_{kin}}\right)\int n_{tot}\,dz \propto N_{tot}.$$

Ω is the solid angle of source or beam and the integral is along the line of sight through the source. In the subthermal regime line fluxes are proportional to (density)^2 ($\equiv$ emission measure) while they are proportional to column density in the thermalized regime. The necessary corrections to Eq. (38) for optically thick lines are discussed below.

In low density gas, the abundance $n_{\rm tot}/n({\rm H_2})$ of a molecule/atom is very important for the emissivity. That is why the low-J mm-transitions of the abundant $^{12}{\rm CO}$ molecule tend to emphasize emission from lower density, extended cloud envelopes. Less abundant species with a high critical density, such as CS or HCN, on the other hand, preferentially sample the high volume and column density cloud cores where the $^{12}{\rm CO}$ emission is optically thick and thermalized.

The extension to a multilevel system is straightforward. The rate equation equivalent to Eq. (36) for level u then becomes

$$ n_{\rm u} \left[\sum_{k \neq u} (n\gamma_{\rm uk} + R_{\rm uk}) + \sum_{k < u} A_{\rm uk} \right] = \sum_{u \neq k} n_{\rm k} (n\gamma_{\rm ku} + R_{\rm ku}) + \sum_{k > u} A_{\rm ku} n_{\rm k}. \quad (39) $$

In Eq. (39) the levels are numbered such that levels with k > u have $E_{\rm k} > E_{\rm u}$ and vice versa. For the purpose of generality we have added in addition to collisional excitation, deexcitation and spontaneous emissions also rates due to stimulated emission and absorption, $R_{\rm uk}$, $R_{\rm ku}$ [s^{-1}], where

$$ R_{\rm uk} = B_{\rm uk} \int_{4\pi} \frac{d\Omega}{4\pi} \left(I_{\rm line} + I_{\rm cont} \right). \quad (40) $$

$I_{\rm line}$, $I_{\rm cont}$ are wavelength/frequency integrated intensities of line and continuum radiation that the molecule/atom sees. This extension takes into account radiative excitations. In optically thin gas, the set of rate equations (39) for $u = 1, \ldots u_{\rm max}$ can be solved for given n, $T_{\rm kin}$, $I_{\rm cont}$ through a matrix inversion. As soon as the line intensity itself contributes to the excitation, and optical depths become appreciable, however, the rate equations become nonlinear and have to be solved iteratively.

Spectral Lines as Probes of Physical Conditions in Interstellar Clouds. Simple considerations involving the upper state energies, wavelengths, and critical densities of various molecular and atomic transitions at radio and infrared wavelengths as given in Tables 1 and 2, already give fairly good first order impressions which lines probe different astrophysical environments (see also discussion by Scoville 1984).

Figure 7 shows the location of the various infrared fine structure lines in a plane of *excitation potential* (energy required to create the ion) and critical density. One way of looking at this diagram is that in a medium with a wide range of densities one empirically expects that a fine structure line of a particular ion X is most sensitive to densities $n \approx n_{\rm crit}$. Atoms with excitation potentials < 13.6 eV sample atomic and molecular clouds. The critical densities of most infrared and submillimeter fine structure lines range between 10^2 and 10^6 cm^{-3}, depending mainly on wavelength ($n_{\rm crit} \sim \lambda^{-3}$). Neutral species, such as O^0, Si^0, C^0, S^0, and ionized species with ionization potentials less that that of hydrogen, such as Si$^+$, C$^+$, and Fe$^+$ sample UV-excited or shock-excited, warm atomic

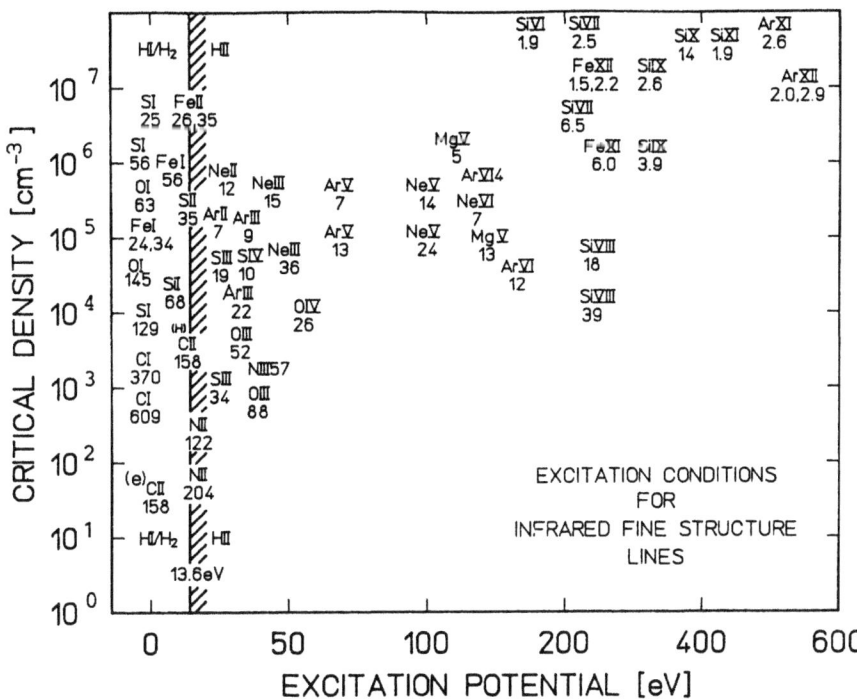

Fig. 7. Infrared fine structure lines as probes of physical conditions in interstellar clouds

gas at cloud surfaces between ionized (HII) zones and fully molecular regions. Ions with ionization potentials between 13.6 and < 40 eV are found in classical HII regions that are photoionized by massive stars. Ions with higher ionization potentials require either very hot stars for photoionization (e.g., planetary nebulae), or come from collisionally ionized, *coronal* gas. Another aspect is to consider different ionic transitions in a given ionization state. It is obvious that line ratios of two lines of a given ionic species but with different wavelengths can give direct estimates of electron densities: [OIII]52/88 μm, [SIII]19/34 μm, [ArIII]9/22 μm, [NeIII]15/36 μm, [ArV]7/13 μm, etc. Similarly, the two ratios of [CII]158 μm/ [OI]63 μm and [OI]145 μm/[OI]63 μm can be used for estimating density and temperature in warm, atomic clouds. As the latter case involves more than one species, the abundance ratio O/C and spatial coexistence have to be assumed, along with low optical depth (see Watson 1984).

Figure 8 shows how the phase space of different physical conditions is probed by different molecular transitions. Rotational lines of CO, for instance, require molecular hydrogen densities of about 4000 J^3[cm^{-3}] and temperatures of 2.7 $J(J+1)$ K for producing "strong" (i.e., thermalized) line emission in the $J \rightarrow J-1$ transition. The $1 \rightarrow 0$ and $2 \rightarrow 1$ millimeter CO transitions are thermalized in gas of molecular hydrogen density 10^3 to 10^4 cm^{-3} and temperatures 10 to 20 K. Mid-J CO submillimeter lines sample gas of hydrogen density a few 10^5 cm^{-3} and temperature ≈ 100 K. Those values span the typ-

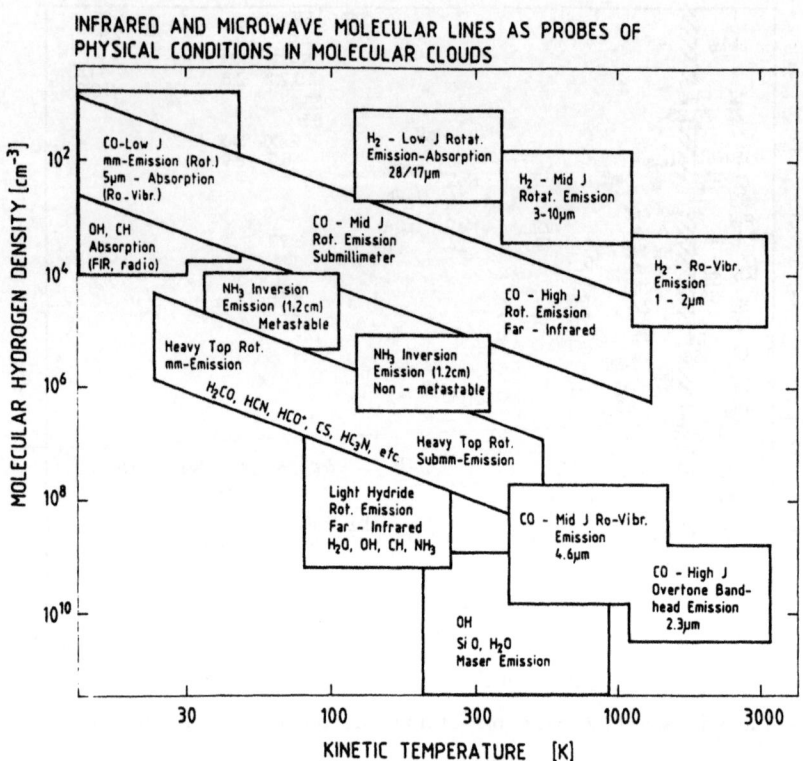

Fig. 8. Molecular lines as probes of physical conditions in interstellar clouds

ical range of physical parameter in molecular clouds. Far-infrared CO lines require a few 10^6 cm^{-3} and 1000 K for thermal population, conditions that are found in interstellar shocks. The same parameter space probed by rotational emission lines is also sampled by ro-vibrational lines in absorption against a hot background continuum source (e.g., the line of sight toward a newly formed star). CO ro-vibrational *emission* lines, on the other hand, require densities of 10^{10} to 10^{11} cm^{-3} and temperatures of a few 10^3 K, conditions that are typical of stellar atmospheres and their envelopes. Other heavy top molecules, such as HCN, CS, SiO, and HCO$^+$ are similar to CO but are less abundant and require typically 10^2 to 10^3 times greater densities for collisional excitation of their rotational states because of their greater electric dipole moments. As mentioned above, the H$_2$ molecule is a special case as it emits only through electric quadrupole transitions. As a result its rotational transitions are thermalized at very low densities $\approx 10^2$ J^3 cm^{-3}, and even the 2 μm ro-vibrational transitions require only densities of about 10^5 cm^{-3}. As the H$_2$ transitions occur in the mid- and near-IR, only hot molecular gas with temperatures between a few hundred and a few thousand K emit observable line radiation. Such physical conditions are typical of shocks or of UV heated gas at the surfaces of molecular clouds. Submillimeter and far-infrared transitions of light hydrides (H$_2$O,

OH, CH, NH_3) are ideal probes of warm ($\approx 10^2$ K) and very dense gas (10^8 to 10^{10} cm^{-3}) that may sample the envelopes and disks of newly formed stars as well as the cooler, downstream zones of shock fronts.

Radiative Excitation. In addition to excitation by collisions, the rate equations (39) contain radiative excitations. Radiative "pumping" by a background continuum source creates only net line photons when pump frequency and line frequency are different. Pumping at the line frequency itself is a scattering process that can only redistribute continuum radiation in frequency space (e.g., P-Cygni profiles). Radiative pumping can thus be important for molecules with many interconnecting transitions (such as asymmetric rotors) and molecules that have strong transmissions in the mid- and far-infrared where continua are intense (e.g., light hydrides).

As an example, consider a molecular cloud illuminated by a thermal background source of temperature T_d and continuum optical depth τ_d that subtends a solid angle Ω_d as seen from the cloud. The solid angle averaged radiative rate in Eq. (40) then becomes

$$R_{ul} = \frac{B_{ul}}{4\pi} \int d\Omega\, I_{back} = \frac{B_{ul}}{4\pi} \left[\frac{2h\nu^3}{c^2}\, \Omega_d\, \frac{1 - \exp(-\tau_d)}{\exp\left(\dfrac{h\nu}{kT_d}\right) - 1} \right]$$

$$\equiv B_{ul} \frac{2k\nu^2}{c^2} \widetilde{T}_{eff}. \tag{41}$$

$\widetilde{T}_{eff}$ is the equivalent Rayleigh-Jeans radiation temperature of the continuum source averaged over all angles. In the limit of a simple two level system the line source function then becomes

$$\Sigma_{line} = \frac{2h\nu^3}{c^2} \left[\frac{g_u n_l}{g_l n_u} - 1 \right]^{-1} = \frac{2h\nu^3}{c^2} \left[\frac{g_u}{g_l} \left(\frac{A_{ul} + R_{ul} + C_{ul}}{R_{lu} + C_{lu}} \right) - 1 \right]^{-1}. \tag{42}$$

The equivalent Rayleigh-Jeans excitation temperature then becomes

$$\widetilde{T}_{ex} = \frac{\widetilde{T}_{eff} + \dfrac{h\nu}{k} \dfrac{n}{n_{crit}}}{1 + \dfrac{h\nu}{k} \dfrac{1}{\widetilde{T}_{kin}} \dfrac{n}{n_{crit}}}. \tag{43}$$

where $\widetilde{T}_{kin} = (h\nu/k)\left(1 - \exp(-h\nu/kT_{kin})\right)^{-1}$. Equation (43) is a general formulation of excitation through radiation and collisions. In the Rayleigh-Jeans limit $\widetilde{T}_{ex} \to T_{ex}$, $\widetilde{T}_{eff} \to T_{eff}$, and $\widetilde{T}_{kin} \to T_{kin}$.

For pure radiative excitation ($n \to 0$)

$$\widetilde{T}_{ex} = \widetilde{T}_{eff}, \tag{44}$$

a fairly obvious result that states that without internal collisions the excitation temperature equilibrates to the effective radiation temperature (e.g., the 3 K cosmic background temperature). Note that for $n = n_{crit}$ and $h\nu/k > \widetilde{T}_{eff}$

$$\tilde{T}_{\mathrm{ex}} = \frac{\tilde{T}_{\mathrm{eff}} + \frac{h\nu}{k}}{1 + \frac{h\nu}{k}\frac{1}{\tilde{T}_{\mathrm{kin}}}} \approx \left(\left(\frac{h\nu}{k}\right)^{-1} + \left(\tilde{T}_{\mathrm{kin}}\right)^{-1}\right)^{-1}. \tag{45}$$

This means that at $n = n_{\mathrm{crit}}$ an infrared transition has a relatively higher excitation temperature than a radio/mm transition. Examples of radiative pumping for line emission in massive star formation regions are the excitation of far-IR transitions of OH (Melnick et al. 1990) and of vibrationally excited HCN (Ziurys & Turner 1986) in Orion-KL.

Line Trapping. The discussion up to now has not taken into account the effect of line optical depth and trapping on the excitation that is also implicitly contained in Eqs. (39) and (40).

In order to qualitatively understand the effect of optical depth, let us consider the simple case of a cloud (volume V) that is emitting line photons in a thermalized transition. If the cloud is optically thin, a photon emitted anywhere within the volume of the cloud will escape without being reabsorbed. In this case the cloud's total line emission per sec is proportional to the number of molecules in the cloud and the spontaneous emission rate, $I_{\mathrm{thin}} \propto n_u V A$ [s^{-1}]. If the optical depth of the cloud is greater than one as the result of larger molecular volume or column densities, or smaller velocity gradients through the cloud, the probability for a once emitted photon to escape without reabsorption will be less than one. Roughly speaking, the escape probability β will scale inversely proportional to the optical depth τ, $\beta \approx 1/\tau$. Once a photon is reabsorbed in our simple model, it will be "destroyed" in a collisional deexcitation and cannot contribute to the net photon emission. The total line emission rate in the optically thick case is then $I_{\mathrm{thick}} \propto n_u V A\beta$. So the effect of "line trapping" in essence scales the spontaneous emission rate by β,

$$A \rightarrow A\beta(\tau) \approx \frac{A}{a\tau}, \tag{46}$$

where a is a numerical factor of order unity. More quantitatively, Eq. (40) can now be written in terms of the escape probability

$$R_{ul} = B_{ul} \int \frac{d\Omega}{4\pi} \int d\nu \, \phi(\nu - \nu_0) \left\{ I_B \exp(-\tau_\nu(z)) + \Sigma \left[1 - \exp(-\tau_\nu(z)) \right] \right\} \tag{47}$$
$$\approx B_{ul} I_B \beta_{\nu_0} + B_{ul} \Sigma (1 - \beta_{\nu_0}).$$

Here we have assumed that Σ does not vary through the cloud. The escape probability in a homogeneous, spherical or plane-parallel cloud with Gaussian line profile and so called "line average" optical depth τ ($\sqrt{\pi}$ times line center optical depth τ_0 (along a radius for spherical clouds)) can be approximated by (de Jong et al. 1980)

$$\beta_{\nu_o} = \int \frac{d\Omega}{4\pi} \int d\nu \, \phi(\nu - \nu_o) \exp\left(-\tau \phi(\nu - \nu_o)\right)$$

$$\approx \begin{cases} \dfrac{1 - \exp(-2.34\,\tau)}{2.34\,\tau}, & \tau < 7 \\[3mm] \dfrac{1}{2\tau \left[\ln\left(\tau/\sqrt{\pi}\right)\right]^{1/2}}, & \tau > 7. \end{cases} \tag{48}$$

Other forms of β depend on the velocity field. For instance in the Sobolev, or *large velocity gradient*, approximation (Castor 1970) one assumes that the radiative interaction length is much smaller than the cloud size. In that case, the escape probability of a plane parallel cloud can be expressed by the so called *Eddington approximation*

$$\beta_{\nu_0} = \frac{1 - \exp(-3\,\tau)}{3\,\tau}, \tag{49}$$

where $\tau = \kappa(\nu_0)\Delta v/(dv/dz)$. Δv is the observed line width and dv/dz the assumed line of sight velocity gradient.

Another effect of line trapping is that it drives a level toward thermalization that would be subthermal in the optically thin case. This can be immediately seen from the fact that for a fixed collision rate C, the ratio of effective spontaneous emission rate to collision rate $A\beta/C = n_{\mathrm{crit}}\beta/n$ is smaller for the optically thick ($\beta < 1$) than for the optically thin ($\beta = 1$) case. The generalization of the equation for line emission from a collisionally excited two level system (Eqs. (37) and (38)) for the optically thick case thus is

$$F_{\mathrm{line}} = \frac{h\nu}{4\pi} A\, \beta(\tau)\, N_{\mathrm{tot}} \, \frac{\dfrac{g_u}{g_1} \exp\left(\dfrac{-h\nu}{kT}\right)}{1 + \dfrac{g_u}{g_1} \exp\left(\dfrac{-h\nu}{kT}\right) + \dfrac{n_{\mathrm{crit}}\,\beta(\tau)}{n}}, \tag{50}$$

N_{tot} is the total column density of molecules (or atoms). For optically thick, thermalized emission ($n_{\mathrm{crit}}\beta/n \ll 1$) the line flux scales with $A\beta$, as discussed above. In the case of subthermal, optically thick emission ($n_{\mathrm{crit}}\beta/n > 1$ and $\beta < 1$), however, the line flux keeps growing linearly with N_{mol} since the β's in the numerator and denominator cancel each other. So an optically thick, but subthermal line behaves much like an optically thin and subthermal line. The difference is that the line trapping results in a greater population in the upper state and a higher excitation temperature T_{ex}. The situation is more complicated for multilevel systems but the basic ideas are much the same.

Line Profiles. Line profiles contain important physical information. Let us first consider the simple case of a homogeneous cloud with a Gaussian profile of FWHM $\Delta v_{\mathrm{local}}$ representing the thermal line width or local random motions (microturbulence). The line profile as a function of velocity shift from line center v can be written as

$$I_{\text{line}} = I_0(T_{\text{ex}})\Big[1 - \exp\big(-\tau(v)\big)\Big], \tag{51}$$

where $\tau(v) = \tau_0 \exp\big(-4\ln 2\,(v/\Delta v_{\text{local}})^2\big)$.

Increasing the line center optical depth τ_0 to greater than 1 then changes the profile from Gaussian to flat-topped with a width that depends logarithmically on τ_0 (cf., Phillips et al. 1979)

$$\Delta v_{\text{FWHM}} = \frac{\Delta v_0}{\sqrt{\ln 2}} \left\{ \ln\left[\frac{\tau_0}{\ln\left(\dfrac{2}{1+\exp(-\tau_0)}\right)}\right]\right\}^{1/2}. \tag{52}$$

A line center optical depth of 10, for instance, results in a FWHM line width 1.96 times as large as in the optically thin case, $\tau_0 = 100$ results in 2.7 and 1000 in 3.24 times the width. With increasing optical depth the lines become more and more rectangular shaped.

The simple single Gaussian model often is not a good description of molecular cloud kinematics. Clouds are highly clumped with large scale motions between clumps significantly larger than the local motions within each clump. This will tend to decrease the dependence of line width on optical depth by approximately the factor $\Delta v_{\text{cloud}}/\Delta v_{\text{local}}$, where Δv_{cloud} is the FWHM width of the large scale motions. Martin et al. (1984) have calculated models of the line profiles in clumpy clouds in the simplified case that large scale motions and local line profiles can both be described by Gaussians. They showed that such models can plausibly fit the ratios of line intensities and widths of various CO isotopic lines in the warm molecular cloud core M17. One of the basic results is that an observed ratio Y of an optically thicker to an optically thinner line with an abundance ratio Z can imply an optical depth of the optically thicker line *much greater* than Z/Y. The reason is that the filling factor of the thicker line is larger than that of the thinner. For instance, in most massive, warm molecular clouds in the disk of our Galaxy one finds a $^{12}\text{CO}/^{13}\text{CO}\ 1\to0$ line ratio of ≈ 5. With an assumed $^{12}\text{CO}/^{13}\text{CO}$ fractional abundance ratio of 50 (Wannier 1980), the simple single Gaussian model of Eq. (51) implies $\tau(^{12}\text{CO}\ 1\to0) \approx 10$ and $\tau(^{13}\text{CO}) < 1$. In the corresponding clumpy cloud model one finds a $\tau(^{12}\text{CO}) \approx 10^2$ and $\tau(^{13}\text{CO}) \geq 1$. One also expects that optically thick line profiles are much more rounded than for the single Gaussian model. This is in fact observed (Martin et al. 1984). More realistic, but also more complicated models of line profiles involve a full solution of the radiative transport with an assumed cloud model (Stenholm 1985).

1.2.f Thermal Dust Emission

The main processes of infrared and radio continuum emission are thermal radiation from small dust grains, free-free radiation in hot plasmas and synchrotron radiation. The first two of these processes are the most relevant for molecular clouds and will be briefly summarized in the following paragraphs. A discussion of synchrotron emission is given, for instance, in Chap. 1 of the book "Galactic and Extragalactic Radio Astronomy" (Verschuur & Kellermann 1988).

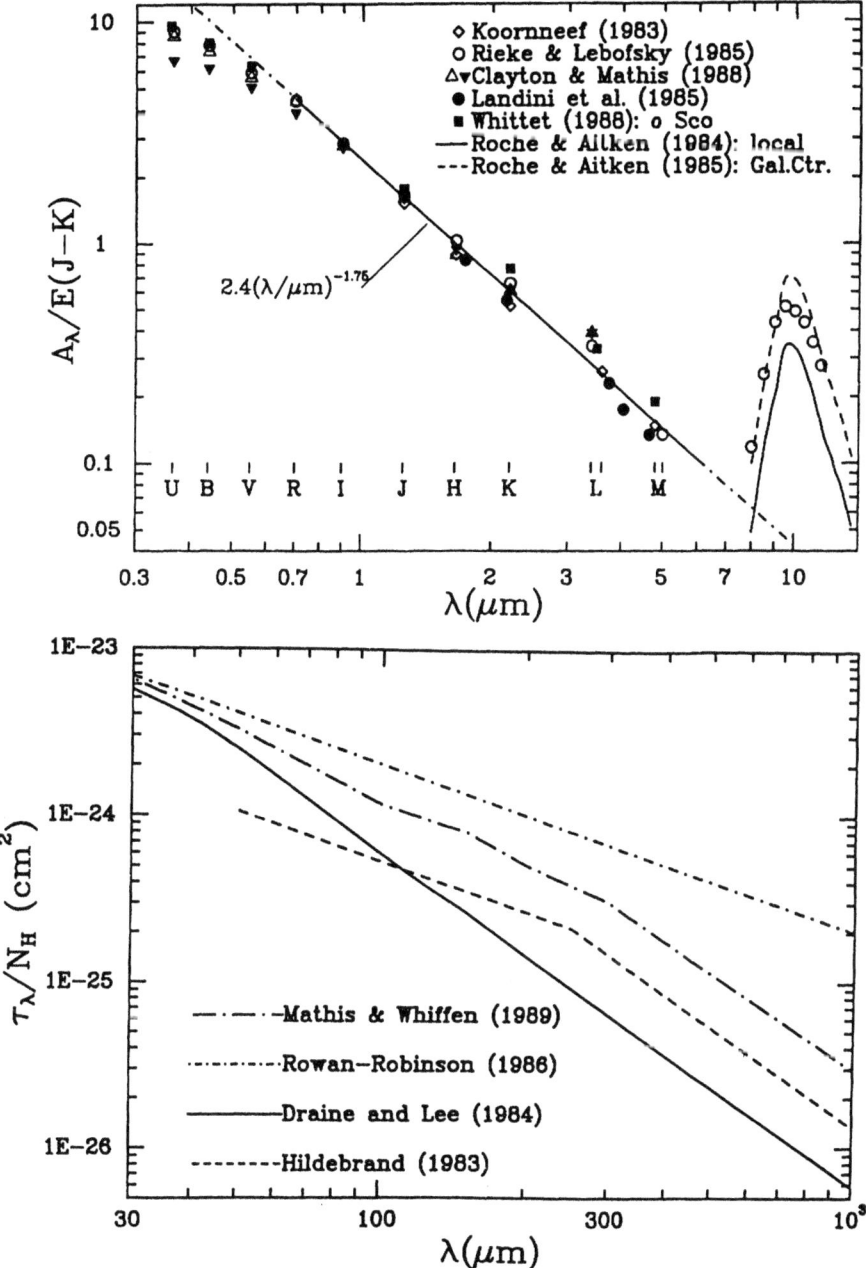

Fig. 9. Extinction curve of interstellar dust in the UV, visible, and infrared (adapted from Draine 1989)

Dust particles in interstellar clouds (diameter $a \approx 0.01$ to $1\,\mu m$) are efficient absorbers of short-wavelength ($\lambda \lesssim a$) radiation. In equilibrium of heating and cooling they reemit a continuous spectrum that for $a > 0.01\,\mu m$ closely resembles a thermal spectrum characterized by a temperature T_d and a smoothly varying absorption/emission efficiency $Q(\nu)$. $Q(\nu) = \sigma(\nu)/\pi a^2$ where $\sigma(\nu)$ is the radiation absorption cross section at ν. The observed flux density S_ν at ν of such a cloud with a line of sight optical depth $\tau_d(\nu)$ and solid angle Ω_ν is given by

$$S_\nu = \frac{2h\nu^3}{c^2} \left[\exp\left(\frac{h\nu}{kT_d}\right) - 1\right]^{-1} \left[1 - \exp(-\tau_d(\nu))\right] \Omega_\nu. \tag{53}$$

Optical depth $\tau_d(\nu)$ and $Q(\nu)$ are directly proportional to each other. The proportionality factor is the line of sight column density of dust particles. For moderately large grains ($\lambda \approx 0.1\,\mu m$), $Q(\nu)$ approaches order of unity in the ultraviolet and decreases steadily with increasing wavelength in the infrared and submillimeter range. Extinction and absorption curves as a function of wavelength are shown in Fig. 9 for dust particles in the solar neighborhood and diffuse clouds ("interstellar"). The resonance at $9.7\,\mu m$ is due to silicates. The value and wavelength dependence of dust opacity in the submillimeter part of the spectrum is poorly known, especially in dense clouds. The dust temperature at any position in the cloud depends on the angle-averaged impinging radiation flux J_ν and is given by an energy balance equation

$$\int_0^\infty Q(\nu) J_\nu \, d\nu = 4\pi \int_0^\infty Q(\nu) B_\nu(T_d) \, d\nu. \tag{54}$$

B_ν is the Planck function. In the very simplified but interesting case of an optically thin cloud at distance R from a short-wavelength source of luminosity L (so that $Q(\nu_{\text{abs}}) \approx O(1) = \text{const}$) and smoothly varying $Q(\nu) = Q_0(\nu/\nu_0)^\beta$ at $\nu \ll \nu_{\text{abs}}$ one can express T_d as

$$T_d(R) = T_0 \left(\frac{L}{L_0}\right)^{2/(4+\beta)} \left(\frac{R}{R_0}\right)^{-2/(4+\beta)}. \tag{55}$$

Observations at far-infrared ($\lambda \approx 100\,\mu m$) wavelengths are consistent with $\beta \approx 1$ and $T_0 \approx 70$ to $95\,K$ for $L_0 = 10^5\,L_\odot$, $R_0 = 3\times10^{17}\,cm$ (e.g., Scoville & Kwan 1976). Equation (55) demonstrates the well established fact that embedded, newly-formed OB stars heat the dust in a $\approx 1\,pc$ region to temperatures of $30\,K$ or more. These dust particles radiate predominantly in the 60 to $300\,\mu m$ band.

Figure 9 also indicates that optical depth effects ($\tau_d \geq 1$) set in at $\lambda \approx 30\,\mu m$ for hydrogen nuclei column densities of $N(H + 2H_2) \geq 3 \times 10^{23}\,cm^{-2}$ and at $\lambda \approx 400\,\mu m$ for $N \geq 10^{25}\,cm^{-2}$ (Hildebrand 1983, Draine 1989). Excepting in the densest regions submillimeter dust emission is thus optically thin and can be used as a tracer of interstellar gas mass once its temperature is known (see Sect. 1.3.b). The $\lambda \leq 100\,\mu m$ emission can, however, become optically thick within massive cloud cores and more detailed radiative transport calculations have to be performed to obtain dust temperature and emergent flux. Such calculations for massive star formation regions have been discussed by Scoville & Kwan (1976), Yorke (1988), and Natta et al. (1981).

The assumption of an average equilibrium dust temperature breaks down for very small dust particles ($\lambda < 0.01\,\mu m$). Their heat capacity is so small that absorption of a single UV photon leads to a large fluctuation in lattice temperature. Evidence for such non-equilibrium heating in very small dust particles, or large molecules, has been found from observations of the 3 to 10 μm continuum and in particular from the $3 \rightarrow 15\,\mu m$ "unidentified" emission features (Sellgren 1984, Puget & Léger 1989). The small dust particles will be discussed below in Sect. 3.2.

1.2.g Free-Free Emission and Recombination Lines

Thermal Bremsstrahlung. Free-free, and free-bound emissions are the dominant continuum emission processes of the ionized gas in HII regions. While HII regions are not the central topic of this discussion, derivation of physical parameters, such as number of Lyman continuum photons, are nevertheless intimately connected to molecular cloud research. The following section will therefore summarize these relevant points. For a detailed discussion of the emission processes the reader is referred to Chap. 5 of Rybicki & Lightman (1979) and Chap. 4 of Osterbrock (1989). The volume emissivity $\epsilon_\nu^{\mathrm{ff}}$ of free-free emission at frequency ν[Hz] and electron temperature T_e [K] can be written as

$$\epsilon_\nu^{\mathrm{ff}} = 6.8 \times 10^{-38}\, Z^2 n_i n_e T_e^{-1/2} \exp\left(\frac{-h\nu}{kT_e}\right) g_{\mathrm{ff}} \; [\mathrm{erg\ cm^{-3}\ s^{-1}\ Hz^{-1}}]. \quad (56)$$

n_i is the volume density of ions of charge Z, n_e is the volume density of electrons, and g_{ff} is the free-free *Gaunt-factor*,

$$g_{\mathrm{ff}} = \begin{cases} 0.551 \ln\left(\dfrac{4.95 \times 10^4\, T_4^{3/2}}{Z\nu_{\mathrm{GHz}}}\right) & \text{for} \quad \dfrac{h\nu}{kT} \ll 1 \\[3mm] 1 & \text{for} \quad \dfrac{h\nu}{kT} > 0.1 \text{ or } \lambda < 10\mu m. \end{cases} \quad (57)$$

T_4 is the electron temperature in units of 10^4 K and ν_{GHz} the frequency in units of 10^9 Hz. The free-free absorption coefficient κ_ν then follows from Kirchoff's law, $B_\nu(T) = \epsilon_\nu/\kappa_\nu$ so that

$$\kappa_\nu^{\mathrm{ff}} = 3.7 \times 10^8\, T_e^{-1/2} Z^2 n_e n_i \nu^{-3} \left[1 - \exp\left(\frac{-h\nu}{kT}\right)\right] g_{\mathrm{ff}}. \quad (58)$$

After line of sight integration, the *optical depth* τ of free-free radiation of a pure hydrogen nebula (charge number $Z = 1$) in the radio/far-IR range ($h\nu \ll kT$) can then be expressed as

$$\begin{aligned} \tau &= 3.01 \times 10^{-2}\, \frac{\mathrm{EM}}{T_e^{1.5}\nu_{\mathrm{GHz}}^2} \left[1.5 \ln T_e + \ln\left(\frac{0.05}{\nu_{\mathrm{GHz}}}\right)\right] \\ &\approx 8.235 \times 10^{-2}\, \mathrm{EM}\, T_e^{-1.35} \nu_{\mathrm{GHz}}^{-2.1}. \end{aligned} \quad (59)$$

Here EM is the linear *emission measure*

$$\text{EM} = \int_{\text{line of sight}} n_e n_i \, dz \quad [\text{cm}^{-6} \text{ pc}]. \tag{60}$$

The free-free continuum emission of HII regions is optically thick (flux density $S_\nu \propto \nu^2$, $T_B \approx T_e$) at frequencies less than the *turnover* frequency

$$\nu_{\text{turn}} = \left[\left(\frac{\text{EM}}{3 \times 10^6 \text{ cm}^{-6} \text{ pc}} \right) T_4^{-1.35} \right]^{0.48} \quad [\text{GHz}]. \tag{61}$$

At $\nu > \nu_{\text{turn}}$ the emission is optically thin and the flux density is proportional to emission measure

$$S_\nu = 2.4 \times 10^{-9} \, T_4^{-0.35} \nu_{\text{GHz}}^{-0.1} \theta_{\text{arcsec}}^2 \text{EM} \quad [\text{Jy}]. \tag{62}$$

θ_{arcsec}^2 is the equivalent solid angle of the source in $(\text{arcsec})^2$. In addition to the linear emission measure, it is useful to define a *volume* emission measure by integrating EM over the area of the entire HII region

$$\text{EM}_V = \int \text{EM} \, dA = 2.9 \times 10^{59} \, T_4^{0.35} \nu_{\text{GHz}}^{0.1} D_{\text{kpc}}^2 S_{\text{Jy}} \quad [\text{cm}^{-3}]. \tag{63}$$

D_{kpc} is the distance to the source in kpc. For an *ionization bounded* HII region, the total number of recombinations to the $k \geq 2$ level of hydrogen per unit time throughout the nebula is balanced by the number of hydrogen ionizing (= Lyman continuum) photons emitted per sec by the central ionizing star(s), $Q(\text{H}^+)$. The total number of recombinations in this so called "case B", pure hydrogen HII region is the multiplication of the volume emission measure with the total recombination coefficient to all levels $k \geq 2$

$$\alpha_B(k \geq 2, \text{H}) = 2.59 \times 10^{-13} \, T_e^{-0.8} \quad [\text{cm}^3 \text{ s}^{-1}], \tag{64}$$

so that

$$Q(\text{H}^+) = 7.5 \times 10^{46} \, T_4^{-0.45} \nu_{\text{GHz}}^{0.1} D_{\text{kpc}}^2 S_{\text{Jy}} f_{\text{He}} \quad [\text{s}^{-1}]. \tag{65}$$

The optically thin radio flux density of an ionization bounded HII region is a direct measure of the number of Lyman continuum photons emitted per unit time. For a pure hydrogen nebula the correction factor for $\text{He}^+ - e$ free-free emission is $f_{\text{He}} = 1$. If $y = n(\text{He}^+)/n(\text{H}^+) > 0$ we have $f_{\text{He}} = (1+yR_0)^{-1}$ where R_0 is the ratio of volumes of the He^+ vs. H^+ zones. The reason for this correction factor is that the free-free emission now has an additional contribution due to interaction between He^+ ions and electrons that should not be counted for calculation of the volume emission measure of hydrogen.

Infrared free-free continuum emission can be estimated from Eq. (57) for $g_{\text{ff}} \approx 1$. In addition, the infrared continuum emission from HII regions includes a significant contribution from free-bound emission. The ratio of free-bound to free-free emissivities is approximately given by

$$\frac{\epsilon_\nu^{\rm fb}}{\epsilon_\nu^{\rm ff}} \approx \frac{\exp\left(\frac{h\nu}{kT}\right)}{g_{\rm ff}}, \qquad (66)$$

where we have "smoothed" over the sharp, series limit resonances. While $(\epsilon^{\rm fb}/\epsilon^{\rm ff})_{\rm radio} \approx 0.2$ the free-bound contribution exceeds the free-free contribution at $\lambda \geq 10\,\mu{\rm m}$. Detection of Bremsstrahlung continuum from ionized gas in the infrared is fairly difficult and rare because it is usually overwhelmed by dust emission within and surrounding the HII region.

Infrared and Radio Recombination Lines. Recombination lines are a measure of emission measure similar to the optically thin Bremsstrahlung continuum. The integrated line intensity of infrared (and optical) recombination lines is conveniently written as (Osterbrock 1989, Wynn-Williams 1984)

$$I_{kk'} = \frac{\gamma_{kk'}(T_e, n_e)}{4\pi}\,(3.08 \times 10^{18})\,\,{\rm EM} \quad [{\rm erg\,s^{-1}\,cm^{-2}\,sr^{-1}}], \qquad (67)$$

where the linear emission measure EM is again in units of $\rm cm^{-6}\,pc$. Values for moderate density $(n_e \geq 10^4\,\rm cm^{-3})$ emission coefficients $\gamma_{kk'}$ at $T_e = 10^4\,\rm K$ are listed in Table 5 for a number of optical/IR hydrogen lines (adapted from Wynn-Williams 1984). To scale to different electron temperatures, multiply $\gamma_{kk'}(10^4)$ by $T_4^{-\alpha}$ ($\alpha \approx 0.9$ for infrared lines). The volume emission measure and number of Lyc-photons are then given by

$$\rm EM_V = 1.19 \times 10^{44}\, D_{kpc}^2\, \frac{\it F_{kk'}}{\gamma_{kk'}} \quad [cm^{-3}], \qquad (68)$$

and

$$Q(\rm H^+) = 3.09 \times 10^{31}\, {\it T}_4^{-0.8}\, {\it D}_{kpc}^2\, \frac{\it F_{kk'}}{\gamma_{kk'}}. \qquad (69)$$

$F_{kk'}\,[\rm erg\,s^{-1}\,cm^{-2}]$ is the recombination line flux. It is often useful to compare radio continuum and near-infrared line estimates. The theoretical, case B, ratio of line flux to radio continuum flux density at ν is

$$\frac{F_{kk'}}{S_{\rm Jy}} = 9.8 \times 10^{13}\, \nu_{\rm GHz}^{0.1}\, \gamma_{kk'}(T_e)\, T_e^{0.35}. \qquad (70)$$

The product of line optical depth τ_1 and FWHM velocity width of a hydrogen *radio* recombination line $\Delta v\,[\rm km\,s^{-1}]$ is

$$\tau_1 \Delta_v = 576\, T_e^{-2.5} {\rm EM}\, \nu_{\rm GHz}^{-1}. \qquad (71)$$

This equation shows that hydrogen recombination lines $(\Delta v \approx 20$ to $30\,\rm km\,s^{-1})$ in the cm-range are optically thin even for very dense, compact HII regions (EM $\approx 10^6$ to 10^7). The line brightness temperature is then given by the product of T_e and τ_1. The ratio of line to adjacent continuum is growing with increasing frequency $(\tau_{\rm cont} < 1)$,

$$\frac{T_1}{T_c} = 0.18\, T_4^{-1.15}\, \nu_{\rm GHz}^{1.1}\, \Delta v^{-1}. \qquad (72)$$

Table 5. Wavelengths and emission coefficients of hydrogen recombination lines [a]

	Balmer $(k'=2)$	Paschen $(k'=3)$	Brackett $(k'=4)$	Pfund $(k'=5)$	Humphreys $(k'=6)$
α $(k=k'+1)$	0.656 (330)	1.875 (34)	4.051 (7.0)	7.46 (2.1)	12.37 (0.78)
β $(k=k'+2)$	0.486 (120)	1.282 (18)	2.625 (4.5)	4.65 (1.5)	7.50 (0.59)
γ $(k=k'+3)$	0.434 (62)	1.094 (11)	2.166 (3.0)	3.74 (1.1)	5.91 (0.44)
δ $(k=k'+4)$	0.410 (36)	1.005 (7.1)	1.945 (2.1)	3.30 (0.77)	5.13 (0.33)
Series Limit $(k=\infty)$	0.365	0.820	1.458	2.28	3.28

a) Adapted from Wynn-Williams (1984), emission coefficients (units $10^{-27}\,\mathrm{erg\,cm^{-3}\,s^{-1}}$) are in parentheses, wavelengths are in μm.

If the HII region contains helium, Eq. (71) has to be multiplied by $(1+Ry)^{-1}$, as in Eq. (65). The ratio of recombination lines to radio continuum can thus be used for estimating the electron temperature. We have discussed above (Sect. 1.2.a) what happens at lower frequency where the line radiation is still optically thin but the continuum becomes optically thick.

Because of the mass dependence of the Rydberg constant in Eq. (1), there are corresponding radio recombination lines of He, C and other heavy elements adjacent to the hydrogen recombination lines and shifted by about -120 to $-160\,\mathrm{km\,s^{-1}}$. Because of the lower abundances of these lines only the He-recombination lines are relatively easy to measure. They give information on the He-abundance and the ionization structure of HII regions (Brown et al. 1978). Recombination lines of carbon are much narrower ($\Delta v \approx 3$ to $10\,\mathrm{km\,s^{-1}}$) and come from photodissociation regions (see Sect. 2.1 below) surrounding the HII-regions. In the infrared, the assumption of hydrogenicity breaks down for these elements so that the wavelengths of the He-recombination lines do not coincide with those of hydrogen (see Osterbrock 1989).

Additional complications due to non-LTE populations, radiative transport or pressure broadening arise for low frequency radio recombination lines (Brown et al. 1978), or infrared recombination lines from dense protostellar envelopes (Simon et al. 1983).

1.3 Estimates of Cloud Column Densities and Masses

The overwhelming fraction of the mass of molecular clouds is almost certainly in form of molecular hydrogen. Atomic hydrogen is the dominant species in diffuse and translucent clouds, but contributes only a fraction to the mass of dense molecular clouds. While atomic hydrogen can be easily detected via its 21 cm hyperfine transition, molecular hydrogen can currently be observed directly only in special circumstances, such as through its infrared, ro-vibrational emission in shocks and photodissociation regions, or via UV-absorption in diffuse clouds. In most cases it is thus necessary to determine the mass of a cloud by less direct indicators, such as the submillimeter/millimeter dust emission, by line emission of CO, the most abundant molecule after H_2, and by MeV γ-ray emission. We now briefly discuss these methods in turn.

1.3.a Direct Measurement of H_2

H_2 has been observed directly by UV absorption in diffuse clouds along the line of sight toward bright stars (e.g., van Dishoeck 1988, Federman et al. 1980). H_2 column densities range between 10^{19} and a few 10^{20} cm^{-2} and correspond to about half of the total column density of hydrogen nuclei in the case of the well investigated star ζOph (van Dishoeck 1988). CO/H_2 fractional abundances range between 5×10^{-8} and 5×10^{-6} in the same diffuse clouds.

H_2 has also been studied in emission by its $2\,\mu$m ro-vibrational lines in many dense clouds with outflows and shocks (Shull & Beckwith 1982). In the case of the hot shocked gas in Orion, Watson et al. (1985) derive a CO/H_2 fractional abundance ratio of 1.2×10^{-4} which is currently the only direct measurement of that quantity in dense clouds. From observations of H_2 $2\,\mu$m lines in absorption against NGC 2024, Black & Willner (1984) find a lower limit of the CO/H_2 abundance of 8×10^{-5} in a cold, dense molecular cloud.

1.3.b Submillimeter Dust Emission

About one percent of the mass of interstellar matter is in form of dust grains. Thermal emission of interstellar dust may be used to determine the dust optical depth and then infer a gas column density. As discussed in Sect. 1.2.f, submillimeter dust emission is usually optically thin and is often on the Rayleigh-Jeans tail of the dust emission spectrum. Recalling the linear relationship between flux density, dust temperature, and dust optical depth $\tau_d(\lambda)$ at wavelengths λ where the Rayleigh-Jeans approximation is fulfilled $\big($Eq. (53)$\big)$, the source averaged dust optical depth at λ, $\langle\tau_d(\lambda)\rangle$, can be calculated in a straightforward manner from the observed flux density and a measured/assumed dust temperature. Measurement of dust optical depth in the submillimeter or millimeter range has the advantage that it likely is a measure of the entire column of dust in contrast to far-infrared observations. IRAS 60 and 100 μm fluxes usually probe only the warmest 10% of the dust. Whitcomb et al. (1981) have related the dust emission at 400 μm to A_V in the reflection nebula NGC 7023 yielding the ratio $\tau(400\,\mu\text{m})/A_V$. It is also typically assumed that the extinction efficiency in the submillimeter scales with wavelength as $\lambda^{-\beta}$ with $\beta \approx 1.5$ to 2 (see Hildebrand 1983, Draine 1989, Cox & Mezger 1989). A_V can then be related to the density of hydrogen nuclei $N(\text{H} + 2\text{H}_2)$ with the calibration of Bohlin, Savage, & Drake (1978) $\big(N(\text{H} + 2\text{H}_2) = 1.9 \times 10^{21}\,A_V\big)$ from UV measurements in diffuse clouds. A gas to dust mass ratio of 100 is equivalent to the Bohlin et al. calibration. The resulting "best" relationship near 400 μm then is (Hildebrand 1983, Cox & Mezger 1989, Draine 1989)

$$\langle N(\text{H} + 2\text{H}_2) \rangle = 1.2 \times 10^{25} \left(Z/Z_\odot \right) \langle \tau_d(\lambda) \rangle \left(\lambda/400\,\mu\text{m} \right)^\beta \quad [\text{cm}^{-2}]. \quad (73)$$

Here $Z/Z_\odot$ is the abundance of heavy elements in units of the solar abundance.

1.3.c CO Line Emission

The CO molecule has transitions in the UV (electronic transitions), at $2.3/4.6\,\mu\text{m}$ (ro-vibrational) and between $50\,\mu\text{m}$ and $2.6\,\text{mm}$ (rotational). It is also the most abundant molecule after H_2. Observations in dense clouds made at infrared, submillimeter and millimeter wavelengths all suggest that the CO/H_2 fractional abundance is close to 10^{-4} (about 30% of carbon in CO). The most common method of determining CO column densities uses measurements of the low-J $(1 \rightarrow 0,\ 2 \rightarrow 1)$ transitions in emission. Unfortunately ^{12}CO transitions in dense molecular clouds are very optically thick $(\tau \approx 100)$.

Tracing Mass with CO Isotopes. One way of avoiding this problem is to observe the less abundant (optically thinner) ^{13}CO, C^{18}O and C^{17}O lines and then to multiply by the appropriate abundance ratio. If only one line is measured, an extrapolation of the column density in the measured levels to all other levels is necessary. Often this is done by assuming LTE population at some temperature. However, the critical density of the excited states increases rapidly with J $(n_{\text{crit}} \sim J^3)$, so that typically only the lowest three or four levels are populated. Excitation/radiative transport calculations show that the integrated intensities of the $2 \rightarrow 1$ transitions of ^{13}CO (if optically thin) or C^{18}O are a good measure of column density and more or less independent of temperature over a reasonably wide parameter range (a few $10^3 < n(\text{H}_2) <$ a few $10^6\,\text{cm}^{-3}$ and $15 < T < 80\,\text{K}$). With $[\text{C}^{18}\text{O}]/[\text{H}_2] = 1.7 \times 10^{-7}$ (Frerking et al. 1982) one then finds

$$N(\text{H}_2) = 3 \times 10^{21}\, I(\text{C}^{18}\text{O}\ 2 \rightarrow 1) \quad [\text{cm}^{-2}/\text{K km s}^{-1}]. \quad (74)$$

Similar relationships hold for optically thin ^{13}CO or C^{17}O lines; one has to just choose the appropriate abundance $(^{13}\text{CO}/\text{C}^{18}\text{O} \approx 8,\ \text{C}^{18}\text{O}/\text{C}^{17}\text{O} \approx 4)$. The $1 \rightarrow 0$ transitions can also be used, although they are more sensitive to temperature; in that case multiply the coefficient on the right by a factor of four. It appears from observations that ^{13}CO is the best mass tracer in dark clouds or the extended gas in giant clouds (e.g., Frerking et al. 1982, Langer et al. 1989) while C^{18}O and C^{17}O probe the mass in dense massive, cloud cores.

Tracing Mass with ^{12}CO $1 \rightarrow 0$. In some cases the isotopic lines are too weak for measurement. A column density is then inferred from the optically thick ^{12}CO line(s) by assuming either that the integrated line flux is proportional to the filling factor of clouds of a constant flux, or by applying the virial theorem (e.g., Dickman et al. 1986, Solomon et al. 1987, Scoville & Sanders 1987). This method is widely used for external galaxies, for instance. In the following, the physical basis and possible pitfalls of the "virial" method are discussed.

Following van Dishoeck & Black (1987) and Dickman et al. (1986), let us consider an ensemble of clouds (or clumps) moving around each other, or around the center of a galaxy with an overall velocity width Δv_{source} (FWHM). Assume that each cloud has an intrinsic velocity width Δv_{cloud} that is determined

Table 6. Temperature and density probes in molecular clouds

METHOD	STRENGTH	WEAKNESS
Temperature		
Brightness of optically thick mm-lines (e.g., ^{12}CO 1→0: $T_R = T_{ex} \approx T_{kin}$)	Easy to measure	Filling factor of emission often less than unity, method may underestimate T_{kin}
Level population as a function of energy from metastable NH_3 inversion lines at $\approx 1.2\,cm$ (also other symmetric tops, such as CH_3CN, CH_3C_2H)	Many energy levels available at about the same wavelength	Optical depth and density correction required, only applicable to fairly dense regions where NH_3, CH_3CN, etc. are abundant
Level population as a function of energy from rotational lines of molecules in submm and infrared (emission, e.g., CO, H_2)	Very sensitive for $E_{ul} > kT_{kin}$, also density information	Measurements need to be made at different wavelengths, must solve the entire excitation problem and take into account optical depth effects, density/ temperature information is coupled
...from ro-vibrational lines in near-IR (absorption)	Same as above, many energy levels at about the same wavelength	Only line of sight toward bright continuum sources
Density		
Average volume density from column density and cloud size (e.g., ^{13}CO or $C^{18}O$)	Easy to measure	Underestimates local densities as volume filling factors $\ll 1$ (clumpiness), strongly depends on abundances and T_{ex}
Detection of "density tracers", such as mm-lines of CS, HCN, and infer $n \approx n_{crit}$	Easy to measure	Rough first order indication, depends strongly on optical depth
Measurement of non-metastable NH_3 inversion lines at $1.2\,cm$	Many energy levels available at about the same wavelength	Optical depth effects and FIR radiative pumping need to be taken into account
Level population as a function of energy from rotational lines above n_{crit} (subthermal regime), especially for optically thin species ($C^{18}O$, $H^{13}CN$, $C^{34}S$, etc.)	Very sensitive to $n(H_2)$, gives also temperature information	Measurements need to be made at different wavelengths, cross sections for all molecules but CO uncertain, must solve the entire excitation problem, density/temperature information is coupled

by virial equilibrium (kinetic energy stabilizes clouds against gravitational collapse). Assume further that at each velocity the (beam) area filling factor of clouds $\Phi(v)$ is smaller than 1 (no shadowing). Each cloud emits optically thick ^{12}CO $1 \to 0$ lines with an equivalent Rayleigh-Jeans excitation temperature $\widetilde{T}_{ex}$, (see Sect. 1.2.a) so that the integrated line flux [K km s^{-1}] for $T_{ex} \gg 3\,$K is given by

$$I(^{12}\mathrm{CO}) = \widetilde{T}_{ex}\, \Phi(v)\, \Delta v_{source} = \widetilde{T}_{ex}\, \Phi\, \Delta v_{cloud}, \tag{75}$$

where $\Phi = \Phi(v)\,\Delta v_{source}/\Delta v_{cloud}$ is the filling factor of clouds irrespective of velocity. The assumption of virialization of each cloud of radius R, mass $M(R)$ and mean (volume averaged) density $\langle n(\mathrm{H}_2)\rangle$ means that

$$\begin{aligned}
\frac{GM(R)}{3R} &= \sigma^2(\mathrm{cloud}) + \frac{kT_{kin}}{\langle m\rangle} \\
&\approx \sigma^2(\mathrm{cloud}) = \left(\frac{\Delta v_{cloud}}{2.35}\right)^2.
\end{aligned} \tag{76}$$

In this equation G is the gravitational constant. We have assumed that the intrinsic cloud motions are significantly larger than thermal. The beam averaged H$_2$ column density can be expressed as

$$\langle N(\mathrm{H}_2)\rangle = \Phi\left[N(\mathrm{H}_2)\right]_{cloud} = \Phi\, 2R\, \langle n(\mathrm{H}_2)\rangle. \tag{77}$$

With $M(R) = (4\pi/3)R^3\langle n(\mathrm{H}_2)\rangle\langle m\rangle$ ($\langle m\rangle = 4.3 \times 10^{-24}\,$g is the mean mass per hydrogen molecule, including heavier elements), Eqs. (40), (41), and (42) can be combined to give the conversion factor, X,

$$X_{\mathrm{CO}} = \frac{\langle N(\mathrm{H}_2)\rangle}{I(^{12}\mathrm{CO})} = 3 \times 10^{20}\, \frac{8\,\mathrm{K}}{\widetilde{T}_{ex}} \left(\frac{\langle n(\mathrm{H}_2)\rangle}{200}\right)^{1/2}. \tag{78}$$

The values for $\widetilde{T}_{ex}$ and $\langle n(\mathrm{H}_2)\rangle$ chosen in Eq. (78) are those quoted by Scoville & Sanders (1987) to be representative for giant molecular clouds in the Galaxy. The conversion factor is the same within a factor of 2 as the ones obtained for galactic disk clouds by several other methods (isotopes, empirical correlation, γ-rays). This forms the basis of the common assumption that the above relationship holds *universally* (Young & Scoville 1991).

Given Eq. (78) and the evidence of much higher excitation temperatures of the ^{12}CO emitting gas near OB star forming regions and in the nuclei of starburst nuclei than in the average disk gas, this assumption has been questioned (Maloney & Black 1988). Yet the fact that mean densities also strongly increase in star forming clouds may "save" the relationship as a first order estimate, albeit with increased uncertainty.

1.3.d Tracing Mass with MeV γ-Rays
High energy γ-rays (30 to 1000 MeV) in interstellar clouds are produced when cosmic rays collide with hydrogen nuclei. The π-mesons produced in this collision decay into γ's. If the cosmic ray density inside clouds in constant, the

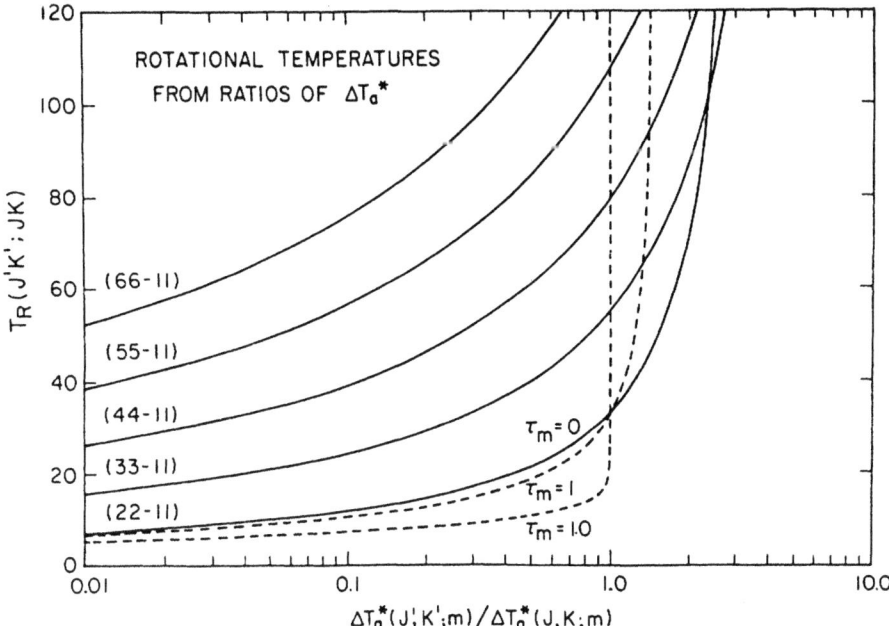

Fig. 10. Rotation temperatures T_R derived from various metastable NH_3 inversion line ratios (adapted from Ho & Townes 1983). The impact of various main satellite line optical depths is shown for the $(2, 2) - (1, 1)$ ratio

γ-ray flux thus measures the number of hydrogen nuclei in a given volume (see Bloemen 1989 for a review). Bloemen et al. (1984) and Strong et al. (1988) have used the COS-B Galactic plane survey to determine masses and the conversion rate between ^{12}CO flux and H_2 column density. The average conversion factor for the Galactic disk corresponds to $X = 2.4 \times 10^{20}$. The conversion factor in the Galactic Center may be up to 10 times smaller (Blitz et al. 1985), supporting cautionary remarks of the last paragraph.

1.4 Measuring Temperature and Density

All basic tools necessary for deriving these basic physical parameters have been discussed in the preceding sections. Different methods for extracting temperature and density information have been applied to the investigation of molecular clouds; the most important ones are listed in Table 6 in increasing order of complexity.

All have advantages and disadvantages. The simplest tools in each category (temperatures from line brightness temperatures and densities from column densities and cloud sizes) are very rough measures only and should be applied with great caution. Even the most sophisticated methods (fitting a non-LTE cloud model to observation of a wide variety of transitions) typically suffer from problems of simplification (single component models), large number of ill-defined parameters (for models with density and temperature gradients and/or

313

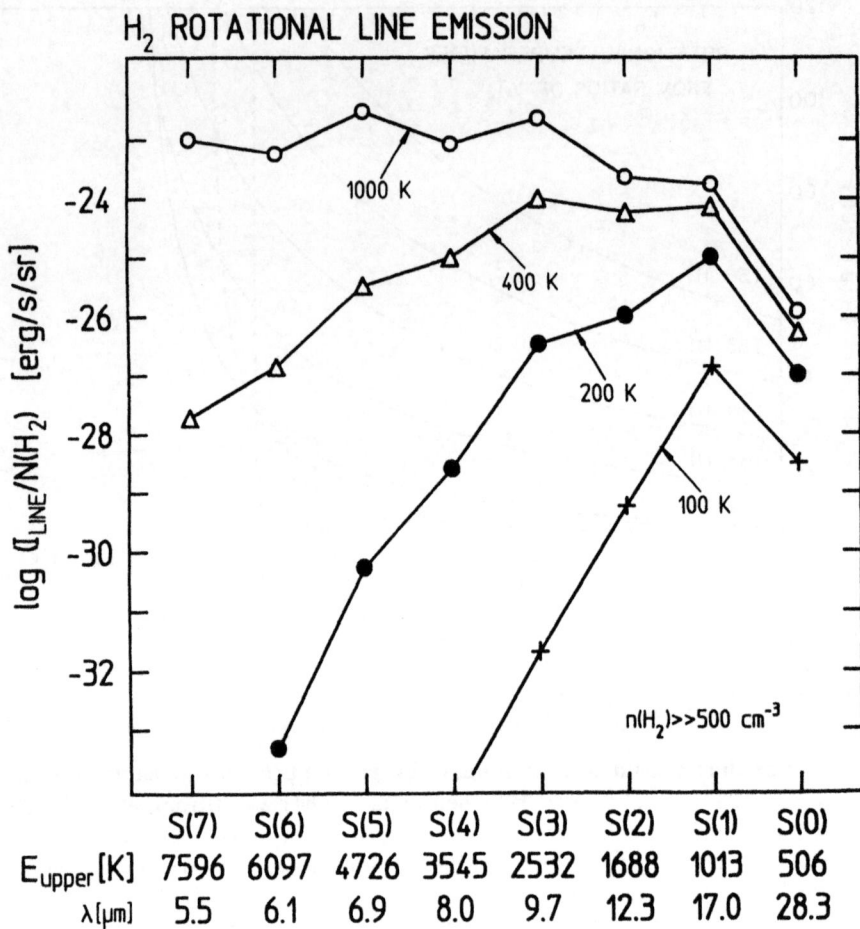

Fig. 11. Line intensities of pure rotational lines of H_2 in the limit of complete thermalization $\left(n(H_2) \gg 500\,\mathrm{cm}^{-3},\ \text{van der Werf, priv. comm.}\right)$

complex velocity fields) and uncertainties in cross sections. If only a small set of lines is considered one has to be aware that they typically sample only a limited range of n and T. Perhaps the best way of getting at cloud temperatures is from the 1.2 cm inversion transitions of NH_3. Figure 10 shows the dependence of metastable NH_3 line ratios on kinetic temperature (Ho & Townes 1983). Another nice indicator of gas temperatures in warmer clouds and shocks is the pure rotational emission of H_2 (Fig. 11). A fairly reliable way of probing density (and temperature) is from line ratios of CO, CS, or HCN rotational transitions. Figure 12 shows the dependence of line radiation temperature and optical depth on density for a number of HCN rotational lines (White 1982). Figure 13 shows the dependence of CO line ratios on temperature and density (from Stutzki 1989).

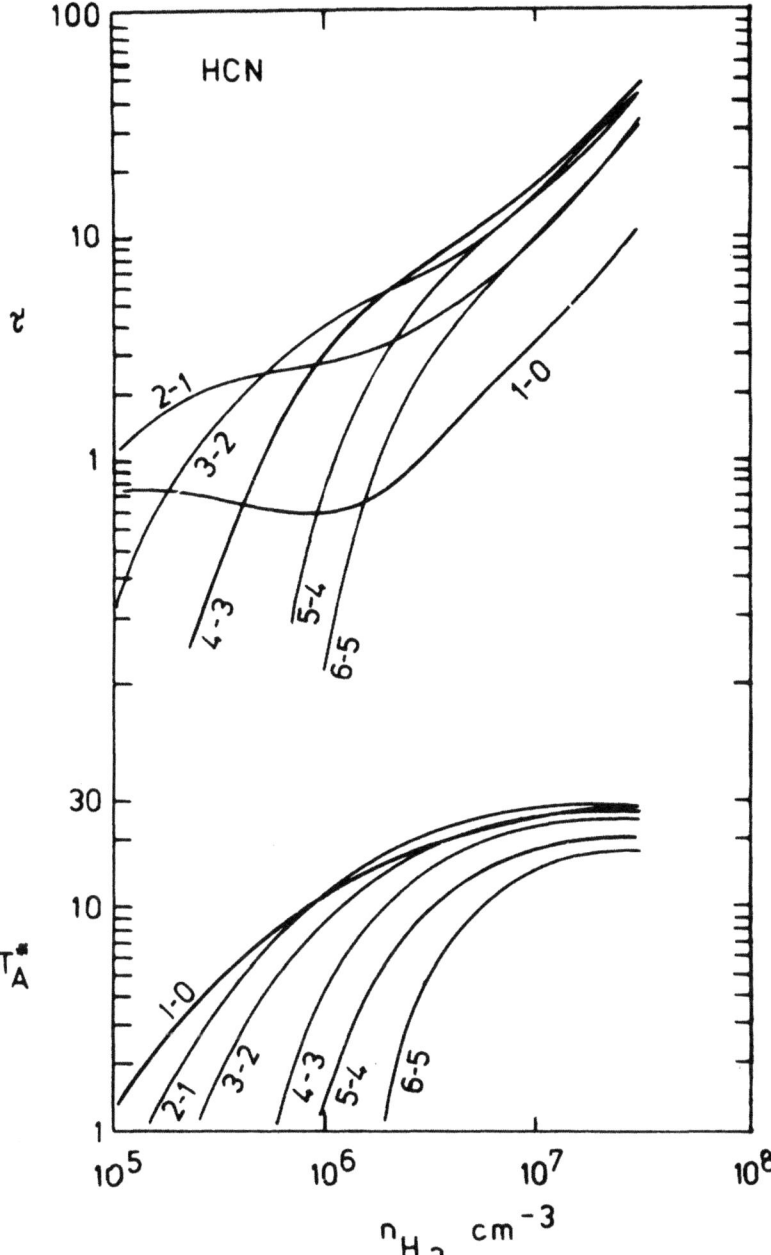

Fig. 12. Dependence of HCN line radiation temperature and optical depth on molecular hydrogen density (adapted from White 1982)

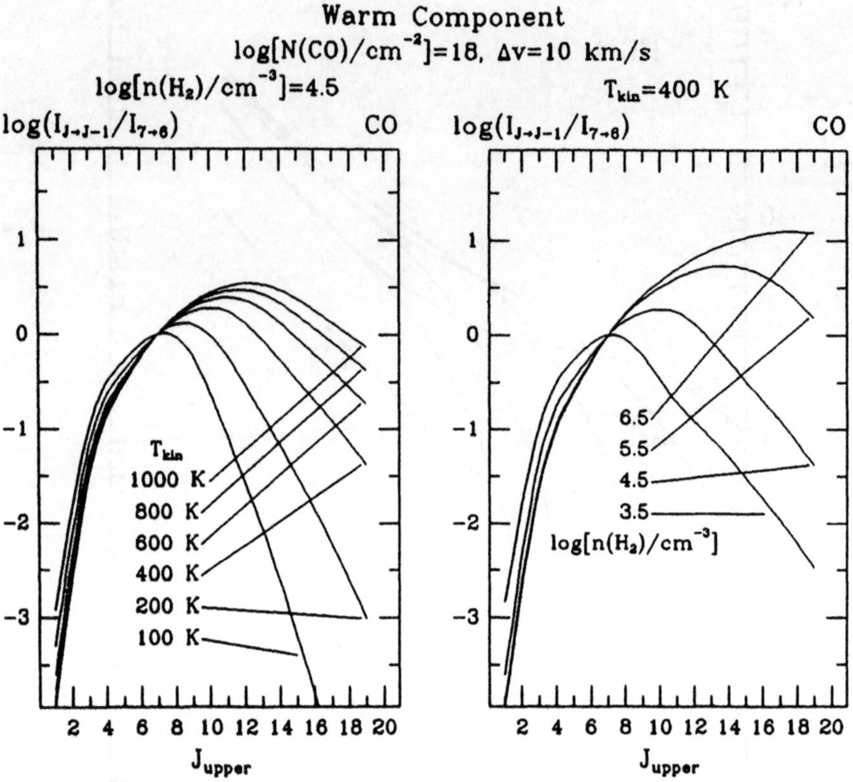

Fig. 13. Dependence of CO line ratios on temperature (left) and density (right) (from Stutzki 1989). Calculations of excitation and radiative transport are for a CO column density of 10^{18} cm^{-2} in a line width of 10 km s^{-1}

A good recent review of the observational situation has been given by Goldsmith (1987), a somewhat older is that of Evans (1980). The basic result is that molecular clouds have a wide range of temperatures and densities, depending on scale size and environment (distance to OB stars, Galactic disk vs. Galactic center). Kinetic temperatures vary from $6 \rightarrow 12$ K (e.g., Myers 1986) in cold, dark clouds to ≥ 50 K in warm cloud cores close to OB star forming sites. Submillimeter and infrared spectroscopy indicates that in these warm cloud cores there is a component of quiescent gas ≥ 100 K with H_2 column densities $10^{22} \rightarrow 10^{23}$ cm^{-2} (Harris et al. 1987, Jaffe et al. 1987, Schmid-Burgk et al. 1989, Graf et al. 1990, Parmar et al. 1991). This warm component represents a significant fraction ($10 \rightarrow 50\%$) of the mass in star forming cores. Gas with temperatures ≥ 50 K is also common and probably dominant throughout the Galactic center molecular clouds (Güsten et al. 1985) and in the nuclei of starburst galaxies (Wild et al. 1992, Harris et al. 1991).

Typical molecular hydrogen densities in molecular clouds range from a few 10^3 cm^{-3} on average in giant clouds, to about 10^4 cm^{-3} in dark cloud cores (Myers 1986) to $\geq 10^5$ cm^{-3} in massive star formation regions (Snell et al.

1984, Stutzki & Güsten 1990). The "hot core" clump (mass $\geq 10\,M_\odot$, size $\approx 0.01\,\mathrm{pc}$) in the BN-KL region has densities of $1 \to 3 \times 10^7\,\mathrm{cm}^{-3}$, Genzel & Stutzki 1989).

References

1.1   Allamandola, L.J. 1984, in *Galactic and Extragalactic Infrared Spectroscopy*, eds. M.F. Kessler, J.P. Phillips (Reidel, Dordrecht), 5

1.2   Black, J.H., Willner, S.P. 1984, *Astrophys. J.* **279**, 673

1.3   Blitz, L., Bloemen, J.B.G.M., Hermsen, W., Bania, T.M. 1985, *Astron. Astrophys.* **143**, 267

1.4   Bloemen, J.B.G.M. 1989, *Ann. Rev. Astron. Astrophys.* **27**, 469

1.5   Bloemen, J.B.G.M. et al. 1984, *Astron. Astrophys.* **139**, 37

1.6   Bohlin, R.C., Savage, B.D., Drake, J.F. 1978, *Astrophys. J.* **224**, 132

1.7   Brown, R.L. Lockman, F.J., Knapp, G.R. 1978, *Ann. Rev. Astron. Astrophys.* **16**, 445

1.8   Castor, J. 1970, *Monthly Notices Roy. Astron. Soc.* **149**, 111

1.9   Cooksy, A.S., Blake, G.A., Saykally, R.J. 1986, *Astrophys. J.* **305**, L89

1.10   Cox, P., Mezger, P.G. 1989, *Astron. Astrophys. Rev.* **1**, 49

1.11   de Jong, T., Dalgarno, A., Boland, W. 1980, *Astron. Astrophys.* **91**, 68

1.12   Dickman, R.L., Snell, R., Schloerb, F.P. 1986, *Astrophys. J.* **309**, 326

1.13   Draine, B.T. 1989, in *Infrared Spectroscopy in Astronomy*, ed. B.H. Kaldeich, *ESA-SP-290*, 93

1.14   Evans, N.J. 1980, in *Interstellar Molecules*, ed. B. Andrew (Reidel, Dordrecht), 1

1.15   Evans, N.J., Lacy, J.H., Carr, J.S. 1991, preprint

1.16   Federman, S.R., Glassgold, A.E., Jenkins, E.B., Shaya, E.J. 1980, *Astrophys. J.* **242**, 545

1.17   Flower, D.R. 1987, in *Interstellar Processes*, eds. D.J. Hollenbach, H.A. Thronson, (Reidel, Dordrecht), 745

1.18   Frerking, M.A., Langer, W.D., Wilson, R.W. 1982, *Astrophys. J.* **262**, 590

1.19   Genzel, R., Stutzki, J. 1989, *Ann. Rev. Astron. Astrophys.* **27**, 41

1.20   Goldsmith, P.F. 1987, in *Interstellar Processes*, eds. D.J. Hollenbach, H.A. Thronson (Reidel, Dordrecht), 51

1.21   Graf, U.U., Genzel, R., Harris, A.I., Hills, R.E., Russell, A.P.G., Stutzki, J. 1990, *Astrophys. J.* **358**, L49

1.22   Güsten, R., Walmsley, C.M., Ungerechts, H., Churchwell, E. 1985, *Astron. Astrophys.* **142**, 381

1.23   Harris, A.I., Stutzki, J., Genzel, R., Lugten, J.B., Stacey, G.J., Jaffe, D.T. 1987, *Astrophys. J.* **322**, L49

1.24   Harris, A.I., Hills, R.E., Stutzki, J., Graf, U.U., Russell, A.P.G., Genzel, R. 1991, *Astrophys. J.* (Ltr.) **382**, L75

1.25   Herzberg, G. 1939, *Molecular Spectra and Molecular Structure I: Diatomic Molecules* (Prentice Hall, New York)

1.26   Herzberg, G. 1945, *Molecular Spectra and Molecular Structure II: Infrared and Raman Spectra of Polyatomic Molecules* (Van Nostrand, Reinhold, New York)

1.27   Hildebrand, R.H. 1983, *Quart. J. Roy. Astron. Soc.* **24**, 267

1.28   Ho, P.T.P., Townes, C.H. 1983, *Ann. Rev. Astron. Astrophys.* **21**, 239

1.29   Jaffe, D.T., Harris, A.I., Genzel, R. 1987, *Astrophys. J.* **316**, 231

1.30   Langer, W.D., Penzias, A.A. 1990, *Astrophys. J.* **357**, 477

1.31   Maloney, P., Black, J.H. 1988, *Astrophys. J.* **325**, 389

1.32   Martin, H.M., Sanders, D.B., Hills, R.E. 1984, *Monthly Notices Roy. Astron. Soc.* **208**, 35

1.33   Melnick, G.J., Stacey, G.J., Genzel, R., Lugten, J.B., Poglitsch, A. 1990, *Astrophys. J.* **348**, 161

1.34   Myers, P. 1986, in *Star Forming Regions*, eds. M. Peimbert, J. Jugaku (Reidel, Dordrecht), 33

1.35 Natta, A., Palla, F., Panagia, N., Preite-Martinez, A. 1981, *Astron. Astrophys.* **99**, 289

1.36 Osterbrock, D.E. 1989, *Astrophysics of Gaseous Nebulae, Active Galactic Nuclei* (Univ. Science Books, Mill Valley)

1.37 Parmar, P., Lacy, J.H., Achtermann, J.M. 1991, *Astrophys. J.* **372**, L25

1.38 Phillips, T.G., Huggins, P.J., Wannier, P.G., Scoville, N.Z. 1979, *Astrophys. J.* **231**, 720

1.39 Puget, J.L., Léger, A. 1989, *Ann. Rev. Astron. Astrophys.* **27**, 161

1.40 Rybicki, G.B., Lightman, A.P., *Radiative Processes in Astrophysics* (Wiley, New York)

1.41 Scoville, N.Z., Kwan, J. 1976, *Astrophys. J.* **206**, 718

1.42 Scoville, N.Z. 1984, in *Galactic and Extragalactic Infrared Spectroscopy*, eds. M.F. Kessler, J.P. Phillips (Reidel, Dordrecht), 167

1.43 Scoville, N.Z., Sanders, D.B. 1987, in *Interstellar Processes*, eds. D.J. Hollenbach, H.A. Thronson (Reidel, Dordrecht), 21

1.44 Sellgren, K. 1984, *Astrophys. J.* **277**, 623

1.45 Shull, J.M., Beckwith, S. 1982, *Ann. Rev. Astron. Astrophys.* **20**, 163

1.46 Simon, M., Felli, M., Cassar, L., Fischer, J., Massi, M. 1983, *Astrophys. J.* **266**, 623

1.47 Solomon, P.M., Rivolo, A.R., Barrett, J., Yahil. A. 1987, *Astrophys. J.* **319**, 730

1.48 Schmid-Burgk, J. et al. 1989, *Astron. Astrophys.* **215**, 150

1.49 Snell, R.L. et al. 1984, *Astrophys. J.* **276**, 625

1.50 Spitzer, L. 1978, *Physical Processes in the Interstellar Medium* (Wiley, New York)

1.51 Stenholm, L.G. 1985, *Astron. Astrophys.* **144**, 179

1.52 Strong, A.W. et al. 1988, *Astron. Astrophys.* **207**, 1

1.53 Stutzki, J. 1989, in *The Physics and Chemistry of Interstellar Molecular Clouds*, eds. G. Winnewisser, J.T. Armstrong (Springer, Berlin), 53

1.54 Stutzki, J., Güsten, R. 1990, *Astrophys. J.* **356**, 513

1.55 Townes, C.H., Schawlow, A.L. 1955, *Microwave Spectroscopy* (McGraw-Hill, New York)

1.56 Verschuur, G.L., Kellermann, K.I. 1988, *Galactic and Extragalactic Radio Astronomy* (Springer, Berlin)

1.57 Wannier, P.G. 1980, *Ann. Rev. Astron. Astrophys.* **18**, 399

1.58 Watson, D.M. 1982, Ph.D. Thesis, Univ. of CA, Berkeley

1.59 Watson, D.M. 1984, in *Galactic and Extragalactic Infrared Spectroscopy*, eds. M.F. Kessler, J.P. Phillips (Reidel, Dordrecht), 195

1.60 Watson, D.M. 1985, *Physica Scripta*, **T11**, 33

1.61 Watson, D.M., Genzel, R., Townes, C.H., Storey, J.W. 1985, *Astrophys. J.* **298**, 316

1.62 Whitcomb, S.E., Gatley, I., Hildebrand, R.H., Keene, J., Sellgren, K., Werner, M.W. 1981, *Astrophys. J.* **246**, 416

1.63 White, G.J. 1982, in *The Scientific Importance of Submillimetre Observations*, eds. T. de Graauw, T.D. Guyenne, *ESA-SP-189*, 5

1.64 Wild, W., Harris, A.I. et al. 1992, in prep.

1.65 Winnewisser, G., Mezger, P.G., Breuer, H.-D. 1974, *Topics in Current Chemistry* **44** (Springer, Berlin)

1.66 Wynn-Williams, C.G. 1984, in *Galactic and Extragalactic Infrared Spectroscopy*, eds. M.F. Kessler, J.P. Phillips (Reidel, Dordrecht), 133

1.67 Young, J.S., Scoville, N.Z. 1991, *Ann. Rev. Astron. Astrophys.* **29**, 581

1.68 Yorke, H.W. 1988, in *Radiation in Moving Gaseous Media* (18th Saas-Fee Course), eds. Y. Chmielewski, T. Lanz, Geneva Observatory, 193

1.69 van Dishoeck, E.F., Black, J.H. 1987, in *Physical Processes in Interstellar Clouds*, eds. G. Morfill, M. Scholer, *NATO-ASI-210*, 241

1.70 van Dishoeck, E.F. 1988, in *Millimetre and Submillimetre Astronomy*, eds. R.D. Wolstencroft, W.B. Burton (Kluwer, Dordrecht), 117

1.71 Ziurys, L.M., Turner, B.E. 1986, *Astrophys. J.* **300**, L19

2 Heating and Cooling Processes

A dense cloud or cloud core in a region of massive star formation is exposed to the intense ultraviolet radiation from newly-formed OB stars that in part ionizes the surrounding cloud and forms *HII regions*. Beyond the penetration depth of Lyman-continuum photons there is a zone where far-UV radiation $(6 \rightarrow 13.6\,\text{eV})$ ionizes species with ionization potentials less than that of hydrogen (C, S, Si, Fe), dissociates molecules and triggers a photon-dominated chemistry. These *photon-dominated* or *photo-dissociation regions* (PDR's) are of particular interest to the following discussion as in these surface zones most of the stellar luminosity is absorbed and reradiated in form of infrared/submillimeter line and continuum radiation. Another important set of heating processes includes the conversion of *mechanical energy (stellar winds and outflows, supernova explosions)*, into thermal energy via *shocks*. Like radiative heating shocks affect cloud surfaces. *Energetic particles (cosmic rays)* and *hard X-rays* penetrate much deeper into clouds and significantly contribute to the heating there. If magnetic fields are present and move (slip) relative to the bulk material of the cloud (e.g., during gravitational collapse), the resulting *frictional heating (ambipolar diffusion)* needs to be taken into account.

These mechanisms contribute a net *heating rate* per volume element $\Gamma(n, T_{\text{kin}})\,[\text{erg s}^{-1}\,\text{cm}^{-3}]$. The cloud cools at a rate $\Lambda(n, T_{\text{kin}})$ that again is a sum of a number of processes (atomic and molecular lines, dust continuum). In *equilibrium* the cloud reaches a kinetic temperature T_{kin} at which heating and cooling balance,

$$\sum_i \Gamma_i(n, T_{\text{kin}}) = \sum_i \Lambda_i(n, T_{\text{kin}}). \tag{79}$$

The cooling rate also controls the *cooling time* during which the cloud can radiate away a significant fraction of its energy

$$\tau_{\text{cool}} = \frac{E_{\text{therm}} + E_{\text{kin}} + E_{\text{turb}}}{\sum_i \Lambda_i(n, T_{\text{kin}})}. \tag{80}$$

The cooling time is often small $\left(O(\text{years})\right)$ compared to other time scales of the system so that in fact thermal equilibrium is reached.

I will now briefly discuss in turn the most important heating and cooling mechanisms for dense, *neutral* clouds (see also Spitzer 1978, Goldsmith & Langer 1987, Black 1987). I will not address fully ionized clouds (HII regions) and refer to the books by Spitzer (1978) and Osterbrock (1989) for an in depth discussion. I begin with the processes affecting the surfaces of dense clouds (PDR's, shocks).

2.1 Photon Dominated/Photodissociation Regions

2.1.a Theory

Theoretical models of the heat balance and chemistry of clouds illuminated by far-UV radiation have been presented by a number of authors (Walmsley 1975, Langer 1976, Gerola & Glassgold 1978, de Jong, Dalgarno & Boland 1980, van Dishoeck & Black 1986, 1988). Tielens & Hollenbach (1985a,b), Sternberg & Dalgarno (1989), Burton et al. (1990b), and Hollenbach et al. (1991) treat the cases of high cloud densities and high UV intensities that are particularly interesting for the present discussion. The basic size scale is given by the penetration depth of far-UV radiation. The most efficient broad-band absorber in this wavelength range is dust, and the maximum thickness of a PDR is given by the condition (Werner 1970) that the UV dust *absorption* optical depth be a few, corresponding to a hydrogen column density of about 6×10^{21} cm^{-2}.

The chemical structure in such a cloud interface displayed in Fig. 14 for a cloud of hydrogen density of 2×10^5 and a UV field that is 10^5 times larger than that of the solar neighborhood (the "standard" model of Tielens & Hollenbach 1985a,b). At $A_V \approx 1$ to 2 from the surface there is a sharp transition from atomic to molecular hydrogen. This is a direct consequence of the fact that photodissociation of H$_2$ requires UV absorption from the $X^1\Sigma_g$ ground electronic state to the $B^1\Sigma_u$ and $C^1\Pi_u$ excited states, followed by a transition into the unbound continuum of the ground state. There is no direct continuum dissociation process. The UV transitions of H$_2$ (the Lyman and Werner bands) rapidly become optically thick and self-shielding sets in (cf., Jura 1974). In this outer layer almost all of the gaseous carbon and oxygen is in form of C$^+$ and O^0. The transition between singly ionized carbon and CO occurs significantly deeper into the cloud ($A_V \approx 5$) with a layer of atomic carbon arising between $A_V \approx 3$ and 8. In Fig. 14 it is assumed that all carbon is in CO as soon as CO can form. Oxygen thus remains atomic deep into the cloud ($A_V \approx 20$) and OI may be the most important carrier of oxygen in interstellar clouds (see Dalgarno 1991). In addition to these most abundant species, the photodissociation zone contains many less abundant ions, such as S$^+$, Mg$^+$, Si$^+$, and Fe$^+$ whose abundances depend heavily on depletion.

Photoelectric heating and UV pumping of H$_2$ are the most important gas heating mechanisms in dense PDR's. The first mechanism involves the absorption of a UV photon by a dust grain, followed by ejection of an electron which then collisionally heats the gas (Draine 1978, de Jong 1977). Per incoming 10 eV photon, the energy per electron available for heating is about 1 eV. The remaining 9 eV go into overcoming the binding energy of the electron to the grain (the "work function", about 6 eV) and the grain's positive electrostatic potential. The potential energy which has to be overcome increases with increasing UV energy density (higher positive charge) and decreases with increasing density (higher rate of recombination). With a "yield" of conversion of incoming photons into photoelectrons of about 0.1 the photoelectric mechanism can plausibly convert a few percent of the UV photons into gas heating ($\eta_{pe} \approx 0.01$ to 0.03). The numerical values are, of course, grain-model dependent and also depend on carbon abundance and grain size distribution. Small dust grains and PAH's

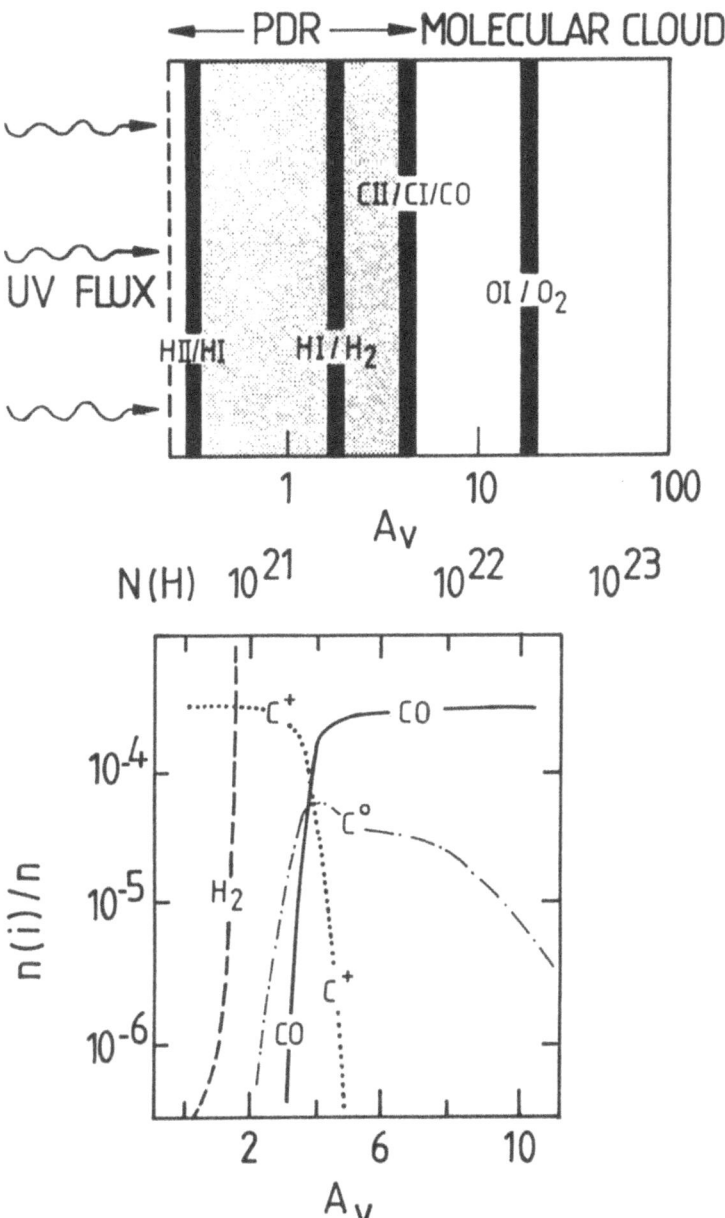

Fig. 14. Schematic of a PDR (top) and chemical abundances of the major species (bottom) in the standard model of Tielens & Hollenbach (1985a,b)

(Sect. 3.2) have a larger photoelectric yield and can heat at higher temperature than bulk dust grains (Verstraete et al. 1990). As a result they may be the dominant heating source in the diffuse ISM (Verstraete et al. 1990).

The net photoelectric heating rate can thus be expressed as (de Jong et al. 1980, Black 1987)

$$\Gamma_{\rm pe} = n_{\rm d}\,\sigma_{\rm d}\,\eta_{\rm pe}\,\chi \approx 4 \times 10^{-26}\left(\frac{\chi}{\chi_0}\right) n_{\rm H} \qquad [{\rm erg\,cm^{-3}\,s^{-1}}]. \qquad (81)$$

Here we have used that the product of volume density $n_{\rm d}$ and cross section $\sigma_{\rm d}$ of dust grains is about $1.5 \times 10^{-21}\,n_{\rm H}$ (Spitzer 1978). χ is the UV flux and $\chi_0 = 2.5 \times 10^{-3}\,{\rm erg\,cm^{-2}\,s^{-1}}$ is the far-UV flux in the solar neighborhood (Draine 1978). χ is of course dependent on the depth into the cloud and scales like $\exp(-\tau_{\rm d})$.

Photoelectric heating by the average interstellar radiation field also likely explain the observed temperatures of moderate to low density HI clouds in the solar neighborhood ($T_{\rm kin} \approx 70 \rightarrow 100\,{\rm K}$). There the main cooling line is the $158\,\mu{\rm m}\ ^2P_{3/2} \rightarrow {}^2P_{1/2}$ transition of C^+. Balancing its cooling rate for $C^+/H = 10^{-4}$ at low densities ($n < n_{\rm crit} \approx 3 \times 10^3\,{\rm cm^{-3}}$) with the photoelectric heating rate from Eq. (81) gives

$$\Gamma_{\rm pe} = 2 \times 10^{-27}\exp\left(\frac{-91\,{\rm K}}{T_{\rm kin}}\right) n_{\rm H}^2\,[{\rm K}], \qquad {\rm and}$$

$$T_{\rm kin} = \frac{-91\,{\rm K}}{\ln\left(\dfrac{20}{n_{\rm H}}\dfrac{\chi}{\chi_0}\right)} \approx 80\,[K] \qquad (82)$$

for $n_{\rm H} = 30\,{\rm cm^{-3}}$ and $\chi = \frac{1}{2}\chi_0$ (solar neighborhood). Within the uncertainties, the photoelectric heating of Eq. (82) also is consistent with the C^+-cooling in the *diffuse* ISM inferred from UV spectroscopy by Pottasch et al. (1979), $\Lambda \approx 10^{-25}\,{\rm erg\,s^{-1}}$ per hydrogen atom.

UV pumping of H_2 molecules can convert a similar fraction of the UV if the hydrogen density is sufficiently high (Sternberg 1988). In this case, electronic excitation of H_2 molecules in the Lyman and Werner bands at $1000\,\text{Å}$ leads to subsequent radiative decay into a vibrationally excited level of the electronic ground state (2 to 3 eV) in 9 out of 10 excitations. At densities greater than the critical density for collisional deexcitation of the vibrational states $\left(n(H_2)_{\rm crit} \approx 7 \times 10^4\,{\rm cm^{-3}}\right)$, this energy can again be converted into gas heating. In the UV transitions of H_2, the conversion efficiency of a single absorbed photon into kinetic energy can be as high as 30%. The total efficiency of converting far-UV photons into heating depends on the fraction of incident radiation absorbed by H_2 molecules throughout the cloud. This fraction depends on the dust to gas ratio and on the ratio of gas density to intensity of the UV field and is usually $\leq 10\%$ (Sternberg 1988, Sternberg & Dalgarno 1989).

At visual extinctions *greater than a few* the gas is predominantly heated through *collisions with warm dust particles* that transfer a fraction of their internal energy in the process. As this heating process involves a collision between

322

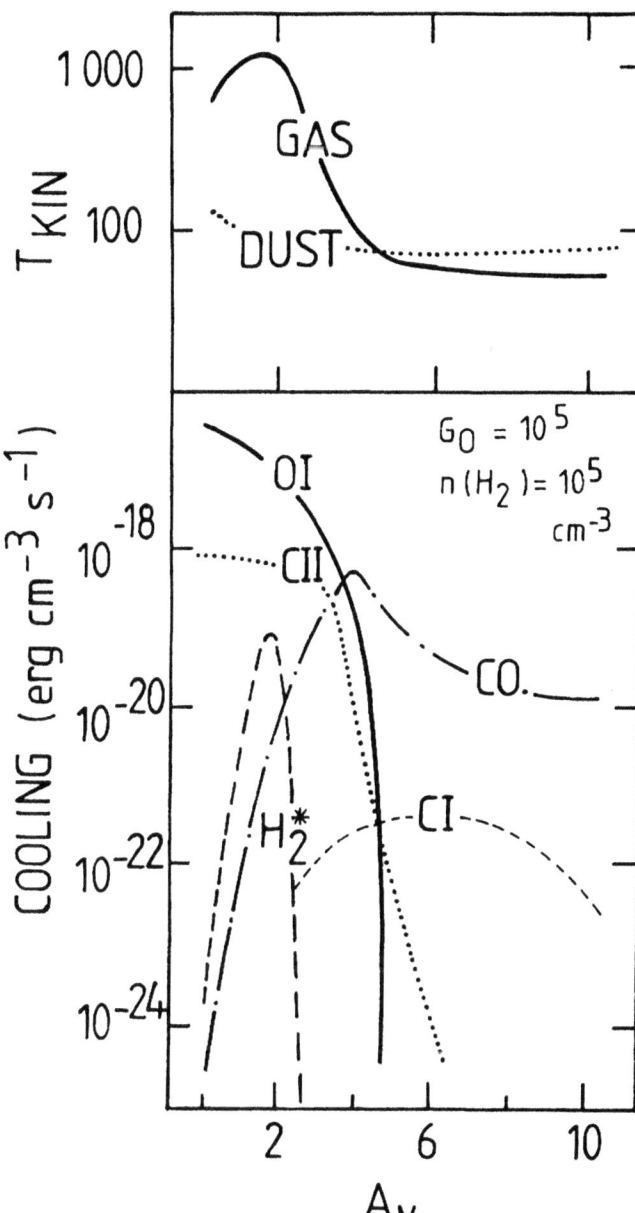

Fig. 15. Kinetic temperature and gas cooling as a function of A_V in the standard model of Tielens & Hollenbach (1985a,b)

two particles it scales proportional to (density)^2 and depends on the composition of the gas through the collision cross sections (Hollenbach & McKee 1979). It also depends on the difference between dust temperature (T_d) and gas kinetic temperature ($T_{kin} \leq T_d$) so that for molecular gas

$$\Gamma_{d-gas} = 2 \times 10^{-33} \, T_{kin}^{1/2} \, (T_d - T_{kin}) \, n_H^2 \qquad [\text{erg s}^{-1} \, \text{cm}^{-3}]. \qquad (83)$$

Clearly the gas is only coupled well to the dust at high densities where the cooling time scale is sufficiently small,

$$t_{d-gas} = 1.5 \times 10^4 \left(\frac{n_H}{10^5 \, \text{cm}^{-3}} \right)^{-1} \left(\frac{T_{kin}}{50 \, \text{K}} \right)^{-1/2} \qquad [\text{y}]. \qquad (84)$$

Gas *cooling* of the PDR at $A_V \leq 5$ is primarily by the fine structure lines of [O I] (63 μm), [C II] (158 μm) and [Si II] (35 μm). For very dense gas the 26 and 35 μm fine structure lines of [Fe II] and the forbidden lines of [O I] and [C I] in the visible may also contribute. Of lesser importance are the H_2 ro-vibrational and rotational lines and the submm fine structure lines of [C I] largely because of the small column densities in the upper state and because of their small A-coefficients.

The major coolants and the resulting thermal structure are depicted in Fig. 15 as a function of depth from the cloud surface, again for the "standard" model of Tielens & Hollenbach (1985a,b). The standard model is representative for dense photodissociation regions within a few tenths of a pc of a massive O star, such as the HII region/cloud interfaces in Orion and M17 or the circumnuclear disk in the Galactic center. The gas temperature in the PDR ranges between about 1000 K at the surface to about 100 K at the C^+/CO transition.

2.1.b Observations of Photon Dominated Regions

The observed infrared and submillimeter line and continuum emission toward the Trapezium region in Orion agree fairly well with the standard Tielens & Hollenbach model (see discussion in Genzel et al. 1989). About 0.6% of the impinging UV radiation is converted into [OI], [SiII], [CII], [CI], H_2, and CO line emission. Recently Stacey et al. (1991) were able to detect the $F = 1 \rightarrow 0$ [^{13}C II] line in the Orion PDR. The large ratio of [^{12}C II]/[^{13}C II] $F = 1 \rightarrow 0$ fluxes (≈ 100) shows that the optical depth of the [^{12}C II] line is moderate ($\tau([^{12}$C II]) $\leq$ a few, see also Boreiko et al. 1988). The [CII] emission must come from gas with temperature ≈ 200 K or more, in good agreement with the theoretical models. The standard model does not predict as much [Si II] emission as is observed (Haas et al. 1986) which may indicate that the assumed gas phase abundance of silicon is too low. More serious is the failure of the model to account for the observed intense submm and far-infrared CO line emission (Harris et al. 1987, Schmid-Burgk et al. 1989, Genzel et al. 1989, Boreiko et al. 1989). Again this is not too surprising, as the models discussed above predict very little CO at temperatures > 100 K that is required to explain the Orion CO data. To overcome this disagreement Burton et al. (1990b) have proposed that the Orion PDR has much higher density than in the standard Tielens

& Hollenbach model ($n_H \approx 10^6$ to 10^7 cm^{-3}). In this case, increased heating (H$_2$ pumping), lower dissociation (shielding), and greater molecular formation rates produce a layer of high temperature molecular gas in the PDR that can account for the submillimeter and far-infrared CO data.

Observations of other Galactic HII region/molecular cloud interfaces (Genzel et al. 1985, 1990, Stutzki et al. 1988, Crawford et al. 1985) and about two dozen nuclei of infrared luminous, external galaxies (Crawford et al. 1985, Lugten et al. 1986, Lord et al. 1992, Stacey et al. 1991) also agree well with the prediction of PDR theory. Self-absorption features in the [CII] lines may indicate substantial optical depths in a number of massive star forming regions (Boreiko et al. 1990). For a detailed discussion we refer to Genzel et al. (1989).

2.2 Shocks in Dense Interstellar Clouds

Interstellar shock waves are generated by supersonic mass motions, such as fast moving clouds, outflows from young stars, stellar winds and supernovae. The kinetic energy of the supersonic motions is converted into thermal energy. In the process, shocks compress and accelerate gas and dust. Shocks radiate primarily in lines from the cooling gas behind the front where the flow motion is converted into random thermal motions (McKee & Hollenbach 1980). With the exception of high velocity shocks ($v_s > 50 \, \mathrm{km \, s^{-1}}$) in dense ($n(H_2) > 10^6$ cm^{-3}) clouds, dust emission is small compared to line emission from hot gas. Shocks in dense molecular clouds often radiate mostly in the infrared and submillimeter lines of molecules, atoms and ions. High velocity, gas dynamic J-shocks dissociate most molecules and ionize atoms. Magneto-hydrodynamic C-shocks create large column densities of moderately warm molecular gas.

2.2.a Theory

J-Shocks. Shock waves in dense interstellar clouds have been recently reviewed in some detail by Shull & Draine (1987) and Hollenbach et al. (1989) who also refer to the many detailed calculations of the last 10 years or so. I summarize here the main results of these theoretical studies. In cold gas (sound speed $< 1 \, \mathrm{km \, s^{-1}}$) with no or weak magnetic fields, a fast ($> 50 \, \mathrm{km \, s^{-1}}$) pressure disturbance creates a *J-shock*, where temperature, density and flow velocity suffer a virtually discontinuous *jump* across the shock front. Figure 16 (top) gives a schematic of the change of the variables across the front. Conservation of mass and momentum (and energy for an adiabatic shock) yield relations between pre-shock variables and downstream variables.

These Rankine-Hugoniot relations are, for instance, discussed in Shull & Draine (1987) and Spitzer's (1978) book (see Elmegreen, this volume). The thermal structure behind the front is shown in Fig. 17, for the case of a $50 \, \mathrm{km \, s^{-1}}$ shock into a $n = 10^5$ cm^{-3} pre-shock density cloud (from Hollenbach & McKee 1989). Immediately behind a fast shock in a molecular cloud the temperature is high,

$$T_s \approx 1.5 \times 10^5 \left(\frac{v_{\mathrm{shock}}}{100 \, \mathrm{km \, s^{-1}}} \right)^2 \quad [\mathrm{K}]. \qquad (85)$$

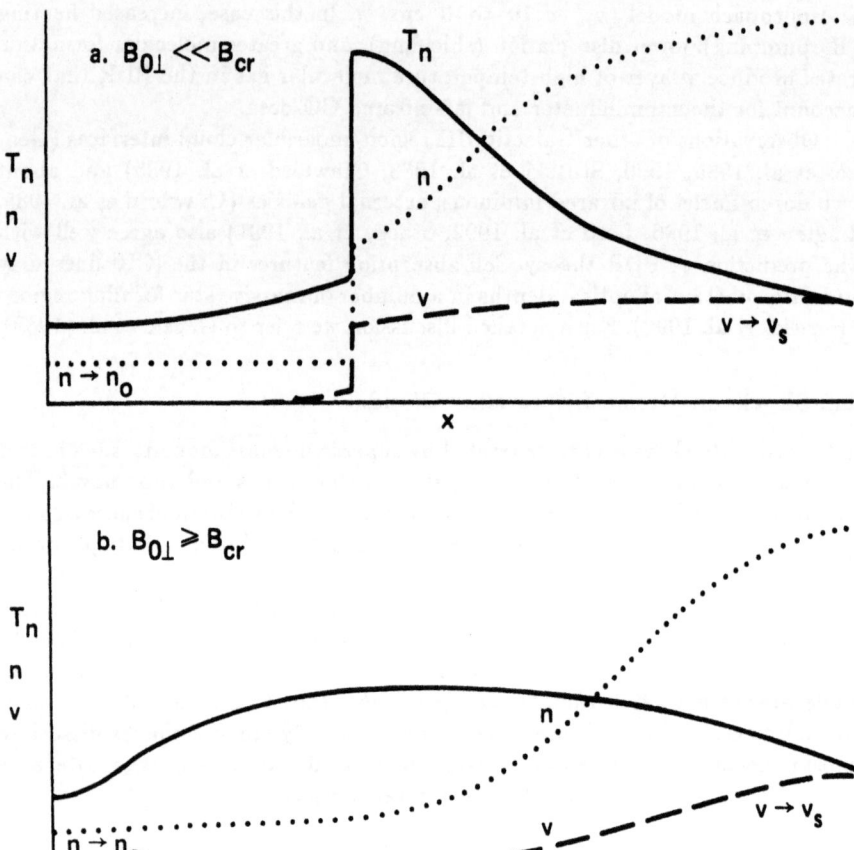

Fig. 16. Schematic structure of J- (top) and C-shocks (bottom, adapted from Hollenbach et al. 1989)

Consequently the gas dissociates and ionizes. It radiates in the UV and visible wavelength bands in resonance, semi-forbidden and forbidden lines of hydrogen, helium and the ions of oxygen, carbon, sulfur, and iron. Further downstream ($N_H \geq 10^{16}$ cm^{-2}), the hydrogen ionizing photons are absorbed, creating a 10^4 K temperature plateau. Hydrogen recombination lines and [Ne II] fine structure lines originate there. Further downstream again ($N_H \approx 10^{20}$ cm^{-2}), the gas recombines and cools rapidly. Molecular formation sets in at temperatures of a few hundred K. As a result of the high temperatures, neutral-neutral chemical reactions proceed rapidly. Carbon is efficiently converted to CO. The remaining oxygen is converted to OH and H$_2$O. In at least moderately dense gas ($n(H_2) \geq 7 \times 10^4$ cm^{-3}), the H$_2$ formation energy is converted into gas heating and creates another temperature plateau at ≈ 400 K (Neufeld & Dalgarno 1989a). The calculations show that much of the infrared emission comes

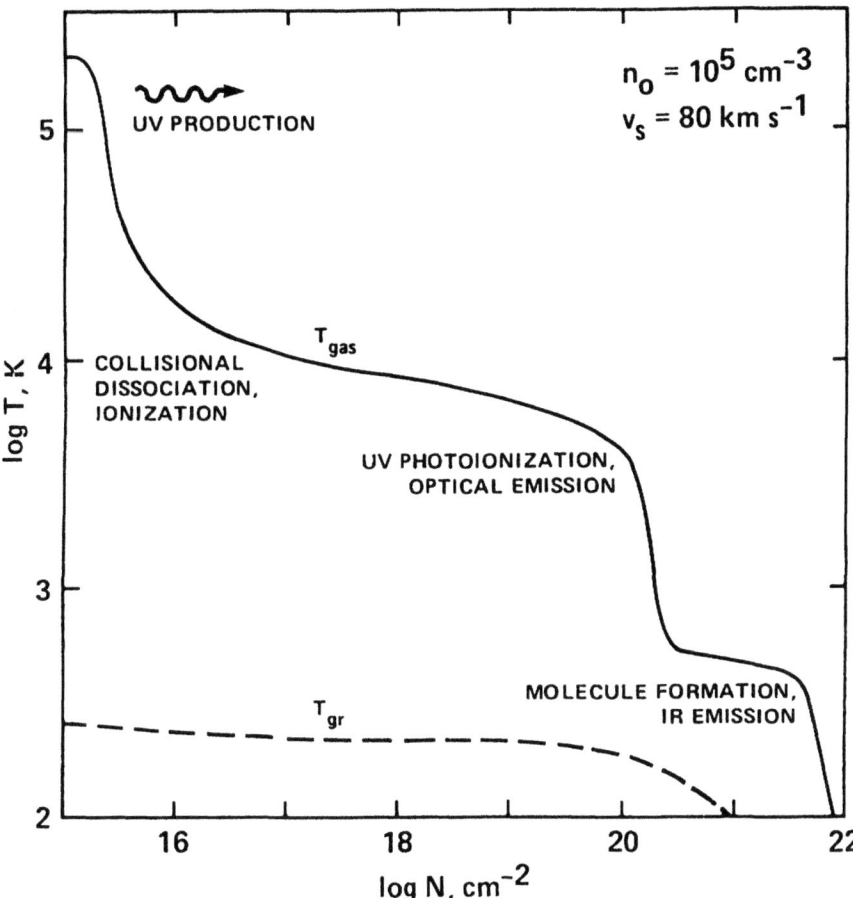

Fig. 17. The post-shock temperature structure of a fast molecular shock ($n_0 = 10^5 \, \mathrm{cm}^{-3}$, $v_s = 80 \, \mathrm{km \, s}^{-1}$ shown here). Three regions are delineated: (1) the hot, $T \approx 10^5$ K, immediate post-shock region, where gas is collisionally dissociated and ionized and UV photons are produced which affect both the pre-shock and post-shock gas; (2) the "recombination plateau" where the Lyman continuum photons are absorbed, maintaining $T \approx 10^4$ K: and (3) the recombining and molecule-forming gas downstream, where chemical energy of H_2 formation can maintain a lower temperature plateau. The column densities of the first two regions are nearly independent of n_0. Note that the grains are weakly coupled to the gas, so that $T_{gr} \ll T_{gas}$ (adapted from Hollenbach & McKee 1989)

from this plateau and from the temperature region $< 10^4$ K where the gas density is about 100 times greater than in the pre-shock gas. Typically 1 to 10% of the shock's energy emerges in infrared lines. Hollenbach & McKee (1989) and Neufeld & Dalgarno (1989a,b) have calculated the chemical and thermal structure and cooling of dissociative shocks with speeds between 40 and $150 \, \mathrm{km \, s}^{-1}$ at pre-shock densities between 10^3 and $10^6 \, \mathrm{cm}^{-3}$. At medium densities ($\approx 10^4 \, \mathrm{cm}^{-3}$) and velocities, the dominant coolants are [O I] 63 μm, [O I] 6300 Å and [C I] 9849 Å, followed by H_2 and CO rotational and ro-vibrational line emission. At densities of 10^6 cm^{-3} or greater the rotational line emission of H_2O and OH dominates and grain cooling becomes important. The [Fe II] lines at $1.3/1.6\,\mu$m, along with vibrational transitions of H_2, are the most prominent emission lines in the near-infrared range. Infrared hydrogen recombination lines of atomic hydrogen are also detectable from J-shocks, but may easily be confused with HII regions along the line of sight. Depending on the gas phase abundance of silicon, [Si II] $35\,\mu$m emission may also be strong. Submillimeter emission of highly excited rotational states of SiO, HCN, CN, SO and NO may be characteristic tracers of J-shocks in dense clouds (Neufeld & Dalgarno 1989b).

C-Shocks. A different type of shock occurs if shock velocities are less than about $40\,$km s^{-1}, if there is a moderately strong magnetic field and if the ionization fraction in the cloud is moderately low (x_e = [electrons]/[hydrogen] $\leq 10^{-6}$). In the presence of a magnetic field disturbances propagate at the magnetosonic, or Alfvén speed v_A, given by

$$v_A = \frac{B}{\sqrt{4\pi\langle m\rangle n}} = 22\left(\frac{B}{1\,\text{mG}}\right)\left(\frac{n}{10^4\,\text{cm}^{-3}}\right)^{-1/2} \quad [\text{km s}^{-1}], \quad (86)$$

where $\langle m\rangle$ and n are the mean mass and number density of the (neutral) gas. Shocks are possible for $v_A < v_{\text{shock}}$, otherwise the pressure disturbance is "communicated" and damped by Alfvén waves. In partially ionized gas, the ions react rapidly to changes in the magnetic field and then communicate those more slowly to the neutrals by ion-neutral collisions. Because of this difference in reaction speed, it is possible to transmit damped Alfvén waves in the ion "fluid" at the ion magnetosonic speed, $v_A(\text{ion}) = \left(n/n(\text{ion})\right)^{1/2}v_A \gg v_A$. For small ion fraction one may then have the situation $v_A < v_{\text{shock}} < v_A(\text{ion})$. In that case magnetic field and ion density $n(\text{ion})$ must vary *continuously* through the shock front. If the neutrals (because of their interaction with the ions) also vary continuously, the shock is called a *"C-shock"* (for continuous, Draine 1980). This situation is depicted in the lower part of Fig. 16. The shock sends a message to the upstream gas via ions and magnetic field. The ions begin to compress and accelerate, so that they drift relative to the neutrals and heat and accelerate them. The neutral gas can rapidly radiate its thermal energy away. In practice this happens if molecules are not dissociated ($T <$ a few 10^3 K) and if the efficient cooling of H_2, CO, OH and H_2O is available.

The detailed structure of a $25\,$km s^{-1} C-shock into a 10^6 cm^{-3} pre-shock density cloud at a magnetic field of $1\,$mG is shown in Fig. 18 (from the models of Draine et al. 1983). C-shocks radiate nearly the entire energy of the shock in many molecular and atomic infrared lines. [O I] $63\,\mu$m emission dominates the cooling for slow ($10\,$km s^{-1}) shocks. Rotational and ro-vibrational line emission of H_2 is the main coolant for higher velocities ($10 < v_{\text{shock}} < 50\,$km s^{-1}), but moderate pre-shock densities. For densities of 10^6 cm^{-3} or greater, H_2O and OH rotational emission become more and more important.

2.2.b Observations of Shocks
One of the best investigated region is the shock in the Orion-KL star forming

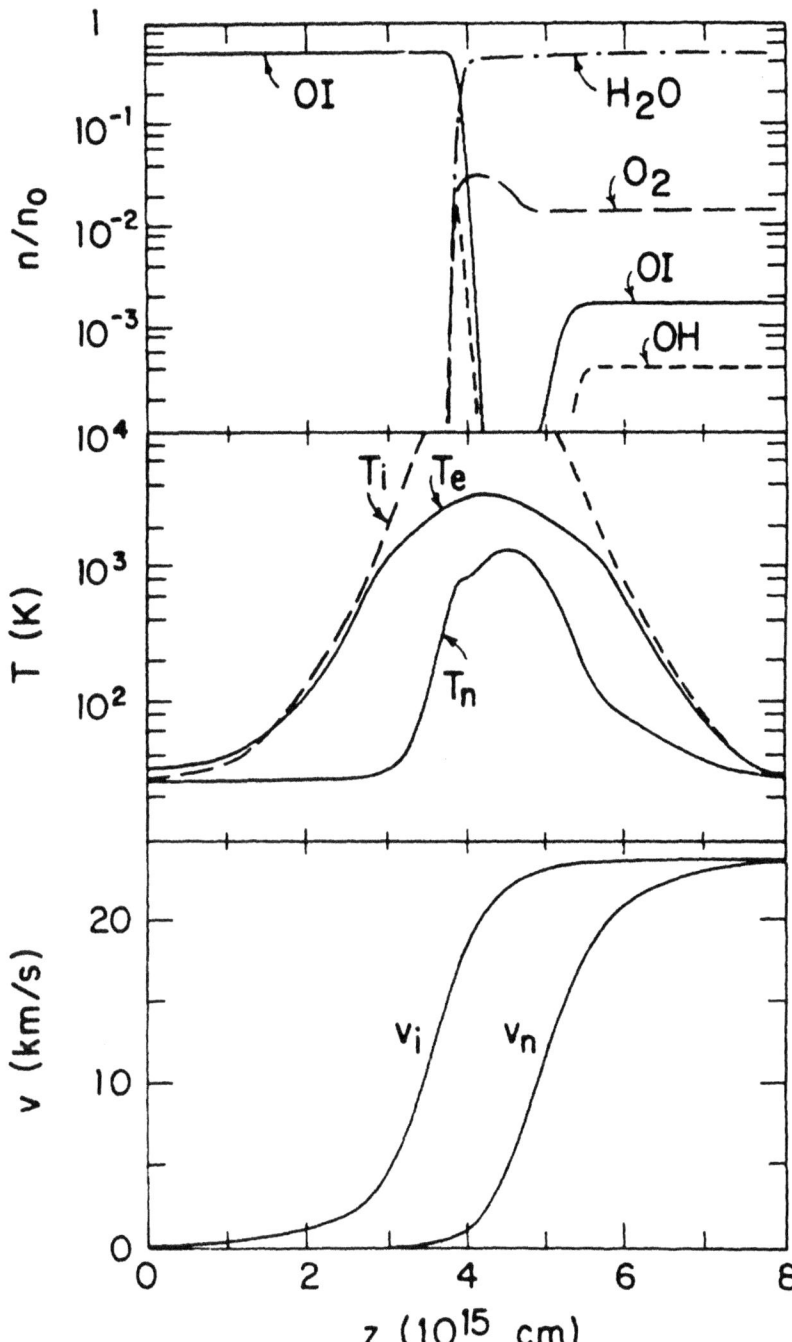

Fig. 18. Structure of a $25\,\mathrm{km\,s^{-1}}$ C-shock in a molecular cloud with $n_0 = 10^6\,\mathrm{cm^{-3}}$, $x_e = 10^{-8}$ and $B_0 = 1\,\mathrm{mG}$ (adapted from Draine et al. 1983)

region. Since the first detection of H_2 ro-vibrational emission by Gautier et al. (1976), many infrared lines have been inferred to come from shock excited gas. (e.g., Watson et al. 1985, Geballe et al. 1986, Geballe & Garden 1987, Brand et al. 1988, Geballe 1990). The shocked zone has a size of about one arcmin (0.15 pc) and is centered on the Orion-KL infrared cluster. The shock is probably created as a (clumpy) molecular outflow emanating from the young stars in the cluster (IRc 2, BN) strikes the surrounding molecular cloud (Sect. 4.4.b). Chernoff et al. (1982) and Draine & Roberge (1982) have successfully modeled most of the emission as coming from a $40\,\mathrm{km\,s^{-1}}$ C-shock, with a magnetic field of about $1\,\mathrm{mG}$ and a pre-shock hydrogen density of 10^5 to $10^6\,\mathrm{cm^{-3}}$. The momentum flux required to drive the shock ($\dot{M}v_{\mathrm{shock}} \approx 10^{30}\,\mathrm{g\,cm\,s^{-2}}$) is quite consistent with other estimates of the mass outflow from IRc 2. Several key observed facts match the theoretical characteristics of C-shocks. H_2 and CO infrared line emission are the dominant coolants ($L(H_2 + CO) \approx 500\,L_\odot$). The H_2 emission comes from thermalized gas at a temperature near 2000 K. There is little ionization. Still there remain questions. The C-shock models cannot explain the strong [Si II] (and probably also the [O I]) emission and an additional J-shock (perhaps the "wind-shock", Hollenbach et al. 1989) may be required. Brand et al. (1988) have argued that the H_2 lines, especially those with large excitation requirements can be better explained by a J-shock. Finally the simple "cloud" C-shock models can also not account for the high velocity wings in the H_2 lines (e.g., Geballe et al. 1986). These high velocity wings may be created in collisions of high velocity cloudlets in the wind. Finally and perhaps most worrisome Wardle (1990) has recently concluded from a theoretical stability analysis that C-shocks are *unstable* to a buckle (Parker) instability of the magnetic fields and the previous time-independent models may not be applicable.

There is an increasing number of infrared spectroscopy studies of shocks in various other sources and astrophysical environments (Geballe 1990). In addition to other star forming regions (e.g., Garden et al. 1986), strong $2\,\mu\mathrm{m}$ H_2, $1.6\,\mu\mathrm{m}$ [FeII], and $63\,\mu\mathrm{m}$ [OI] emission has been detected from shocks associated with the supernova remnants IC 443 and RCW 103 (Burton et al. 1990a, Oliva et al. 1989). For IC 443 Burton et al. (1990a) conclude that both C- and J-shocks are required for an explanation of the available data. J-shocks in low density gas with velocities $> 30\,\mathrm{km\,s^{-1}}$ may account for the extended $158\,\mu\mathrm{m}$ [CII] emission found in IC 443 (Haas, private communication, Poglitsch, private communication). In the Cygnus loop supernova remnant Graham et al. (1991) find extended H_2 emission that may originate in a J-shock with a magnetic precursor that is running into low density ($n_H \approx 10\,\mathrm{cm^{-3}}$) gas. Strong H_2 emission associated with Herbig-Haro objects HH 7-11 may originate in bow shocks at the interface between clumps in the wind and the surrounding cloud (Garden et al. 1988). Finally, recent imaging spectroscopy of H_2 line emission in the merging galaxy pair NGC 6240 indicates that the infrared line emission comes from an extended arc that is located between the two nuclei (Van der Werf et al. 1992). This result is consistent with the original proposal of a large scale shock of the ISM due to the collision of the two galaxies (Harwit et al. 1987).

J-Shocks vs. C-Shocks vs. PDR's. How can J- and C-shocks be observationally distinguished and how can shocks be distinguished from photodissociation regions? In comparison to J-shocks, C-shocks have essentially no infrared emission from ionized species ([CII], [Ne II], [Fe II], [Si II], hydrogen recombination lines). C-shocks have much larger column densities of moderately high temperature (10^3 K) molecular gas than J-shocks, resulting in a relatively high ratio of far-infrared CO to near-infrared H_2 cooling. In principle one can easily distinguish J- from C-shocks by the line profiles. In J-shocks, the cooling gas has been fully accelerated and thus emits a narrow line displaced from the velocity of the quiescent cloud (by the shock velocity). In C-shocks, there is a smooth, broad emission profile centered on the velocity of the quiescent cloud, as most of the emission occurs when the gas is just being accelerated. In practice, however, the geometry of the shocked zone may be complex with several shocks along the line of sight seen at different angles.

Perhaps the most definite indicators of a photodissociation region vs. an (ionizing) J-shock are the absolute and relative intensities of the [C II] 158 μm and [O I] 63 μm fine structure lines. In shocks one finds weak [C II] emission (or no [C II] emission in C-shocks), but strong [O I] emission $\left(I([O\,I])/I([C\,II]) \geq 10\right)$, while that ratio is between 1 and 10 in PDR's. The 1.6 μm [FeII] fine structure line seems to correlate with fast shocks associated with supernova remnants (Oliva et al. 1990, Greenhouse et al. 1991), and radio jets (Blietz et al. 1992). Another commonly used criterion is the ratio of H_2 $v = 2 \rightarrow 1$ to $v = 1 \rightarrow 0$ $S(1)$ line fluxes. If that ratio is small (≤ 0.1), the emission is ascribed to shocks. If the ratio is ≈ 0.5, it is ascribed to a PDR. This is not generally a good indicator, however, as a small ratio merely indicates thermalized emission at a temperature not much exceeding $\approx 10^3$ K. A small $v = 2 \rightarrow 1/1 \rightarrow 0$ ratio can also occur in a dense $\left(n(H_2) \gg 7 \times 10^4 \text{ cm}^{-3}\right)$ PDR (Sternberg 1988, Sternberg & Dalgarno 1989).

2.3 Heating by Energetic Particles

Energetic particles (e.g., $1 \rightarrow 100$ MeV cosmic rays) and hard X-rays (> 250 eV) contribute to the heating throughout molecular clouds by collisional impact ionization. A cosmic ray (usually a proton) hitting a neutral hydrogen atom, for instance, produces a proton and a fast electron,

$$H + CR(1 \rightarrow 100 \text{ MeV}) \longrightarrow H^+ + e^- (E_{kin} \approx 35 \text{ eV}) + CR. \qquad (87)$$

If atomic hydrogen is replaced by molecular hydrogen the molecular ion H_2^+ is created, a key ion in the ion-molecule chemistry chain (Chap. 3). The primary electron can ionize again or it can contribute to non-ionizing heating of the gas. In dense neutral gas $\left(x_p = n(H^+)/n(H) \leq 10^{-3}\right)$ about $E_{nh} = 7$ eV is fed into gas heating (including secondary processes, Cravens & Dalgarno 1978). The total heating rate then is the product of hydrogen density, ionization rate per hydrogen atom/molecule ζ_{CR} (s^{-1}) and E_{nh}. The primary ionization rate in the solar neighborhood can in principle be directly determined from in situ measurements of the cosmic ray flux. The principal difficulty of this direct method has so far been the fact that the measured cosmic ray flux in the vicinity of the Earth needs to be corrected by a large and uncertain factor involving the interaction of cosmic rays with the interplanetary magnetic field and the solar wind.

The best determination of ζ_{CR} in the solar neighborhood currently comes from an analysis of the ionization and chemical abundances in interstellar clouds, both being strongly dependent on the primary ionization rate. From an analysis of diffuse clouds van Dishoeck & Black (1986) find $\zeta_{CR} \approx 4 \times 10^{-17}\,s^{-1}$. With this value we have

$$\Gamma_{CR} = 4.7 \times 10^{-28} \left(\frac{\zeta_{CR}}{4 \times 10^{-17}\,s^{-1}} \right) \left(\frac{E_{nh}}{7\,eV} \right) n(H, H_2) \quad [erg\,s^{-1}\,cm^{-3}]. \quad (88)$$

Heating by cosmic rays with $\zeta_{CR} \approx 10^{-17}$ to $10^{-16}\,s^{-1}$ can plausibly account for the kinetic temperatures of the bulk of cold, non-star forming molecular clouds, $T \approx 10\,K$ (e.g., Solomon et al. 1987). At molecular hydrogen densities between a few 10^2 and $10^4\,cm^{-3}$ and kinetic temperatures between 5 and 30 K the cooling function is dominated by ^{12}CO and ^{13}CO rotational emission and can be approximated as

$$\Lambda_{CO} \approx 10^{-24} \left(\frac{T_{kin}}{10\,K} \right)^{2.75} \left(\frac{n(H_2)}{10^3\,cm^{-3}} \right)^{0.3} \left(\frac{\Delta v / \Delta l}{1\,km\,s^{-1}\,pc^{-1}} \right) \quad (89)$$
$$[erg\,s^{-1}\,cm^{-3}].$$

At the default values for the input parameters Eqs. (88) and (89) imply kinetic temperatures between 7 and 13 K. Cosmic ray heating can also explain higher kinetic temperatures in the vicinity of strong cosmic ray sources, such as supernova remnants.

X-rays photoionize and heat gas in much the same manner as cosmic rays (Krolik & Kallman 1983, Lepp & McCray 1983). In the solar neighborhood the ionization rate due to cosmic rays dominates so that X-ray heating plays only a minor role in the average interstellar medium. The X-ray luminosity of OB stars is about $10^{-7}\,L_{tot}$ and T-Tau stars emit about $10^{-4}\,L_{tot}$ in X-rays (Rosner et al. 1985). The total X-ray luminosity of the central region around θ^1 Ori is about 3×10^{33} to $10^{34}\,erg\,s^{-1}$ (Ku & Chanan 1979). Assuming that the stellar X-ray spectrum is thermal with $kT_* \approx 500\,eV$, Krolik & Kallman (1983) find an overall X-ray heating as a function of hydrogen column density N_H into the cloud and distance R from the X-ray sources of

$$\Gamma_X \approx 3 \times 10^{-26} \left(\frac{N_H}{10^{22}\,cm^{-2}} \right)^{-1.6} \left(\frac{L_X}{10^{34}\,erg\,s^{-1}} \right) \left(\frac{R}{pc} \right)^{-2} n(H_2) \quad (90)$$
$$[erg\,s^{-1}\,cm^{-3}].$$

The assumed X-ray optical depth scales with photon energy E approximately as $\tau_X \sim N_{22} E_{keV}^{-2.5}$ (Cruddace et al. 1974) where $N_{22} = (N_H/10^{22} \, \text{cm}^{-2})$. Equation (90) shows that near OB and T-Tau associations X-ray heating can significantly contribute to cloud heating and ionization (Krolik & Kallman 1983).

X-ray heating of molecular clouds may be of great importance in the vicinity of luminous AGN's and QSO's. Krolik & Lepp (1989) calculate that dense parsec scale circum-nuclear tori of molecular gas in Seyfert galaxies would emit primarily in far-infrared and submillimeter CO lines. Between 6×10^{-4} and 6×10^{-3} of the X-ray luminosity may emerge in the $2 \, \mu\text{m}$ $S(1)$ H_2 line (Lepp & McCray 1983, Draine & Woods 1990).

2.4 Heating by Ambipolar Diffusion

In addition to shocks, "mechanical" heating processes involving the interaction of the gas with surrounding and embedded magnetic fields, or magnetic waves may make a significant contribution to cloud heating (Black 1987). In the following we will consider in particular the heating that results from a slow drift between magnetic field and neutral gas (Scalo 1977). This "ambipolar diffusion" heating may occur during cloud collapse in dense gas. Following Spitzer (1978) and Zweibel (1987) let us consider a magnetic field of strength B drifting through a neutral cloud of density n_n and ionization fraction x_c ($n_c = x_c n_n$). Recall that the charged particles in the cloud are closely coupled to the magnetic field while the neutrals interact with the field only indirectly via collisions with the charged particles on the "slow down" time scale

$$t_{cn} = \frac{1}{n_n \langle \sigma u \rangle_{cn}} \approx \frac{10^9}{n_n} \quad [\text{s}]. \tag{91}$$

$\langle \sigma u \rangle_{cn}$ is the average product of cross section and differential velocity in collisions between neutral and charged particles. The change in momentum of the drift motion due to these collisions is

$$\frac{dp_d}{dt} = -\frac{\mu \, n_c \, w_{drift}}{l_{cn}}. \tag{92}$$

w_{drift} is the drift velocity and $\mu = m_c m_n/(m_c + m_n)$ is the reduced mass. For a constant drift velocity to occur this momentum change must be balanced by the force of a magnetic field gradient (magnetic field assumed to be straight and perpendicular to the drift velocity) over scale length L,

$$\frac{dp_d}{dt} = -\left| \frac{\nabla B^2}{8\pi} \right| \approx -\frac{B^2}{8\pi L}. \tag{93}$$

The drift velocity can then be expressed as

$$w_{drift} \approx \frac{B^2}{8\pi L \, n_n^2 \, x_c \, \mu \, \langle \sigma u \rangle_{cn}} \approx 8 \frac{B_{\mu G}^2}{L_{pc} \, n_n^2 \, x_c \, (\mu/m_H)} \quad [\text{cm s}^{-1}]. \tag{94}$$

$B_{\mu G}$ is the magnetic field in units of μG and L_{pc} is in units of pc. The resulting ambipolar diffusion heating rate then is

$$\Gamma_{\text{amb}} = \frac{\mu \, n_c \, w_{\text{drift}}^2}{t_{cn}} \approx 10^{-31} \frac{B_{\mu G}^4}{L_{pc}^2 \, n_n^2 \, x_c \, (\mu/m_H)} \qquad [\text{erg cm}^{-3} \, \text{s}^{-1}]. \qquad (95)$$

If B and L are free variables, Γ_{amb} scales as n_n^{-2} and is most important for the diffuse interclump medium (Pérault et al. 1985).

From observations of clouds over a wide range of density and scale size, mean density and scale size appear to be correlated inversely to each other ("Larson's" relation, Larson 1981, Sect. 4.2.b),

$$n_n L_{pc} \approx N_0 \approx 2 \times 10^3 \qquad (96)$$

(see, for example, Scalo 1990). Furthermore it is theoretically expected that magnetic field strengths increase with increasing density, $B \sim n_n^\alpha$ with $\alpha \approx 1/2$ (Mouschovias 1987). From the measured values of magnetic fields in diffuse clouds and dense dark clouds (Troland & Heiles 1986) a reasonable estimate may be

$$B_{\mu G} \approx 200 \left(\frac{n_n}{10^4 \, \text{cm}^{-3}} \right)^{1/2} = 2 \left(\frac{B_0}{2\mu G} \right) n_n^{1/2} \quad [\mu G]. \qquad (97)$$

With Eqs. (96) and (97) substituted into Eq. (95) one finds for molecular hydrogen gas ($\mu = 0.67 \, m_H$)

$$\Gamma_{\text{amb}} \approx 10^{-29} \left(\frac{B_0}{2\,\mu G} \right)^4 \left(\frac{10^{-7}}{x_c} \right) \left(\frac{5 \times 10^{21}}{N_0(H_2)} \right)^2 n^2(H_2) \quad [\text{erg cm}^{-3} \, \text{s}^{-1}]. \qquad (98)$$

With these assumptions ambipolar heating is of particular importance in dense, clumps of small size, low column density, and low ionization fraction. Ambipolar heating can be neglected in the outer layers of clouds where the ionization fraction may approach a few 10^{-3} due to the far-UV photoionization of carbon.

An interesting question is whether ambipolar heating can contribute significantly to the heating of dense $\left(n(H_2) \geq 10^5 \, \text{cm}^{-3} \right)$, warm $(T \geq 40 \, \text{K})$ cloud cores with column densities of $N(H_2) \geq 5 \times 10^{22} \, \text{cm}^{-2}$. For these conditions we use the total line cooling function given by Goldsmith & Langer (1978)

$$\Lambda_{\text{tot}} \approx 7 \times 10^{-30} \, T^{2.2} \, n(H_2) \qquad [\text{erg cm}^{-3} \, \text{s}^{-1}]. \qquad (99)$$

With Eqs. (98) and (99) we find

$$T_{\text{kin}} \approx 27 \left(\frac{B_0}{2\,\mu G} \right)^{1.8} \left(\frac{x_c}{10^{-7}} \right)^{-0.45} \left(\frac{N(H_2)}{5 \times 10^{22}} \right)^{-0.9} \left(\frac{n(H_2)}{10^5} \right)^{0.45} \quad [\text{K}]. \qquad (100)$$

Given the considerable uncertainties of all the parameters going into Eq. (100), ambipolar heating may well be an interesting mechanism for explaining temperatures near 50 K (see Scalo 1977). It is less likely, however, that this mechanism can account for substantial molecular column densities at temperatures $\geq 100 \, \text{K}$ that are found near PDR's (Graf et al. 1990).

References

2.1   Black, J.H. 1987, in *Interstellar Processes*, eds. D.J. Hollenbach, H.A. Thronson (Reidel, Dordrecht), 731

2.2   Blietz, M. et al. 1992, in prep

2.3   Boreiko, R.T., Betz, A.L., Zmuidzinas, J. 1988, *Astrophys. J.* **325**, L47

2.4   Boreiko, R.T., Betz, A.L., Zmuidzinas, J. 1989, *Astrophys. J.* **337**, 332

2.5   Boreiko, R.T., Betz, A.L., Zmuidzinas, J. 1990, *Astrophys. J.* **353**, 181

2.6   Brand, P.W.J.L., Moorhouse, A., Burton, M.G., Geballe, T.R., Bird, M., Wade, R. 1988, *Astrophys. J.* **334**, L103

2.7   Burton, M.G., Hollenbach, D.J, Haas, M.R., Erickson, E.F. 1990a, *Astrophys. J.* **355**, 197

2.8   Burton, M.G., Hollenbach, D.J., Tielens, A.G.G.M. 1990b, *Astrophys. J.* **365**, 620

2.9   Chernoff, D.G.F., Hollenbach, D.J., McKee, C.F. 1982, *Astrophys. J.* **259**, L97

2.10   Cravens, T.E., Dalgarno, A. 1978, *Astrophys. J.* **219**, 750

2.11   Crawford, M.K., Genzel, R., Townes, C.H., Watson, D.M. 1985, *Astrophys. J.* **291**, 755

2.12   Cruddace, R., Paresce, F., Bowyer, S., Lampton, M. 1974, *Astrophys. J.* **187**, 497

2.13   Dalgarno, A. 1991, in *Chemistry in Space*, eds. J.M. Greenberg, V. Pironello, *NATO-ASI-323* (Kluwer, Dordrecht), 71

2.14   de Jong, T. 1977, *Astron. Astrophys.* **55**, 137

2.15   de Jong, T., Dalgarno, A., Boland, W. 1980, *Astron. Astrophys.* **91**, 68

2.16   Draine, B.T. 1978, *Astrophys. J. Suppl. Ser.* **36**, 595

2.17   Draine, B.T. 1980, *Astrophys. J.* **241**, 1021

2.18   Draine, B.T., Roberge, W.G. 1982, *Astrophys. J.* **259**, L91

2.19   Draine, B.T., Roberge, W.G., Dalgarno, A. 1983, *Astrophys. J.* **264**, 485

2.20   Draine, B.T., Woods, D.T. 1990, *Astrophys. J.* **363**, 464

2.21   Garden, R., Geballe, T.R., Gatley, I., Nadeau, D. 1986, *Monthly Notices Roy. Astron. Soc.* **220**, 203

2.22   Garden, R., Russell, A.P.G., Burton, M.G. 1990, *Astrophys. J.* **354**, 232

2.23   Gautier, T., Fink, U., Treffers, R., Larson, H. 1976, *Astrophys. J.* **207**, L129

2.24   Geballe, T.R., Persson, S.E., Simon, T., Lonsdale, C.J., McGregor, P.J. 1986, *Astrophys. J.* **302**, 500

2.25   Geballe, T.R., Garden, R. 1987, *Astrophys. J.* **317**, L107

2.26   Geballe, T.R. 1990, in *Molecular Astrophysics*, ed. T. Hartquist (Univ. Press, Cambridge), 345

2.27   Genzel, R., Watson, D.M., Crawford, M.K., Townes, C.H. 1985, *Astrophys. J.* **297**, 766

2.28   Genzel, R., Harris, A.I., Stutzki, J. 1989, in *Infrared Spectroscopy in Astronomy*, ed. B.H. Kaldeich, *ESA-SP-290*, 115

2.29   Genzel, R., Stacey, G.J., Harris, A.I., Townes, C.H., Geis, N., Graf, U.U., Poglitsch, A., Stutzki, J. 1990, *Astrophys. J.* **356**, 160

2.30   Gerola, H., Glassgold, A. 1978, *Astrophys. J. Suppl. Ser.* **37**, 1

2.31   Goldsmith, P.F., Langer, W.D. 1978, *Astrophys. J.* **222**, 881

2.32   Graf, U.U., Genzel, R., Harris, A.I., Hills, R.E., Russell, A.P.G., Stutzki, J. 1990, *Astrophys. J.* **358**, L49

2.33   Graham, J.R., Wright, G.S., Hester, J.J, Longmore, A.J. 1991, *Astron. J.* **101**, 175

2.34   Greenhouse, M.A., Woodward, C.E., Thronson, H.A., Rudy, R.J., Rossano, G.S., Erwin, P., Puetter, R.C. 1991, *Astrophys. J.* **383**, 164

2.35   Haas, M.R., Hollenbach, D.J., Erickson, E.F. 1986, *Astrophys. J.* **301**, L57

2.36   Harris, A.I., Stutzki, J., Genzel, R., Lugten, J.B., Stacey, G.J., Jaffe, D.T. 1987, *Astrophys. J.* **322**, L49

2.37   Harwit, M., Houck, J.R., Soifer, B.T., Palumbo, G.G. 1987, *Astrophys. J.* **315**, 28

2.38   Hollenbach, D.J., McKee, C.F. 1979, *Astrophys. J. Suppl. Ser.* **41**, 555

2.39   Hollenbach, D.J., Chernoff, D.F., McKee, C.F. 1989, in *Infrared Spectroscopy in Astronomy*, ed. B.H. Kaldeich, *ESA-SP-290*, 245

2.40   Hollenbach, D.J., McKee, C.F. 1989, *Astrophys. J.* **342**, 306

2.41   Hollenbach, D.J., Takahashi, T., Tielens, A.G.G.M. 1991, *Astrophys. J.* **377**, 192

2.42 Jura, M. 1974, *Astrophys. J.* **191**, 375
2.43 Krolik, J.H., Kallman, T.R. 1983, *Astrophys. J.* **267**, 610
2.44 Krolik, J.H., Lepp, S. 1989, *Astrophys. J.* **347**, 179
2.45 Ku, W.H.M., Chanan, G.A. 1979, *Astrophys. J.* **234**, L59
2.46 Langer, W.D. 1976, *Astrophys. J.* **206**, 699
2.47 Larson, R.B. 1981, *Monthly Notices Roy. Astron. Soc.* **194**, 809
2.48 Lepp, S., McCray, R. 1983, *Astrophys. J.* **269**, 560
2.49 Lord, S. et al. 1992, in prep
2.50 Lugten, J.B., Watson, D.M., Crawford, M.K., Genzel, R. 1986, *Astrophys. J.* **311**, L51
2.51 McKee, C.F., Hollenbach, D.J. 1980, *Ann. Rev. Astron. Astrophys.* **18**, 219
2.52 Mouschovias, T. 1987, in *Physical Processes in Interstellar Clouds*, eds. G. Morfill, M. Scholer (Reidel, Dordrecht), 453
2.53 Neufeld, D.A., Dalgarno, A. 1989a, *Astrophys. J.* **340**, 869
2.54 Neufeld, D.A., Dalgarno, A. 1989b, *Astrophys. J.* **344**, 251
2.55 Oliva, E., Moorwood, A.F.M., Danziger, I.J. 1989, *Astron. Astrophys.* **214**, 307
2.56 Osterbrock, D.E. 1989, *Astrophysics of Gaseous Nebulae and Active Galactic Nuclei* (Univ. Science Books, Mill Valley)
2.57 Pérault, M., Falgarone, E., Puget, J.L. 1985, *Astron. Astrophys.* **152**, 371
2.58 Pottasch, S.R., Wesselius, P., van Duinen, R. 1979, *Astron. Astrophys.* **74**, L15
2.59 Rosner, R., Golub, L., Vaiana, G.S. 1985, *Ann. Rev. Astron. Astrophys.* **23**, 413
2.60 Scalo, J.M. 1977, *Astrophys. J.* **213**, 705
2.61 Scalo, J.M. 1990, in *Physical Processes in Fragmentation and Star Formation*, eds. R. Capuzzo-Dolcetta, C. Chiosi, A. deFazio (Kluwer, Dordrecht), 151
2.62 Schmid-Burgk, J. et al. 1989, *Astron. Astrophys.* **215**, 150
2.63 Shull, M., Draine, B.T. 1987, in *Interstellar Processes*, eds. D.J. Hollenbach, H.A. Thronson (Reidel, Dordrecht), 283
2.64 Solomon, P.M., Rivolo, A.R., Barrett, J., Yahil, A. 1987, *Astrophys. J.* **319**, 730
2.65 Spitzer, L. 1978, *Physical Processes in the Interstellar Medium* (Wiley, New York)
2.66 Stacey, G.J., Geis, N., Genzel, R., Lugten, J.B., Poglitsch, A., Sternberg, A., Townes, C.H. 1991, *Astrophys. J.* **373**, 423
2.67 Stacey, G.J., Townes, C.H., Poglitsch, A., Madden, S.C., Jackson, J.M., Herrmann, F., Genzel, R., Geis, N. 1991, *Astrophys. J.* **382**, L37
2.68 Sternberg, A. 1988, *Astrophys. J.* **332**, 400
2.69 Sternberg, A., Dalgarno, A. 1989, *Astrophys. J.* **338**, 197
2.70 Stutzki, J., Stacey, G.J., Genzel, R., Harris, A.I., Jaffe, D.T., Lugten, J.B. 1988, *Astrophys. J.* **332**, 379
2.71 Tielens, A.G.G.M., Hollenbach, D.J. 1985a, *Astrophys. J.* **291**, 722
2.72 Tielens, A.G.G.M., Hollenbach, D.J. 1985b, *Astrophys. J.* **291**, 747
2.73 Troland, T.H., Heiles, C. 1986, *Astrophys. J.* **301**, 339
2.74 van der Werf, P., et al. 1992, in prep
2.75 van Dishoeck, E.F., Black, J.H. 1986, *Astrophys. J. Suppl. Ser.* **62**, 109
2.76 Verstraete, L., Léger, A., d'Hendecourt, L., Dutuit, O., Défourneau, D. 1990, *Astron. Astrophys.* **237**, 436
2.77 Wardle, M. 1990, *Monthly Notices Roy. Astron. Soc.* **246**, 98
2.78 Walmsley, C.M. 1975, in *HII Regions and Related Topics*, eds. T.L. Wilson, D. Downes (Springer, New York), 17
2.79 Watson, D.M., Genzel, R., Townes, C.H., Storey, J.W.V. 1985, *Astrophys. J.* **298**, 316
2.80 Werner, M.W. 1970, *Astrophys. Letters* **6**, 81
2.81 Zweibel, E.G. 1987, in *Interstellar Processes*, eds. D.J. Hollenbach, H.A. Thronson, (Reidel, Dordrecht), 195

3 Interstellar Chemistry

3.1 Chemistry of the Gas

This chapter gives a brief discussion of interstellar chemistry and of the composition of gas and dust clouds. For recent detailed discussions of interstellar chemistry I refer to Duley & Williams (1984), Herbst (1987), van Dishoeck (1988), Prasad et al. (1987), and Dalgarno (1991).

3.1.a Neutral Gas Phase Reactions

At the low temperatures and densities of quiescent interstellar clouds, neutral gas phase chemistry is generally too slow to explain observed abundances of any but a few molecules (e.g., Herbst 1987). For two body *radiative association* of species X and Y,

$$X + Y \rightarrow XY + h\nu, \tag{101}$$

to be successful radiative processes have to be very fast as the interaction time is fairly short ($\approx 10^{-12}$ s). With the exception of a few molecules like CH, this is not the case, however. For example, the time scale for H_2 formation through radiative association reactions of two hydrogen atoms is about $3 \times 10^{15}/n_H$ [y]. The reason for this prohibitively long time scale is that H_2 lacks a permanent dipole moment. Clearly, the large abundance of interstellar H_2 cannot be explained by neutral gas phase reactions.

In other cases, neutral-neutral reactions have an activation barrier $\left(\text{endothermic reactions with } \Delta E \sim O(eV)\right)$ that cannot be overcome at $T \approx 10$ K. Three body collisions are much too rare to be of importance. Neutral-neutral, gas phase chemistry thus cannot explain the chemistry in dark cloud cores or diffuse clouds.

Neutral-neutral reactions can, however, be very important in shock fronts or warm cloud cores where temperatures exceed 10^2 K. Shock/high temperature chemistry may explain the large abundances of OH, H_2O, SiO, SO, and SO_2 in warm cloud cores (e.g., Iglesias & Silk 1978, Hartquist et al. 1980, Mitchell 1984, Hartquist 1988, and Glassgold et al. 1991).

3.1.b Grain Surface Chemistry

Formation of molecules on grain surfaces is another important process (see Duley 1989 for a recent review). Assume that molecule or atom X strikes a dust grain whose surface is already occupied by molecule or atom Y. Call y the probability (yield) that X is adsorbed on the surface, migrates to the location of Y, binds to Y and forms XY which then evaporates from the surface. The rate of formation of XY through the process

$$X + Y : g \rightarrow XY + g, \tag{102}$$

then is given by

$$R_g = n(X) \, n(g) \, \sigma_g \, y \, \Delta v \qquad [\text{cm}^{-3} \, \text{s}^{-1}]. \tag{103}$$

In the case of formation of H_2 molecules on the surfaces of graphite or silicate grains Hollenbach & Salpeter (1971) inferred a large yield, $y \approx 1$ so that

$$R_g(H_2) = 3 \times 10^{-18} \, T^{1/2} n_H^2 \qquad (104)$$

The time scale for H_2 formation on dust grains at $\approx 10\,K$ is

$$t_g(H_2) \approx \frac{3 \times 10^7}{n_H} \qquad [y], \qquad (105)$$

still fairly long but 8 orders of magnitude faster than for radiative association. Several other molecules, in particular saturated species like CH_4, NH_3, or H_2O may be formed on grain surfaces as well and may be released back to the gas phase in "hot cores" heated by protostars (Williams 1984, Tielens & Allamandola 1987, Duley 1989, Millar 1989, and Brown et al. 1988).

Depletion of gas phase molecules on dust grains by adsorption has a time scale

$$t_{ads} \approx \frac{10^9}{n_H} \qquad [y], \qquad (106)$$

which, for denser clouds ($n_H \approx 10^4 \, cm^{-3}$), can be comparable to the time scale of gas phase reactions and shorter than the lifetime of a molecular cloud ($\geq 10^7$ y, see van Dishoeck 1988). The question thus arises whether gas phase molecules are severely depleted in cold clouds and why they are observed at all. "Ice" mantles on dust grains mode of various molecules are in fact observed in such clouds (see Sect. 3.2.a below). As molecules are observed to exist in the gas phase even in the densest cloud cores (but see evidence for depletion found by Mezger et al. 1986, 1988), there must be efficient mechanisms returning absorbed molecules to the gas phase. Possible mechanisms are shocks (Williams & Hartquist 1984) or grain "explosions" (d'Hendecourt et al. 1982).

3.1.c Ion-Molecule Chemistry

If an interstellar cloud contains *molecular ions* as well as neutrals, fast ion-molecule reactions can proceed even at low temperatures. The long range polarization forces due to the polarization of a neutral by an ion result in large reaction coefficients that are approximately given by temperature independent Langevin coefficient,

$$R_1 = 2\pi e \left(\frac{\alpha}{\mu} \right)^{1/2} \approx 10^{-9} \qquad [cm^{+3} \, s^{-1}] \qquad (107)$$

(Langevin 1905). The parameter α is the polarizability. Ion-molecule reactions thus proceed on a time scale

$$t_{IM} \approx \frac{30}{n_{ion}} \approx \frac{30 \to 10^9}{n_H} \qquad [y], \qquad (108)$$

that is, orders of magnitude faster at low-temperature than neutral-neutral or neutral-grain reactions.

In a classic paper Herbst & Klemperer (1973) proposed that *cosmic-ray induced ionization of H_2* in turn could lead to a fast ion-molecule chemistry network that could explain the basic features of the chemistry in dark, quiescent clouds (for more recent discussions see Herbst 1987, Dalgarno 1991). The key ions in this reaction scheme are the ions H_2^+ and H_3^+ that are formed in the reactions

$$H_2 + \text{cosmic ray} \rightarrow H_2^+ + e + \text{cosmic ray}$$
$$H_2^+ + H_2 \rightarrow H_3^+ + H. \tag{109}$$

The H_3^+ ions undergo *proton transfer* reactions with a wide range of species X,

$$H_3^+ + X \rightarrow XH^+ + H_2. \tag{110}$$

More complex, neutral species can then be built up by a series of *hydrogen abstraction* reactions, followed by rapid *dissociative recombination*.

$$
\begin{aligned}
XH_{n-1}^+ + H_2 &\rightarrow XH_n^+ + H \\
XH_n^+ + e &\rightarrow XH_{n-1} + H \\
&\rightarrow XH_{n-2} + H_2.
\end{aligned} \tag{111}
$$

In this simple scheme, OH and H_2O are formed by the reactions

$$
\begin{aligned}
H_3^+ + O &\rightarrow OH^+ + H_2 \\
OH^+ + H_2 &\rightarrow H_2O^+ + H \\
H_2O^+ + H_2 &\rightarrow H_3O^+ + H \\
H_3O^+ + e &\rightarrow H_2O + H \\
&\rightarrow OH + H_2.
\end{aligned} \tag{112}
$$

H_2^+ and H_3^+ have not yet been directly observed in interstellar space, but the ions HCO^+, N_2H^+, and probably H_2D^+ have been found in high abundance, confirming the basic scheme of ion-molecule chemistry. The hydronium ion (H_3O^+) probably has also been recently seen in several of its 850 μm inversion transitions (Wootten et al. 1991).

While radiative association reactions are usually slow for neutrals, they can be rapid in ion-molecule reactions and can lead to large molecules, such as methanol or acetylene (Dalgarno 1991),

$$
\begin{aligned}
CH_3^+ + H_2O &\rightarrow CH_3OHH^+ + h\nu \\
CH_3OHH^+ + e &\rightarrow CH_3OH + H \\
C^+ + CH_4 &\rightarrow C_2H_3^+ + H \\
C_2H_3^+ + e &\rightarrow C_2H_2 + H.
\end{aligned} \tag{113}
$$

In ion-molecule chemistry the most abundant molecule after H_2, CO, is made by (dissociative) recombination of HCO^+ or CO^+.

Quantitative models of interstellar chemistry require computation of complicated reaction networks with $O(10^3)$ coupled nonlinear kinetic equations (e.g., Prasad & Huntress 1980, Graedel et al. 1982, Leung et al. 1984). In early models, only the equilibrium solution was determined, while more recently time

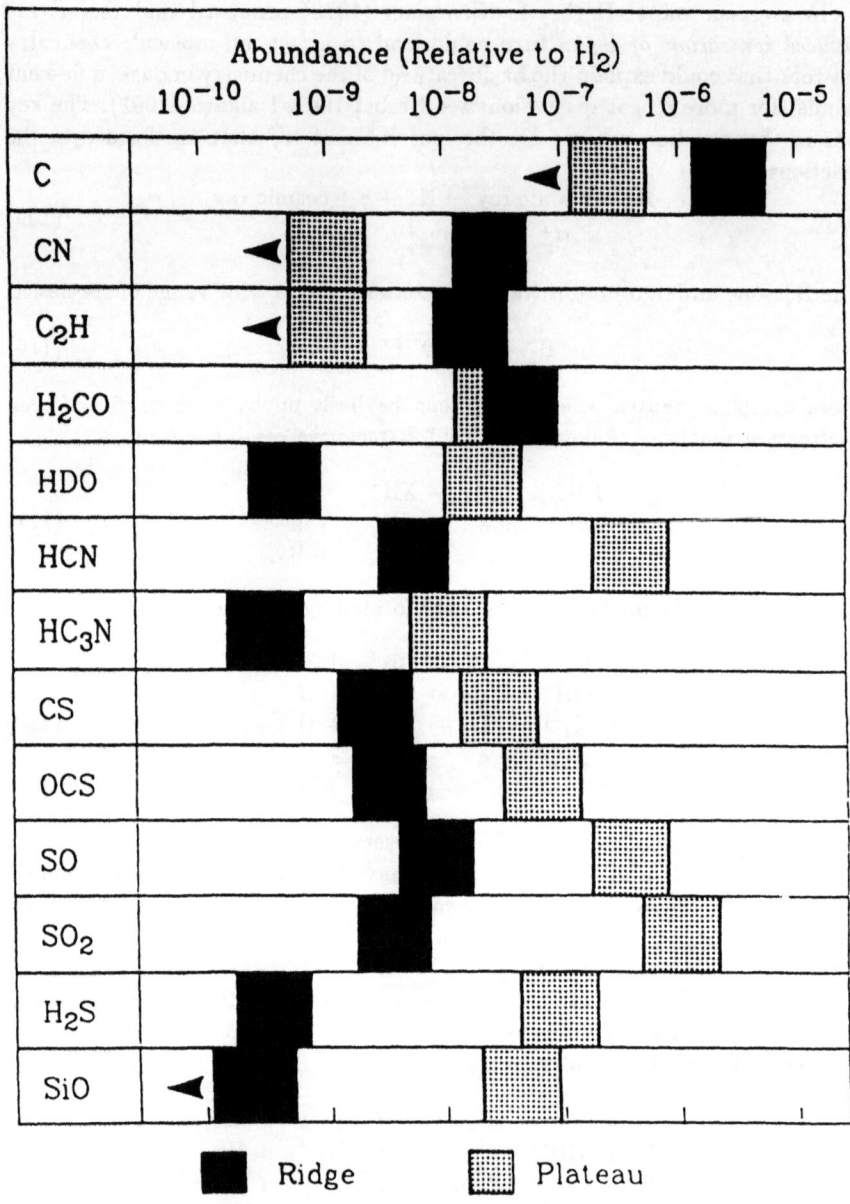

Fig. 19. Molecular abundances of a number of molecules in the "ridge" and "plateau" components of the core of Orion (adapted from Blake et al. 1987)

dependent solutions have been investigated as well. These show large time abundance variations on time scales of 10^5 to 10^6 years for a number of molecules (e.g., Herbst & Leung 1986).

On the whole the models are fairly successful in accounting for the observed abundances of many small molecular species, both in diffuse and in dense clouds (see van Dishoeck 1988 for an excellent review). The chemistry of dense clouds in strong radiation fields and photon-dominated regions has already been discussed in Sect. 2.1. There are also some obvious failures of the models in explaining, for instance, the high CH^+ abundance in diffuse clouds (van Dishoeck 1988), or the high C^0 abundance in dense clouds (e.g., Phillips & Huggins 1981). Discussions of the latter can be found in Keene et al. (1985), van Dishoeck (1988), and Genzel et al. (1989). A possible solution of the high C^0 abundance in GMC's may be parsec scale penetration of ultraviolet radiation and the resulting presence of PDR's fairly deep inside clumpy molecular clouds (Stutzki et al. 1988, Genzel et al. 1988).

3.1.d Observations: Chemistry in Different Environments

Reviews of the observed composition of dense clouds have been given by Irvine (1987, 1991). The sensitive dependence of the relevant chemistry on physical environments has been beautifully demonstrated by a detailed chemical analysis of the center of the Orion-KL cloud core/star formation region (Johansson et al. 1984, Sutton et al. 1985, Blake et al. 1986, 1987, White et al. 1986). Blake et al. (1987) were able to show that different spatial/physical components of interstellar gas in this region also have different chemical characteristics, along the lines of the discussion in the last paragraphs. This is shown in more detail in Fig. 19 for the quiescent molecular gas in the extended "ridge" in comparison to the dynamically active "plateau" gas (Sect. 4.4.b).

Cooler gas in the quiescent, large scale "ridge" shows chemical abundances typical for a cosmic ray induced ion-molecule chemistry. This chemical composition in the ridge is similar to that found in other dark, quiescent clouds (Irvine 1987, 1991). Warm and hot gas in outflows from the central infrared star cluster has chemical abundances typical for a shock/high temperature chemistry. Finally, very dense gas in the immediate vicinity of the young massive stars has chemical abundances that is best described by recent (10^3 to 10^4 y) evaporation of molecules from grain surfaces (Brown et al. 1988).

The chemistry in the massive Galactic center molecular SgrB2 has been studied by Cummins et al. (1986), Turner (1989, 1991), and Sutton et al. (1992). Again substantial differences in chemical composition of different physical regions in the cloud are found. The chemical compositions in circum-stellar shells as a probe of stellar nucleosynthesis is another important area. The reader is referred to Omont (1991) for a recent in-depth discussion.

3.1.e Abundance Gradients in the Galaxy

Measurements of elemental and isotopic abundances across the Galaxy are in principle important indicators of large scale chemical gradient and stellar nucleosynthesis (Wannier 1980). While optical spectroscopy of HII regions does not reach into the central 5 kpc of the Galaxy (Shaver et al. 1983), both radio spectroscopy of molecular lines (Wannier 1980) and far-infrared spectroscopy of atomic fine structure lines (Lester et al. 1987, Herter 1989) have been used to study large scale abundance gradients. Despite substantial efforts, however, only a few undisputed trends have emerged. The ^{12}C/^{13}C isotopic ratio in the Galactic center (20 to 30) is significantly lower than in the disk (40 to 80) but there appears to be no clear gradient in the disk (Wannier 1980, Shaver et al. 1983). Unfortunately different methods result in different values for a given cloud, thus potentially masking any small variations. For instance, Langer & Penzias (1990) derive from mm-spectroscopy a ^{12}C/^{13}C ratio of 67 for the Orion cloud, while Hawkins & Jura (1987) find a ratio of 43 from UV spectroscopy. Another, fairly clear result is that the ^{16}O/^{18}O ratio is about 500 in the Galactic disk, and 250 in the Galactic center (Wannier 1980). ^{14}N/^{15}N is about a factor of 3 to 4 greater in the Galactic center than in the disk (270). Any large scale variations in deuterium appear to be masked by fractionation effects (Sect. 3.1.g). Lester et al. (1987) find that the N^{++}/O^{++} abundance in the Galactic center is 3 to 5 greater than in the solar neighborhood and interpret this result in terms of a much increased oxygen abundance and secondary nucleosynthesis of nitrogen in the nucleus. The inferred increase of oxygen is also consistent with the gradient in electron temperature of radio HII regions (e.g., Wink et al. 1983). However, the excitation potentials of N^{++} and O^{++} are different by 5eV so that a gradient can occur for constant N/O if there is a gradient in the temperatures of the exciting stars.

3.1.f Electron Abundance in Dense Clouds

In diffuse clouds and at the surfaces of dense clouds the electron abundance is controlled by photodissociation of carbon, silicon, iron, etc. and is comparable to the gas phase carbon abundance, $[e]/[H] = x_e \approx [C]/[H] \approx 3 \times 10^{-4}$. Beyond the range of UV penetration, the electron abundance in dense clouds is controlled by the cosmic ray ionization rate ζ_{CR} [s^{-1}] on the one hand and the electron recombination rate of various molecular ions on the other hand (Langer 1985),

$$
\begin{aligned}
\zeta_{CR}: \quad & H_2 && \to H_2^+, H_3^+ \\
& H_3^+ &+ e \quad & \to H_2 && + H \\
& H_3^+ &+ HD \quad & \to H_2D^+ && + H_2 \\
& H_2D^+ &+ e \quad & \to HD && + H \\
& H_3^+ &+ CO \quad & \to HCO^+ && + H_2 \\
& HCO^+ &+ e \quad & \to CO && + H_2.
\end{aligned} \tag{114}
$$

Given this fairly complex network of reaction channels with partially unknown rate coefficients Langer (1985) has shown that lower and an upper bounds to x_e are

$$
10^{-8} \approx \frac{[HCO^+]}{[H_2]} \leq x_e \leq \zeta_{CR} \left[k_e \left(HCO^+ \right) n(H_2) \frac{[HCO^+]}{[H_2]} \right]^{-1} \approx 3 \times 10^{-7} \tag{115}
$$

where $k_e(HCO^+)$ is the rate coefficient for dissociative recombination of HCO$^+$ and $\zeta_{CR} = 4 \times 10^{-17}$ s^{-1} (Sect. 2.3). Wootten (1989) derives an upper limit of $x_e \leq 5 \times 10^{-8}$ from the H$_2$D$^+$ observations of Phillips et al. (1985). The electron abundance in clouds is a key parameter that determines, for instance, the ambipolar diffusion/drift of magnetic fields (Sects. 2.4, 4.2) and the propagation speed of Alfvén waves in the ionized component (see C-shocks in Sect. 2.2).

3.1.g Fractionation

Because of significant differences in zero-point, vibrational energies ($\Delta E/k \approx$ $10 \rightarrow 300$ K), *fractionation* of isotopes plays an important role in interstellar clouds. For example, the reaction

$$^{13}C^+ + {}^{12}CO \rightarrow {}^{12}C^+ + {}^{13}CO + 35 \, [K], \tag{116}$$

can lead to fractionation of CO at the surfaces of clouds; it is counteracted by isotope selective dissociation (Bally & Langer 1982, van Dishoeck & Black 1988). The most dramatic fractionation occurs for deuterated molecules, as in the reaction

$$H_3^+ + HD \rightarrow H_2D^+ + H_2 + 200 \, [K]. \tag{117}$$

In dark clouds, the ratio of abundances of deuterated to normal species (e.g., [DCN]/[HCN]) can be greater by a factor of $10^2 \rightarrow 10^3$ than the [D]/[H] ratio, consistent with the fractionation model (Penzias 1979, Wootten 1989).

Walmsley et al. (1987) found a hundred fold fractionation of deuterated molecules in the Orion-KL hot-core, despite a gas temperature of $\approx$ 200 K. This result cannot be explained by gas phase fractionation reactions and probably requires recent grain evaporation of molecules that have formed and fractionated in an originally cold environment (Brown et al. 1988, Walmsley et al. 1987).

3.2 Composition of Dust Grains
3.2.a Bulk Grains

Our knowledge about the chemical composition and size distribution of interstellar dust grains has substantially improved in the last decade, mainly due to the availability of detailed infrared spectrophotometry (see Allamandola 1984, Roche 1989, Allamandola & Tielens 1989). The overall $0.1\,\mu m$ to $1\,mm$ dust extinction curve (Sect. 1.2.f) is reasonably fit by a mixture of (bulk) graphite and silicate grains with a power law size distribution (Mathis et al. 1977, Draine & Lee 1984)

$$\frac{dN(a)}{da} \sim a^{-3.5} \quad [100\text{Å} \leq a \leq 3000\text{Å}], \tag{118}$$

a is the grain size and $dN(a)$ gives the number of dust particles between a and $a + da$.

In dense clouds these grains acquire volatile mantles due to freeze out of H_2O, CH_4, NH_3, CH_2OH, etc. from the gas phase. The composition of the mantles can be studied by near- and mid-IR absorption along the line of sight of bright compact IR sources. If the IR source is behind or embedded in a dense cloud, characteristic absorption features can be observed due to various vibrational modes of the mantle molecules (see Fig. 6). As an example, Fig. 20 shows the near-IR spectrum of the deeply embedded "protostars" W33 A along

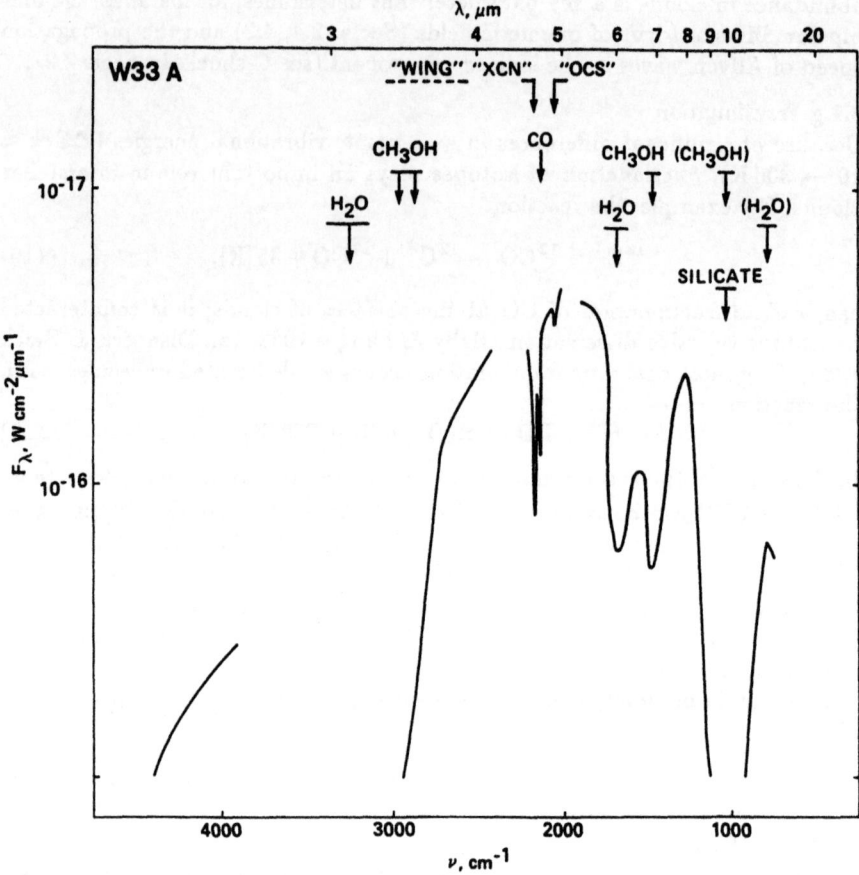

Fig. 20. Schematic 2 to 20 μm spectrum of the protostellar source W33A, along with identifications of the most prominent absorption features (adapted from Tielens 1989)

with the most probable identifications of the various absorption features (from
Tielens 1989). Deep absorption features of H_2O ice (3.1 μm), solid CO (4.8 μm),
hydrocarbons (6.0 and 6.8 μm) and silicates (9.7 μm) are clearly discernible.
Differences in shape of the absorption features from source to source may be
understood as a consequence of different amounts of thermal or UV processing
and resulting amorphousness (Greenberg 1991). For a more detailed discussion
of recent observations and their interpretation the reader is referred to the
reviews of Allamandola (1984), Roche (1989), Tielens (1989), and Greenberg
(1991).

3.2.b Very Small Dust Grains and Large Molecules: The PAH Model
In the early and mid-1980's several lines of evidence started to indicate that
the interstellar medium contains much smaller dust particles (10 to 100 Å)
than predicted by the Mathis-Rumpl-Nordsieck (1977) model. From 2 − 5 μm
photometric observations of reflection nebulae Sellgren (1984) inferred the ex-

istence of a component of hot dust ($T_{\text{dust}} \approx 500 \to 1000\,\text{K}$) that is heated by a non-equilibrium process. If the dust grains are small enough, absorption of a single UV photon is sufficient to raise their temperature substantially. Mid- and far-IR photometry of galactic HII regions and molecular cloud complexes also indicated the presence of small, hot dust grains (Andriesse 1978, Price 1981, Pajot et al. 1986). Finally, Léger & Puget (1984) identified IR emission features at 3.3, 6.2, 7.7, 8.6, and 11.3 μm (Gillett et al. 1973) as vibrational bands of C–H bonds on *polycyclic aromatic hydrocarbons* (PAH's), similar to coronene or naphtalene. A second, similar proposal involves aromatic side-groups on larger hydrogenated amorphous carbon (HAC) particles (Duley & Williams 1988). Very recently Webster (1991) has proposed that aromatic groups on the surface of Fullerenes (C_{60}, C_{70}, etc.) could also play a role. The PAH-model (see review of Puget & Léger 1989) predicts a smooth transition from (large) molecules to (very small) dust grains and has so far been fairly successful in explaining the available observations.

The photometric observations are in good agreement with the PAH-model. For a single 10 eV-photon to raise the lattice temperature to ≥ 500 K the number of atoms has to be $\approx 10^2$, approximately the number of atoms in a medium-size PAH. Giard et al. (1989) showed that the large scale spatial distributions of the 3.3 μm feature and of the 10 μm diffuse IRAS emission correlate very well. There are several interesting consequences of the PAH-model (see Puget & Léger 1989). First, for a fit of the measurements about 20% of interstellar carbon has to be in form of PAH's. Second, the extinction curve in the far-UV ($\lambda > 2000$Å) may well be dominated by PAH's. This prediction, based on laboratory data, is consistent with the fact that about 20% of the diffuse IRAS luminosity of the Galaxy emerges in the 10 $\to$ 20 μm band (Puget & Léger 1989). PAH's may be destroyed in strong UV radiation fields due to Coulomb explosions following multiple ionization events. As a result one ought to expect an anticorrelation of the the mid-IR emission and the far-UV rise in the extinction. Such an anticorrelation has in fact been seen toward the O star σ-Sco, but is less convincing or even lacking in other hot stars. Third, Verstraete et al. (1990) have shown that the photoelectric yield of PAH's is substantially greater than that of bulk grains and that they can contribute to photoelectric heating at high temperature. Photoelectrons from PAH's may, therefore, be the dominant heating source in the diffuse interstellar medium. Finally, Desert et al. (1991) find the somewhat surprising result that PAH's may also contribute significantly to the dust emissivity at *very long* (e.g., submm) wavelengths. The reason is that for planar particles the emissivity decreases more slowly with wavelength than for the bulk grains that dominate the far-infrared emissivity ($Q(\lambda) \sim \lambda^{-1}$ instead λ^{-2}).

3.2.c Origin of Dust Grains

Gehrz (1989) & Sedlmayr (1989) have given recent reviews on the origin and formation of dust grains. It is fairly certain from the observational material that mass losing, late type stars are significant sources of interstellar dust grains. Oxygen rich Miras, OH-IR stars, and carbon stars may contribute equally

(Gehrz 1989). A much smaller source of dust are supernovae, novae, or hot stars (Wolf-Rayet stars, etc.). In the cool outflows of the late type stars molecules form larger clusters which can then be the nuclei of growing dust grains (Sedlmayr 1989). It is presently not certain whether the stars are also the only source of dust grains or whether grains also form directly in interstellar clouds by accretion.

References

3.1 Allamandola, L.J. 1984, in *Galactic and Extragalactic Infrared Spectroscopy*, eds. M.F. Kessler, J.P. Phillips (Reidel, Dordrecht), 5

3.2 Allamandola, L.J., Tielens, A.G.G.M. (eds.) 1989, *Interstellar Dust* (Kluwer, Dordrecht)

3.3 Andriesse, C.D. 1978, *Astron. Astrophys.* **66**, 169

3.4 Bally, J., Langer, W.D. 1982, *Astrophys. J.* **255**, 143

3.5 Blake, G.A., Sutton, E.C., Masson, C.R., Phillips, T.G. 1986, *Astrophys. J. Suppl. Ser.* **60**, 357

3.6 Blake, G.A., Sutton, E.C., Masson, C.R., Phillips, T.G. 1987, *Astrophys. J.* **315**, 621

3.7 Brown, P.D., Charnley, S.B., Millar, T.J. 1988, *Monthly Notices Roy. Astron. Soc.* **231**, 409

3.8 Cummins, S.E., Linke, R.A., Thaddeus, P. 1986, *Astrophys. J. Suppl. Ser.* **60**, 819

3.9 Dalgarno, A. 1991, in *Chemistry in Space*, eds. J.M. Greenberg, V. Pironello, *NATO-ASI-323* (Kluwer, Dordrecht), 71

3.10 Desert F.X., Boulanger, F., Puget, J.L. 1990, *Astron. Astrophys.* **237**, 215

3.11 d'Hendecourt, L.B., Allamandola, L.J., Baas, F., Greenberg, J.M. 1982, *Astron. Astrophys.* **109**, L12

3.12 Draine, B.T., Lee, H.M. 1984, *Astrophys. J.* **285**, 89

3.13 Duley, W.W., Williams, D.A. 1984, *Interstellar Chemistry*, (Academic Press, London)

3.14 Duley, W.W., Williams, D.A. 1988, *Monthly Notices Roy. Astron. Soc.* **96**, 269

3.15 Duley, W.W. 1989, in *The Physics and Chemistry of Interstellar Molecular Clouds*, eds. G. Winnewisser, J.T. Armstrong (Springer, Berlin), 353

3.16 Gehrz, R. 1989, in *Interstellar Dust*, eds. L.J. Allamandola, A.G.G.M. Tielens (Kluwer, Dordrecht), 445

3.17 Genzel, R., Harris, A.I., Jaffe, D.T., Stutzki, J. 1988, *Astrophys. J.* **332**, 1049

3.18 Genzel, R., Harris, A.I., Stutzki, J. 1989, in *Infrared Spectroscopy in Astronomy*, ed. B.H. Kaldeich, *ESA-SP-290*, 115

3.19 Giard, M., Pajot, F., Lamarre, J.M., Serra, G., Caux, E. 1989, *Astron. Astrophys.* **215**, 92

3.20 Gillett, F.C., Forrest, W.J., Merrill, K.M. 1973, *Astrophys. J.* **183**, 87

3.21 Glassgold, A., Mamon, G.A., Huggins, P.J. 1991, *Astrophys. J.* **373**, 254

3.22 Graedel, T.E., Langer, W.D., Frerking, M.A. 1982, *Astrophys. J. Suppl. Ser.* **48**, 321

3.23 Greenberg, J.M. 1991, in *Chemistry in Space*, eds. J.M. Greenberg, V. Pironello, *NATO-ASI-323* (Kluwer, Dordrecht), 227

3.24 Hartquist, T.W., Oppenheimer, M., Dalgarno, A. 1980, *Astrophys. J.* **236**, 180

3.25 Hartquist, T.W. 1988, in *Millimetre and Submillimetre Astronomy*, eds. R.D. Wolstencroft, W.B. Burton (Kluwer, Dordrecht), 165.

3.26 Hawkins, I., Jura, M. 1987, *Astrophys. J.* **317**, 926

3.27 Herbst, E., Klemperer, W. 1973, *Astrophys. J.* **185**, 505

3.28 Herbst, E., Leung, C.M. 1986, *Astrophys. J.* **310**, 378

3.29 Herbst, E. 1987, in *Interstellar Processes*, eds. D.J. Hollenbach, H.A. Thronson (Reidel, Dordrecht), 611

3.30 Herter, T. 1989, in *Infrared Spectroscopy in Astronomy*, ed. B.H. Kaldeich, *ESA-SP-290*, 403

3.31 Hollenbach, D.J., Salpeter, E.E. 1971, *Astrophys. J.* **163**, 155

3.32 Iglesias, E.R., Silk, J. 1978, *Astrophys. J.* **226**, 851
3.33 Irvine, W.M., Goldsmith, P.F., Hjalmarson, Å. 1987, in *Interstellar Processes*, eds. D.J. Hollenbach, H.A. Thronson (Reidel, Dordrecht), 561
3.34 Irvine, W.M. 1991, in *Chemistry in Space*, eds. J.M. Greenberg, V. Pironello, *NATO-ASI-323* (Kluwer, Dordrecht), 89
3.35 Johansson, L.E.B., Andersson, C., Elldér, J., Friberg, P., Hjalmarson, Å. et al. 1984, *Astron. Astrophys.* **130**, 227
3.36 Keene, J., Blake, G.A., Phillips, T.G., Huggins, P.J., Beichman, C. 1985, *Astrophys. J.* **299**, 967
3.37 Langer, W.D. 1985, in *Protostars and Planets II*, eds. D. Black, M. S. Mathews, (University of Arizona Press, Tucson), 650
3.38 Langer, W.D., Penzias, A.A. 1990, *Astrophys. J.* **357**, 477
3.39 Langevin, P. 1905, *Ann. Chem. Phys.*, **5**, 245
3.40 Léger, A., Puget, J.L. 1984, *Astron. Astrophys.* **137**, L5
3.41 Lester, D.F., Dinerstein, H.L., Werner, M.W., Watson, D.M., Genzel, R., Storey, J.W.V. 1987, *Astrophys. J.* **320**, 573
3.42 Leung, C.M., Herbst, E., Huebner, W.F. 1984, *Astrophys. J. Suppl. Ser.* **56**, 231
3.43 Mathis, J.S., Rumpl, W., Nordsieck, K.H. 1977, *Astrophys. J.* **217**, 425
3.44 Mitchell, G.M. 1984, *Astrophys. J. Suppl. Ser.* **54**, 81
3.45 Mezger, P.G., Chini, R., Kreysa, E., Gemünd, H.P. 1986, *Astron. Astrophys.* **160**, 324
3.46 Mezger, P.G., Chini, R., Kreysa, E., Wink, J.E., Salter, C.J. 1988, *Astron. Astrophys.* **191**, 44
3.47 Millar, T.J. 1989, in *Infrared Spectroscopy in Astronomy*, ed. B.H. Kaldeich, *ESA-SP-290*, 109
3.48 Omont, A. 1991, in *Chemistry in Space*, eds. J.M. Greenberg, V. Pironello, *NATO-ASI-323* (Kluwer, Dordrecht), 171
3.49 Pajot, F., Boissé, P., Gispert, R., Lamarre, J.M., Puget, J.L., Serra, G. 1986, *Astron. Astrophys.* **157**, 393
3.50 Penzias, A.A. 1979, *Astrophys. J.* **228**, 430
3.51 Phillips, T.G., Huggins, P.J. 1981, *Astrophys. J.* **251**, 533
3.52 Phillips, T.G., Blake, G.A., Keene, J., Woods, R.C., Churchwell, E. 1985, *Astrophys. J.* **294**, L45
3.53 Prasad, S.S., Huntress, W.T. 1980, *Astrophys. J. Suppl. Ser.* **43**, 1
3.54 Prasad, S.S., Tarafdar, S.P., Villere, K.R., Huntress, W.T. 1987, in *Interstellar Processes*, eds. D.J. Hollenbach, H.A. Thronson (Reidel, Dordrecht), 631
3.55 Price, S.D. 1981, *Astron. J.* **86**, 193
3.56 Puget, J.L., Léger, A. 1989, *Ann. Rev. Astron. Astrophys.* **27**, 161
3.57 Roche, P.F. 1989, in *Infrared Spectroscopy in Astronomy*, ed. B.H. Kaldeich, *ESA-SP-290*, 79
3.58 Sedlmayr, E. 1989, in *Interstellar Dust*, eds. L.J. Allamandola, A.G.G.M. Tielens (Kluwer, Dordrecht), 467
3.59 Sellgren, K. 1984, *Astrophys. J.* **277**, 623
3.60 Shaver, P.A., McGee, R.X., Newton, L.M., Danks, A.C., Pottasch, S.R. 1983, *Monthly Notices Roy. Astron. Soc.* **204**, 53
3.61 Stutzki, J., Stacey, G.J., Genzel, R., Harris, A.I., Jaffe, D.T., Lugten, J.B. 1988, *Astrophys. J.* **332**, 379
3.62 Sutton, E.C., Blake, G.A., Masson, C.R., Phillips, T.G. 1985, *Astrophys. J. Suppl. Ser.* **58**, 341
3.63 Sutton, E.C., Jaminet, P.A., Danchi, W.C., Blake, G.A. 1992, *Astrophys. J.* in press
3.64 Tielens, A.G.G.M., Allamandola, L.J. 1987, in *Interstellar Processes*, eds. D.J. Hollenbach, H.A. Thronson (Reidel, Dordrecht), 397
3.65 Tielens, A.G.G.M. 1989, *In Interstellar Dust*, eds. L.J. Allamandola, A.G.G.M. Tielens (Kluwer, Dordrecht), 239
3.66 Turner, B.E. 1989, *Astrophys. J. Suppl. Ser.* **70**, 539
3.67 Turner, B.E. 1991, *Astrophys. J. Suppl. Ser.* **76**, 617
3.68 van Dishoeck, E.F. 1988, in *Millimetre and Submillimetre Astronomy*, eds. R.D. Wolstencroft, W.B. Burton (Kluwer, Dordrecht), 117
3.69 van Dishoeck, E.F., Black, J.H. 1988, *Astrophys. J.* **334**, 711
3.70 Verstraete, L., Léger, A., d'Hendecourt, L., Dutuit, O., Défourneau, D. 1990, *Astron. Astrophys.* **237**, 436

3.71 Walmsley, C.M., Hermsen, W., Henkel, C., Mauersberger, R., Wilson, T.L. 1987, *Astron. Astrophys.* **172**, 311
3.72 Wannier, P.G. 1980, *Ann. Rev. Astron. Astrophys.* **18**, 399
3.73 Webster, A. 1991, *Nature* **352**, 412
3.74 White, G.J., Monteiro, T.S., Richardson, K.J., Griffin, M.J., Rainey R. 1986, *Astron. Astrophys.* **162**, 253
3.75 Williams, D.A. 1984, in *Galactic and Extragalactic Infrared Spectroscopy*, eds. M.F. Kessler, J.P. Phillips (Reidel, Dordrecht), 59
3.76 Williams, D.A., Hartquist, T.W. 1984, *Monthly Notices Roy. Astron. Soc.* **210**, 141
3.77 Wink, J.E., Wilson, T.L., Bieging, J.H. 1983, *Astron. Astrophys.* **127**, 211
3.78 Wootten, A. 1989, in *Astrochemistry*, eds. M.S. Vardya, S.P. Tarafdar, (Reidel, Dordrecht), 311
3.79 Wootten, A., Mangum, J.G., Turner, B.E., Bogey, M., Boulanger, F., Combes, F., Encrenaz, P.G., Gerin, M. 1991, *Astrophys. J.* **380**, L79

4 Structure and Dynamics of Molecular Clouds

4.1 The Spatial Structure of Molecular Clouds

4.1.a Molecular Line Maps

About half of the mass of neutral interstellar gas in our Galaxy is contained in a few thousand giant molecular clouds (GMC's) of mass 10^5 to 10^7 $M_\odot$ (Scoville & Sanders 1987, Solomon et al. 1987, Blitz 1991, Thaddeus 1991). It was realized as early as the mid-seventies that molecular clouds have substantial structure on scales smaller than the arcmin beam sizes of typical single dish telescopes (e.g., Barrett et al. 1977). Overwhelming evidence for the clumpiness of molecular clouds on all accessible scale sizes (0.003 to 30 pc) has recently come from large scale maps of clouds with moderate size and large single dish millimeter telescopes, as well as from the VLA and millimeter interferometers (cf. Wilson & Walmsley 1989). Much of the recent progress is based on mapping sufficiently large areas that contain many resolution elements and on selecting optically thin lines that give much more intensity contrast than the very optically thick ^{12}CO transitions.

A good illustration of the complex and highly inhomogeneous distribution of the gas in molecular clouds is Fig. 21 which gives molecular line maps of the Orion A molecular cloud over four orders magnitude in scale. The left part of Fig. 21 shows a ^{13}CO $1 \rightarrow 0$ map between 6.5 to 7.5 km s^{-1} LSR of the entire Orion A molecular cloud at a resolution of 90" (0.19 pc at 450 pc) from the work of Bally et al. (1987).

There are dense clumps and filaments, as well as bubbles and cavities. Bally et al. conclude that the massive stars of the Orion OB associations as well as embedded lower mass stars have a strong dynamical effect on the cloud structure. Bally (1989) points out a possible correlation between a kinematic "twist" motion in the southern part of the cloud and a helical structure of the magnetic field wrapping around the cloud that may be inferred from HI Zeeman and optical polarization measurements.

The middle part of Fig. 21 gives a 3 mm CS $2 \rightarrow 1$ interferometer map of the central "ridge" of OMC1 in the dense northern part of Orion A (Mundy et al. 1988) The CS data have a resolution of 7.5" (0.016 pc), or less than one tenth of the Bally et al. (1987) data. The CS map corresponds to only 3 linear resolution elements of the ^{13}CO map on the left and clearly has significant substructure down to the scale of the interferometer beam. Finally the right part of Fig. 21 displays a 1" resolution VLA map of the 1.2 cm NH$_3$ (3,2) inversion line toward the "hot core", the most prominent condensation of the CS ridge (Migenes et al. 1989). The Orion-KL hot core region again shows considerable structure to the resolution limit of the VLA images (2.3×10^{-3} pc, Genzel et al. 1982, Pauls et al. 1983, Migenes et al. 1989). It is clear from Fig. 21 that the molecular gas in Orion A has structure from the largest to the smallest observable scales, with no preferred size scale. While the Orion molecular cloud is not typical of the average conditions in GMC's, as it is associated with past and very recent OB star formation, much the same is found qualitatively also in dark clouds

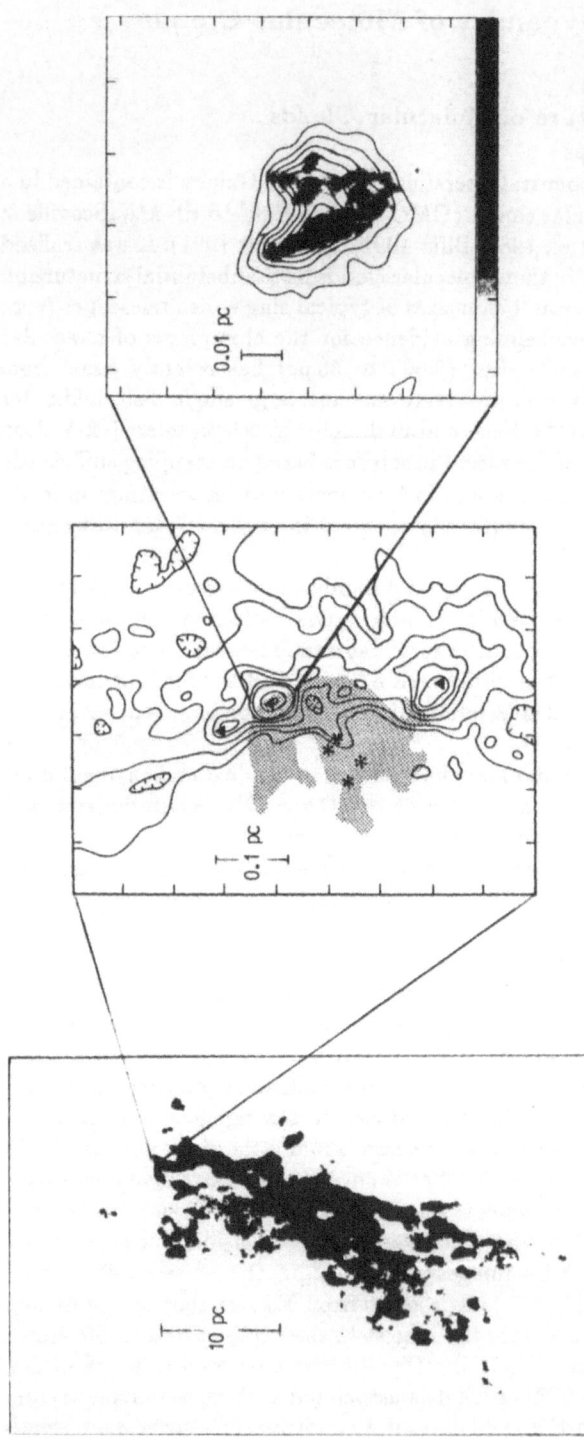

Fig. 21. Molecular line maps of Orion A over four orders of magnitude in spatial scale. Left: ^{13}CO $1 \rightarrow 0$ map between LSR 6.5 and 7.5 km s^{-1} (Bally et al. 1987, Bell Labs telescope, 90″ beam). Middle: CS $2 \rightarrow 1$ map of the OMC1 ridge between LSR -0.2 and $+18$ km s^{-1} (Mundy et al. 1988, OVRO mm interferometer, 7.5″ beam). Asterisks and stippled region mark the location of the θ^1C Trapezium OB stars and the Orion A HII region. Right: NH$_3$ (3,2) map of the Orion-KL hot core between LSR 7.4 and 8.6 km s^{-1} (gray-scale, Migenes et al. 1989, VLA, 1.2″ beam). Contours represent a velocity integrated NH$_3$ (3,2) map, smoothed to 2″

(Cernicharo & Guélin 1987, Loren 1990, Scalo 1990, Falgarone & Puget 1988). The main quantitative difference appears to be that a much larger fraction of the mass of the Orion cloud ($\approx 50\%$, similarly in M17, Stutzki & Güsten 1990) is found in high density "clumps".

It is presently still debated whether these clumps represent stable, physical entities or whether they are temporary fluctuations in an ever shifting dynamical evolution. In favor of the first interpretation speaks the fact that the clumps are not just column density fluctuations in projection on the sky but that they appear to be well defined also in the velocity domain (Stutzki & Güsten 1990, Blitz & Stark 1986). The individual clumps visible on column density maps can be easily separated from other velocity components, and thus represent high-contrast condensations in three dimensional phase space. Furthermore, Stutzki & Güsten (1990) find in M17 that an increase in local line intensity is also accompanied by an increase in local line width suggesting that gravity plays a role and that the clumps are actual mass concentrations. It is plausible from the various observations that the line widths follow virial equilibrium, to within the measurement accuracy (about a factor ± 2 in mass). There can, however, be significant differences in the spatial distributions of different molecules suggesting that spatial variations in chemical abundances play a role (cf. Swade 1989, Goldsmith et al. 1992). For this reason, it is probably preferable to investigate molecular cloud structure in species whose abundances are not expected to vary much with environment, such as CO and its isotopes, or in submillimeter dust emission. Finally, molecules may freeze out on dust grains and deplete the gas phase in cold and very dense condensations (Mezger et al. 1986, 1988).

Scalo (1990) has challenged the interpretation that intensity maxima on molecular line maps can be interpreted as stable physical entities, such as "cores", clumps of fragments. He points out that limited spatial resolution and dynamic range, selection bias and the natural tendency of astronomers to put things in categorized boxes will conspire and lead to false, quasi-static evolutionary models. He shows that the spatial structure of the Taurus cloud as derived from a large scale IRAS $100\,\mu m$ map has some features of a random, fractal structure. He also points out that uncertainty and spread of the data in the velocity width vs. size relationship are too large to deduce virial equilibrium. Falgarone & Phillips (1990) find that the velocity field in molecular clouds with a wide range of physical conditions can be characterized by turbulence with intermittency. Compressible turbulence with moderately large Mach numbers can perhaps account for the observed filamentary structures. Falgarone et al. (1992) have shown that molecular line contour maps over four magnitudes in spatial scale follow a power law relationship between perimeter and area, with a fractal dimension of 0.68. This indicates a large surface to volume ratio on all accessible scales.

4.1.b What is Between Clumps?
Important criteria for assessing the reality of individual emission peaks as stable physical objects are the spatial correlation between clumps and recently formed stars and the density contrast between clumps and the medium in be-

tween them. Observations of both GMC's and dark clouds indicate an intimate relationship between dense cloud cores and star formation sites. T-Tau stars or embedded IRAS point sources in Taurus are often located in or close to NH_3 cores (Myers 1987, Emerson 1987). The densest part of the ρ Oph molecular cloud contains a cluster of embedded stars (Wilking & Lada 1983). Lada (1990) finds clusters of embedded stars associated with four of the five major CS $2 \rightarrow 1$ emission knots in the Orion B molecular cloud. It is likely that IRc 2, a very young and luminous star in the Orion-KL star forming region, has formed in the Orion-KL hot core (right part of Fig. 21), the most prominent gas concentration in the Orion A ridge.

The clump to interclump density contrast in the warm and dense clouds that form O and B stars appears to be remarkably large. Blitz & Stark (1986) infer that ratio to be 10 or larger from ^{13}CO and ^{12}CO $1 \rightarrow 0$ observations of the Rosette molecular cloud. Stutzki & Güsten (1990) derive a density contrast of at least 20 from their $C^{18}O$ $2 \rightarrow 1$ maps of the M17 interface region. The clump/interclump contrast is significantly smaller in dark clouds (Pérault et al. 1985, Falgarone & Puget 1988).

Clump to interclump contrasts of up to a factor of 100 are inferred from the large observed spatial extent of the C^+/C^0 regions around OB stars. Far-infrared observations of the 158 μm [CII] line in a number of Galactic clouds (Genzel et al. 1985, Stutzki et al. 1988a, Genzel et al. 1989, Howe et al. 1991, and Stacey et al. 1992) show that the far-infrared line emission is clearly more extended than the radio continuum emission and appears to penetrate several parsecs from the central luminosity sources into the surrounding molecular clouds. Similar results are found in the 609 and 371 μm [CI] lines (Phillips & Huggins 1981, Keene et al. 1985, and Genzel et al. 1988). The large extent of the [CII] emission region is interpreted as a large penetration depth of the UV radiation in the context of the PDR models of Sect. 2.1. The average column density of neutral gas, as measured from isotopic CO emission, between the central OB stars in the HII regions and the point where the [CII] emission has dropped significantly corresponds to $A_V \approx 50$ to 300 in W3 and M17 (Stutzki et al. 1988a, Howe et al. 1991). Without obscuration, the diluted UV flux from the central OB stars corresponds to a few $10^3 \chi_0$ at a distance of 2 parsec, not much larger than the UV flux necessary to account for the observed [CII] flux at that radius. Hence the average visual extinction toward the central OB stars must be less than $A_V \approx$ a few. This is in contradiction with the above mentioned measured column density unless most of the column density is concentrated in dense clumps with a clump to interclump contrast between 10 and 100. Numerical calculations confirm this argument and exclude homogeneous cloud models (Stutzki et al. 1988a, Boissé 1990, Howe et al. 1991). Assuming that geometry effects do not fool us in deducing large penetration depths in all clouds observed, there appears presently to be only two ways around this argument. If the clouds contain a distributed population of B stars, enhancing the far-UV radiation throughout, clouds could be more homogeneous. Alternatively, it is conceivable that the UV dust absorptivity per hydrogen nucleus is lower in interclump gas than it is in the dense clumps. This could occur if PAH's are responsible for a significant fraction of the far-UV extinction and also have a much lower abundance in the interclump medium, due to photodestruction by UV radiation. The detection of faint non-Gaussian emission components in the wings of ^{12}CO line profiles may constitute direct evidence for an interclump medium (Blitz & Stark 1986, Falgarone & Phillips 1990).

In summary then, keeping Scalo's (1990) comments in mind as an important warning and caveat it appears quite convincing that many of the molecular column density peaks are self-gravitating physical entities with a large density contrast relative to the interclump medium. This finding puts strong constraints on molecular cloud models. Next we have to address the clump's mass spectrum and stability.

4.1.c Mass Spectrum of Clumps

Stutzki & Güsten (1990) fitted Gaussian clumps to the data cube of $C^{18}O$ spectra in the M17 interface. They decompose the emission into 179 clumps of size ≤ 10" to 60", FWHM velocity width 0.5 to 3 km s^{-1} and mass 10 to 10^3 $M_\odot$. The clumps have a molecular hydrogen density between 10^5 and 10^6 cm^{-3} and fill about 30% of the volume. The derived clump mass spectrum follows a power law with an exponent of about 1.7 $\left(dN(M)/dM \propto M^{-1.7} \right)$. Similar power law spectra with exponents between 1.1 and 1.6 have been found for the Rosette cloud by Blitz (1987), for ρ Oph by Loren (1989) and for the general mass spectrum of giant clouds by Casoli et al. (1984) and Sanders et al. (1985).

A power law spectrum with an exponent of ≈ 1.5 is the plausible result of an equilibrium of coagulation and fragmentation (Spitzer 1982). It may also be consistent with a Salpeter type stellar mass spectrum $\left(dN(M)/dM \propto M^{-2.35} \right)$ as the eventual outcome of that fragmentation if the fraction of the clump's mass that ends up in the star decreases with increasing mass (Zinnecker 1989).

4.2 Energy Balance of Molecular Clouds
4.2.a Stability of Clumps

One of the long standing problems of molecular cloud research has been the question of how the clouds and clumps are stabilized against immediate gravitational collapse. Assuming that the clouds are indeed self-gravitating on all scales (Solomon et al. 1987), they should collapse on a few times the free fall time scale, $t_{\text{ff}} \approx (3G\rho)^{-1/2}$, or less than a few 10^6 years. The result would be a star formation rate much larger than that observed in the Galaxy. A very likely explanation is that magnetic fields are approximately in balance with gravity (e.g., Myers & Goodman 1988) and that gravitational collapse is prevented by magnetic pressure until the field has slowly diffused out by ambipolar diffusion (Mestel & Spitzer 1956, Shu et al. 1987). Star formation will then proceed (with low efficiency) on a time scale (Zweibel 1987, Shu et al. 1987)

$$t_{\text{amb}} \approx 2 \times 10^7 \left(\frac{x_e}{10^{-7}} \right) \approx 10 \, t_{\text{ff}} \quad [\text{y}]. \tag{119}$$

x_e is the fractional electron abundance. It is presently not established whether a uniform component of the field or magnetic waves (Alfvén waves) dominate the

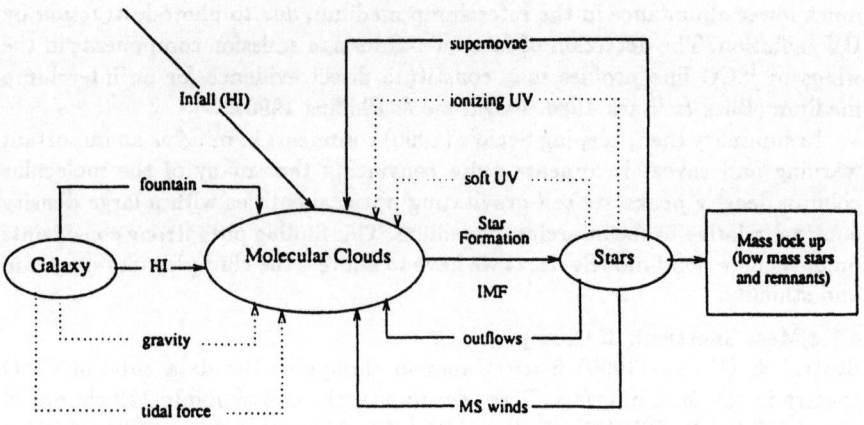

Fig. 22. Energy pathways in molecular clouds (adapted from Bally 1989)

magnetic pressure locally. In most cases one finds large "non-thermal" local line widths that are on the one hand sufficient to stabilize the clouds against free fall collapse, but on the other hand are supersonic. Dissipation of kinetic energy by shocks in cloud-cloud or clump-clump collision would then take place on a time scale comparable to the free-free collapse or dynamical time scales $t_{ff} \approx t_{dyn} = R/v$. Many authors have therefore argued that the turbulence is supersonic but sub-Alfvénic (Shu et al. 1987). In this case magnetic fields mediate clump-clump collisions with a low rate of dissipation. The observed "turbulence" may just be the result of magnetic waves (Arons & Max 1975). Alfvén waves and turbulence could then be replenished on a much larger time scale. Supernovae, Galactic tidal shear, UV radiation of massive stars and stellar winds may all contribute at some level to feeding of the "turbulent cascade" (Falgarone & Puget 1988, Wilson & Walmsley 1989). Promising local sources of kinetic energy that could create the observed turbulence are outflows from young stars (Norman & Silk 1980). Fukui et al. (1986) and Margulis & Lada (1986) have carried out unbiased surveys for outflows in Orion and Monoceros. In the surveyed regions of the Orion molecular cloud alone there are more than 20 flows, approximately evenly distributed with a mean size of about 1 pc and a separation of 5 pc (Fukui 1989). Fukui (1989) estimates that all stars more massive than about 1 $M_\odot$ go through a few 10^4 year period of pre-main sequence mass loss and that mass outflows can significantly contribute to the turbulence and cloud support. However, the mass outflow time scales are highly uncertain and estimates of dissipation time scale (Scalo & Pumphrey 1982) sensitively depend on the assumptions in numerical models. Clouds without outflows have line widths not very much less than clouds which do contain outflow sources.

The high mass clumps observed in Orion (Fig. 21 middle and right) and other OB star forming regions may be in a different regime as their masses are likely greater than the mass that can be supported by magnetic fields. This critical mass is given by (Mouschovias & Spitzer 1976, Shu et al. 1987)

$$M_{cr} = 25 \left(\frac{B}{300 \, \mu G} \right) \left(\frac{R}{0.1 \, pc} \right)^2 \quad [M_\odot],$$

or

(120)

$$N(H_2)_{cr} \approx 4 \times 10^{22} \left(\frac{B}{300 \, \mu G} \right) \quad [cm^{-2}].$$

The likely consequence is that most clumps in Orion-KL and M17 must collapse, and are in the process of forming a high density stellar system with high efficiency. Support for this expectation comes from the work by Herbig & Terndrup (1986), McCaughrean et al. (1992), and Lada et al. (1992) who find that the most prominent molecular condensations in Orion A and B are associated with young star clusters whose density is comparable to that of the parent cloud.

It is fairly clear that an explanation of the energy balance of molecular clouds and of the time evolution of clumps must involve a number of factors. Figure 22 (from Bally 1989) is an attempt to sketch the complicated network of energy pathways in molecular clouds.

4.2.b Larson's Correlations

Larson (1981) was the first to point out that average molecular hydrogen volume density $\langle n \rangle$ and velocity width Δv appear to correlate with scale size R via simple power laws: $\langle n \rangle \propto R^{-\gamma}$ and $\Delta v \propto R^\delta$. Recent evaluations give $\gamma \approx 1$ and $\delta \approx 0.5$ (Scalo 1990). These two relationships are consistent with just two intrinsic conditions; virial equilibrium and an approximately constant column density, $N(H_2) \approx$ a few 10^{21} cm^{-2}, independent of scale size. There are currently four competing explanations for the approximately constant column densities:

1. Most clouds have column densities near, but below the critical value for support by magnetic fields (Eq. (120), Shu et al. 1987). Higher column densities of clumps near OB stars are the result of super-critical collapse and/or much higher fields at higher densities.

2. Molecular clouds are in equilibrium with external pressure (Maloney 1988, Elmegreen 1989). A simple estimate of hydrostatic equilibrium gives

$$\rho \frac{d \left(\frac{GM(R)}{R} \right)}{dR} = \frac{dP}{dR}$$

or, at the surface

(121)

$$P_{ext} \approx \frac{GM^2}{R^4} \propto N^2.$$

Maloney (1988) and Elmegreen (1989) show that the observed correlations can be met by virialized clouds and their atomic envelopes that are in virial equilibrium with the general pressure of the ISM, $P_{ext} \approx 4000$ cm^{-3} K.

The higher column densities of the gas in OB star forming regions then correspond to higher pressures. Since the gas pressure of the atomic gas is directly coupled to the external radiation field, the higher column and volume densities of these clumps should then scale with the square root of the density of the UV radiation field $\chi^{1/2}$. This is in reasonable agreement with the observations and may explain why the densest and most massive clumps are typically seen in the immediate vicinity of luminous O stars.

3. The third explanation by McKee (1989) rests on the fact that the time scale for ambipolar diffusion is proportional to the fractional electron abundance x_e (Eq. (119)). Low mass star formation (sub-critical in the sense of Eq. (120)) is only possible in regions of low electron abundance, that is, regions where UV radiation cannot photoionize carbon ($A_V \geq$ a few, or $N(H_2) \geq$ a few $10^{21}\,cm^{-2}$). In the scenario of "photoionization-regulated" star formation proposed by McKee an initially diffuse cloud contracts quickly to $A_V \approx 4$. At that point its central parts are shielded from external UV radiation, the ambipolar diffusion time drops rapidly to the value determined by cosmic rays alone and star formation commences. Outflows from newly-formed low mass stars then stabilize the cloud and prevent further collapse. On average most clouds are then expected to have column densities near the "onset" value of a few $10^{21}\,cm^{-2}$.

4. The fourth explanation argues that there is *no* physical basis of the correlation (Kegel 1989, Scalo 1990). The apparently constant column density is a direct consequence of observational bias and sensitivity limits (intensity limit = column density limit).

4.3 An Example of a Star Forming Cloud: Orion

4.3.a Global Structure

Figure 23a gives an overview of the Orion molecular cloud complex (adapted from Maddalena et al. 1986). The complex is made up of several sub-complexes (Orion A and B clouds, etc.) and has a size of about 120 pc (15° at a distance of 450 pc). The complex's mass is estimated to be a few $10^5\,M_\odot$ from observations of CO $1 \rightarrow 0$ line emission (Maddalena et al. 1986). This corresponds to a mean hydrogen density of $100\,cm^{-3}$ or more. The Orion cloud system thus is one of the less massive of ≈ 4000 giant molecular clouds (GMC's) in the Galaxy (Solomon et al. 1987, Scoville & Sanders 1987).

In this cloud OB stars ($M \geq 10\,M_\odot$) have been forming for the past $10^7\,y$ or so. Four distinct OB associations can be distinguished in visible studies (Blaauw 1964). The three older ones are marked by hatched circles/ellipses in Fig. 23a. The youngest is associated with the Orion nebula. In addition to the associations which are in front of or near the cloud the most recent star forming activity (within the last 10^6 years) has occurred within the cloud itself. This includes the Orion nebula, NGC 2023, NGC 2024, NGC 2068, and NGC 2071. All major star forming sites in Orion can be easily traced by their intense $12 \rightarrow 120\,\mu m$ far-infrared continuum emission that is shown, for instance, on the beautiful IRAS image on Fig. 23b (Beichman 1988). It is clear that star formation in the Orion cloud currently is taking place at a high rate.

The most prominent and luminous of these star forming regions is the Orion nebula itself. It contains the θ^1 (Trapezium) OB association ($L = 10^5 L_\odot$), as well as a remarkable concentration of lower mass stars surrounding it (Herbig & Terndrup 1986, McCaughrean et al. 1992). About 0.1 pc behind the HII region that is ionized by the Trapezium stars is the Orion-KL region. The KL region contains a luminous ($10^5 L_\odot$) cluster of compact infrared sources near the center of the densest core of the molecular cloud, with the first infrared "protostars" named after they discoverers, the Becklin-Neugebauer object (BN, 1967) and the Kleinmann-Low nebula (KL, 1967).

4.3.b The Relation Between Dense Gas and Star Formation

I had already pointed out in context with Fig. 21 that star formation and dense cloud cores are well correlated. E. Lada (1990) has recently presented a survey of molecular and near-infrared emission in the Orion B cloud (Fig. 23) that probably represents a key result for the understanding of star formation in molecular clouds. Her study shows that the number of $2\,\mu m$ sources (as a measure of content of young embedded stars) has four peaks coincident with or very close to the locations of four to five prominent CS mm-emission peaks (as a measure of column density of dense gas). This close correlation is shown in Fig. 24.

After statistical correlation for background stars, E. Lada finds that up to 96% of all young stars are related to the five CS cores that comprise about 50% of the mass of the entire cloud. As in the Orion A, Trapezium area discussed above (Herbig & Terndrup 1986) the high density (> a few hundred $M_\odot$ pc^{-3}) of the young stellar associations suggests efficient stellar formation there. There is no other concentration of stars elsewhere in the Orion B cloud. Evans & Lada (1991) conclude that the young stellar clusters formed in this "cluster mode" of star formation must disperse later on, as less that 10% of all stars in the Galaxy reside in gravitationally bound, open clusters.

A possible interpretation is then that in Orion (external?) triggering has created large, dense molecular cores that form star clusters with high efficiency. In the rest of the cloud, however, as in dark clouds ("Taurus star formation mode") and on average in other GMC's, star formation is proceeding only at low (few %) efficiency. The idea of external triggering of (massive) star formation by radiation, ionization fronts and winds had been previously proposed (also in context with Orion) by Elmegreen & Lada (1977). It is tempting to speculate that starburst galaxies form stars efficiently exactly because they have managed to concentrate very dense molecular clouds at their nuclei.

4.3.c The Magnetic Field

Optical polarization measurements of magnetically aligned dust grains indicate that the magnetic field component in the plane of the sky is within 30° of the long axis of the Orion A molecular cloud (Vrba et al. 1988). Hence, the Orion A cloud cannot have contracted to its present elongated or flattened shape along magnetic field lines. Rather, the magnetic field may support the

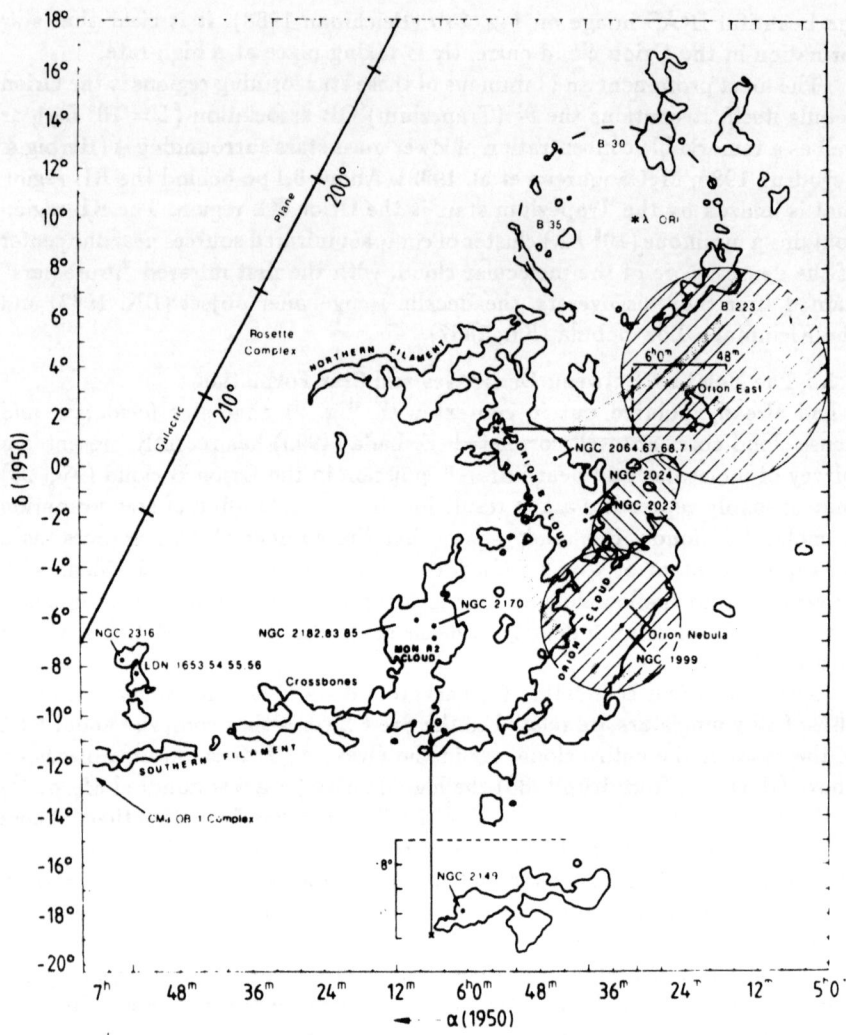

Fig. 23. (a): Schematic of the giant Molecular Cloud in Orion, along with locations of OB associations and visible nebulae (adapted from Maddalena et al. 1986)

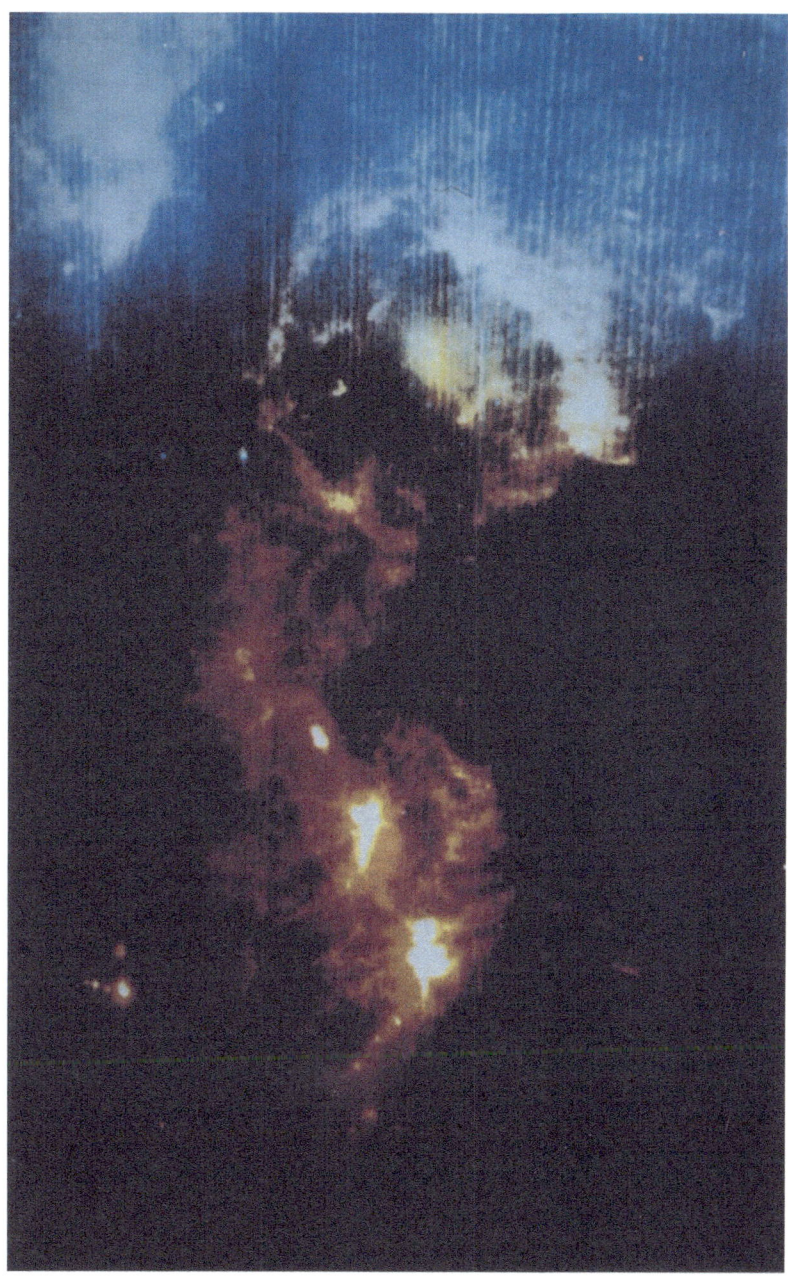

Fig. 23. (b): IRAS 12 → 120 μm image of the same area (adapted from Beichman 1988)

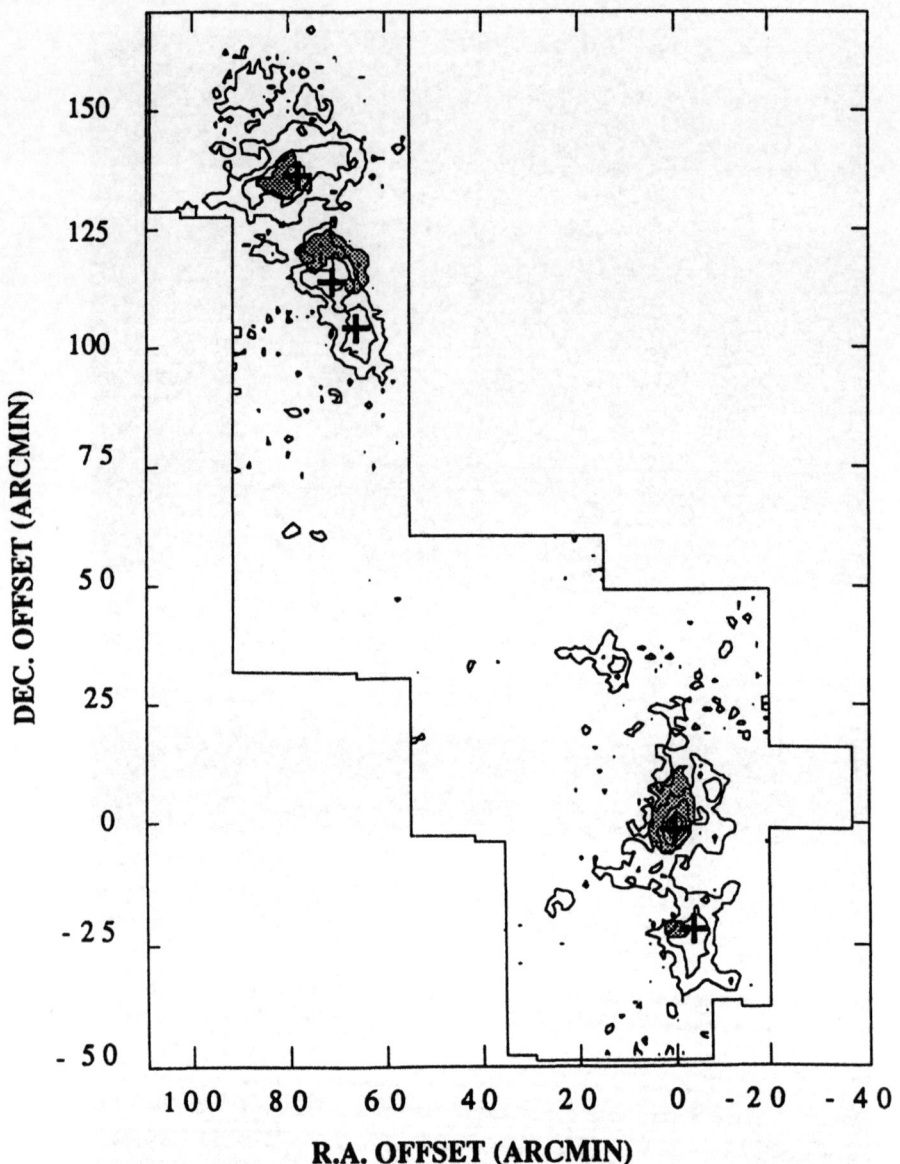

Fig. 24. Dense molecular cores in CS 2 → 1 line emission (contours, crosses) and concentrations of stars (2 μm, stippled) in Orion B molecular cloud (E. Lada 1990, E. Lada et al. 1992)

elongated structure of the Orion A cloud. Heiles & Stevens (1986) find that the line of sight magnetic field component, as determined from radio OH and HI Zeeman observations, reverses its direction across the cloud, pointing toward the Sun on the side of the cloud towards the Galactic plane and away from the Sun at lower galactic latitudes. A possible explanation is a helical geometry in which the magnetic field wraps around the cloud (Bally 1989) with its largest component along the cloud's long axis. As indicated by the large scatter of polarization position angles, the orientation of the magnetic field on small scales does not necessarily trace the large-scale average orientation discussed above. Near-IR and far-IR polarization measurements in BN-KL indicate that the magnetic field is approximately along the long axis of the local outflow (Dyck & Lonsdale 1979, Dragovan 1986). On the other hand, the directions of several optical jets/outflows are aligned with the large-scale field (Strom et al. 1986) which suggests that the field plays an important role in the presently ongoing star formation. Line of sight field strengths for several positions range from 50 to $125\,\mu$G in the large beam OH/HI observations (see Genzel & Stutzki 1989 for references). The field strength inferred for the OH masers in BN-KL is a few mG (Hansen et al. 1977) and $40\,$mG in the H_2O maser spots (Fiebig & Güsten 1989).

4.4 The Orion-KL Star Forming Core

4.4.a The BN-IRc2 Infrared Cluster

Following the discovery of the Becklin-Neugebauer object and of the cooler, extended Kleinmann-Low nebula 10" south of it, Rieke et al. (1973) found that at higher resolution the BN-KL infrared nebula splits up into a number of compact sources of different color temperatures. Figure 25 shows a contour map of a recent $19\,\mu$m camera image at 1.4" resolution (Cameron et al. 1992), superposed on a VLA map of $1.2\,$cm NH_3 (3.2) emission of similar resolution by Genzel et al. (1982). This image and other high-resolution, 2 to 30 μm maps show half a dozen compact sources embedded in extended emission, with standard designations marked on Fig. 25. The KL nebula is composed of the sources IRc 2 − 5 and surroundings.

Are these sources self-luminous, and what is their evolutionary state? Until the mid-80's the common interpretation had been that the infrared nebula is a cluster of newly formed stars in a very early stage of evolution that is, protostars in Larson's (1969) sense. In the following, we discuss how measurements of the past ten years have changed this picture considerably.

The total mid- and far-infrared luminosity of the complex is indicating that the young stars in the cluster are luminous and massive. Only about 10% of that luminosity emerges at $\lambda \leq 30\,\mu$m, where large ground-based telescopes allow the separation into individual sources. This fact also makes clear that radiative transport effects at near- and mid-infrared wavelengths are very important.

Rieke et al. (1973) first noted that the integrated depth of the 9.7 μm silicate absorption feature is stronger in the sources of the KL nebula than in BN itself. Measurements at higher spatial and spectral resolution show that the silicate

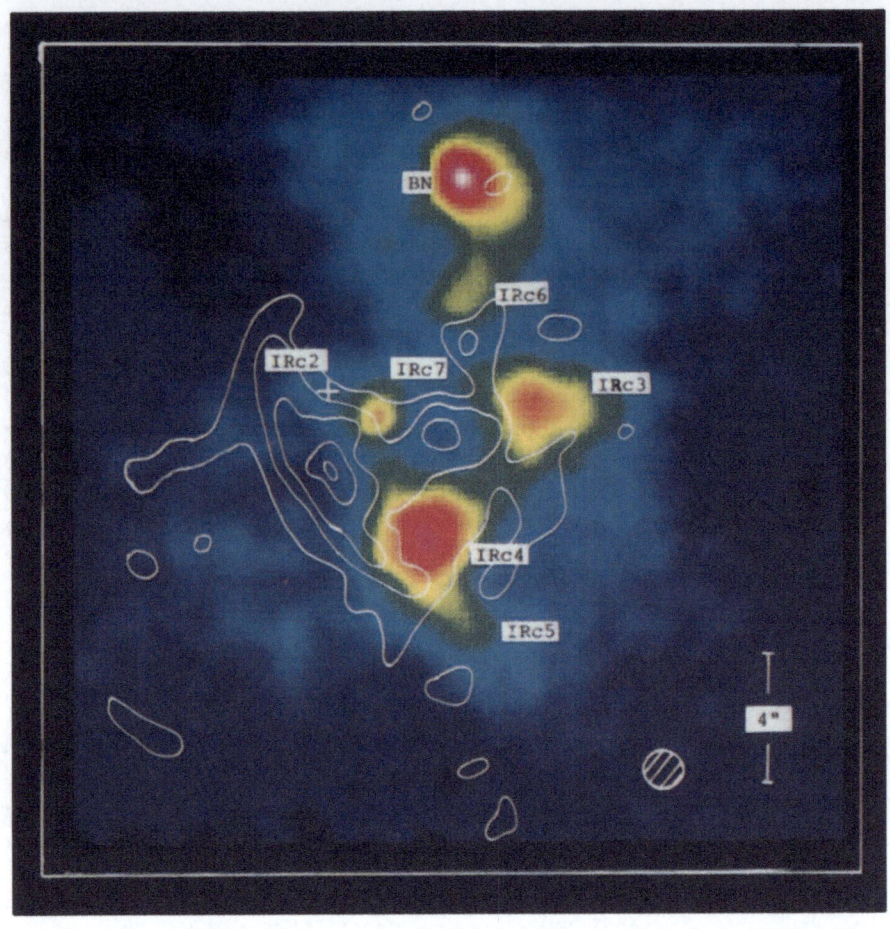

Fig. 25. 19 μm continuum image of the BN-KL region, taken with MIRACLE camera on UKIRT (1.4" resolution, Cameron et al. 1992), superposed on contours of 1.2 cm NH₃ (3,2) integrated line emission (1.2", VLA, Genzel et al. 1982)

absorption depth varies greatly from source to source, with IRc 2, 3, and 4 having the deepest features. Correction for the overlying absorption implied by the silicate feature then leads to the conclusion that the sources in the KL nebula, although much less conspicuous than BN at most near/mid-infrared wavelengths are intrinsically at least as luminous as BN. Downes et al. (1981) proposed that IRc 2, then already known to be intimately connected to the massive outflows in the BN-KL region is responsible for most of the region's luminosity of $10^5 L_\odot$.

For the present interpretation of the infrared nebula, the 3.8 μm polarization observations by Werner et al. (1983) and the high-resolution 2–30 μm mapping by Lee et al. (1983) and Wynn-Williams et al. (1984) are particularly relevant. Briefly, these observations show that:

1. The nebula's radiation between 2 and 4 μm is highly polarized. The high degree of polarization (up to 50%), its wavelength dependence, and in particular, its large-scale systematic spatial pattern unambiguously demonstrate that the polarized radiation is due to scattered light from a few central sources. BN and IRc 2 are the main sources illuminating this reflection nebula.

2. The most prominent 8 – 12.5 and 12.5 – 20 μm color temperature peaks are BN and IRc 2 suggesting that they contain the warmest dust and are close to heating sources. Elsewhere in the region, the color temperature is relatively constant.

3. The zones of strong scattering at 2.2 μm and 3.8 μm (e.g., IRc 3, 4, 5, 7) coincide with the most intense 20 and 30 μm peaks implying that their emission at the longer wavelengths is largely thermal reradiation of photons from BN and IRc 2 absorbed at 2 – 20 μm.

4. Column density maxima in molecular gas coincide with peaks of 20 – 30 μm dust opacity and intensity minima at 8 – 30 μm. This clearly demonstrates the presence of patchy extinction. The anticorrelation of infrared and molecular emission is apparent on the overlay in Fig. 25.

Werner et al. (1983) and Wynn-Williams et al. (1984) propose a self-consistent model of the central 30" of the nebula in which the geometry of the BN-KL region is that of a clumpy cavity (diameter 10^{17} cm), rather than that of a number of isolated objects. The relatively low average infrared opacity through the cavity probably results from the clearing by mass outflow from IRc 2. The cavity is centered on or near IRc 2, which is the source of most of the luminosity of the region. IRc 2 is intrinsically a powerful 2 – 30 μm source; however, it is heavily obscured by the edge of an optically thick gas and dust ridge centered about 2" south of the source $(N(\mathrm{H}_2) = 10^{24}\,\mathrm{cm}^{-2}$ or $\tau(10 - 30\,\mu m) = 10 - 20)$. BN $(L = 10^4\,L_\odot)$ and IRc 9 $(L \geq 200\,L_\odot)$ are the other two sources in the region that clearly are self-luminous.

The model that best fits all the infrared and radio data implies that most of the other peaks in the Kleinmann-Low nebula (IRc 3, 5, 7, etc.) are irregularities in the material within, at the edge of, and surrounding the cavity, rather that major self-luminating sources, (Wynn-Williams et al. 1984). It is not excluded, however, that IRc 3 and 4 contain self-luminous sources at a level of a few $10^3 L_\odot$.

4.4.b The Orion-KL Cloud Core

Multi-wavelength infrared and microwave spectroscopy, high-resolution line mapping and detailed measurements of molecular abundances now give a fairly coherent physical picture of the molecular gas in terms of spatially, kinematically, and chemically distinct regions. The measurements demonstrate that the violent mass outflows from IRc 2/BN have substantially affected the structure and physical conditions of the cloud core. Shocks and grain mantle evaporation in the warm circumstellar environment have substantially altered the chemistry.

Ridge of Quiescent Gas. The core of the quiescent molecular cloud ridge (Fig. 21 middle and right) contains several clumps or condensations with a total gas mass of about $150\,M_\odot$ within 0.1 pc of IRc 2. Typical parameters are a temperature of 70 K, a molecular hydrogen volume density of $10^5\,\mathrm{cm}^{-3}$, and a column density of $3 \times 10^{23}\,\mathrm{cm}^{-2}$. Chemically this region is characterized by standard gas phase, ion-molecule chemistry, with an abundance of carbon-rich species (e.g., CS, CN, CCH), and a lack of oxygen-rich molecules (Fig. 19).

Generally considered separate from the extended ridge is a 15" size condensation of quiescent gas at $v_{LSR} = 8\,\mathrm{km\,s}^{-1}$, located at the southwestern edge of the KL nebula. This "compact ridge", first noted spectrally in the 3 mm Onsala survey (Johansson et al. 1984), is warmer ($T = 100 - 150\,\mathrm{K}$) than the rest of the ridge and has a significantly different chemistry. Complex oxygen-rich molecules, such as CH_3OH, $HCOOCH_3$, or CH_3OCH_3, are highly abundant. The compact ridge also contains several CH_3OH masers (Menten et al. 1988). Blake et al. (1987) propose that mixing of the outflows with ridge gas in this region releases oxygen into the gas phase, which is then incorporated into the large oxygen-rich molecules by radiative association reactions. The proposed interaction is also made plausible by the spatial coincidence between the compact ridge and low-velocity H_2O masers in the same region. The masers probably are the locations of clump-cloud shocks (Sect. 5.3).

There is a velocity difference of about $2\,\mathrm{km\,s}^{-1}$ between the ridge gas just southwest and northeast of BN-KL. Ho & Barrett (1978) proposed that two clouds are colliding, whereas others (e.g., Hasegawa et al. 1984, Vogel et al. 1985) interpret the velocity gradient as evidence for rotation of the inner ridge. The ridge velocity appears to diverge approaching IRc 2 from either side, a signature consistent with Keplerian rotation about a central object. Vogel et al. (1985) derive a central mass of $25\,M_\odot$.

Hot Core. The "hot core" is apparent in Fig. 25 as a 10" diameter, elongated gas and dust ridge centered 2" south of IRc 2. Its hydrogen column density is at least $10^{24}\,\mathrm{cm}^{-2}$, as estimated from millimeter and far-infrared ^{13}CO measurements and from millimeter and sub-millimeter dust continuum observations (e.g., Masson et al. 1984). The total gas mass is $10\,M_\odot$ or more.

Because of the large column densities radiative trapping of the line radiation is often very important in the hot core. Current best estimates of the molecular hydrogen density are about $2 \times 10^7 \, \mathrm{cm}^{-3}$ (e.g., Hermsen et al. 1988) and temperatures range between 150 and 200 K. Pumping by far-infrared continuum radiation is important for a number of molecules for excitation of rotational levels in the ground vibrational state (e.g., OH, Betz & Boreiko 1989, Melnick et al. 1990). Mid-infrared pumping of the vibrational states has been found to account for millimeter rotational transitions of molecules in vibrationally excited states, such as HC_3N, CH_3CN, HCN, and NH_3 (Goldsmith et al. 1983, Ziurys & Turner 1986). Radiation from IRc 2 almost certainly is responsible for the pumping.

The hot core ridge appears to consist of a turbulent ensemble of clumps with sizes of 1" ($7.2 \times 10^{15} \, \mathrm{cm}$) or smaller on VLA maps of 1.2 cm NH_3 inversion transitions (Figs. 21, 25). Its velocity centroid is identical to the centroid of the SiO masers on IRc 2 which probably is a reliable indicator of the stellar velocity. Hence, IRc 2 and the hot core are probably physically related. The hot core may be the remnant of the dense parental cloud out of which IRc 2 has formed and that now interacts with its outflow and radiation (Masson & Mundy 1988).

The chemistry of the hot core is characterized by unusually high abundances of hydrogen-saturated molecules, such as NH_3, H_2O (observed in its isotopic form HDO), CH_3CN and CH_2CH_3CN. A likely interpretation is that, analogous to comets, the gas phase is enriched with fully hydrogenated species by evaporation of grain mantles that have been heated by radiation from IRc 2 (Brown et al. 1988).

Outflows. As has been mentioned several times already, the Orion-KL is at the center of high velocity outflows. In fact Orion was the first source in which high velocity wings in CO were discovered (Kwan & Scoville 1976, Zuckerman et al. 1976). The dynamically active gas is sometimes separated into the "low-velocity" (or "18 km s^{-1}" or "expanding doughnut") flow, the "high velocity" flow (or "plateau"), and the "shock-excited gas" because of differences in kinematics, spatial distribution, and excitation. The flows represent gas streaming away from the center of the BN-IRc 2c cluster. This has been demonstrated by measurements of proper motions of H_2O masers (Genzel et al. 1981) and of Herbig-Haro objects (Axon & Taylor 1983). Spectroscopy at near-infrared (Hall et al. 1978) and far-infrared wavelengths (Betz & Boreiko 1989, Melnick et al. 1990) show the blueshifted, high-velocity gas in absorption against the dust continuum, again suggesting an overall outflow of the high-velocity gas. The center of the outflows is within a few arcsec of IRc 2. Their dynamical age is a few times 10^3 y. The inferred mass loss rate, as derived from that age and the total mass involved, is about $10^{-3 \pm 0.3} \, M_\odot \, \mathrm{y}^{-1}$ (Genzel et al. 1981).

A low-velocity flow is recognized in the SiO, OH, and low-velocity H_2O masers, as well as in thermal millimeter and submillimeter rotational transitions of several heavy top molecules (SiO, HCN, SO_2, SO, cf. Plambeck et al. 1982). This flow can be traced from within 0.2" of IRc 2 to about 20" from the star, with an approximately constant velocity range ($\Delta v \approx 35 \, \mathrm{km \, s}^{-1}$). The double peaked appearance of the H_2O and SiO "shell" masers strongly suggests a spherically or axially symmetric outflow, where the greatest maser gain is along the front and back sides of the expanding shell. The flowing gas probably is clumped, as strong emission lines from molecular transitions are seen that require very high densities $(n(H_2) \approx 10^7 \, cm^{-3}$ for their excitation (Stutzki et al. 1988b)). The observations fit a simple model whereby the initially free outflow from the star plunges into the dense molecular ridge at $R = 10^{17} \, cm$. In the shocks at this interface, low-velocity H_2O maser are created as a thin, but very dense, zone of molecular species that require shock chemistry for their formation. The inner, relatively low-density, free outflow zone is identical with the "infrared cavity" discussed in the last section. The observational data are in good agreement with model calculations of the chemistry and the formation of masers in such shocks (Elitzur et al. 1989).

A second flow of larger, but less well-defined, velocity range $(\Delta v \leq 250 \, km \, s^{-1})$ is approximately perpendicular to the low-velocity flow and the ridge. This high-velocity flow has a weakly bipolar velocity structure, similar to those found in many other regions of star formation (Erickson et al. 1982). As in the case of the low-velocity flow, the SiO molecule traces the high-velocity gas closest to the center, which is within a few arcsec of IRc 2.

The fact that the low-velocity flow is mostly along the dense ridge, whereas the higher velocity gas appears to stream mainly perpendicular to it, has been interpreted as the channeling of one single, high-velocity flow by the surrounding cloud (Canto et al. 1981). However, the two flows are distinct to within about 2" (10^{16} cm) of IRc 2, far inside the outflow cavity. Furthermore, the mean gas density in the outflow cavity is not sufficient for dynamic channeling of the high-velocity flow. Hence, either the two flows emerge from different stars, or else the outflowing gas has to be channeled in the immediate environment of the star, possible by a circumstellar disk.

In addition to the main outflow centered on the BN-KL cluster there are several streamers, jet-like features and Herbig-Haro objects (Hasegawa 1987, Axon & Taylor 1983, and Schmid-Burgk et al. 1990) that are likely connected with other outflows in the vicinity. A group of very compact, partially ionized globules in the neighborhood of θ^1 may be the result of the interaction of the radiation and winds from the Trapezium OB stars with the surrounding cloud or with nearby circum-stellar disks (Garay et al. 1987, Churchwell et al. 1987).

High-Velocity Shocked Gas. The high-velocity flow plunges into the surrounding cloud about 30" from the dynamical center, creating high-velocity H_2O masers and a zone of shocked molecular gas cooling in infrared and submillimeter lines. The shocked zone has been studied in detail in the near-infrared lines of excited molecular hydrogen (Beckwith et al. 1978, Geballe et al. 1986, and Geballe & Garden 1987) and in the far-infrared lines of CO, OH, [OI], and [SiII] (Watson et al. 1985, Melnick et al 1990, Haas et al. 1991). The interpretation of this component in terms of J- and C-shocks has already been discussed in Sect. 2.2.

4.5 Outflows in Star Forming Regions

The phenomenon of molecular outflows that was discussed in the last section for Orion-KL has also been found in almost all other regions of recent star formation (Genzel & Downes 1982, Bally & Lada 1983, Lada 1985). Evidence is now rather firm that most if not all stars $\geq 1 M_\odot$ go through a period of mass outflow during their formation phase. The outflows have velocities of about 20 to 300 km s^{-1}, and have an ionized (Panagia 1991), a neutral atomic (Lizano et al. 1988, Natta & Giovanardi 1991), and a molecular component (Bally & Lane 1991). The outflows can be traced in a number of signposts, including optical jets and Herbig-Haro objects (Mundt 1988, Reipurth 1991), infrared recombination lines and radio continuum, broad wings in thermal mm-emission lines (CO, HCN, SiO, etc.), H_2O, OH and SiO masers and infrared emission lines from shocks.

Most of the mass is in the lower velocity, molecular component and the derived mass loss rates range from 10^{-8} $M_\odot$ y^{-1} for T-Tau stars to $> 10^{-2}$ $M_\odot$ y^{-1} for massive star-forming clouds, like W49 and W51. The momentum transported in the mass outflows is substantially larger than what can be accounted for by single scattering radiation pressure and scales approximately with source luminosity (e.g., Bally & Lada 1983).

The molecular flows usually have bipolar morphology, with a fairly broad opening angle (e.g., Snell 1987). In contrast, the optical jets close to the exciting stars in T-Tau star forming regions are extremely well collimated (e.g., Reipurth 1991). The lifetime of the outflow phase appears to be $\approx 10^5$ y (Genzel & Downes 1979, Bally & Lada 1983, Lada 1985) comparable to the pre-main sequence time of massive stars. It is now widely assumed that mass outflow and mass accretion in the formation phase of stars are closely related, either through centrifugal ejection of matter from magnetized disks (Pudritz et al. 1991) or through ejection from the surfaces of the newly-formed stars (Shu et al. 1987). Mass ejection is probably a direct consequence of the fact that most of the angular momentum of the accreting material has to be removed before it can fall onto the forming star.

References

4.1   Arons, J., Max, C.E. 1975, *Astrophys. J.* **196**, L77
4.2   Axon, K., Taylor, K. 1983, *Monthly Notices Roy. Astron. Soc.* **207**, 4
4.3   Bally, J., Lada, C.J. 1983, *Astrophys. J.* **265**, 824
4.4   Bally, J., Langer, W.D., Stark, A.A., Wilson, R.W. 1987, *Astrophys. J.* **312**, L45
4.5   Bally, J. 1989, in *Low Mass Star Formation and Pre-Main Sequence Objects*, ed. B. Reipurth, ESO Conf. Proc. **33**, 1
4.6   Bally, J., Lane, A.P. 1991, in *The Physics of Star Formation and Early Evolution*, eds. C.J. Lada, N.D. Kylafis, *NATO-ASI-342*, (Kluwer, Dordrecht), 471
4.7   Barrett, A.H., Ho, P.T.P., Myers, P.C. 1977, *Astrophys. J.* **211**, L39
4.8   Becklin, E.E., Neugebauer, G. 1967, *Astrophys. J.* **147**, 799

4.9   Beckwith, S., Persson, S.E., Neugebauer, G., Becklin, E.E. 1978, *Astrophys. J.* **223**, 464

4.10 Beichman, C.A. 1988, *Astrophys. Ltr. Commun.*, **27**, 67

4.11 Betz, A.L., Boreiko, R.T. 1989, *Astrophys. J.* **346**, L101

4.12 Blake, G.A., Sutton, E.C., Masson, C.R., Phillips, T.G. 1987, *Astrophys. J.* **315**, 621

4.13 Blitz, L., Stark, A.A. 1986, *Astrophys. J.* **300**, L89

4.14 Blitz, L. 1987, in *Physical Processes in Interstellar Clouds*, eds. G. Morfill, M. Scholer (Reidel, Dordrecht), 35

4.15 Blitz, L. 1991, in *The Physics of Star Formation and Early Stellar Evolution*, eds. C.J. Lada, N.D. Kylafis (Kluwer, Dordrecht), 3

4.16 Boissé, P. 1990, *Astron. Astrophys.* **228**, 483

4.17 Brown, P.D., Charnley, S.B., Millar, T.J. 1988, *Monthly Notices Roy. Astron. Soc.* **231**, 409

4.18 Cameron, M. et al. 1992, in prep

4.19 Canto, J., Rodriguez, L.F., Barral, J.F., Carral, P. 1981, *Astrophys. J.* **244**, 102

4.20 Casoli, F., Combes, F., Gerin, M. 1984, *Astron. Astrophys.* **133**, 99

4.21 Cernicharo, J., Guélin, M. 1987, *Astron. Astrophys.* **176**, 299

4.22 Churchwell, E., Felli, M., Wood, D.O., Massi, M. 1987, *Astrophys. J.* **321**, 516

4.23 Downes, D., Genzel, R., Becklin, E.E., Wynn-Williams, C.G. 1981, *Astrophys. J.* **244**, 869

4.24 Dragovan, M. 1986, *Astrophys. J.* **308**, 270

4.25 Dyck, H.M., Lonsdale, C.J. 1979, *Astron. J.* **84**, 1339

4.26 Elitzur, M., Hollenbach, D.J., McKee, C.F. 1989, *Astrophys. J.* **346**, 983

4.27 Elmegreen, B.G., Lada, C.J. 1977, *Astrophys. J.* **214**, 725

4.28 Elmegreen, B.G. 1989, *Astrophys. J.* **338**, 178

4.29 Emerson, J. 1987, in *Star Forming Regions*, eds. M. Peimbert, J. Jugaku (Reidel, Dordrecht), 19

4.30 Erickson, N.R., Goldsmith, P.F., Snell, R.L., Berson, R.L., Huguenin, G. R. et al. 1982, *Astrophys. J.* **261**, L103

4.31 Evans, N.J., Lada, E. 1992, in press

4.32 Falgarone, E., Puget, J.L. 1988, in *Galactic and Extragalactic Star Formation*, eds. R. Pudritz, M. Fich (Kluwer, Dordrecht), 195

4.33 Falgarone, E., Phillips, T.G. 1990, *Astrophys. J.* **359**, 344

4.34 Falgarone, E., Phillips, T.G., Walker, C. 1992, *Astrophys. J.* in press

4.35 Fiebig, D., Güsten, R. 1989, *Astron. Astrophys.* **214**, 333

4.36 Fukui, Y., Sugitami, K., Takaba, H., Iwata, T., Mizuno, A., Ogawa, H., Kawabata, K. 1986, *Astrophys. J.* **311**, L85

4.37 Fukui, Y. 1989, in *Low Mass Star Formation and Pre-Main Sequence Objects*, ed. Bo Reipurth, ESO Proceedings **33**, 95

4.38 Garay, G., Moran, J.M., Reid, M.J. 1987, *Astrophys. J.* **314**, 535

4.39 Geballe, T.R., Persson, S.E., Simon, T., Lonsdale, C.J., McGregor, P.J. 1986, *Astrophys. J.* **302**, 500

4.40 Geballe, T.R., Garden, R. 1987, *Astrophys. J.* **317**, L107

4.41 Genzel, R., Downes, D. 1979, *Astron. Astrophys.* **72**, 234

4.42 Genzel, R., Reid, M.J., Moran, J.M., Downes, D. 1981, *Astrophys. J.* **244**, 884

4.43 Genzel, R., Downes, D. 1982, in *Regions of Recent Star Formation*, eds. R.S. Roger, P.E. Dewdney (Reidel, Dordrecht), 251

4.44 Genzel, R., Downes, D., Ho, P.T.P., Bieging, J.H. 1982, *Astrophys. J.* **259**, L103

4.45 Genzel, R., Watson, D.M., Crawford, M.K., Townes, C.H. 1985, *Astrophys. J.* **297**, 766

4.46 Genzel, R., Harris, A.I., Jaffe, D.T., Stutzki, J. 1988, *Astrophys. J.* **332**, 1049

4.47 Genzel, R., Harris, A.I., Stutzki, J. 1989, in *Infrared Spectroscopy in Astronomy*, ed. B.H. Kaldeich, *ESA-SP-290*, 115

4.48 Genzel, R., Stutzki, J. 1989, *Ann. Rev. Astron. Astrophys.* **27**, 41

4.49 Goldsmith, P.F., Krotkov, R., Snell, R.L., Brown, R.D., Godfrey, P. 1983, *Astrophys. J.* **274**, 184

4.50 Goldsmith, P.F., Margulis, M., Snell, R.L., Fukui, Y. 1992, *Astrophys. J.* in press

4.51 Haas, M.R., Hollenbach, D.J., Erickson, E.F. 1991, *Astrophys. J.* **374**, 555

4.52 Hall, D.N.B., Kleinmann, S.G., Ridgway, S.T., Gillett, F.C. 1978, *Astrophys. J.* **223**, L47

4.53   Hansen, S.S., Moran, J.M., Reid, M.J., Johnston, K.J., Spencer, J.H., Walker, R.C. 1977, *Astrophys. J.* **218**, L65
4.54   Hasegawa, T., Kaifu, N., Inatani, J., Morimoto, M., Chikada, Y. et al. 1984, *Astrophys. J.* **283**, 117
4.55   Hasegawa, T. 1987, *In Star Forming Regions*, eds. M. Peimbert, J. Jugaku (Reidel, Dordrecht), 123
4.56   Heiles, C., Stevens, M. 1986, *Astrophys. J.* **301**, 331
4.57   Herbig, G.H., Terndrup, D.M. 1986, *Astrophys. J.* **307**, 609
4.58   Hermsen, W., Wilson, T.L., Bieging, J.H. 1988, *Astron. Astrophys.* **201**, 276
4.59   Ho, P.T.P., Barrett, A.H. 1978, *Astrophys. J.* **224**, L23
4.60   Howe, J.E., Jaffe, D.T., Genzel, R., Stacey, G.J. 1991, *Astrophys. J.* **373**, 158
4.61   Johansson, L.E.B., Andersson, C., Elldér, J., Friberg, P., Hjalmarson, Å. et al. 1984, *Astron. Astrophys.* **130**, 227
4.62   Keene, J., Blake, G.A., Phillips, T.G., Huggins, P.J., Beichman, C. 1985, *Astrophys. J.* **299**, 967
4.63   Kegel, W.H. 1989, *Astron. Astrophys.* **225**, 517
4.64   Kleinmann, D.E., Lo, F.J. 1967, *Astrophys. J.* **149**, L1
4.65   Kwan, J., Scoville, N.Z. 1976, *Astrophys. J.* **210**, L39
4.66   Lada, C.J. 1985, *Ann. Rev. Astron. Astrophys.* **23**, 267
4.67   Lada, E. 1990, Ph.D. Thesis, Univ. of Texas
4.68   Lada, E., de Poy, D.L., Evans, N.J., Gatley, I. 1992, *Astrophys. J.* in press
4.69   Larson, R.B. 1981, *Monthly Notices Roy. Astron. Soc.* **194**, 809
4.70   Lee, T.J., Beattie, D.H., Geballe, T.R., Pickup, D.A. 1983, *Astron. Astrophys.* **127**, 417
4.71   Lizano, S., Heiles, C., Rodriguez, L.F., Koo, B.C., Shu, F.H., Hasegawa, T., Hayashi, S., Mirabel, I.F. 1988, *Astrophys. J.* **328**, 763
4.72   Loren, R.B. 1989, *Astrophys. J.* **338**, 902
4.73   Maddalena, R.J., Morris, M., Moscovitz, J., Thaddeus, P. 1986, *Astrophys. J.* **303**, 375
4.74   Maloney, P. 1988, *Astrophys. J.* **334**, 761
4.75   Margulis, M., Lada, C.J. 1986, *Astrophys. J.* **309**, L87
4.76   Masson, C.R., Berge, G.L., Claussen, M.J., Heiligman, G.M., Leighton, R.B. et al. 1984, *Astrophys. J.* **283**, L37
4.77   Masson, C.R., Mundy, L.G. 1988, *Astrophys. J.* **324**, 538
4.78   McCaughrean, M.J. et al. 1992, in press
4.79   McKee, C.F. 1989, *Astrophys. J.* **345**, 782
4.80   Melnick, G.J., Stacey, G.J., Genzel, R., Lugten, J.B., Poglitsch, A. 1990, *Astrophys. J.* **348**, 161
4.81   Menten, K.M., Walmsley, C.M., Henkel, C., Wilson, T.L. 1988, *Astron. Astrophys.* **198**, 253
4.82   Mestel, L., Spitzer, L. 1956, *Monthly Notices Roy. Astron. Soc.* **116**, 503
4.83   Mezger, P.G., Chini, R., Kreysa, E., Gemünd, H.P. 1986, *Astron. Astrophys.* **160**, 324
4.84   Mezger, P.G., Chini, R., Kreysa, E., Wink, J.E., Salter, C.J. 1988, *Astron. Astrophys.* **191**, 44
4.85   Migenes, V., Johnston, K.J., Pauls, T.A., Wilson, T.L. 1989, *Astrophys. J.* **347**, 294
4.86   Mouschovias, T., Spitzer, L. 1976, *Astrophys. J.* **210**, 326
4.87   Mundt, R. 1988, in *Formation and Evolution of Low Mass Stars*, eds. A.K. Dupree, M.T.V.T. Lago, *NATA-ASI-241*, (Kluwer, Dordrecht), 257
4.88   Mundy, L.G., Cornwell, T.J., Masson, C.R., Scoville, N.Z., Baath, L.B., Johansson, L.E.B. 1988, *Astrophys. J.* **325**, 382
4.89   Myers, P. 1987, in *Star Forming Regions*, eds. M. Peimbert, J. Jugaku (Reidel, Dordrecht), 33
4.90   Myers, P.C., Goodman, A.A. 1988, *Astrophys. J.* **329**, 392
4.91   Natta, A., Giovanardi, C. 1991, in *The Physics of Star Formation and Early Evolution*, eds. C.J. Lada, N.D. Kylafis, *NATO-ASI-342*, (Kluwer, Dordrecht), 595
4.92   Norman, C., Silk, J. 1980, *Astrophys. J.* **238**, 158
4.93   Panagia, N. 1991, in *The Physics of Star Formation and Early Evolution*, eds. C.J. Lada, N.D. Kylafis, *NATO-ASI-342* (Kluwer, Dordrecht), 565
4.94   Pauls, T.A., Wilson, T.L., Bieging, J.H., Martin, R.N. 1983, *Astron. Astrophys.* **124**, 23
4.95   Pérault, M., Falgarone, E., Puget, J.L. 1985, *Astron. Astrophys.* **152**, 371

4.96   Phillips, T.G., Huggins, P.J. 1981, *Astrophys. J.* **251**, 533
4.97   Plambeck, R.L., Wright, M.C.H., Welch, W.J., Bieging, J.H., Band, B. et al. 1982, *Astrophys. J.* **259**, 617
4.98   Pudritz, R., Pelletier, G., Gomez de Castro, A. in *The Physics of Star Formation and Early Evolution*, eds. D.J. Lada, N.D. Kylafis, *NATO-ASI-342*, (Kluwer, Dordrecht), 539
4.99   Reipurth, B. 1991, in *The Physics of Star Formation and Early Evolution*, eds. C.J. Lada, N.D. Kylafis, *NATO-ASI-342*, (Kluwer, Dordrecht), 497
4.100 Rieke, G.H., Low, F.J., Kleinmann, D.E. 1973, *Astrophys. J.* **186**, L7
4.101 Sanders, D.B., Scoville, N.Z., Solomon, P.M. 1985, *Astrophys. J.* **289**, 373
4.102 Scalo, J., Pumphrey, W.A. 1982, *Astrophys. J.* **258**, L29
4.103 Scalo, J. 1990, in *Physical Processes in Fragmentation and Star Formation*, eds. R. Capuzzo-Dolcetta, C. Chiosi, A. deFazio (Kluwer, Dordrecht), 151
4.104 Schmid-Burgk, J., Güsten, R., Mauersberger, R., Schulz, A., Wilson, T.L. 1990, *Astrophys. J.* **362**, L25
4.105 Scoville, N.Z., Sanders, D.B. 1987, in *Interstellar Processes*, eds. D.J. Hollenbach, H.A. Thronson (Reidel, Dordrecht), 21
4.106 Shu, F.H., Adams, F.C., Lizano, S. 1987, *Ann. Rev. Astron. Astrophys.* **25**, 23
4.107 Snell, R.L. 1987, in *Star Forming Regions*, eds. M. Peimbert, J. Jugaku (Reidel, Dordrecht), 213
4.108 Solomon, P.M., Rivolo, A.R., Barrett, J., Yahil, A. 1987, *Astrophys. J.* **319**, 730
4.109 Spitzer, L. 1982, *Searching Between Stars* (Yale Univ. Press), 148
4.110 Stacey, G.J. et al. 1992, in press
4.111 Strom, K.M., Strom, S.E., Wolff, S.C., Morgan, J., Wenz, M. 1986 *Astrophys. J. Suppl. Ser.* **62**, 39
4.112 Stutzki, J., Genzel, R., Harris, A.I., Herman, J., Jaffe, D.T. 1988b *Astrophys. J.* **330**, L125
4.113 Stutzki, J., Stacey, G.J., Genzel, R., Harris, A.I., Jaffe, D.T., Lugten, J.B. 1988a, *Astrophys. J.* **332**, 379
4.114 Stutzki, J., Güsten, R. 1990, *Astrophys. J.* **356**, 513
4.115 Swade, D 1989, *Astrophys. J. Suppl. Ser.* **71**, 219
4.116 Thaddeus, P. 1991, in *Molecular Clouds*, eds. R.A. James, T.J. Millar, (Cambridge Univ. Press. Cambridge), 1
4.117 Vogel, S.N., Bieging, J.H., Plambeck, R.L., Welch, W.J., Wright, M.C.H. 1985, *Astrophys. J.* **296**, 600
4.118 Vrba, F.J., Strom, S.E., Strom, K.M. 1988, *Astron. J.* **96**, 680
4.119 Watson, D.M., Genzel, R., Townes, C.H., Storey, J.W.V. 1985, *Astrophys. J.* **298**, 316
4.120 Werner, M.W., Dinerstein, H.L., Capps, R.W. 1983, *Astrophys. J.* **265**, L13
4.121 Wilking, B., Lada, C.J. 1983, *Astrophys. J.* **274**, 698
4.122 Wilson, T.L., Walmsley, C.M. 1989, *Astron. Astrophys. Rev.* **1**, 141
4.123 Wynn-Williams, C.G., Genzel, R., Becklin, E.E., Downes, D. 1984, *Astrophys. J.* **281**, 271
4.124 Zinnecker, H. 1989, in *Evolutionary Phenomena in Galaxies*, ed. J. Beckman (Cambridge Univ. Press, Cambridge), 115
4.125 Ziurys, L.M., Turner, B.E. 1986, *Astrophys. J.* **300**, L19
4.126 Zuckerman, B., Kuiper, T., Rodriguez-Kuiper, E. 1976, *Astrophys. J.* **209**, L137
4.127 Zweibel, E.G. 1987, in *Interstellar Processes*, eds. D.J. Hollenbach, H.A. Thronson, (Reidel, Dordrecht), 195

5 Circumstellar Clouds and Masers

5.1 A Close-Up View of BN and IRc 2

Grasdalen's (1976) discovery of HI Brα emission from BN was the first strong evidence that the infrared source (size ≈ 0.1" or 46 AU, Dougados et al. 1992) contains a central star of temperature $\geq 10^4$ K. Radio observations demonstrate that the circumstellar HII region is very small (≤ 20 AU) with electron densities of $\geq 10^7\,\mathrm{cm}^{-3}$ (Moran et al. 1983). Weak radio emission has also been detected from IRc 2 which supports the view that this source contains a hot central star as well (Churchwell et al. 1987, Garay et al. 1987).

High-resolution near-infrared spectroscopy has given detailed information on the circumstellar environment of BN. Scoville et al. (1983) identify four distinct regimes in the circumstellar and line-of-sight gas. First there is the ultradense HII region discussed above with an ionized gas mass of $3 \times 10^{-6}\,M_\odot$. High-velocity wings in the infrared HI recombination lines ($\Delta v = 200\,\mathrm{km\,s}^{-1}$) probably imply supersonic ionized gas flows. Second, $2.3\,\mu$m high-J band-head emission of CO indicates the presence of a highly confined region of a very hot ($T \approx 3500$ K) and very dense $\left(n(\mathrm{H_2}) = 10^{12}\,\mathrm{cm}^{-3}\right)$ molecular gas that may be located directly adjacent to the ionized region. Third, emission in 2.3 and $4.6\,\mu$m ro-vibrational CO lines shows the presence of a molecular circumstellar envelope with $n(\mathrm{H_2}) = 10^7\,\mathrm{cm}^{-3}$ and temperature 600 K. This envelope is probably identical with the infrared "dust photosphere". Finally, there is lower excitation gas at $T \approx 150$ K from the molecular ridge and plateau in front of BN, apparent in CO absorption.

In the model of Scoville et al. BN is a hot, zero-age main sequence star with a UV emission rate equivalent to a B0.5 star ($T_{\mathrm{eff}} = 26,000$ K, $L = 10^4\,L_\odot$). Ionized gas is flowing out from the star at $\geq 100\,\mathrm{km\,s}^{-1}$, with an inferred mass outflow rate of $4 \times 10^{-7}\,M_\odot\,\mathrm{y}^{-1}$. The compact HII region in turn is surrounded by a dense circumstellar envelope or disk. The CO band-head emission comes from the inner rim (thickness a few AU) of this neutral envelope at a distance of about 20 AU from the star. A very interesting and puzzling result of the spectroscopic observations is the fact that BN's systemic velocity is $21\,\mathrm{km\,s}^{-1}$ LSR which implies a motion of $12\,\mathrm{km\,s}^{-1}$ relative to the molecular cloud. If BN is not in a binary system, it will leave the cloud core on a time scale of only a few thousand years. A similar conclusion also applies to IRc 2, whose systemic velocity of $5\,\mathrm{km\,s}^{-1}$ differs by $4\,\mathrm{km\,s}^{-1}$ from that of the cloud core.

Less is known about the nature of IRc 2, other than that it has a luminosity approaching $10^5\,L_\odot$ and a circumstellar HII region similar to that of BN. The main reason is that, owing to the combined effects of line-of-sight extinction (through the hot core) and the large size of the infrared photosphere (≈ 1"; Chelli et al. 1984, Lester et al. 1985), no detailed circumstellar infrared spectroscopy has been possible up to now. Recent $5\,\mu$m speckle imaging by Dougados et al. (1992) indicates that IRc 2 consists of several components separated spatially by less than 1". The spatial distribution and velocities of the

SiO masers on a scale of ≈ 0.2" are fit by a model of a rotating and expanding disk that is inclined at $\approx 45°$ with respect to the line of sight (Plambeck et al. 1990).

From modeling of the SiO maser associated with IRc 2, Elitzur (1982a) has derived a scenario for the circumstellar environment. In his model, the outflow starts at a radius of a few times 10^{14} cm from the star. The SiO masers occur at $R = 5 \times 10^{14}$ cm, where the outflow (driven by radiation pressure on grains) has fully developed. In order to pump the masers, the molecular hydrogen density must be just under 10^{12} cm^{-3}. Strong turbulent motions heat the gas above the temperature of the dust, a necessary condition for maser pumping. Elitzur derives a mass outflow rate of about 10^{-3} $M_\odot$ y^{-1}, consistent with the other estimates given above.

5.2 Circumstellar Disks in Low Mass Star Formation Regions

Evidence for circumstellar disks surrounding protostars and young stellar objects has been increasing steadily over the past few years, especially for low mass star forming regions (see Lada 1991). The very broad spectral energy distributions of many young stellar objects with comparable contributions to the luminosity in the submm, far-IR, mid-IR, near-IR, and visible wavebands can be best understood as a result of young pre-main sequence stars that are surrounded by passive or active disks (Adams et al. 1987, Beckwith et al. 1990). Perhaps the most convincing evidence for such disks comes from optical/infrared spectroscopy of FU Ori. Hartmann & Kenyon (1987) find that the velocity widths of optical lines are greater than those of infrared lines by an amount that is consistent with that expected for a Keplerian disk with a radial temperature gradient. The size of the disks inferred from the infrared/submillimeter dust emission is $\leq 10^2$ AU. In HL-Tau Beckwith & Sargent (1991) find a much larger (2000 AU) flattened structure in ^{13}CO $1 \to 0$ emission that appears to be in differential rotation about the star. The ^{13}CO disk may be an extension of the inner dust disk to larger radii. The observed circumstellar disks can be plausibly understood in the framework of current theoretical models of the formation of solar mass stars from dark cloud cores (Shu et al. 1987, Shu 1991).

5.3 Interstellar Masers

5.3.a Summary of Observations

Molecular interstellar masers, along with molecular outflows and Herbig-Haro objects are the most spectacular signposts of recent stellar formation. They are predominantly observed in high mass star formation regions. Since the discovery of OH and H_2O masers in the late 60's, several tens of cm, mm, and submm transitions of at least 7 molecules have been shown to require amplification by stimulated emission for explanation of their observed high intensities and narrow line profiles. There are many excellent reviews of observations and

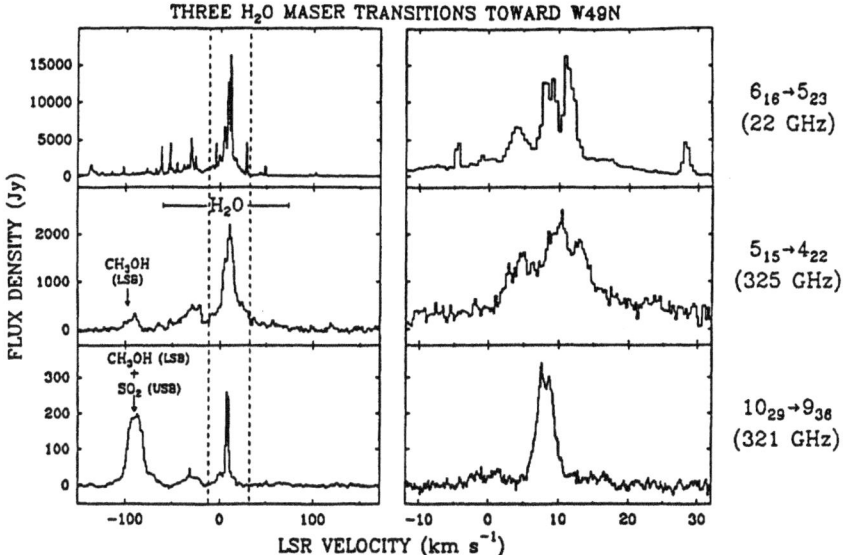

Fig. 26. Three different H_2O maser transitions toward the luminous star forming region W49. The left panels show a velocity range of $\pm 150\,\mathrm{km\,s^{-1}}$ while the right panels display the strongest features near the central velocity (adapted from Menten et al. 1990)

theory in the literature (for details see Reid & Moran 1981, 1988, Elitzur 1982b, Genzel 1986, Elitzur 1992).

Of the 5 strongly masing molecules with brightness temperatures exceeding 10^3 K (H_2O, OH, SiO, CH_3OH, H_2CO) the 22 GHz H_2O and 1.7 GHz OH maser transitions are the brightest ($T_B \approx 10^8 \rightarrow 10^{15}$ K), followed by CH_3OH masers. H_2O and OH masers have been detected toward several hundred Galactic and several ten extragalactic sources. The number of methanol masers has also been increasing rapidly. In the Galaxy strong masers correlate closely with signposts of recent OB star formation, such as compact HII regions (size $\leq$ 0.1 pc) and compact mid-IR sources. In the case of the 22 GHz H_2O masers there are many sharp features in their spectra that are spread over several tens and sometimes several hundred $\mathrm{km\,s^{-1}}$ (Fig. 26). These features are well explained by compact ($\approx 10^{13}$ cm), dense ($n(H_2) \approx 10^8$ to 10^{10} cm^{-3}) cloudlets moving at high velocity in mass outflows from a recently formed, massive star. This kinematic model was first discussed in detail by Strelnitskii & Sunyaev (1973). Final observational proof came with VLBI proper motion studies (Genzel et al. 1981a,b, Reid 1989). The three dimensional velocity field of the masers and their time variability imply a fast, clumpy wind that smashes into dense clumps in the surrounding cloud, and as a result creates very dense cloudlets, sheets or filaments. These compressed cloudlets are the maser spots. They have approximately planetary mass ($10^{26} \rightarrow 10^{29}$ g) and move outward and disperse on a dynamical time scale (≈ 1 to 10 years). Very recently, several other mm- and submm-transitions of H_2O have been shown to be masing in the same sources where 22 GHz masers are observed (Fig. 26, Menten et al. 1990 a,b, Cernicharo et al. 1990).

While it is difficult enough to theoretically account for the pumping of the strongest Galactic H_2O masers, such as W49 $(L(H_2O) \approx 0.1 \to 1\,L_\odot$ see below), a plausible explanation of the even more spectacular extragalactic H_2O masers $(\geq 10^2\,L_\odot)$ is presently lacking (see discussion of Henkel et al. 1991).

H_2O maser spots typically occur in 0.1 pc size clusters that are located close to but at some distance $(0.1-1\,\mathrm{pc})$ from compact HII regions. In contrast OH maser clusters tend to coincide with the HII regions. It is likely that the OH masers (size $\geq 10^{14}$ cm) are part of an (expanding) shell directly outside of, and perhaps compressed by the expanding shock/ionization front of the HII region (Elitzur & de Jong 1978, Norris & Booth 1981, Bloemhof et al. 1989). The same general explanation may also apply to Class II CH_3OH and the rarer H_2CO masers (Menten 1991, Elitzur 1992). The very luminous extragalactic OH "megamasers" $(L_{OH} \approx 1 \to 10^4\,L_\odot)$ are located in the nuclear regions of luminous IRAS galaxies and likely amplify the intense radio continuum background radiation (Baan 1991, Henkel et al. 1991).

The statistics of OH and H_2O masers in the Galaxy suggests that the maser phase coincides with the mass outflow/compact HII region phase of O and B (and perhaps lower mass) stars and lasts about 10^5 years (Genzel & Downes 1979).

5.3.b Theory

Interstellar masers are one pass, *travelling wave masers* without feedback (there are no end mirrors in interstellar clouds). If there is a population inversion between upper and lower levels $(\Delta n = n_u/g_u - n_l/g_l > 0)$, a radiation input entering the maser on one side (either a background photon or a photon created by spontaneous emission) is amplified through the maser cloud. Equations (11) and (12) can then be used again, but with the difference that the absorption coefficient is now negative. Its absolute value is called the *maser gain*,

$$g(x) = -\kappa(x) \propto (n_u - n_l) \propto \frac{\Delta P}{P}\, n_{\mathrm{mol}} \left[\frac{P}{P + B\bar{J}(x)}\right]. \qquad (122)$$

On the right side of Eq. (122) we have made the simplifying assumption that a "pump" source (radiation or collisions) is cycling the molecules through the various rotational levels at a rate $P\,[\mathrm{s}^{-1}]$. This mechanism is further assumed to pump more efficiently to the upper than to the lower maser level $(\Delta P = P_u - P_l > 0)$, thus creating the required population inversion. The maser pump has to compete with the rate of stimulated radiative downward transitions, given in Eq. (123) by $B\bar{J}$. Here B is the Einstein coefficient, and $\bar{J}(x)$ is the angle averaged line intensity at x. For simplicity we have averaged all quantities over the line profile. For a *beamed* maser of beam opening solid angle $\Omega(x)$ at x we have $\bar{J}(x) = I(x)\Omega(x)/4\pi$ where $I(x)$ is the peak line intensity at beam center.

If the pump rate is faster than the stimulated emission rate $(P > B\bar{J})$ the maser gain is independent of position, the intensity grows exponentially

$(I(x) \propto \exp(gx))$ and the maser is called *unsaturated*. Weak masers, such as the 3 GHz CH transitions (Rydbeck et al. 1976, Stacey et al. 1987), or the extragalactic OH megamasers are very likely unsaturated and amplify the continuum background (cf. Henkel et al. 1991). If $P < B\bar{J}$ the maser amplification is entirely controlled by the pump rate. The maser gain decreases with increasing x and the intensity only grows with a power law in x,

$$I(x) \propto \frac{\Delta P}{P}\, n(\mathrm{H_2O})\, x^3. \tag{123}$$

The dependence on x^3 is a result of the fact that for a cylinder or sphere $\Omega(x)$ scales proportional to x^2 (Reid & Moran 1988). A maser obeying Eq. (123) is called *saturated*. As the cycling/pumping rate cannot be faster than the rate for spontaneous emission in far-infrared rotational transitions ($A_{\mathrm{FIR}} \approx 10^{-2}$ to $1\,\mathrm{s}^{-1}$ for OH and $\mathrm{H_2O}$), saturation sets in at a brightness temperature of

$$T_B^{\mathrm{SAT}} \leq \left(\frac{A_{\mathrm{FIR}}}{A_{\mathrm{mas}}}\right)\left(\frac{h\nu_{\mathrm{mas}}}{k}\right)\left(\frac{4\pi}{\Omega}\right). \tag{124}$$

A_{mas} and ν_{mas} are Einstein coefficient and frequency of the maser transition. For the 1.7 GHz OH and 22 GHz $\mathrm{H_2O}$ masers $\left(A(\mathrm{OH}) = 7 \times 10^{-11}, A(\mathrm{H_2O}) = 2 \times 10^{-9}\,\mathrm{s}^{-1}\right)$ one then finds $T_B^{\mathrm{SAT}} \leq 10^{10}$ to 10^{11} K for $(\Omega/4\pi) \geq 10^{-3}$. Observed brightness temperatures are $10^{12\pm1}$ K for OH masers and $10^{13\pm2}$ K for $\mathrm{H_2O}$ masers. Hence, *strong* interstellar masers are almost certainly *saturated*.

Several lines of arguments suggest that strong masers are also strongly beamed: $(\Omega/4\pi) \approx 10^{-2}$ to 10^{-3}. For cylindrical $\mathrm{H_2O}$ masers of diameter d and length l on has $\Omega = (d/l)^2$, for instance, and a lower limit to the molecular hydrogen volume density can be derived from the observed maser intensity and reasonable upper limits to the efficiency of any pumping mechanism (Eq. (123), Genzel 1986, Elitzur 1992). This results in

$$n(\mathrm{H_2}) \geq 10^{14}\, N_{45}\, \Omega^{3/2}. \tag{125}$$

Here N_{45} is the emission rate of maser photons in units of $10^{45}\,\mathrm{s}^{-1}$ deduced from the observed line flux F_1 and source distance D *assuming isotropic emission* $(N = F_1\, 4\pi D^2/h\nu)$.

In many standard pumping schemes the hydrogen density has to be less than $\approx 10^{11}\,\mathrm{cm}^{-3}$, the critical density for thermalization of the $\mathrm{H_2O}$ rotational levels. For an average $\mathrm{H_2O}$ maser, $N_{45} \approx 1$, and one finds $\Omega \leq 10^{-2}$. The strongest galactic $\mathrm{H_2O}$ masers $N_{45} \geq 10^2$ imply even more extreme beaming. Other arguments in favor of beaming come from the observed clustering size and intensity ratio of masers at low and high radial velocities (Genzel 1986). Filamentary/sheet like structures are also fully consistent with the wind/cloud interaction model that was mentioned in the last paragraph.

Each maser photon requires at least one pump photon, leading to the so called Manley-Rowe relationship

$$T_{\mathrm{pump}} \geq \left(\frac{\Omega}{\Omega_{\mathrm{p}}}\right)\left(\frac{\nu_{\mathrm{mas}}}{\nu_{\mathrm{pump}}}\right) T_B^{\mathrm{MAS}}. \tag{126}$$

$\Omega_{\rm p}$ is the solid angle subtended by the maser cloudlet as seen from a possible external pump source, $\nu_{\rm pump}$ is the mean pump frequency and $T_{\rm pump}$ is the Rayleigh-Jeans brightness temperature of that pump source. It is now clear for H_2O masers that the maser cloudlets are far from external radiation sources ($\Omega_{\rm p} < 10^{-3}$). Together with the constraints on the maser beaming solid angle and the observed brightness temperatures, Eq. (126) then leads to the conclusion that H_2O *masers must be pumped internally* and cannot be pumped by external radiation sources.

Many detailed pumping models have been presented over the past twenty years since the discovery of the maser phenomenon. In the case of H_2O masers, *collisional* pumping by H_2 molecules *can* invert the 22 GHz transition (Chandra et al. 1985, Elitzur et al. 1989, Kylafis & Norman 1991) *as well* as the other recently discovered transitions (Neufeld & Melnick 1990) and is a natural result of the level structure of the H_2O molecule. The upper levels of all observed maser transitions lie along the "backbone" rotational ladder (the equivalent of the metastable ($J = K$) levels in a symmetric top molecule) that is more easily collisionally populated than the rest of the levels. The model of Elitzur et al. (1989) is based on shock excitation and compression which is also in good agreement with the other observational facts, as mentioned above. H_2O masers are located in the 400 K H_2 formation plateau of J-shocks (Sect. 2.2.a, Fig. 17). As is apparent from the simple considerations leading to Eq. (125), however, models involving collisions with H_2 fail to explain the most powerful masers unless very small values for the beaming solid angle are adopted. To overcome the problem of thermalization Strelnitskii (1984) proposed that H_2O molecules collide with two types of particles (H_2 and e) at different temperatures. Kylafis & Norman (1987) showed that this "two stream" model can account for strong masers in environments where the neutral gas is cooler than the electrons, such as in magnetic precursors of C-shocks (Sect. 2.2). More recently the two stream pump model has been questioned, however, as Anderson & Watson (1990) find that with the newest $H_2O - H_2$ collision rates (Palma et al. 1988) inversion does not occur.

In the case of 1665 MHz OH masers, collisional pumping alone is unlikely (Kylafis & Norman 1990). In a recent calculation Cesaroni & Walmsley (1991) find that most observational features of the OH maser emission in W3(OH) can be accounted for by combination of collisions, infrared continuum radiation and infrared line overlap.

Class I methanol masers (Menten 1991) are perhaps excited by collisions (Walmsley et al. 1988) while the pumping of class II methanol is currently not understood (Elitzur 1992).

References

5.1 Adams, F.C., Lada, C.J., Shu, F.H. 1987, *Astrophys. J.* **321**, 788
5.2 Anderson, N., Watson, W.D. 1990, *Astrophys. J.* **348**, L69
5.3 Baan, W.A. 1991, in *Proceedings of 3rd Haystack Conference*, eds. A.D. Haschick, P.T.P. Ho (Astr. Soc. Pac., San Francisco), 45
5.4 Beckwith, S.V.W., Sargent, A.I., Chini, R.S., Güsten, R. 1990, *Astron. J.* **99**, 924
5.5 Beckwith, S.V.W., Sargent, A.I. 1991, *Astrophys. J.*
5.6 Bloemhof, E.E., Reid, M.J., Moran, J.M. 1989, in *Physics and Chemistry of Interstellar Molecular Clouds*, eds. G. Winnewisser, J.T. Armstrong (Springer, Berlin), 228
5.7 Cernicharo, J., Thum, C., Hein, H., John D., Garcia, P., Mattioco, F. 1990, *Astron. Astrophys.* **231**, L15
5.8 Cesaroni, R., Walmsley, C.M. 1991, *Astron. Astrophys.* **241**, 537
5.9 Chandra, S., Kegel, W.H., Warshalovich, D.A. 1985, *Astron. Astrophys.* **148**, 145
5.10 Chelli, A., Perrier, C., Léna, P. 1984, *Astrophys. J.* **280**, 163
5.11 Churchwell, E., Felli, M., Wood, D.O., Massi, M. 1987, *Astrophys. J.* **321**, 516
5.12 Dougados, C., Léna, P., Ridgway, S., Christou, J. 1992, in *High Resolution Imaging by Interferometry*, ed J. Beckers, ESO Reports, in press
5.13 Elitzur, M., de Jong, T. 1978, *Astron. Astrophys.* **67**, 323
5.14 Elitzur, M. 1982a, *Astrophys. J.* **262**, 189
5.15 Elitzur, M. 1982b, *Rev. Mod. Phys.* **54**, 1225
5.16 Elitzur, M., Hollenbach, D.J., McKee, C.F. 1989, *Astrophys. J.* **346**, 983
5.17 Elitzur, M. 1992, *Ann. Rev. Astron. Astrophys.* **30**, in press
5.18 Garay, G., Moran, J.M., Reid, M.J. 1987, *Astrophys. J.* **314**, 535
5.19 Genzel, R., Downes, D. 1979, *Astron. Astrophys.* **72**, 234
5.20 Genzel, R., Reid, M.J., Moran, J.M., Downes, D. 1981a, *Astrophys. J.* **244**, 884
5.21 Genzel, R. et al. 1981b, *Astrophys. J.* **247**, 1039
5.22 Genzel, R. 1986, in *Masers, Molecules, and Mass Outflows in Star Forming Regions*, ed. A.D. Haschick (Haystack Observatory), 233
5.23 Grasdalen, G.L. 1976, *Astrophys. J.* **205**, L83
5.24 Henkel, C., Baan, W.A., Mauersberger, R. 1991, *Astron. Astrophys. Rev.* **3**, 47
5.25 Hartmann, L., Kenyon, S.J. 1987, *Astrophys. J.* **323**, 714
5.26 Kylafis, N.D., Norman, C. 1987, *Astrophys. J.* **323**, 346
5.27 Kylafis, N.D., Norman, C. 1990, *Astrophys. J.* **350**, 209
5.28 Kylafis, N.D., Norman, C. 1991, *Astrophys. J.* **373**, 525
5.29 Lada, C.J. 1991, in *The Physics of Star Formation and Early Stellar Evolution*, eds. C.J. Lada, N.D. Kylafis, *NATO-ASI-342*, (Kluwer, Dordrecht), 329
5.30 Lester, D.F., Becklin, E.E., Genzel, R., Wynn-Williams, C.G. 1985, *Astron. J.* **90**, 2331
5.31 Menten, K.M., Melnick, G.J., Phillips, T.G. 1990a, *Astrophys. J.* **350**, L41
5.32 Menten, K.M., Melnick, G.J., Phillips, T.G., Neufeld, D.A. 1990b, *Astrophys. J.* **363**, L27
5.33 Menten, K.M. 1991, in *Proceedings of 3rd Haystack Conference*, eds. A.D. Haschick, P.T.P. Ho (Astr. Soc. Pac, San Francisco), 119
5.34 Moran, J.M., Garay, G., Reid, M.J., Genzel, R., Wright, M.C.H., Plambeck, R.L. 1983, *Astrophys. J.* **271**, L31
5.35 Neufeld, D.A., Melnick, G.J. 1990, *Astrophys. J.* **368**, 215
5.36 Norris, R.P., Booth, R.S. 1981, *Monthly Notices Roy. Astron. Soc.* **195**, 213
5.37 Palma, A., Green, S., deFrees, D.J., McLean, A.D. 1988, *Astrophys. J. Suppl. Ser.* **68**, 287
5.38 Plambeck, R.L., Wright, M.C.H., Carlstrom, J.E. 1990, *Astrophys. J.* **348**, L65
5.39 Reid, M.J., Moran, J.M. 1981, *Ann. Rev. Astron. Astrophys.* **19**, 231
5.40 Reid, M.J., Moran, J.M. 1988, in *Galactic and Extragalactic Radio Astronomy*, eds. G.L. Verschuur, K.I. Kellermann (Springer, Berlin), Chap. 6
5.41 Reid, M.J. 1989, in *The Center of the Galaxy*, ed. M. Morris (Kluwer, Dordrecht), 37
5.42 Rydbeck, O.E.H., Kollberg, E., Hjalmarson, Å., Sume, A., Elldér, J., Irvine, W.M. 1976, *Astrophys. J.* **31**, 333
5.43 Scoville, N.Z., Kleinmann, S.G., Hall, D.N.B., Ridgway, S.T. 1983, *Astrophys. J.* **275**, 201

5.44   Shu, F.H., Adams, F.C., Lizano, S. 1987, *Ann. Rev. Astron. Astrophys.* **25**, 23

5.45   Shu, F.H. 1991, in *The Physics of Star Formation and Early Stellar Evolution*, eds. C.J. Lada, N.D. Kylafis, *NATO-ASI*-**342** (Kluwer, Dordrecht), 365

5.46   Stacey, G.J., Lugten, J.B., Genzel, R. 1987, *Astrophys. J.* **313**, 859

5.47   Strelnitskii, V.S., Sunyaev, R.A. 1973, *Sov. Astron.* **16**, 579

5.48   Strelnitskii, V.S. 1984, *Monthly Notices Roy. Astron. Soc.* **207**, 339

5.49   Walmsley, C.M., Batrla, W., Matthews, H.E., Menten, K.M. 1988, *Astron. Astrophys.* **197**, 271

6 Molecular Clouds in the Galactic Center and External Galaxies

Most of the molecular gas in the disk of our Galaxy is contained in cool $(T \approx 10\,\mathrm{K})$ giant molecular clouds of volume averaged hydrogen density $\approx 200\,\mathrm{cm}^{-3}$ (Scoville & Sanders 1987). These clouds appear to be forming stars with low average efficiency. A global view of this low excitation interstellar medium has been recently possible through the beautiful results of the COBE mission (Wright et al. 1991). A schematic of the Galactic line and continuum infrared spectrum as seen by COBE is shown in Fig. 27. In addition to the dust continuum from the atomic and molecular medium peaking at 100 to $200\,\mu\mathrm{m}$ (with a secondary peak by PAH's between 10 and $20\,\mu\mathrm{m}$), one recognizes CO emission from the $\approx 10\,\mathrm{K}$ molecular cloud medium, [CI] emission from cloud interfaces and [CII] emission from a combination of PDR's, diffuse HI and diffuse HII gas (Wright et al. 1991, Gry et al. 1991). Of particular interest is the strong [NII] emission which indicates the ubiquitous presence of a low excitation, diffuse HII medium (Wright et al. 1991, Gry et al. 1991). The characteristics of the gas in the central few hundred parsecs of our Galaxy and the nuclei of luminous external galaxies is quite different (see review of Henkel et al. 1991).

6.1 Our Galaxy: The Central 500 pc

There is a substantial concentration of interstellar material near the Galactic nucleus. Within about 500 pc of the Galactic center, the mass of molecular material is about $10^8\,M_\odot$ (Güsten 1989). This corresponds to about 10% of the entire amount of molecular gas in the Galaxy and an *average* hydrogen density of $\approx 10^2\,\mathrm{cm}^{-3}$ (the average density in the disk is about $1\,\mathrm{cm}^{-3}$). This concentration is very obvious in surveys of $60/100\,\mu\mathrm{m}$ dust emission or CO emission (Burton, this volume).

Most of the gas in the Galactic center is in giant clouds ($10^5 \rightarrow 10^7\,M_\odot$) that are significantly denser ($\langle n(H_2)\rangle \approx 10^3 - 10^4\,\mathrm{cm}^{-3}$) and warmer ($\geq 40\,\mathrm{K}$) than in the disk (Güsten et al. 1989). While high density tracing CS $1 \rightarrow 0/2 \rightarrow 1$ emission in the disk typically only arises in small regions near the cloud cores (see Fig. 24), it is seen throughout the clouds in the Galactic center (Tsuboi et al. 1989, Stark et al. 1989). The derived gas temperatures ($40 \rightarrow 150\,\mathrm{K}$) are significantly higher than the temperature of the dust grains ($< 30\,\mathrm{K}$) thus excluding dust-gas collisions as the main heating mechanism (Güsten 1989, Genzel et al. 1990).

What is the reason for these different physical properties? Higher cosmic ray ionization/heating (about two orders of magnitude greater than in the disk!), greater UV heating through enhanced massive star formation (as in starburst galaxies, see below) or increased ambipolar diffusion heating can all be excluded with some certainty. The COS-B γ-ray observations toward the center show, if anything, a *decrease* in inferred cosmic ray density (Blitz et al. 1985). As is

FIR/SUBMM EMISSION OF THE GALAXY

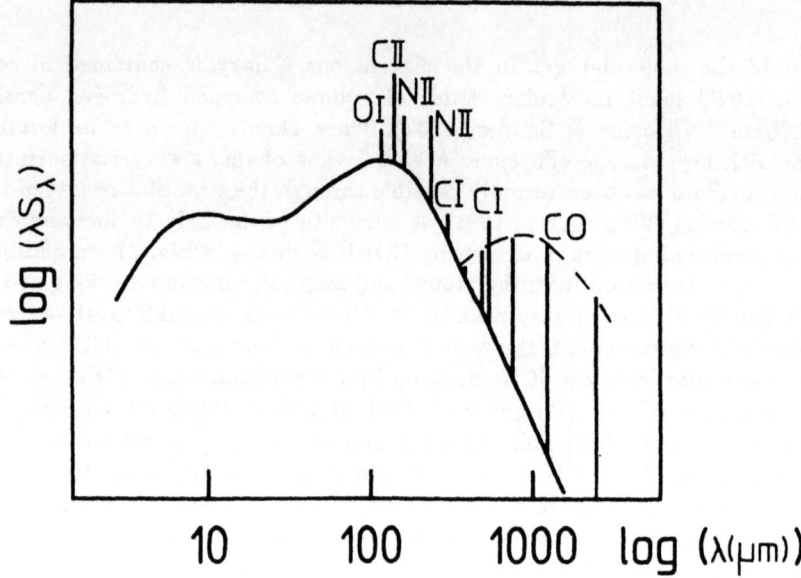

Fig. 27. Schematic of the FIR/submm line and continuum emission as seen by the COBE satellite (Wright et al. 1991)

evident from the fairly low dust temperature the *average* star forming activity in the central 500 pc is fairly low at present (Morris 1989, Cox & Laureijs 1989). Finally, despite greater magnetic fields in the central 10^2 pc ($10^{2\pm1}$ μG), magnetic field slippage and thus ambipolar heating is reduced because of the greater densities. Hence gas temperatures much in excess of 10 K cannot be explained by ambipolar heating.

Perhaps a key is that the Galactic center clouds, in addition to being denser and warmer, also have greater velocity widths than disk clouds ($\Delta v_{\mathrm{FWHM}} = 15$ to 30 km s^{-1} viz. 5 to 10 km s^{-1}). This fact suggests dissipation of clump-clump turbulence created by the strong differential rotation as a promising heating mechanism (cf. Güsten 1989). The turbulent volume heating rate is

$$\Gamma_{\mathrm{tur}} = \tfrac{1}{2}\left(\frac{\rho\,\Delta v_t^3}{L}\right) \qquad [\mathrm{erg\,s^{-1}\,cm^{-3}}], \tag{127}$$

where ρ is the density and L is the dissipation scale length of the turbulence. With $\Delta v_t \approx 10$ km s^{-1} and $L = 30$ pc one then obtains a gas temperature of 40 K from the heating-cooling balance (Chap. 2). For this mechanism to work, the time scale for cloud shearing due to differential rotation t_{shear} at galactocentric radius R must be smaller than the turbulent dissipation time scale t_{diss},

$$t_{shear} = \frac{R}{v_{rot}} \leq \frac{L}{\Delta v_t}. \tag{128}$$

In the Galactic center this condition probably applies. For $R = 100\,\mathrm{pc}$ and $v_{rot} = 150\,\mathrm{km\,s^{-1}}$, Eq. (128) gives $t_{shear} = 7 \times 10^5\,\mathrm{y} < t_{diss} = 1.5 \times 10^6\,\mathrm{y}$. In the disk at $R = 5\,\mathrm{kpc}$, however, $t_{shear} = 2 \times 10^7\,\mathrm{y} > t_{dis} = 5 \times 10^6\,\mathrm{y}$.

A plausible explanation for the high gas densities of the clouds in the center may come from the requirement of stability against tidal disruption (Stark et al. 1989). The well known "Roche"-criterion applied to Galactic center clouds requires

$$n(H_2) \geq \frac{3\,v_{rot}^2}{2\pi\,m(H_2)\,GR^2} \approx 5 \times 10^3 \left(\frac{R}{10^2\,\mathrm{pc}}\right)^2 \quad [\mathrm{cm^{-3}}], \tag{129}$$

consistent with the finding that the identifiable Galactic center clouds have gas densities of $\approx 10^4\,\mathrm{cm^{-3}}$ (Stark et al. 1989).

A significant fraction of the molecular/atomic gas in the Galactic nucleus appears to move on non-circular orbits (Burton, this volume). Binney et al. (1991) show in a recent detailed theoretical analysis that the observed gas motions can be well accounted for by non-circular, box, cusp, and polar orbits in a bar potential. In their model gas is streaming into the central $10^2\,\mathrm{pc}$ as a result of dispersion of energy through cloud-cloud collisions and shocks.

The observational evidence is also increasing that there is a sizeable mass influx into the central few parsecs ($\approx 10^{-3}$ to $10^{-2}\,M_\odot\,\mathrm{y^{-1}}$, Lo & Claussen 1983, Jackson et al. 1991, Genzel 1989). Massive molecular clouds located only a few arcmin away from the center may feed gas into the circum-nuclear environment (Ho et al. 1991, Jackson et al. 1991). The ionized central 1.5 pc is surrounded by a dense ring or disk-like structure consisting predominantly of warm molecular material: the "circum-nuclear disk" (Genzel 1989). While the dominant motion of the circum-nuclear gas is rotation, non-circular streaming motions are clearly detected and are quite prominent for the ionized/atomic gas within the inner rim of the circum-nuclear disk (Serabyn & Lacy 1985, Lacy et al. 1991, Jackson et al. 1991).

What is happening to the gas approaching the central few parsecs? Clearly the probability for collisions between clouds or clumps is increasing with decreasing radius, probably allowing a fraction of the material to loose angular momentum and then falling in yet closer to the center. The evidence for formation of $\approx 10 \rightarrow 100$ massive stars no longer than a few million years ago is now rather compelling (Rieke et al. 1989, Allen et al. 1990, Krabbe et al. 1991), suggesting that a sizeable fraction of the infalling gas is rapidly (and perhaps efficiently) converted to newly formed stars.

6.2 Molecular Gas in Starburst Galaxies

Because of substantial improvements in sensitivity in the past ten years studies of molecular line emission in external galaxies have become widely possible. The reader is referred, for example, to the recent reviews by Young & Scoville (1991) and Henkel et al. (1991) for summaries of the large amount of information now available. In the following three paragraphs I will highlight three cases, namely studies of the starburst galaxy M82, of PDR's in external galaxies, and of the nuclear region of the Seyfert galaxy NGC 1068.

Figure 28 shows a composite infrared and radio spectrum of the central 700 pc of the nearby irregular galaxy M82. This spectrum is significantly different from the low excitation spectrum of our own Galaxy (Fig. 27) and is very reminiscent of that of the Orion star forming region. The available observations are well explained by the original proposal of Rieke et al. (1980) that the central kpc of M82 is a giant star forming site, a "starburst" nucleus. In the starburst scenario most of M82's $\approx 2 \times 10^{10} L_{\odot}$ far-infrared luminosity comes from OB stars that have formed within the last 10^7 years. A large central concentration of young giants and supergiants evident from $2\,\mu$m observations suggests that the active star formation phase of M82 started about 10^8 years ago (Rieke et al. 1980). The current star formation rate within 500 pc is $\approx 5\,M_{\odot}\,y^{-1}$, or about ten times that of the nuclear region in our own Galaxy.

From comparison of star formation rate, number of giants and total number of stars derived from the rotation curve Rieke et al. (1980) also found that predominantly massive stars ($> 3\,M_{\odot}$) are formed during the burst phase. While the exact value of the lower mass cutoff is uncertain because of uncertainties in the $2\,\mu$m extinction (Scalo 1990), the qualitative conclusion of a lower mass cutoff in the initial mass function of starburst galaxies seems plausible (see Scalo 1990 for a critical discussion).

The intensities of the various spectral lines in Fig. 28 also show that the neutral interstellar medium in M82 is dense (intense [OI] and excited rotational lines of HCN, HCO$^+$, etc.), warm (excited rotational lines of CO) and there is a a lot of it (strength of ^{13}CO, C^{18}O lines). About 1 to $2 \times 10^8\,M_{\odot}$ of molecular material is concentrated within the central 1' (≈ 1 kpc). The distribution of ^{12}CO mm emission is double-peaked and centered on the near-infrared nucleus (Fig. 29). The two peaks have been interpreted as evidence for a $R \approx 200$ pc ring or torus of molecular material within which the interstellar material has been removed by the intense star formation activity (Nakai et al. 1987, Sofue 1988). Recent molecular line and submillimeter dust continuum mapping, however, show a prominent peak within 10" of the nucleus suggesting that the central 500 pc are not devoid of molecular material (Loiseau et al. 1988, Carlstrom 1989, Wild et al. 1991, Smith et al. 1990).

The coincidence of the central gas concentration with the starburst zone is evident from Fig. 30 which shows a high resolution radio continuum map (Kronberg et al. 1985) superposed on a gray-scale representation of the ^{12}CO map (from Lo et al. 1987). The compact, non-thermal radio continuum sources are identified as young, expanding radio supernova remnants that are formed at a rate of 0.1 to $0.3\,y^{-1}$ (Kronberg et al. 1985). The ratio of gas to total stellar mass in the central few hundred parsec of M82 is very high, ≈ 0.2 as required by and consistent with the starburst scenario.

The high temperature of the molecular ISM in M82 is evident from the brightness temperature of ^{12}CO emission lines at a few arcsec resolution (15 −

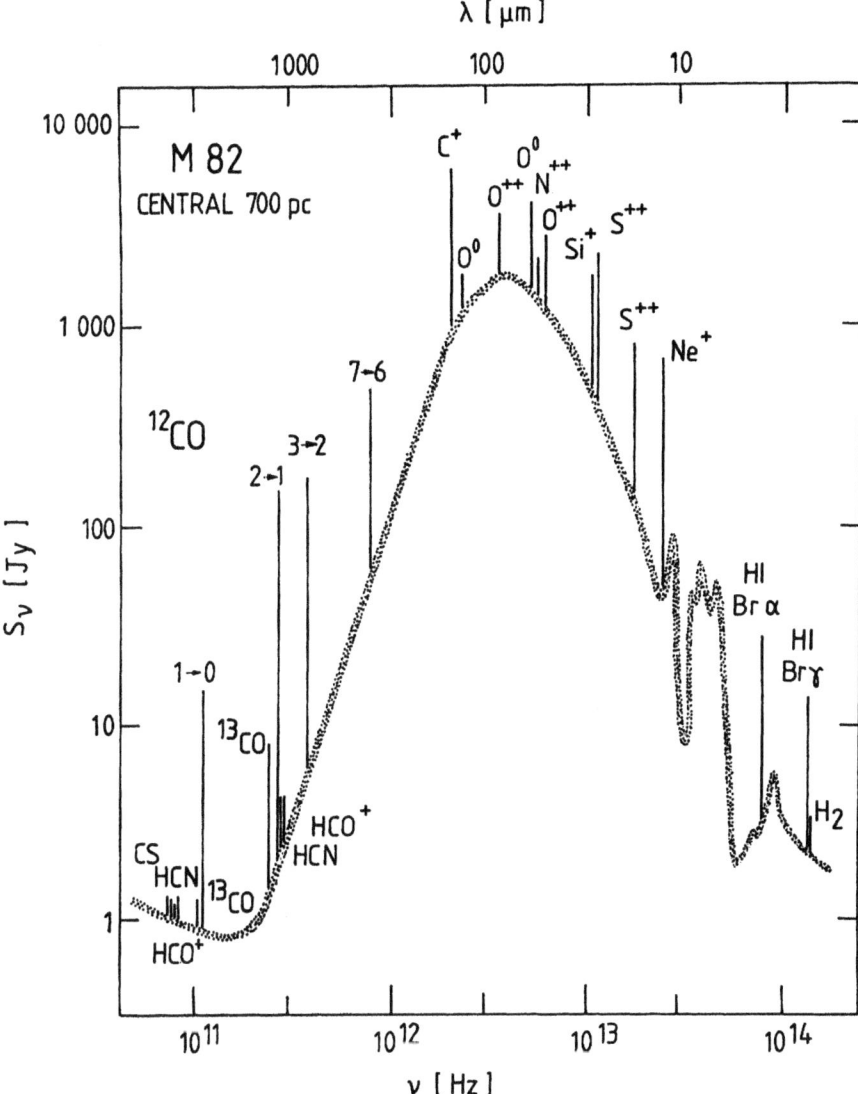

Fig. 28. Composite infrared and radio spectrum of the starburst galaxy M82

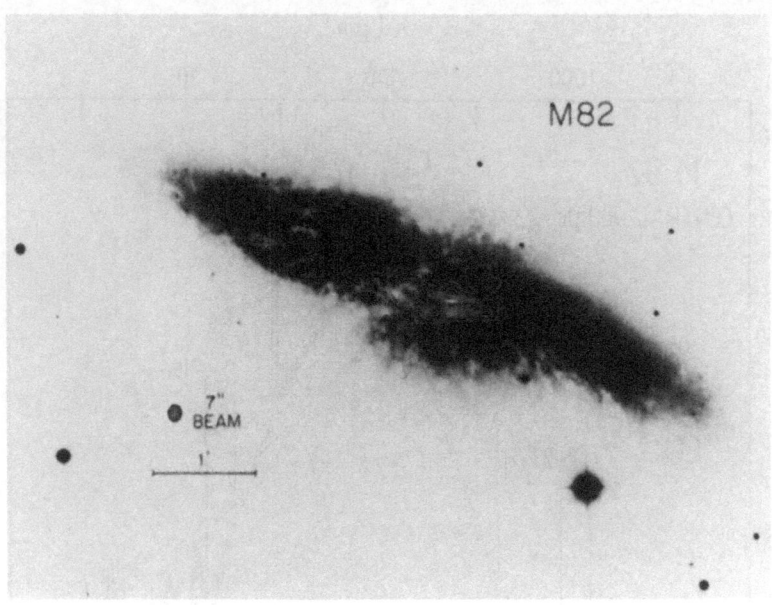

Fig. 29. CO $1 \rightarrow 0$ aperture synthesis map (resolution 7"), superposed on a optical photograph of M82 (adapted from Lo et al. 1987)

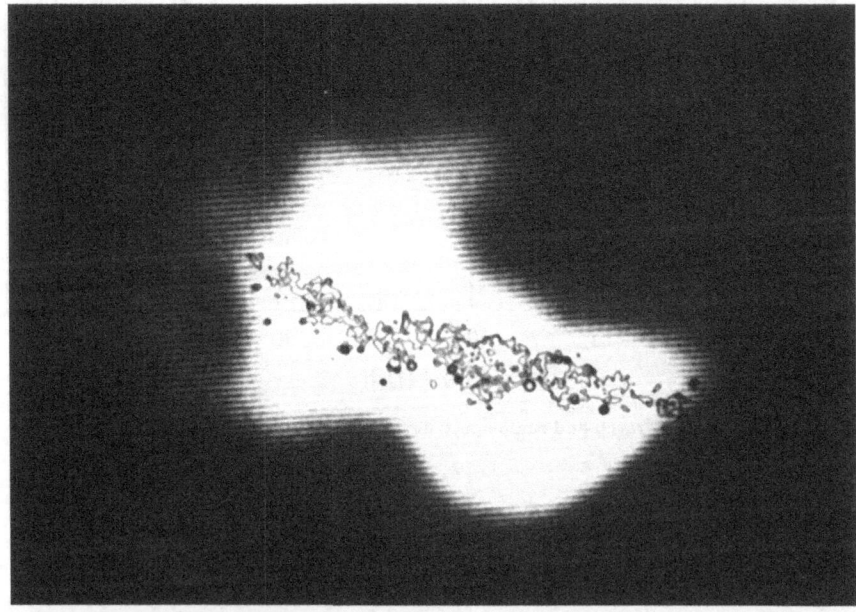

Fig. 30. High resolution 5 GHz radio continuum map (Kronberg et al. 1985) superposed on a gray-scale presentation of the ^{12}CO map of Fig. 29 (adapted from Lo et al. 1987)

MOLECULAR ISM IN GALAXIES

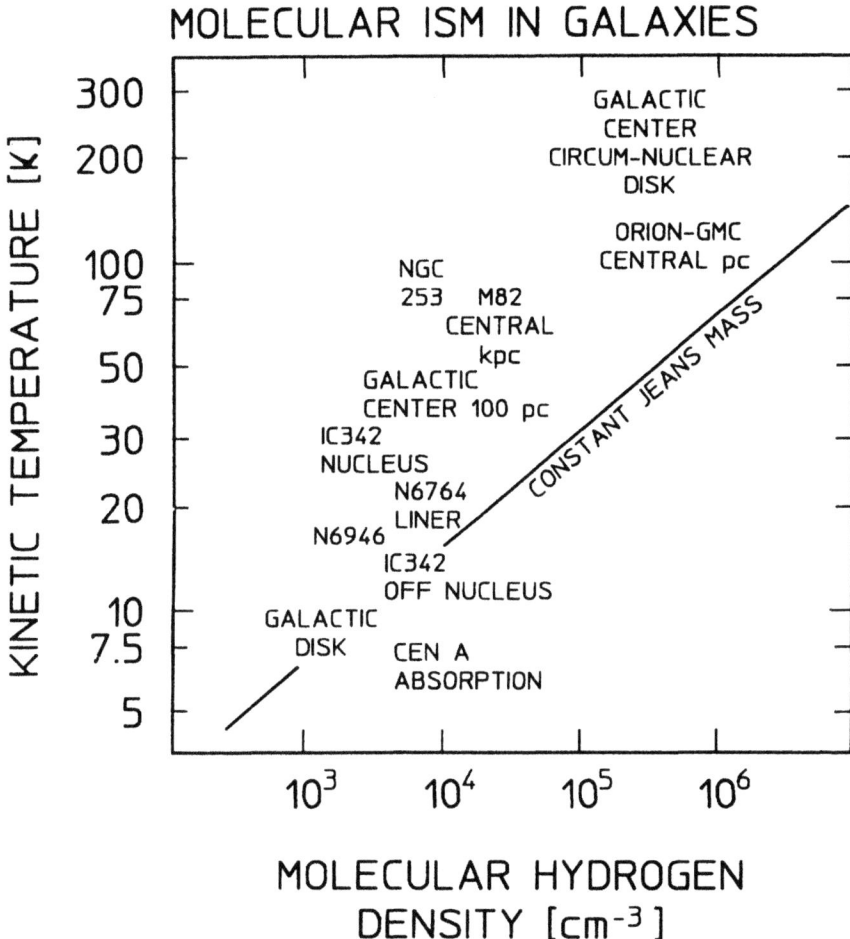

Fig. 31. Summary of physical properties derived for the molecular ISM's of several IR luminous galaxies, in comparison to our own Galaxy. The density-temperature points are derived from multi-line analysis and refer to the nuclei, unless noted otherwise (Eckart et al. 1990a,b, 1991a,b, Genzel et al. 1985, Genzel 1990, Güsten 1989, Henkel et al. 1991)

20 K, Lo et al. 1987, Carlstrom 1989) and in particular, from the strength of recently observed ^{12}CO $6 \rightarrow 5$ emission (Harris et al. 1991). A single component analysis of the ^{12}CO lines observed to date indicates bulk cloud temperatures of at least 40 K.

Similarly high temperatures are also implied by measurements of CH_3CN, and CH_3CCH lines (Mauersberger et al. 1991). A multi-line excitation analysis of isotopic CO emission fits a single-component model with $T \approx 40 \rightarrow 50$ K and $n(H_2) \approx 10^4$ cm^{-3} (Wild et al. 1991). Intense rotationally excited emission of HCN, CS, HC$_3$N, etc. suggest that a substantial fraction ($\approx 10\%$) of the gas is at even higher density, $10^5 \rightarrow 10^6$ cm^{-3} (Wild et al. 1991, Henkel et al. 1991). While single temperature, single density models are clearly not a realistic description of the interstellar medium, these estimates nevertheless indicate that the molecular cloud medium of M82 is much more excited than that in the disk of our Galaxy. These conditions are plausibly connected to the intense star forming activity, but we have seen above that the ISM in the central few hundred parsecs of our Galaxy is also warmer and denser than in the disk, *not* as a result of star formation, but much more likely as a consequence of tidal shearing in the steep gravitational potential.

A number of detailed excitation studies in other starburst galaxies (NGC 253, IC 342, NGC 6764) have recently become available and generally confirm the conclusions for M82 (see Henkel et al. 1991 for a review, Eckart et al. 1990a,b, 1991a,b). Other starburst nuclei recently investigated by detailed mm interferometry of molecular emission are IC 342 (Ishizuki et al. 1990), NGC 253 (Canzian et al. 1988, Carlstrom et al. 1990), and NGC 3690/Arp 299 (Sargent & Scoville 1991). An overview of the derived average physical conditions in several well investigated galaxies in comparison to our own Galaxy is given in Fig. 31. A wide range in the parameters of the interstellar media of galaxies is already indicated from this small selection. If the Jeans mass criterion is a guide, the large implied variation in physical parameters of the interstellar gas may be reflected in a comparable variation of the initial mass function.

6.3 Neutral Gas in Photon Dominated Regions

Advances in infrared spectrometers have made external galaxies accessible for study in several of the prominent far-infrared cooling lines. Crawford et al. (1985) studied 7 nearby, infrared luminous galaxies (including M82) and found that the 158 μm [CII] line is bright toward the nuclei and typically contains about 0.5% of the bolometric luminosity. Crawford et al. (1985) concluded that the [CII] emission in these galaxies comes from fairly dense photon dominated regions at the surfaces of molecular clouds that are created by embedded or nearby OB stars.

Stacey et al. (1991) extended the [CII] observations to about 20 galaxies, with a wider range of characteristics. As shown in Fig. 32 they find — as did Crawford et al. (1985) — that the [CII] and CO $1 \rightarrow 0$ line emission in infrared luminous spiral galaxies correlate well and can be explained in terms of dense PDR's created by OB stars (cf. Wolfire et al. 1989). In starburst galaxies, up to 30% of the ISM can be contained in such PDR's. In contrast, more normal galaxies have a significantly smaller [C II]/CO ratio (Fig. 32). Stacey et al. (1991) propose that the [CII]/CO ratio can be used as a qualitative measure of star formation activity in galaxies.

Madden et al. (1992) have recently presented the first detailed, two dimensional [CII] map of a galaxy. Figure 33 shows the [CII] distribution in NGC 6946, a fairly low excitation, nearby Sc galaxy. In addition to [CII] emission from the nucleus and the spiral arms, there is also extended ($\pm 3'$ or ± 9 kpc) [CII] emission that dominates the overall [CII] luminosity. Madden et al. (1991) conclude that the extended [CII] emission originates in the HI medium and extended low

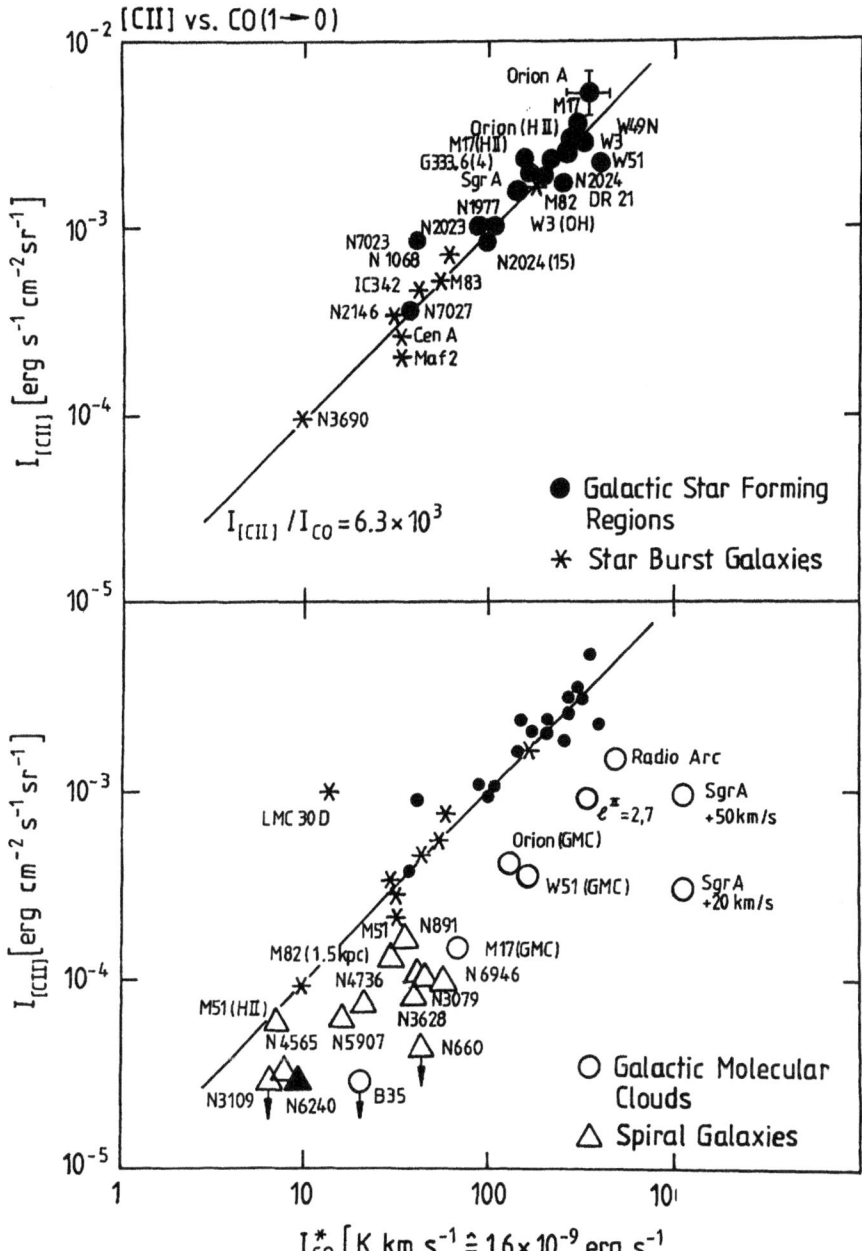

Fig. 32. Correlation between 158 μm [CII] and 2.6 mm CO 1→0 line intensities for starburst galaxies and Galactic star formation regions (top), and for other spiral galaxies and Galactic GMC's (bottom, adapted from Stacey et al. 1991)

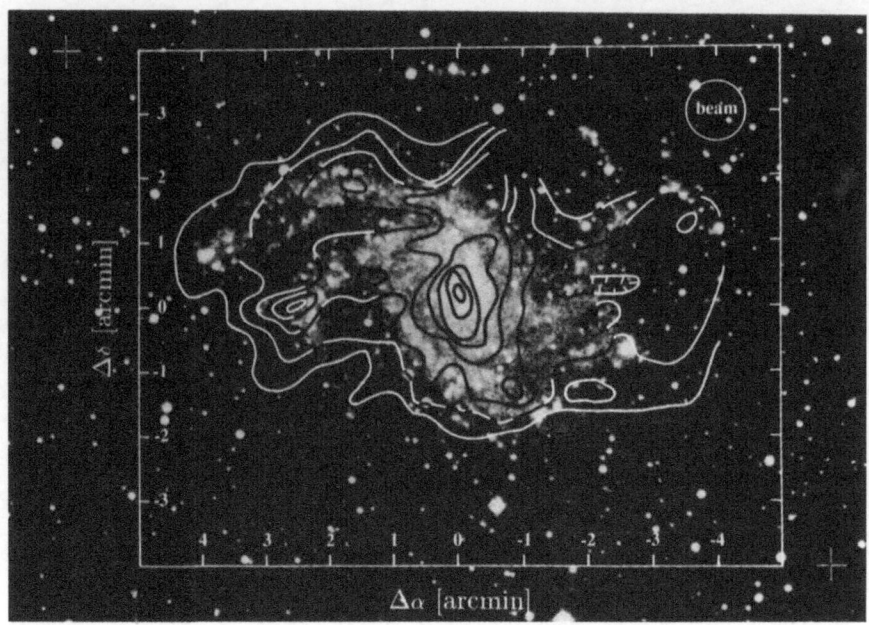

Fig. 33. Map of 158 μm [CII] line emission from NGC 6946 (adapted from Madden et al. 1992)

density HII regions, in agreement with the COBE results for our Galaxy. Madden et al. deduce a pressure of about $5 \rightarrow 10 \times 10^3 \, \mathrm{K \, cm^{-3}}$ for the HI medium, similar to our own Galaxy, and infer that the atomic ISM of NGC 6946 has to be fairly clumpy or filamentary, perhaps similar to the "cirrus" in our Galaxy.

6.4 Molecular Gas in AGN's

As a last example in this brief review of our knowledge of the neutral ISM in external galaxies, I want to comment on recent work in AGN's. Evidence is increasing that Seyfert galaxies and QSO's have large amounts of molecular gas in their nuclei (Sanders et al. 1986, Young & Scoville 1991). Interferometric observations of ^{12}CO $1 \rightarrow 0$ show fairly intense emission within $5 - 10$" of the nuclei of a number of Seyfert galaxies (Meixner et al. 1990, Planesas et al. 1991). Strong molecular emission is also found toward the nucleus of the archetypal superluminous IRAS galaxy Arp 220 (Scoville et al. 1991, Radford et al. 1991). Barvainis et al. (1989) infer a molecular mass of $\approx 10^{10} \, M_\odot$ in the host galaxy of the QSO I Zw1.

Figure 34 shows a ≈ 1" resolution image of $2 \, \mu m$ H_2 vibrational emission toward the nucleus of the Seyfert 2 galaxy NGC 1068 (Blietz et al. 1992). There is a few arcsec diameter, double-lobed H_2 source centered on the central radio source that is coincided with the active nucleus. The central H_2 molecular concentration is also apparent on the CO $1 \rightarrow 0$ interferometer map of Planesas

388

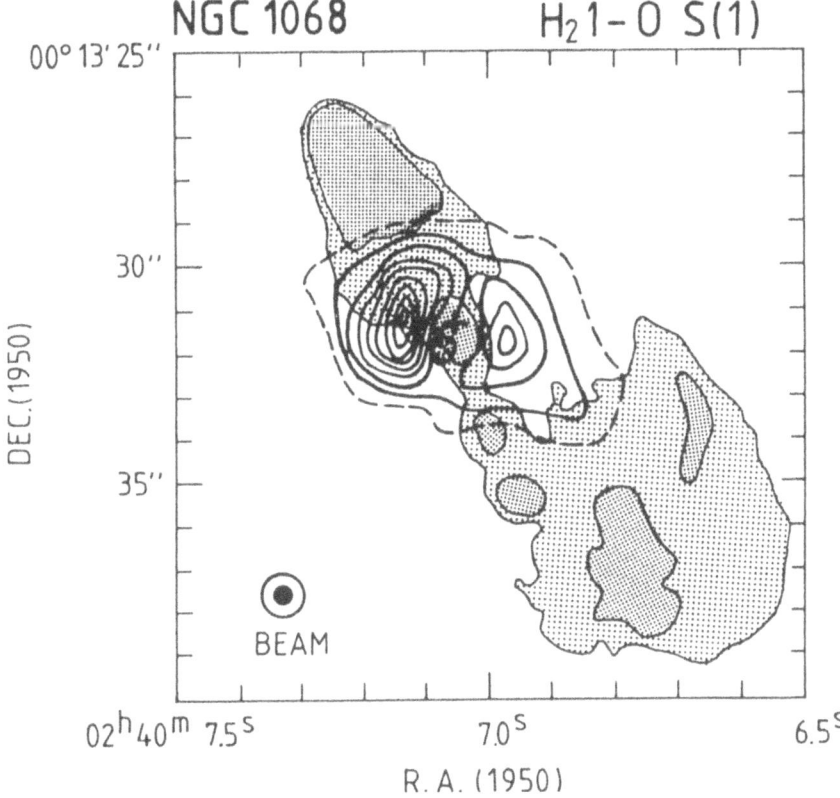

Fig. 34. $2\,\mu m$ H_2 $S(1)$ emission at 0.8" resolution in the nucleus of the Seyfert galaxy NGC 1068, superposed on a VLA map of the radio continuum (Rotaciuc et al. 1991, Blietz et al. 1992)

et al. (1991). The H_2 torus appears to be spatially anticorrelated with the radio jet (Wilson & Ulvestad 1983). The V-shaped structure of the H_2 emission north of the nucleus correlates with a cone structure whose existence is inferred from observations of narrow-line clouds in the visible (Hofmann et al. 1989, Ebstein et al. 1989, Evans et al. 1991). The H_2 emission probably comes from a few 10^2 pc size disk or ring of $\approx 10^7 \rightarrow 10^8\,M_\odot$ of molecular material surrounding the nucleus. It is likely that the H_2 torus is the long sought far confining medium for jet and expanding narrow line clouds (Cecil et al. 1990). Rotaciuc et al. (1991) propose that the molecular gas is heated by X-rays and UV radiation from the AGN. H_2 emission zones of a few arcsec diameter have recently been found in a number of very luminous galaxies and AGN. They clearly show molecular gas exposed to an environment that is very different from that in our own Galaxy.

Acknowledgements. I am very grateful to Barbara Schnelle who put in many long hours for creating and revising the text of this manuscript. I also thank M. Blietz, M. Cameron, A. Krabbe, D. Lutz, and P. van der Werf for letting me use recent results prior to publication, and J. Stutzki for communicating references for collision cross sections.

References

6.1 Allen, D.A., Hyland, A.R., Hillier, D.J. 1990, *Monthly Notices Roy. Astron. Soc.* **244**, 706

6.2 Barvainis, R., Alloin, D., Antonucci, R. 1989, *Astrophys. J.* **337**, L69

6.3 Binney, J., Gerhard, O.E., Stark, A.A., Bally, J., Uchida, K.I. 1991, *Monthly Notices Roy. Astron. Soc.* **252**, 210

6.4 Blietz, M. et al. 1992, in prep

6.5 Blitz, L., Bloemen, J.B.G.M., Hermsen, W., Bania, T.M. 1985, *Astron. Astrophys.* **143**, 267

6.6 Canzian, B., Mundy, L.G., Scoville, N.Z. 1988, *Astrophys. J.* **333**, 157

6.7 Carlstrom, J.E. 1989, Ph.D. Thesis, Univ. of California, Berkeley

6.8 Carlstrom, J.E., Jackson, J.M., Ho, P.T.P., Turner, J.L. 1990, in *The Interstellar Medium in External Galaxies*, eds. H.A. Thronson, D.J., Hollenbach, NASA Conference Proceedings, 337

6.9 Cecil, G., Bland, J., Tully, R.B. 1990, *Astrophys. J.* **355**, 70

6.10 Cox, P., Laureijs, R. 1989, in *The Center of the Galaxy*, ed. M. Morris (Kluwer, Dordrecht), 121

6.11 Crawford, M.K., Genzel, R., Townes, C.H., Watson, D.M. 1985, *Astrophys. J.* **291**, 755

6.12 Ebstein, S.M., Carleton, N.P., Papaliolios, C. 1989, *Astrophys. J.* **336**, 103

6.13 Eckart, A., Downes, D., Genzel, R., Harris, A.I., Jaffe, D.T., Wild, W. 1990a, *Astrophys. J.* **348**, 434

6.14 Eckart, A., Cameron, M., Rothermel, H., Wild, W., Zinnecker, H., Rydbeck, G., Olberg, M., Wiklind, T. 1990b, *Astrophys. J.* **363**, 451

6.15 Eckart, A., Cameron, M., Genzel, R., Jackson, J.M., Rothermel, H., Stutzki, J., Rydbeck, G., Wiklind, T. 1991a *Astrophys. J.* **365**, 522

6.16 Eckart, A., Cameron, M., Jackson, J.M., Genzel, R., Harris, A.I. 1991b, *Astrophys. J.* **372**, 67

6.17 Evans, I.N., Ford, H.C., Kinney, A.L., Antonucci, R.R.J., Armus, L., Caganoff, S. 1991, *Astrophys. J.* **369**, L27

6.18 Genzel, R., Watson, D.M., Crawford, M.K., Townes, C.H. 1985, *Astrophys. J.* **297**, 766

6.19 Genzel, R. 1989, in *The Center of the Galaxy*, ed. M. Morris (Kluwer, Dordrecht), 393

6.20 Genzel, R., Stacey, G.J., Harris, A.I., Townes, C.H., Geis, N., Graf, U.U., Poglitsch, A., Stutzki, J. 1990, *Astrophys. J.* **356**, 160

6.21 Gry, C., Lequeux, T., Boulanger, F. 1991, *Astron. Astrophys.* in press

6.22 Güsten, R. 1989, in *The Center of The Galaxy*, ed. M. Morris (Kluwer, Dordrecht), 89

6.23 Harris, A.I., Hills, R.E., Stutzki, J., Graf, U.U., Russell, A.P.G., Genzel, R. 1991, *Astrophys. J.* **382**, L75

6.24 Henkel, C., Baan, W.A., Mauersberger, R. 1991, *Astron. Astrophys. Rev.* **3**, 47

6.25 Ho, P.T.P., Ho., L.C., Szczepanski, J.C., Jackson, J.M., Armstrong, J.T., Barrett, A.H. 1991, *Nature* **350**, 309

6.26 Hofmann, K.H., Mauder, W., Weigelt, G. 1989, in *Extranuclear Activity in Galaxies*, eds. E.J.A. Meurs, R.A.E. Fosbury, ESO Conference Proceedings **32**, 35

6.27 Ishizuki, S., Kawabe, R., Ishiguro, M., Okumura, S.K., Morita, K.I., Chikada, Y., Kasuga, T. 1990, *Nature* **344**, 224

6.28 Jackson, J., Geis, N., Genzel, R., Harris, A.I., Madden, S.C., Poglitsch, A., Stacey, G.J., Townes, C.H. 1991, *Astrophys. J.* submitted

6.29 Krabbe, A., Genzel, R., Drapatz, S., Rotaciuc, V. 1991, *Astrophys. J.* **382**, L19

6.30 Kronberg, P.P., Biermann, P., Schwab, F.R. 1985, *Astrophys. J.* **291**, 693

6.31 Lacy, J.H., Achtermann, J.M., Serabyn, E. 1001, *Astrophys. J.* **380**, L71

6.32 Lo, K.Y., Claussen, M.J. 1983, *Nature* **306**, 647

6.33 Lo, K.Y., Cheung, K.W., Masson, C.R., Phillips, T.G., Scott, S.L., Woody, D.P. 1987, *Astrophys. J.* **312**, 574

6.34 Loiseau, N., Reuter, H.-P., Wielebinski, R., Klein, U. 1988, *Astron. Astrophys.* **200**, L1

6.35 Madden, S.C., Geis, N., Genzel, R., Herrmann, F., Jackson, J.M., Poglitsch, A., Stacey, G.J., Townes, C.H. 1992, *Astrophys. J.* in prep

6.36 Mauersberger, R., Hakel, C., Walmsley, C.M., Sage, L.J., Wiklind, T. 1991, *Astron. Astrophys.* **247**, 307

6.37 Meixner, M., Puchalsky, R., Blitz, L., Wright, M.C.H., Heckman, T. 1990, *Astrophys. J.* **354**, 158

6.38 Morris, M. 1989, in *The Center of the Galaxy*, ed. M. Morris (Kluwer, Dordrecht), 171

6.39 Nakai, N., Hayashi, M., Handa, T., Sofue, Y., Hasegawa, T., Sasaki, M. 1987, *Publ. Astron. Soc. Japan* **39**, 685

6.40 Planesas, P., Scoville, N.Z., Myers, S.T. 1991, *Astrophys. J.* **369**, 364

6.41 Radford, S.J.E. et al. 1991, in *Dynamics of Galaxies and their Molecular Cloud Distribution*, eds. F. Combes, F. Casoli, (Kluwer, Dordrecht), 303

6.42 Rieke, G.H., Lebofsky, M.J., Thompson, R.I., Low, F.J., Tokunaga, A.T. 1980, *Astrophys. J.* **238**, 24

6.43 Rieke, G.H., Rieke, M.J., Paul, A.E. 1989, *Astrophys. J.* **336**, 752

6.44 Rotaciuc, V., Krabbe, A., Cameron, M., Drapatz, S., Genzel, R., Sternberg, A., Storey, J.W.V. 1991, *Astrophys. J.* **370**, L23

6.45 Sanders, D.B., Scoville, N.Z., Young, J., Soifer, B.T., Danielson, G. 1986, *Astrophys. J.* **305**, L45

6.46 Sargent, A.I., Scoville, N.Z. 1991, *Astrophys. J.* **366**, L1

6.47 Scalo, J. 1990, in *Windows on Galaxies* eds. G. Fabbiano et al. (Kluwer, Dordrecht), 125

6.48 Scoville, N.Z., Sanders, D.B. 1987, in *Interstellar Processes*, eds. D.J. Hollenbach, H.A. Thronson (Reidel, Dordrecht), 21

6.49 Scoville, N.Z., Sargent, A.I., Sanders, D.B., Soifer, B.T. 1991, *Astrophys. J.* **366**, L5

6.50 Serabyn, E., Lacy, J.H. 1985, *Astrophys. J.* **293**, 445

6.51 Smith, P.A., Brand, P.W.J.L., Puxley, P.J., Mountain, C.M., Sakai, N. 1990, *Monthly Notices Roy. Astron. Soc.* **243**, 97

6.52 Sofue, Y. 1988, in *Galactic and Extragalactic Star Formation*, eds. R.E. Pudritz, M. Fich (Kluwer, Dordrecht), 409

6.53 Stacey, G.J., Geis, N., Genzel, R., Lugten, J.B., Poglitsch, A., Sternberg, A., Townes, C.H. 1991, *Astrophys. J.* **373**, 423

6.54 Stark, A.A., Bally, J.II., Wilson, R.W., Pound, M.W. 1989, in *The Center of the Galaxy*, ed. M. Morris (Kluwer, Dordrecht), 129

6.55 Tsuboi, M., Handa, T., Inoue, M., Inatani, J., Ukita, N. 1989, in *The Center of the Galaxy*, ed. M. Morris (Kluwer, Dordrecht), 135

6.56 Wild, W., Harris, A.I., Eckart, A., Genzel, R., Graf, U.U., Jackson, J.M., Jaffe, D.T., Stutzki, J. 1991, in prep

6.57 Wilson, A.S., Ulvestad, J.S. 1983, *Astrophys. J.* **275**, 8

6.58 Wolfire, M.G., Hollenbach, D.J., Tielens, A.G.G.M. 1989, *Astrophys. J.* **344**, 770

6.59 Wright, E. et al. 1991, preprint

6.60 Young, J.S., Scoville, N.Z. 1991, *Ann. Rev. Astron. Astrophys.* **29**, 581

Subject Index

gamma, 63, 69
 stray, 9–11, 133, 136
 synchrotron, 63, 277
 UV, 63
radiative association, 337
radiative transport, 287, 289
radio continuum survey, 86
raisin pudding model, 29
Rayleigh-Jeans approximation, 20, 288
Rayleigh-Jeans regime, 291
RCW103, 330
reaction
 ion-molecular, 338
 ion-molecule, 339
 neutral-grain, 337
 neutral-neutral, 337
recombination, 209
red giant, 216
reddening, 157
relation
 Manley-Rowe, 375
 P-L, 38
 Ranking-Hugoniot, 325
 size linewidth, 47, 48, 50
resonance, 224
Roche criterion, 381
Rosette, 352, 353
rotation
 circular, 101
 curve, 36
 galactic, 2, 36, 72
rotation curve, 39, 41, 42, 51–53, 59, 61, 67,
 92, 100–103, 121, 127, 129, 133,
 137, 236, 237, 239, 264, 267, 269
 bump, 51, 61
Rydberg constant, 308

Sagittarius complex, 100, 105
saturation, 29, 31
scale length
 stellar disk, 92
Scutum Crux feature, 90
self-absorption, 12, 16, 21, 29–33, 107–109
self-similarity, 167
Sgr A, 118
Sgr A*, 105
Sgr B, 118
Sgr B2, 38, 341
shadowing, 45, 48, 50, 69
shearing, 137
sheet, 373
shell, 50, 164, 167, 212, 215, 216, 218, 264,
 341
shock, 191, 193, 198–200, 202–204, 206, 208,
 212, 215, 216, 219, 224, 232–235,

238, 319, 325, 327, 328, 331, 337,
 338, 341, 364, 367, 376
 bow, 216, 219, 330
 C-shock, 325, 326, 328–331, 376
 chemistry, 366
 front, 157, 183, 191, 192, 194
 J-shock, 325, 326, 328, 330, 331, 376
 magnetic, 195, 208
 post-shock condition, 195
 pre-shock condition, 195
 spiral arm, 257
sidelobe, 9–11, 133
silicate, 93, 304, 343
source function, 288
spectroscopy
 absorption, 292
 infrared, 277
 radio, 277
spin statistic, 283
spiral arm, 60, 61, 65, 159, 224, 236–245,
 257, 258, 265, 272
 Carina, 90
 Norma, 90
 shock, 257
standard solar motion, 8
star
 AGB, 122
 carbon, 42, 345
 Cepheid, 38, 42
 KM, 64
 massive, 341
 Mira, 39, 345
 O, 216, 219, 345
 OB, 39, 42, 63, 85, 142, 180, 214, 250,
 304, 312, 316, 319, 332, 350, 352,
 354–356, 366, 373, 374, 382, 386
 OH-IR, 123, 345
 RR Lyrae, 39, 40
 T-Tau, 332, 352, 367
 Wolf-Rayet, 214, 346
star formation, 63, 70, 91, 126, 147, 164,
 169, 181, 216, 218, 230, 243–245,
 260, 264–267, 269–273, 316, 319,
 341, 349, 352, 353, 356, 357, 361,
 367, 372, 379, 382, 386, 387
 rate, 72, 382
starburst, 275
 galaxy, 272, 273, 316, 357, 381, 386
 nucleus, 382
Strömgren radius, 213
stretch vibration, 285
stretching mode, 284
structure
 hierarchical, 167, 168, 170

Printing: Mercedesdruck, Berlin
Binding: Buchbinderei Lüderitz & Bauer, Berlin